Complete legend to cover illustration:
The observatory of German amateur Max Lammerer at Lichtenfels-Köttel has a cylindrical dome with two slides to open it. It houses an 18-inch Newtonian telescope, as seen to the left in the dome slit, and a 12-inch long focus (planetarian) Newtonian, pictured to the right, a 5-inch rich-field telescope and a 6-inch Schmidt-camera, altogether on a single mounting. The observatory has the usual equipment for visual and photographic observation, a spectrograph and a CCD-camera. The stars in the background are in the constellation of Cassiopeia.

Photograph by M. Lammerer

Günter Dietmar Roth (Ed.)

Compendium of Practical Astronomy

Volume 1:
Instrumentation and
Reduction Techniques

Translated and Revised by
Harry J. Augensen and Wulff D. Heintz

With 153 Figs., some in color, and 51 Tables

Springer-Verlag
Berlin Heidelberg New York
London Paris Tokyo
Hong Kong Barcelona Budapest

Dipl.-Kfm. Günter Dietmar Roth
Ulrichstraße 43, Irschenhausen, D–82057 Icking/Isartal, Germany

Dr. Harry J. Augensen
Department of Physics and Astronomy, Widener University, Chester,
PA 19013, USA

Professor Dr. Wulff-D. Heintz
Department of Physics and Astronomy, Swarthmore College, Swarthmore,
PA 19081, USA

Completely Revised and Enlarged Translation of the 4th German Edition of the title "Roth (Ed.), Handbuch für Sternfreunde, Vols. 1 and 2".

ISBN 3-540-53596-9 Springer-Verlag Berlin Heidelberg New York
ISBN 0-387-53596-9 Springer-Verlag New York Berlin Heidelberg

ISBN 3-540-56273-7 Volumes 1, 2, and 3
ISBN 0-387-56273-7 Volumes 1, 2, and 3

Library of Congress Cataloging-in-Publication Data
Handbuch für Sternfreunde. English
Compendium of practical astronomy / Günter Roth, ed. :
translated by Harry J. Augensen and Wulff D. Heintz.
  p. cm.
Rev. translation of: Handbuch für Sternfreunde (4th ed.).
Includes index.
Contents: v. 1. Instrumentation and reduction techniques -- v.
2. Earth and solar system -- v. 3. Stars and stellar systems.
ISBN 0-387-56273-7 (New York). -- ISBN 3-540-56273-7 (Berlin)
1. Astronomy--Handbooks, manuals, etc. I. Roth, Günter Dietmar.
QB64.H3313 1993 520--dc20 93-27023

This work is subject to copyright. All rights are reserved, whether the whole or part of the material is concerned, specifically the rights of translation, reprinting, reuse of illustrations, recitation, broadcasting, reproduction on microfilm or in any other way, and storage in data banks. Duplication of this publication or parts thereof is permitted only under the provisions of the German Copyright Law of September 9, 1965, in its current version, and permission for use must always be obtained from Springer-Verlag. Violations are liable for prosecution under the German Copyright Law.

© Springer-Verlag Berlin Heidelberg 1994
Printed in Germany

The use of general descriptive names, registered names, trademarks, etc. in this publication does not imply, even in the absence of a specific statement, that such names are exempt from the relevant protective laws and regulations and therefore free for general use.

Cover Design: Erich Kirchner, Heidelberg
Typesetting: Data conversion by Lewis & Leins Buchproduktion, Berlin
Production: PRODUserv Springer Produktions-Gesellschaft, Berlin
SPIN 10018869   55/3020 - 5 4 3 2 1 0 – Printed on acid-free paper

Contributing Authors to Volume 1

Altenhoff, Wilhelm J., Dr.
Max-Planck-Institut für Radioastronomie, Auf dem Hügel 69, D–53121 Bonn, Germany

Augensen, Harry, J., Dr.
Department of Physics, Widener University, One University Place, Chester, PA 19013, USA

Duerbeck, Hilmar W., Dr.
Astronomisches Institut der Universität Münster, D–48149 Münster, Germany

Häfner, Reinhold, Dr.
Universitäts-Sternwarte, Scheinerstrasse 1, D–81679 München, Germany

Heintz, Wulff-Dieter, Prof. Dr.
Department of Physics and Astronomy, Swarthmore College, Swarthmore, PA 19081, USA

Hoffmann, Martin, Dr.
Alter Weg 7, D–54570 Weidenbach, Germany

Koch, Bernd, Dipl.-Phys.
Fliederweg 10, D–42699 Solingen, Germany

Nicklas, Harald, Dipl.-Phys.
Universitäts-Sternwarte, Geismarlandstrasse 11, D–37083 Göttingen, Germany

Roth, Günter Dietmar, Dipl.-Kfm.
Ulrichstrasse 43, Irschenhausen, D-82057 Icking/Isartal, Germany

Schmeidler, Felix, Prof. Dr.
Mauerkircherstrasse 17, D–82057 München, Germany

Sommer, Norbert
Hammer Dorfstrasse 198, D–40221 Düsseldorf, Germany

Ziegler, Herwin G., El.-Ing.
Ringstrasse 1a, CH–5415 Nussbaumen, Switzerland

Preface to the Second English Edition

It has been a particular pleasure to produce this revised English edition of the German *Handbuch für Sternfreunde*, thus making it available to a wider readership. I should like to express my gratitude to the authors and translators, who have contributed invaluably to the process of revision and translation.

I am deeply indebted to Prof. W.D. Heintz and Prof. H.J. Augensen, who not only made the translation but also assisted in the critical reading and improvement of various chapters. Moreover, Prof. Augensen has written for Vol. 1 a chapter which says a great deal about "Astronomy Education and Instructional Aids". This contains full information with regard to the situation in the United Kingdom, Canada, and for the United States. It is an extremely helpful survey for both teachers and students in colleges and high schools, and the staff of planetaria and public observatories.

Welcomed as a new author is R. Kresken, who wrote the chapter "Artificial Earth Satellites" especially for this English edition.

Last, but not least, I gratefully acknowledge the helpfulness of Springer-Verlag, Heidelberg, where Prof. W. Beiglböck always gave every possible consideration to the translators' and my suggestions.

Irschenhausen *Günter D. Roth*
Summer 1994

Translators' Preface

It is a pleasure to present this work, which has been well received in German-speaking countries through four editions, to the English-speaking reader. We feel that this is a unique publication in that it contains valuable material that cannot easily—if at all—be found elsewhere. We are grateful to the authors for reading through the English version of the text, and for responding promptly (for the most part) to our queries. Several authors have supplied us, on their own initiative or at our suggestion, with revised and updated manuscripts and with supplementary English references. We have striven to achieve a translation of *Handbuch für Sternfreunde* which accurately presents the qualitative and quantitative scientific principles contained within each chapter while maintaining the flavor of the original German text. Where appropriate, we have inserted footnotes to clarify material which may have a different meaning and/or application in English-speaking countries from that in Germany.

When the first English edition of this work, *Astronomy: A Handbook* (translated by the late A. Beer), appeared in 1975, it contained 21 chapters. This new edition is over twice the length and contains 28 authored chapters in three volumes. At Springer's request, we have devised a new title, *Compendium of Practical Astronomy*, to more accurately reflect the broad spectrum of topics and the vast body of information contained within these pages. It should be noted that, while much of this information is directed toward the "amateur," it is equally applicable to the professional astronomer or physicist who teaches at a small college and is searching for suitable astronomical projects to give to his or her students.

The *Compendium of Practical Astronomy* is structured somewhat differently from its German counterpart. The former consists of three volumes, the latter two. Volume 1 is essentially the same as the corresponding volume in the *Handbuch für Sternfreunde*, but the chapters have been reordered in a sequence which, we feel, presents the topics in more homogeneous groups. In addition, "Astronomy Education and Instructional Aids" (H.J. Augensen) has been added as the last chapter. Volume 2 contains chapters covering the Earth and Solar System, including "The Terrestrial Atmosphere and Its Effects on Astronomical Observations" (F. Schmeidler), which appeared in Vol. 1 of the *Handbuch*. Also, "Artificial Earth Satellites" (R. Kresken) replaces the corresponding chapter by Petri in the German edition. Volume 3 is devoted to topics of stellar, galactic, and extragalactic astronomy. In the fourth German edition, these latter two volumes appear as one very thick second volume.

At the end of each volume, we have included a "Supplemental Reading List" for each of the chapters in that volume. We have prepared a new appendix, "Educational Resources in Astronomy" (Appendix A in Vol. 1), as a supplement to Chap. 12, and have also updated some tables, primarily those which constitute Appendix B, "Astronomical Data," in Vol. 3. Recognizing the fact that many readers will want to utilize computer techniques in the various reduction procedures presented in this *Compendium*, it is our hope that the references on computers and programmable calculators given in Sect. 12.4.4 of Vol. 1 will prove helpful.

The superb guidance provided by Prof. W. Beiglböck and the able assistance of his secretary Ms. S. Landgraf (Springer-Verlag, Heidelberg) has been invaluable in the completion of this project. We are also indebted to Mark Seymour at Springer for his thorough proofreading of the entire manuscript, and to Dr. Fred Orthlieb of Swarthmore College for his help in translating several technical terms. Finally, we gratefully acknowledge the unwavering assistance of Computing Services at Widener University, in particular Barry Poulson, James Connalen, John Neary, Kim Stalford, Lynn Pollack, and David Walls, who provided expert advice on the preparation of this entire manuscript in T$_E$X.

One of us (WDH) still cherishes his acquaintance with A. Güttler, W. Jahn, R. Kühn, R. Müller, and K. Schütte—the now-deceased authors of the first edition, whose enthusiasm helped Günter Roth's project off to a good start over 30 years ago.

Swarthmore/Chester   *Wulff D. Heintz*
June 1994   *Harry J. Augensen*

Preface to the Fourth German Edition

The ability to employ objective techniques to appreciate the wide variety of cosmic phenomena quantitatively is not restricted to professional astronomers. It is the principal aim of the *Handbuch für Sternfreunde* to provide the astronomically interested public—amateur observers as well as teachers—with instruction and guidance in practical astronomical activities. This goal has remained unchanged since the first edition, which appeared in 1960.

What has changed is the technical content and organizational structure in various areas. Larger and more effective telescopes are currently within the reach of non-professionals. The professional accessories used in photography, photometry, and spectroscopy are now being operated by amateurs; schools and private observatories own electronic equipment. Thus equipped, the amateur can now engage in observational tasks ranging from photoelectric photometry of planets and variable stars to studies of high-resolution photographs of distant galaxies.

These developments are reflected in every chapter of the present work. The presentation of new tools, techniques, and tasks has required a significant expansion, so that the *Handbuch für Sternfreunde* now appears in two volumes.

Volume 1 provides the technical basis for astronomical observations and measurements with amateur equipment. These include fundamental methods for recording and processing light intensities in photography, photometry, and spectroscopy. The optical range is augmented by radio-astronomical observations, whose instrumental basics are described. Also included within this volume are instructions on how to organize astronomical observations and subsequently process the results by mathematical methods, a guide to the literature, and a brief history of astronomy.

The following chapters in Vol. 1 have been completely rewritten or newly added for the fourth edition: "Optical Telescopes and Instrumentation" (H. Nicklas), "Telescope Mountings, Drives, and Electrical Equipment" (H.G. Ziegler), "Astrophotography" (B. Koch, N. Sommer), "Principles of Photometry" (H.W. Duerbeck, M. Hoffmann), and "Historical Exploration of Modern Astronomy" (G.D. Roth).

Volume 2 presents the objects of astronomical study in detail, commenting on the execution and evaluation of observational tasks. The topics covered include, among others, the various Solar System bodies, the stars, the

Milky Way, and extragalactic objects. This volume contains an expanded section of tables, general literature references, and the cumulative subject index for both volumes. Also included as an appendix is the contribution "Instructional Aids in Astronomy" (A. Kunert).

The following chapters in Vol. 2 have been completely rewritten or newly added for the fourth edition: "The Sun" (R. Beck et al.), "Lunar Eclipses" (H. Haupt), "Noctilucent Clouds, Polar Aurorae, and the Zodiacal Light" (C. Leinert), "Stars" (T. Neckel), "Variable Stars and Novae" (H. Drechsel, T. Herczeg), "The Milky Way and the Objects Composing It" (T. Neckel), and "Extragalactic Objects" (J.V. Feitzinger).

I also wish to thank all the authors on this occasion for their successful collaboration. Welcomed as new contributors are: Dr. R. Beck and his coworkers V. Gericke, H. Hilbrecht, C.H. Jahn, E. Junker, K. Reinsch, and P. Völker (the "Sun" Working Group of the Vereinigung der Sternfreunde); Dr. H. Drechsel (Bamberg), Dr. H. Duerbeck (Münster), Prof. J.V. Feitzinger (Bochum), Prof. H. Haupt (Graz), Prof. T.J. Herczeg (Norman, OK), Dr. M. Hoffmann, B. Koch (Düsseldorf), Dr. C. Leinert (Heidelberg), Dr. T. Neckel (Bochum), Dr. H. Nicklas (Göttingen), and N. Sommer (Düsseldorf).

The preparation of the fourth edition was aided by the valuable advice given by Prof. F. Schmeidler (Munich) and Dr. H.J. Staude, managing editor of the magazine *Sterne und Weltraum*. Dr. W. Kruschel (Konstanz) supported the preparation of the tables in Vol. 2 by providing updated material. Photographs and tables were made available by C. Albrecht (Freiburg), H. Haug and coworkers at the Wilhelm-Foerster-Sternwarte (Berlin), and J. Meeus (Erps-Kwerps, Belgium).

As the representative of Springer-Verlag, Prof. W. Beiglböck has attended to this large project with care and provided numerous suggestions. The somewhat tedious task of preparing the manuscripts for printing was in the capable hands of Mrs. C. Pendl, to whom the editor and the authors are very grateful.

Irschenhausen *Günter D. Roth*
Summer 1989

Contents of Volume 1

1	**Introduction to Astronomical Literature and Nomenclature** by W.D. Heintz	1
1.1	Astronomy and the Observer	1
1.2	The Astronomy Library	2
1.3	Catalogues and Maps	4
1.4	Almanacs	6
1.5	Reduction of Observations	7
2	**Fundamentals of Spherical Astronomy**	9
	by F. Schmeidler	9
2.1	Introduction	9
2.2	The Coordinates	9
2.2.1	Geographic Coordinates	9
2.2.2	Horizontal Coordinates	11
2.2.3	The Equatorial System, Vernal Equinox, and Sidereal Time	13
2.2.4	Transformation of Horizontal Coordinates into Equatorial Coordinates and Vice Versa	14
2.2.5	Other Coordinate Systems	15
2.3	Time and the Phenomena of Daily Motion	17
2.3.1	True and Mean Solar Time	17
2.3.2	The Relation Between Sidereal Time and Mean Time	18
2.3.3	Other Phenomena of Diurnal Motion	20
2.4	Changes in the Coordinates of a Star	21
2.4.1	Proper Motion	21
2.4.2	Precession and Nutation	21
2.4.3	Aberration	23
2.4.4	Parallax	24
2.4.5	Reduction of Mean Position to Apparent Position	25
2.5	Problems of the Calendar and Time Zones	26
2.5.1	The Calendar and the Measurement of Years	26
2.5.2	The Length and Beginning of the Year	27

2.5.3	The Julian Date and the Beginning of the Mean Day	28
2.5.4	Time Zones and the Date Line	29
2.6	The Variability of the Time Systems	29
2.6.1	Various Causes for Changes in the Length of the Day	29
2.6.2	Astronomical Effects of the Variable Rotation of the Earth	30
2.7	Spherical Trigonometry	32
2.7.1	Fundamental Equations	32
2.7.2	Derived Equations	32
2.7.3	The Right-Angled Spherical Triangle	34
	References	35

3 Applied Mathematics and Error Theory
by F. Schmeidler 37

3.1	Introduction	37
3.1.1	Pocket Calculators	37
3.1.2	Computers	38
3.2	The Theory of Errors	39
3.2.1	Direct Observations	39
3.2.2	Indirect Observations	41
3.3	Interpolation and Numerical Differentiation and Integration	45
3.4	Photographic Astrometry	49
3.5	Determination of the Position and Brightness of Planets and of the Planetographic Coordinates	50
3.6	The Reduction of Stellar Occultations	56
	References	57

4 Optical Telescopes and Instrumentation
by H. Nicklas 59

4.1	Introduction	59
4.2	Basics of Optical Computation	60
4.2.1	Ray Tracing and Sign Convention	60
4.2.2	The Cardinal Points of an Optical System	62
4.2.3	Pupils and Stops	64
4.3	Imaging Errors	66
4.3.1	Seidel Formulae	66
4.3.2	The Primary Aberrations	70
4.3.3	Chromatic Aberrations	75
4.4	Methods of Optical Testing	77

4.4.1	Surface Manufacturing of Optical Quality Surfaces	77
4.4.2	Determination of Focal Length	79
4.4.3	The Hartmann Test	81
4.4.4	Foucault's Knife-Edge Test	83
4.4.5	Interferometric Tests	84
4.5	Telescope Systems	86
4.5.1	Refractors	86
4.5.2	The Newtonian Reflector	88
4.5.3	The Cassegrain Telescope	91
4.5.4	The Ritchey–Chrétien System	93
4.5.5	The Schiefspiegler	94
4.5.6	The Schmidt Camera	95
4.5.7	Schmidt–Cassegrain Systems	98
4.5.8	Maksutov Systems	99
4.5.9	Instruments for Solar Observations	101
4.6	Telescope Performance	104
4.6.1	Resolving Power	104
4.6.2	Magnification and Field of View	107
4.6.3	Image Brightness and Limiting Magnitude	109
4.7	Accessories	113
4.7.1	Eyepieces	113
4.7.2	The Barlow Lens	115
4.7.3	Tube and Dewcap	118
4.7.4	Finding and Guiding Telescopes	120
4.7.5	Eyepiece Micrometers	121
4.7.6	The Photometer	122
4.7.7	The Spectrograph and the Spectroscope	123
4.7.8	Sun Projection Screens	124
4.7.9	The Clock	125
4.8	Visual Observations	126
4.8.1	The Eye	126
4.8.2	Binoculars	127
4.9	Photographic Plates and Photoelectric Detectors	129
4.9.1	Astrophotographs and Their Limitations	129
4.9.2	Photomultipliers	132
4.9.3	Charge-Coupled Devices—CCDs	132
4.10	Services for Telescopes and Accessories	133
	References	134
5	**Telescope Mountings, Drives, and Electrical Equipment** *by H.G. Ziegler*	137
5.1	Introduction	137
5.2	Basic Types of Telescope Mountings	138

5.2.1	The Alt-Azimuthal or Horizontal Mounting	138
5.2.2	The Parallactic or Equatorial Mount	138
5.3	General Design Criteria	140
5.3.1	The General Mechanical Context and the Design Catechism	140
5.3.2	Basic Criteria for Telescope Mountings Regarding Statics, Kinetics, and Kinematics	145
5.4	Static Criteria of Telescope Mountings	147
5.4.1	Stiffness as the Static Characteristic of Telescope Mountings	147
5.4.2	Stiffness as a Design and Layout Quantity	150
5.4.3	The Modulus of Elasticity	153
5.5	Shafts and Bearings	153
5.5.1	Stiffness of Bearings	153
5.5.2	Stresses on the Declination Axis	159
5.5.3	Friction Bearings	162
5.5.4	Antifriction Bearings	162
5.5.5	Stiffness of Antifriction Bearings	163
5.5.6	Stiffness and Distance Between Bearings	164
5.6	The Foundation and Stability Against Tilting	164
5.7	Joining Elements	166
5.8	Measuring the Stiffness	167
5.9	Telescope Vibrations	168
5.9.1	Principles of Mechanical Oscillations	168
5.9.2	The Mounting as Oscillator Chain and Mechanical Low-Pass Filter	169
5.10	Kinematic Aspects of Telescope Mountings	172
5.10.1	General Criteria and Instrument Errors	172
5.10.2	The Accuracy of Manufacture of Mechanical Parts	175
5.11	Drives in Right Ascension and Declination	176
5.11.1	General Considerations	176
5.11.2	The Mechanics of Drives	177
5.11.3	Drive Motors	180
5.11.4	The Control Circuit of Telescope Drives	184
5.11.5	Control Electronics for Synchronous Motor Drives	185
5.11.6	Control Electronics for Stepping Motor Drives	188
5.11.7	Circuitry for Microstep Operation	189
5.11.8	Control Electronics for DC Operation	193
5.12	Photoelectric Guiding Systems	195
5.12.1	Direct and Off-Axis Guiding	195
5.12.2	The Guiding Head	196
5.12.3	Light Receivers	198
5.12.4	Signal Processing and Control Electronics	200

5.13	Adjust Elements and Telescope Alignment on the Celestial Pole	201
5.13.1	Adjustment Elements	201
5.13.2	Adjustment of the Axis System by the Scheiner Method	202
5.14	The Setting Circles and Their Adjustment	203
5.14.1	Setting Circles	203
5.14.2	Adjustment of the Circles	204
5.14.3	Digital Position Displays	204
5.15	Electrical Equipment	206
5.15.1	Power Sources and Safety Aspects	206
5.15.2	Batteries: Properties and Hazards	207
5.15.3	Power Supplies for Electronic Circuits	209
5.15.4	Illumination	212
5.15.5	The Dewcap and Its Heating	214
5.16	Comments on Literature Cited Here and in the Supplemental Reading List	215
	References	216

6 Astrophotography
by B. Koch and N. Sommer 217

6.1	Introduction	217
6.2	Cameras and Lenses	217
6.2.1	Single-Lens Reflex Cameras	217
6.2.2	Medium and Large-Sized Cameras	218
6.2.3	Other Cameras	219
6.2.4	Refractors	219
6.2.5	Reflectors	219
6.2.6	Special Optics for Astrophotography	220
6.2.7	Video Cameras, Image Intensifiers, CCDs	220
6.3	General Considerations	221
6.3.1	Focusing	221
6.3.2	Modification of the Primary Focal Length	224
6.3.3	Guiding	224
6.3.4	Polar Alignment	228
6.3.5	Miscellaneous	230
6.4	Stationary Cameras	230
6.4.1	Trails, Constellations, Planetary Conjunctions	230
6.4.2	Atmospheric Phenomena	232
6.4.3	Meteors	232
6.4.4	Eclipses	233
6.4.5	Artificial Satellites	234
6.5	Tracking Cameras	235

6.5.1	Focal Lengths $f \leqq 500$ mm	235
6.5.2	Lunar Halos	235
6.5.3	Planetary Moons	235
6.5.4	Comets and Minor Planets	236
6.5.5	Eclipses	237
6.5.6	Deep Sky	238
6.5.7	Spectrography	239
6.5.8	Zodiacal Light and Gegenschein	242
6.6	Long-Focus Astrophotography	242
6.6.1	Instrumental Requirements	242
6.6.2	Moon, Sun, and Planets	243
6.6.3	Moons of Planets	245
6.6.4	Comets	245
6.6.5	Deep Sky	245
6.7	Films for Astrophotography	246
6.7.1	Film Formats	246
6.7.2	Physical Composition of the Film and Formation of the Latent Image	246
6.7.3	The Characteristic Curve	248
6.7.4	Film Sensitivity	250
6.7.5	Astrophotography with Filters	250
6.7.6	Low-Intensity Reciprocity Failure (LIRF)	254
6.7.7	Resolving Power	256
6.7.8	Recommended Films	258
6.7.9	Storage of Films	258
6.7.10	Film Development	259
6.7.11	Sensitization of Films	259
6.8	Advanced Darkroom Techniques	269
6.8.1	Darkroom Equipment	269
6.8.2	Black-and-White Paper	269
6.8.3	Color Processing	271
6.8.4	Special Techniques	271
6.9	Photographic Limiting Magnitude	275
6.9.1	Recorded Photographic Limiting Magnitude m	275
6.9.2	Maximum Photographic Magnitude Limit m_{lim}	276
6.9.3	Maximum Exposure Time t_{max}	278
6.9.4	The Standard Sequence for Determining the Limiting Magnitude	278
6.10	Further Reading	280
	References	281
7	**Fundamentals of Spectral Analysis** by R. Häfner	287
7.1	Introduction	287

7.2	The Theory of Spectra	287
7.2.1	The Laws of Radiation	287
7.2.2	The Line Spectrum	289
7.2.3	Excitation and Ionization	292
7.3	The Objects of Spectral Studies	293
7.3.1	Stars	293
7.3.2	The Sun	297
7.3.3	Planets and Moons	298
7.3.4	Comets	299
7.3.5	Meteors	299
7.4	Spectroscopic Instruments	300
7.4.1	The Methods of Spectral Dispersion	300
7.4.2	The Design of a Spectral Instrument	303
7.4.3	Radiation Detectors	306
7.4.4	Designs, Hints for Operation, and Accessories	307
7.5	The Analysis	313
7.5.1	Spectral Classification	314
7.5.2	Line Changes	314
7.5.3	Radial Velocities	314
7.5.4	Color Temperatures	316
7.5.5	Equivalent Widths and Line Profiles	316
	References	318
8	**Principles of Photometry**	
	by H.W. Duerbeck and M. Hoffmann	319
8.1	Introduction	319
8.1.1	General Description and Historical Overview	319
8.1.2	Units of Brightness	320
8.1.3	The Receivers	321
8.2	Limits and Accuracies of Photometric Measurements	327
8.3	Astronomical Color Systems	329
8.4	The Technique and Planning of Observations	332
8.4.1	Point Photometry and Surface Photometry	334
8.4.2	Visual Photometry: Differential Observations	334
8.4.3	Photoelectric Photometry: Differential Observations	337
8.4.4	Absolute Photoelectric Photometry	338
8.4.5	Photometry of Occultations	339
8.4.6	Visual, Photographic, and Electronic Surface Photometry	341
8.5	Reduction Techniques	342
8.5.1	Reduction of Photometric Measurements—Generalities	342
8.5.2	Reduction of Photographic Observations—Generalities	345
8.5.3	Reduction of Digital Images—Generalities	346
8.5.4	The Reduction of Time: the Heliocentric Correction	347

8.5.5	Determination of Minimum Light Times and of Periods	351
8.6	Photometry of Different Astronomical Objects – Generalities	356
8.6.1	Photometry of Solar System Objects	356
8.6.2	Stellar Photometry	365
8.6.3	Surface Photometry	368
8.7	Construction or Purchase of Receivers and Equipment for Reductions	369
8.7.1	Advice on the Purchase of Photometers	369
8.7.2	Advice on the Construction of Photometers	370
8.8	General Literature on Photometry	375
	References	376

9 Fundamentals of Radio Astronomy
by W.J. Altenhoff 381

9.1	Introduction	381
9.2	Radio Radiation	384
9.2.1	Thermal Radiation	385
9.2.2	Nonthermal Radiation	388
9.3	Atmospheric Influences	388
9.3.1	The Ionosphere	388
9.3.2	The Troposphere	390
9.3.3	Artificial Interference and Protected Frequencies	390
9.4	Instrumentation	391
9.4.1	Antennas	391
9.4.2	Radio Receivers	395
9.4.3	The Computer	397
9.5	The Objects of Observation	399
9.5.1	Continuous Radiation	400
9.5.2	Spectral Lines	407
9.6	Tested Observing Systems	408
9.6.1	Solar Flare Monitoring	409
9.6.2	Jovian Bursts	409
9.6.3	A Model Interferometer	409
9.6.4	FM and Television Receivers	409
9.6.5	Satellite Receivers	410
9.6.6	Evaluation	410
9.7	Amateur Radio Astronomy Groups	411
	References	411

10 Modern Sundials
by F. Schmeidler 413

10.1	Introduction	413
10.2	The Equinoxial Dial	414
10.3	Horizontal Dials and Vertical East–West Dials	414
10.3.1	Computations	414
10.3.2	Graphical Construction	415
10.4	The Vertical Deviating Dial	417
10.4.1	Determination of the Azimuth of the Wall	417
10.4.2	Calculation of the Dial at the Wall	418
10.4.3	Construction of the Calculated Dial Face at the Wall	419
10.4.4	Inserting the Style	420
10.5	Designs for Higher Accuracy	422
10.5.1	Correction for Geographic Longitude	422
10.5.2	Correction for the Equation of Time	422
	References	424
11	**An Historical Exploration of Modern Astronomy** *by G.D. Roth*	425
11.1	Introduction	425
11.2	The Heliocentric System	425
11.3	Evolution of the Theory of Motions	426
11.4	Cataloguing the Stellar Sky	427
11.5	Astrophysics	428
11.5.1	Stellar Photometry	428
11.5.2	Spectroscopy of the Sun and Stars	429
11.5.3	Astronomical Photography	429
11.5.4	Large Telescopes	430
11.6	Stellar Evolution and Stellar Systems	431
11.6.1	Stellar Evolution	431
11.6.2	Stellar Systems	431
11.7	Observations at Invisible Wavelengths and Space Exploration	432
11.8	Research in Historical Astronomy	433
11.8.1	Research Problems	433
11.8.2	Sources	433
11.8.3	Processing	434
	References	435
12	**Astronomy Education and Instructional Aids** *by H.J. Augensen*	437
12.1	Introduction	437

12.2	Formal Astronomy Education	438
12.2.1	The Education of Professional Astronomers: An Overview	439
12.2.2	Pre-College/University Preparation	440
12.2.3	Astronomy at the College/University Level	441
12.2.4	Astronomy at the Graduate Level	445
12.2.5	Postdoctoral Positions	448
12.2.6	Vocational Opportunities	448
12.2.7	Interdisciplinary Approaches to Astronomy	450
12.2.8	Comments on Astrology	450
12.3	Facilities and Services Available to Schools and the General Public	451
12.3.1	Planetariums, Museums, and Exhibits	451
12.3.2	Observatories and Research Laboratories	452
12.3.3	Lectures	453
12.3.4	Workshops and Other Programs	454
12.3.5	Where to Find Information on Astronomy as a Career	455
12.3.6	Films on Astronomy as a Career	455
12.3.7	Astronomical Societies and Clubs	456
12.3.8	Inns and Travel Tours	457
12.3.9	Astronomy Books for the Visually Handicapped	457
12.4	Educational Resources in Astronomy	458
12.4.1	Mechanical Models and Exhibit Items	459
12.4.2	Audio-visual Media	466
12.4.3	Broadcasting and Communications	470
12.4.4	Computers and Software	472
12.4.5	Printed Material	474
12.4.6	Games	476
12.4.7	Music	477
	References	478

Appendix A: Educational Resources in Astronomy 481

A.1	Planetariums, Museums, and Exhibits	481
A.2	Observatories and Research Laboratories	488
A.3	Astronomical Societies and Clubs	494
A.4	General List of Sources for Mechanical Models and Exhibit Items	498
A.5	General List of Sources for Audio-Visual Aids	500
A.6	Telescopes and Observing Equipment	503
A.7	Printed Materials	513

Supplemental Reading List for Vol. 1 527

Index 535

Contents of Volume 2

13 The Sun
*by R. Beck, V. Gericke, H. Hilbrecht, C. Jahn, E. Junker
K. Reinsch, P. Völker, Fachgruppe Sonne der
Vereinigung der Sternfreunde e.V.*

14 Observations of Total Solar Eclipses
by W. Petri

15 The Moon
by G.D. Roth

16 Lunar Eclipses
by H. Haupt

17 Occultations of Stars by the Moon
by W.D. Heintz

18 Artificial Earth Satellites
by R. Kresken

19 Observations of the Planets
by G.D. Roth

20 Comets
by R. Häfner

21 Meteors and Bolides
by F. Schmeidler

22 Noctilucent Clouds, Polar Aurorae, and the Zodiacal Light *by C. Leinert*

23 The Terrestrial Atmosphere and Its Effects on Astronomical Observations *by F. Schmeidler*

Supplemental Reading List for Vol. 2

Index

Contents of Volume 3

24 **The Stars**
   by T. Neckel

25 **Variable Stars**
   by T.J. Herczeg and H. Drechsel

26 **Binary Stars**
   by W.D. Heintz

27 **The Milky Way Galaxy and the Objects Composing It**
   by T. Neckel

28 **Extragalactic Objects**
   by J.V. Feitzinger

Appendix B: Astronomical Data

Supplemental Reading List for Vol. 3

Index

1 Introduction to Astronomical Literature and Nomenclature

W.D. Heintz

1.1 Astronomy and the Observer

Astronomy involves the exploration of all phenomena external to and including the Earth. Its realm is the whole of space, out to the remotest distances, and also all of time back to the origin of the Universe as deduced from stellar dating and cosmological theory. Apart from meteorites and the very nearest celestial bodies which can be reached directly by spacecraft, astronomical researchers are spatially separated from the objects of their study and therefore cannot experiment with them at will. They must make observations at the times and under the conditions that Nature prescribes. Furthermore, the apparent diameters, separations, and motions which must be measured in astrometric work as well as the amount of light available for astrophysical analysis are usually so small that an appreciable degree of natural uncertainty is inherent in most observational measurements. The analysis and removal of this uncertainty has itself become a primary focus of astronomical research. As with many other mathematical methods, Gauss's *theory of errors* was originally developed to meet the stringent requirements which arise in the processing of astronomical measurements, and thus serves as a compliment to the thoroughness of our astronomical forefathers. In this way, from the computational conquest of the motions of celestial objects, and on the basis of volumes of observed data, and with a careful consideration of all possible error sources, the term "astronomical" was once a proverbial byword for accuracy (as it is now for enormous size).

Astronomy is closely related to mathematics and other "exact" sciences in that its methods and results are usually quantitative. The reduction of the observations, that is, the processing of raw data to obtain a result, often requires lengthy computations and the use of extensive tables; the astronomical theories that depend on the observations frequently use the most complex mathematical methods and high-speed computers.

While the domain of the amateur astronomer is remarkably wide, it would not be desirable to waste precious observing time on astronomical problems which either have already been solved or can only be advanced by means of state-of-the-art equipment. With regard to higher mathematics, many amateur observers encounter difficulties which perhaps pose even more of a barrier than any limitations in the use of expensive and specialized instruments (spectrographs, CCD detectors, etc.). While they seldom have the interest or the leisure time to concentrate on the theoretical foundations of their telescopic observations, such constraints need not interfere with the success of the observations and the pleasure derived from them.

Nevertheless, without some numerical work and simple calculation precious little can be achieved in this field. Surely even one's first look at the starry sky through binoculars simply for pleasure immediately leads to quantitative questions—How large and how bright is the object? How do we know its distance and its motion? To what percentage are these numbers accurate? Astronomy teachers in particular must continually ask themselves the questions: How much mathematics can we afford to present to the students without dampening their enthusiasm for astronomy? Which relations can be made logically plausible, can be quantitatively explained, and can then perhaps be implanted firmly in the mind? How can the quantitative facts be presented in a continuous, logical fashion and without burdening the memory?

Astronomers, amateurs and professionals alike, yearn to understand the *modus operandi* of the heavens. This handbook is written to meet the needs of those who wish to master the elementary astronomical, physical, and mathematical concepts which are the foundation stones of the vast field of observational astronomy; it is not intended for those whose fervent ambition is, for example, to comprehend theories of black holes or discuss relativistic models of the universe.

The first volume of this work gives the observer a survey of the tools, instruments, and basic knowledge available. The basic computational methods are presented in Chaps. 2 and 3. We begin in the next section with comments on the astronomical literature.

1.2 The Astronomy Library

Every observer is certainly acquainted with at least some of the basic books which cover the entire length (but certainly not breadth) of astronomy in a single volume; there is no shortage of such books. Celestial science has always had a fascination of its own, and progress in the knowledge of space has aroused the public interest. The market for the comprehensive introductory and general astronomy texts is particularly large in the U.S. because most of its many colleges and universities offer at least an introductory class in the subject for students of all disciplines. While these books may be somewhat unequal in the breadth, depth, and quality of the material covered, instructionally they are all quite valuable, and from them every instructor can glean useful hints for a smooth presentation of the subject. Most of the available texts clearly distinguish between solid knowledge and less dependable or speculative views, and they are also less inclined toward sensational presentations than are the popular, purely nontechnical books. It should be noted, however, that the results of astronomical research in foreign countries are often underrepresented in U.S. textbooks.

The periodicals which in nontechnical form report on selections of papers and keep the reader up to date are an important link to research. Some of these magazines, such as *Astronomy*, *Astronomy Now*, and *Sky & Telescope*, focus in particular on the interests of amateur observers, whereas *Mercury* sometimes deals with issues in the teaching of astronomy.

The situation in the technical literature of all disciplines is in general more complex. Specialization and the short lifetime of the flood of publications are most certainly not restricted to astronomy. In the past there existed certain "definitive" books which lasted through several generations of students, but today monographs which present a transition between the introductory works and research literature are rather thin and quickly become outdated. In many subjects, it is virtually impossible to locate a publication whose content is not at least half-outdated. Even more ephemeral and disjointed is information extracted from the presentations of the many special conferences and symposia; such publications report what is going on only in various specialized areas.

Several more hints may serve to help the reader orient him- or herself in a large specialist library and to track down original sources. Several classification schemes have evolved over the past century. The *Dewey Decimal System* divides library holdings into ten broad categories, with subheadings to three digits (520 = Astronomy), and further subdivision, if needed, by digits following a decimal point, zero being reserved for books of a general nature. A preferable modification, widely seen in Europe, is the *Universal Decimal Classification*, offering an easy-to-use, more subject- than method-oriented subdivision (UDC 52 = Astronomy). U.S. libraries often use the *Library of Congress System*, characterized by two subject letters (QB = Astronomy), but its substructure is most ambiguous. Surprisingly, books (in the narrow sense) comprise only a small portion of the accessions to the astronomical library.

Technical journals or periodicals contain the major part of new literature. They fill the shelves quickly; the *Astrophysical Journal*, for instance, annually publishes twelve thick volumes plus several supplement volumes. *Astronomy and Astrophysics*, the merger of western European technical journals, is equally voluminous. The so-called "synoptic literature," which assembles numerous individual contributions on one subject into survey reports and evaluates them critically, is much less widespread in astronomy than it is, say, in chemistry. The volumes *Annual Review of Astronomy and Astrophysics* and the serial *Scientific American* are good examples of synoptic literature. Few libraries can afford to subscribe to all technical volumes; new journals, some of which will quickly be terminated, are continually being founded. Many papers of astronomical interest appear also in journals on geophysics, planetology, technology, instrumentation, mathematics, data processing, and other subjects.

For this reason, bibliographies are the key to using a library efficiently. *Astronomy and Astrophysics Abstracts* are published in Heidelberg semiannually and offer a complete survey of the literature. Every astronomy student and every observer should be acquainted with this publication. The predecessor was the *Astronomische Jahresbericht* dating back to 1899. The Russian periodical *Referatiwny Journal* also serves to provide quick information.

Via a library exchange, a copy of inaccessible literature can often be obtained from large libraries at nominal cost. Authors in institutes often have reprints of their papers available, and are willing to send them to interested persons as long as the supply lasts. It is also possible to subscribe to some journals on microfilm; out-of-print volumes are microfilm copied, which has helped in some ways to curb the expansion of library space. Many astronomical institutes issue their own series of publications which go via international exchange to other observatories and not to commercial channels. This branch of publications has lost some importance because of the spreading of commercial periodicals, but it still serves as a source of copies from journals with limited circulation and also for extensive catalogue material which cannot appear in journals. A large fraction of Soviet publications still appears in institutes' series. Many extensive series of observations are contained in older publications so that these volumes have irreplaceable value.

For observations which require rapid circulation, the *IAU Circulars* are airmailed from the Smithsonian Astrophysical Observatory in Cambridge, Massachusetts, U.S.A. Many observatories also subscribe to a telegraph service in order to be fully apprised of comet and nova discoveries.

A sense of the kinds of research done at the various institutes can be gained by looking through their annual reports. If advice on a special question is desired or if a student needs information regarding particular areas of study, the reports in the *Bulletin of the American Astronomical Sociey*, the *Quarterly Journal of the Royal Astronomical Society*, and similar publications by societies in other countries presenting such reports will prove to be useful sources of information. Astronomers are less enthusiastic about bibliographical search services which keep literature citations according to keyword categories ready for computer access by subscribers; such services leave much to be desired, both with respect to completeness and with respect to tagging or keywording. A very short survey of recent research can be obtained from the *Transactions of the International Astronomical Union*, which includes triannual reports of the research commissions. Volume 12C of the *Transactions, Astronomers Handbook* (1966) contained, among other things, a list of bibliographic abbreviations, a manual on how to prepare and proofread scripts, and the telegram code. The style manual was revised by the IAU in 1991.

1.3 Catalogues and Maps

Astronomical catalogues present large amounts of observed results and other numerical data, and they are a much-used section in observatory libraries. Compilatory catalogues on, for instance, variable stars, cometary orbits, and X-ray sources, are reedited at suitable times so that they can provide reliable information on the current state of the subject. Such information is especially reliable if it is based on complete data, but this is not always a safe assumption. Some catalogues, like the *Bonner Durchmusterung*, are available commercially, but can also be purchased from the issuing institutes or, in some cases, from second-hand bookshops.

Catalogues and astronomical data in the literature are also collected at data centers and are ready for computer retrieval, especially at the Centre de Données Stellaires (11 rue de l'Université, Strasbourg, France). This center's general database is called SIMBAD (an acronym for Set of Identifications, Measurements, and Bibliography for Astronomical Data). It includes a Bibliographical Stellar Index (BSI) and Catalogue of Stellar Identifications (CSI). All special catalogues for which there is demand are separately stored. This is also true of those which, because of their length or preliminary nature, do not appear in print; the material can be accessed via the International Data Networks and is available at nominal cost on magnetic tape and microfiche and in small extracts in printed form (hard copy) or via telex.

Corrections and additions to catalogues are not the task of data centers but rather of the original catalogue editors. Thus, for reasons of documentation and copyright laws, a new catalogue superseding a previous one is loaded into the database in its entirety, and the old catalogue will be withdrawn from the use index and archived when demand has diminished.

Every observer has undoubtedly made an early acquaintance with celestial atlases, especially those which show the naked-eye stars. Maps for fainter stars are used at the telescope, for example the *BD* which reaches to about magnitude $9.^{m}7$. The multivolume *Astrographic Catalogue (Carte du Ciel)* reaches to about two magnitudes fainter, but it has only to a minimal degree appeared in maps, and the printed catalogue gives only rectangular plate coordinates. Small regional maps for identifying very faint objects are often excerpted from photographs, especially from the Palomar Sky Survey. Where did the variety of star names and numbers found in celestial atlases originate?

While the division of the sky into constellations was begun many centuries ago, it was completed and homogenized only in the 18th and 19th centuries, especially for the southern sky. At present, 90 celestial regions are recognized: 88 constellations with their Latin names and abbreviations (see Table B.27 in Vol. 3) with their boundaries following the meridians and parallels, and the two Magellanic Clouds, LMC and SMC.

For bright stars, the designations by Greek letters after Bayer are used: for example, α Leonis = alpha in the constellation of the Lion (= Regulus). In each constellation, the letters generally stand in order of magnitude, but within a magnitude class according to position in the constellation. Flamsteed numbered each of the stars down to about fifth magnitude within each constellation in order of right ascension (e.g., 61 Cygni).

A corresponding scheme in the southern sky was introduced by Gould, and is indicated by the notation "G." following the number (for instance 38G. Puppis). Older numberings (Hevelius, Bode) and Latin letter designations are to be avoided. Popular names, usually from the Arabic, and frequently garbled, are known for about 130 of the brightest stars, predominantly in the northern sky. Sometimes the same name was used for several stars, and some stars had several names or spellings thereof; thus only a few of these names, such as Sirius, Arcturus, and Antares, can be used without confusion since they are well known.

The running number in one of the large compilation catalogues is one of the most widely used designations of stars. The best-known "inventory lists" of the sky include the *Bonner Durchmusterung (BD)*, the *Henry Draper Catalogue (HD)*, and the *General Catalogue (GC)*; for instance, star BD $+75°752$ (star 752 in the declination zone $+75°$) = HD 197433 = GC 28804. South of $-23°$ declination the *BD* is supplemented by the *Cordoba-* and *Cape-Durchmusterungs*. Also the *Smithsonian Catalog for Positions (SAOC)* and the *Catalogue of Bright Stars (BS* or *HR = Harvard Revised)* are often referred to for magnitudes, spectra, and so on. Star clusters and nebulae are numbered according to a list given by Messier (M) or in the *New General Catalogue* (NGC) by Dreyer, where, for instance, M 31 = NGC 224 = the Great Andromeda Nebula. Variable stars have their own designations in the constellation in which they are located with one or two capital letters (e.g., UU Gem, RR Lyr), or with a "V" number (e.g., V 444 Cyg). For the brightest of them which are already in Bayer's catalogue (δ Cep, β Lyr), this designation is retained.

Catalogue designations and abbreviations have recently proliferated almost chaotically. There are also difficulties with the identification of the objects within other units (galaxies and clusters), with extended objects or with radio sources whose position may even depend on the wavelength of observation. The IAU has therefore, through its Commission 5 (Documentation and Astronomical Data), been striving toward homogeneous guidelines of nomenclature, a task in which it has also been joined by the majority of technical journals. To this end, the *First Dictionary of the Nomenclature of Celestial Objects* (1983) is available; the second edition, plus a supplement, is now in preparation at Meudon Observatory. For nomenclature dealing with objects within the planetary system, a special working group (WGPSN) is in charge.

Next to its name, a star is characterized by its position in the sky (right ascension and declination). Identifiers which are formed from coordinates have recently become preferred to running numbers. They are admittedly clumsier but have the advantage of not changing with each new catalogue. PSR 0531 +21 for the pulsar in the Crab Nebula is a good example of an unambiguous designation.

The position of any celestial object changes significantly through precession; sometimes the proper motion is also noticeable. Therefore, it is important to know what epoch (the so-called Equinox) the position refers to. The *BD* maps are valid for 1855.0, the *HD* for 1900.0, and the latest issue of the *Variable Star Catalogue* for 1950.0; for the present it is recommended to use the Equinox 2000.0; for conversion refer to Appendix Table B.10 in Vol. 3.

In order to avoid misidentifications and misrecordings of data owing to typographical errors or misprints, it is advisable to designate objects by two independent identifiers, for instance the approximate coordinates together with a catalogue number.

The lists of objects in Appendix B in Vol. 3 will be sufficient for most purposes; recent specialized catalogues have been referenced in the Supplemental Reading List for Vol. 1.

1.4 Almanacs

Tabulated information on all the changing phenomena are to be found in the astronomical almanacs: the courses of the Sun, Moon, and planets, eclipses, data for planetary observations (central meridians, orientation of the axes, illumination, positions of the satellites), sidereal time, and others. *The Astronomical Ephemeris*[1] contains extensive calculations of higher precision than generally needed by observers. The familiar celestial calendars based on the same ephemeris bases are just as dependable. However, the reorganization of the *Ephemeris* in recent years has not exactly contributed to its ready use, and so the reader must therefore frequently refer to the Appendix with its brief explanations. The *Ephemeris* also contains tables whose use is not tied to

---

[1] published jointly since 1981 by the Nautical Almanac Office and the U.S. Naval Observatory and available in the United States and Canada through the Superintendent of Documents, U.S. Government Printing Office, Washington D.C.; or in the British Commonwealth through Government Bookshop, P.O. Box 276, London SW8 5DT.

a particular year, for instance twilight tables and lists of photometric standard stars, classified bright galaxies, and radio sources.

For earlier years, the predessessors of the *Astronomical Ephemeris*—the *Nautical Almanac* and the *American Ephemeris*—can be consulted. There exist special annual publications, for instance, for minor planets (*Efemeridy malych planet*, St. Petersburg) and variable stars (*Rocznik*, Krakow).

1.5 Reduction of Observations

Observers with research interests who perform series of observations will have already consulted with experienced colleagues and will have explored in the literature what instruments and what accuracy are required for the stated results and conclusions to pass muster. Similarly, they will explore which data center is in charge of the material, whether the material should be published in abridged form, which basic data the manuscript should contain, and similar questions. Results from many observers will often have to be compiled in a later evaluation, and the deductions obtained would thus certainly be impaired had a significant and useful part of the available material been omitted.

Essential for the evaluation of the observations are data on the instrument: aperture, magnification, focal length, etc. Records on atmospheric conditions at the time of observation are retained in the observing journal for later reference although they may or may not appear in the final observing report. The two conditions usually reported are the *transparency* of the sky, which can vary from a perfectly clear, dark background to haze and fog to cloudiness, and the image quality, as measured by the *seeing* (see Chap. 4), which can be clear and sharp in the optimal case, but fuzzy and unstable in the worst case, the latter resulting in a star appearing to "dance" about in the field of view. An estimated scale is used to quantify these two different influences which the beginner can easily confuse. The time of observation is later converted to universal time (UT) or, in some cases, to fractions of a day. Sometimes a numerical estimate of the uncertainty can be made using the theory of errors: an average or *mean* is formed from several measurements, and the *standard deviation* is calculated to serve as a guide for the uncertainties in the individual measurements. Some observing programs (of stellar occultations by the Moon, for instance) employ standard formats which prescribe the contents and arrangement of the observing report. All technical publications, by the way, use the metric system (SI units unless otherwise specified), which is also used throughout this book.

If the manuscript is intended for a central office or an editorial office, it should contain the results and related data in a concise but complete form. Unclear or messy scripts often result in many printing errors which are guaranteed to evoke the editor's displeasure; additional unnecessary costs are incurred when the author pencils into the galley proofs those additions and annotations which should have been contained in the original submitted draft. Subsequently published errata sheets and supplements are at best an ineffective means of correcting mistakes since they will invariably be overlooked by readers.

In the rare case of an important unexpected event—a nova, a comet, or perhaps even the fall of a meteor—the nearest observatory should be informed, but only after the observer is convinced of the reality of the event. While the frequent and valuable support by amateur astronomers and the public should not be devalued, a lot of unnecessary work may be created in clearing up the confusion resulting from simple mistakes. It is incumbent upon the amateur observer to obtain sufficiently precise data that the observatory can confirm the event if it is real, and in any case prepare a sufficiently clear report for further investigation. The immediate follow-up on the discovery observations with suitable instruments can be very important, as was recently demonstrated by the discovery of the supernova SN1987A by observers in the southern hemisphere.

2 Fundamentals of Spherical Astronomy

F. Schmeidler

2.1 Introduction

As its name suggests, the subject of spherical astronomy purports to describe, using the language of mathematics, the positions and motions of phenomena which occur on the celestial sphere. This is a field that today is often neglected, and yet it is still the foundation of many branches of astronomy for observations and research, even if this is not always clearly evident. Every observer is confronted with problems involving spherical astronomy in practical work; for example, to observe a certain astronomical object, it is necessary to find its location, or coordinates, on the celestial sphere. The reduction or processing of observed data also often involves methods which are based on spherical astronomy.

The aim of this chapter is to provide the reader with a brief survey of the fundamentals of spherical astronomy. Because of restricted space in this book, several engaging subjects have been excluded. Also, derivations of the formulae were necessarily omitted; the interested reader can find them in textbooks on spherical astronomy (see the Supplemental Reading List for Vol. 1). The reader is assumed to have some knowledge of numerical calculations, to be familiar with pocket calculators (or logarithm tables) and trigonometric functions, and to understand the basics of the *theory of errors* and the *method of least squares* (see Sect. 3.2 in Chap. 3). Some tables relating to spherical astronomy and references on more extensive literature are to be found in Appendix B in Vol. 3 and the Supplemental Reading List for Vol. 1, respectively.

2.2 The Coordinates

A point on a sphere is defined by two coordinates, assuming that the coordinate system itself is uniquely defined. There are several kinds of coordinate systems in spherical astronomy, and the most important ones will be dealt with in this chapter.

2.2.1 Geographic Coordinates

With the exception of space-based observations, observers work from a point on the Earth's surface defined by three coordinates:

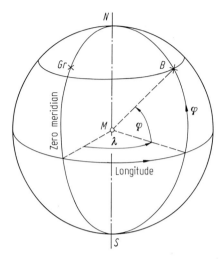

Fig. 2.1. Spherical coordinates.

Geographic latitude, φ
Geographic longitude, λ
Altitude above sea level, H

The geographic latitude is measured from 0° at the Earth's equator up to 90° at its poles, positive to the north, negative to the south. All circles parallel to the Earth's equator are called *parallels of latitude*, and all are smaller than the equator itself.

Any plane which cuts through the imaginary polar axis of the Earth is perpendicular to the equator and intersects the Earth along a *meridian*, which is used to measure geographic longitude (see Fig. 2.1). By international agreement, the meridian through the Royal Observatory, Greenwich, England, has been taken as the *zero meridian*. Longitude is measured from 0° to 180°, positive to the west of Greenwich and negative to the east.[2]

The determination of the geographic coordinates and the shape and size of the Earth is one of the tasks of astronomy and geodesy. In reality the Earth is not precisely spherical but rather is shaped somewhat like a flattened, or oblate, ellipsoid. For practical reasons, the so-called *reference ellipsoid* was introduced as a standard surface. This is chosen so that the directions of its normals agree as closely as possible with the directions of a plumb line at points on the Earth's surface. Also by international agreement, the following dimensions have been accepted for this reference

---

[2] *Translators' Note*: The current (since 1984) definition by the International Astronomical Union reverses these signs of longitude.

ellipsoid:

$$a = \text{semimajor axis} = \text{equatorial radius} = 6378.140 \text{ km},$$

$$f = \text{flattening} = \frac{1}{298.257}.$$

From the relation

$$f = \frac{a-b}{a}, \tag{2.1}$$

the value b is obtained:

$$b = \text{semiminor axis} = \text{polar radius} = 6356.755 \text{ km}.$$

The polar radius is therefore 21.4 km smaller than the equatorial radius. The physical surface of the Earth as a rule deviates somewhat from the reference ellipsoid. On the other hand, the term *geoid* refers to an equipotential surface defined by the average sea level, and whose position over the continents can be imagined as being represented by the water level in channels connected with the oceans.

As a consequence of the ellipsoidal shape of the Earth, there is a difference between the *geographic latitude* φ, defined by the angle between the equatorial plane and a plumb line at that point, and the *geocentric latitude* φ' given by the angle between the equatorial plane and the direction to the center of the Earth. This difference is

$$\varphi - \varphi' = 695\rlap{.}''65 \sin 2\varphi - 1\rlap{.}''17 \sin 4\varphi + \cdots. \tag{2.2}$$

Correspondingly, the distance r' of a point at sea level from the center of the Earth is given by

$$r' = a(0.998\,320 + 0.001\,684 \cos 2\varphi - 0.000\,004 \cos 4\varphi \ldots). \tag{2.3}$$

2.2.2 Horizontal Coordinates

If one imagines a plane passing through the place of observation at right angles to the vertical, it will then intersect the apparent sphere in a circle called the *horizon*, or, more precisely, the *apparent horizon*. Vertically above the observer is the *zenith*, and vertically downward is the *nadir* (see Fig. 2.2).

At any given point on the horizon, one can imagine a circle drawn perpendicular to it. Such a circle is called an *altitude circle*, and all such circles intersect at the zenith and below the horizon at the nadir. The altitude of a star above the horizon is measured along the altitude circle from 0° to 90° at the zenith. All altitudes measured above the apparent horizon are called *apparent altitudes* a; owing to atmospheric refraction, they are slightly greater than the true altitudes. The *zenith distance* $z = 90° - a$ is the complement of the altitude and is frequently given instead of a.

A plane through the center of the Earth, running parallel to the apparent horizon, intersects the celestial sphere at the *true horizon*. Altitudes referred to it are called *true altitudes*. Since any celestial body is located at a finite distance from Earth, its apparent altitude will be smaller, sometimes substantially so, than the true one; in the case of the Moon, the difference can amount to about 1°. On the other hand, astronomical refraction caused by the deflection of the rays from a celestial object

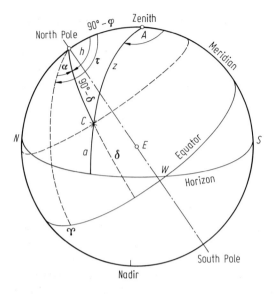

Fig. 2.2. The horizon coordinate system.

in the Earth's atmosphere elevates its apparent position with respect to the horizon; therefore the apparent altitude is increased with respect to the true one. All celestial observations must be corrected for the effects of refraction (see Chap. 23).

The second coordinate in the horizon system is defined by the the direction of the point where the altitude circle through the star cuts the horizon and is called the *azimuth A*. In astronomy, it is measured south to west, north, and east, from 0° to 360°. In geodesy, on the other hand, the measurement starts from the north and proceeds in the same sense. The four points on the horizon having astronomical azimuths 0°, 90°, 180°, and 270° are called the *south point*, *west point*, *north point*, and *east point*, respectively. The altitude circle passing through the south point, the zenith, and the north point is called the *local meridian*; the altitude circle perpendicular to it through the west point, the zenith, and the east point is called the *prime vertical*.

Because of the rotation of the Earth, each celestial body traverses the meridian twice every day, once in its highest position, or *upper transit*, which for observers in the northern hemisphere is usually in the south, and the second time in its lowest position, or *lower transit* in the north. For some stars both transits take place above the horizon; such stars are called *circumpolar*. The condition for a star with declination δ to be circumpolar is given by the relation

$$|\delta + \varphi| \geqq 90°, \tag{2.4}$$

where φ is the geographic latitude of the observer. In the northern hemisphere, $\varphi \geqq 0$ and the circumpolar condition becomes simply that $\delta \geqq 90° - \varphi$.

All stars that do *not* fulfill the circumpolar condition of Eq. (2.4) must rise and set, provided that they can be seen at all. The positions on the horizon where the star rises and sets and the star's motion above the horizon all depend on the geographic latitude of the observer as well as on the declination of the star.

As viewed from a point on the Earth's equator, all stars rise perpendicularly to the horizon. The further the place of observation is from the equator the smaller is the angle between the horizon and the direction of the star's rising or setting. When viewed from the Earth's North or South Poles, all stars move in circles parallel to the horizon and therefore never rise or set; the Sun, Moon, planets, and comets, however, because of their own intrinsic motions in declination, can in fact rise and set as seen from these two specific locations.

2.2.3 The Equatorial System, Vernal Equinox, and Sidereal Time

The major disadvantage of using the horizon system is that the coordinates of each star change continuously during the course of the day, and also that they differ from place to place. This disadvantage is removed in the *equatorial system*. If one imagines an extension of the plane of the Earth's equator, it intersects the celestial sphere at the *celestial equator*. The imaginary rotational axis of the Earth is perpendicular to this plane and intersects the celestial sphere at the north and south celestial poles. In a system thus defined, it is possible to define more permanent coordinates of the star, analogous with the geographical coordinates of a point on the Earth's surface.

The first equatorial coordinate is obtained by measuring the angular distance of a star from the celestial equator toward the north or south from $0°$ at the celestial equator to $90°$ at the celestial pole; this quantity is called the *declination*, δ. It is counted positive toward the north and negative toward the south. Any great circle through the star and perpendicular to the equator is called a *circle of declination*.

The other coordinate is counted along the equator in analogy to the geographic longitude. Here the zero point is the *vernal equinox*, also called the *first point of Aries* (Υ), which is one of the two points of intersection between the celestial equator and the plane of the annual solar motion (the *ecliptic*). When the Sun arrives at this point, spring begins. The diametrically opposite point is called the *autumnal equinox*. The second coordinate of the star corresponds to the geographic longitude and is measured by the angular distance from the vernal equinox to the point of intersection of the circle of declination with the equator; the possible range is from $0°$ to $360°$, and the sense of the angle is from west to east, that is, in the sense of the annual motion of the Sun. This coordinate is called the *right ascension*, α, or simply RA. The position of a star on the celestial sphere is uniquely determined by the two coordinates of right ascension and declination.

The vernal equinox itself therefore has the coordinates

$$\alpha = 0^h 0^m 0^s \quad \text{and} \quad \delta = 0°0'.$$

Because of the slow displacement of the vernal equinox, to which reference will be made again later, the coordinates α and δ are subject also to slow, progressive, and periodic variations (see Sect. 2.4).

It is not difficult to find a relation between the coordinate systems referred to the horizon and to the equator if the concept of the *hour angle h* is introduced. This quantity is the angular distance between the circle of declination through the star and the upper meridian. It is measured from south through west, north, and east from 0° to 360°, or from 0^h to 24^h. Since a whole revolution of 360° takes place in 24^h, the angles can always be expressed either in time *or* in degrees by means of the following conversions:

$$1^h = 15°$$
$$1^m = 15'$$
$$1^s = 15''.$$

At upper transit, the star has the hour angle $0°(0^h)$, and at the lower transit $180°(12^h)$. Of course the vernal equinox, just like a star, also has a certain hour angle. This special quantity is known as the *sidereal time* τ. When the vernal equinox is in upper transit the local sidereal time is $\tau = 0^h 0^m 0^s$. There exists a very simple relationship between sidereal time, right ascension, and hour angle (see also Fig. 2.2):

$$\tau = h + \alpha \quad \text{or} \quad h = \tau - \alpha. \tag{2.5}$$

This relation is of greatest practical importance when, for instance, one wishes to locate an object in the sky: if $\tau > \alpha$, the hour angle is west; if $\tau < \alpha$ the hour angle is east.

2.2.4 Transformation of Horizontal Coordinates into Equatorial Coordinates and Vice Versa

The transformation from horizontal coordinates into equatorial ones and vice versa is achieved by applying the formulae of spherical trigonometry (refer to Sect. 2.7 and texts on spherical astronomy for derivations). To convert horizontal into equatorial coordinates, the following outline can be used:

Given: $z = 90° - a, \varphi, A$
Find: h, δ

The necessary equations are:

$$\cos \delta \sin h = \sin z \sin A, \tag{2.6a}$$
$$\sin \delta = \sin \varphi \cos z - \cos \varphi \sin z \cos A, \tag{2.6b}$$
$$\cos \delta \cos h = \cos \varphi \cos z + \sin \varphi \sin z \cos A. \tag{2.6c}$$

The inverse problem of converting equatorial coordinates into horizontal ones is solved analogously:

Given: $\varphi, \delta, h = \tau - \alpha$
Find: A, z

The formulae are:

$$\sin z \sin A = \cos \delta \sin h, \qquad (2.7a)$$
$$\cos z = \sin \varphi \sin \delta + \cos \varphi \cos \delta \cos h, \qquad (2.7b)$$
$$\sin z \cos A = -\cos \varphi \sin \delta + \sin \varphi \cos \delta \cos h. \qquad (2.7c)$$

Example: A star has declination $\delta = +10°36'\!.3$ and hour angle $h = +2^h12^m8$. What is the zenith distance z and the azimuth A if the geographic latitude of the observer is $\varphi = +48°8\!.^m9$?

For computational work, it is usually convenient to convert the angles into pure degree units. Expressed in degrees, the hour angle is $h = +33°12'\!.0 = +33°\!.200$. Converting arcminutes to degree fractions, the other two angles become $\delta = +10°\!.605$ and $\varphi = +48°\!.148$. Formula (2.7b) then gives

$$\cos z = \sin(+48°\!.148)\sin(10°\!.605) + \cos(48°\!.148)\cos(10°\!.605)\cos(33°\!.200)$$
$$= +0.685845.$$

Thus, the zenith distance is

$$z = +46°\!.698 = +46°41\!.^m9.$$

Next, using Eq. (2.7a),

$$\sin A = \frac{\cos(10°\!.605)\sin(+33°\!.200)}{\sin(+46°\!.698)} = 0.739556.$$

The azimuth follows as

$$A = 47°\!.6936 = 47°41\!.^m6,$$

which completes the calculation.

The general principle underlying coordinate transformations can be found in Schütte [2.1].

2.2.5 Other Coordinate Systems

Apart from the systems of coordinates referred to the horizon and to the equator, there are two other coordinate systems which are used in astronomy: *ecliptic coordinates* and *galactic coordinates*.

2.2.5.1 The System of the Ecliptic. In the ecliptic coordinate system the fundamental plane is the ecliptic itself. The *ecliptic latitude* β is measured at right angles from the ecliptic to the north and the south, running from $0°$ to $\pm 90°$, and the *ecliptic longitude* λ starts at the vernal equinox and runs from $0°$ to $360°$ in the same eastward sense as the right ascension. Since ε, the *obliquity*, or inclination of the ecliptic plane, is known, the equatorial coordinates can easily be converted into ecliptic coordinates, and vice versa. The obliquity is slowly variable and given by the formula

$$\varepsilon = 23°26'21''\!.448 - 46''\!.82T - 0''\!.0006T^2 + 0''\!.0018T^3, \qquad (2.8)$$

with T in Julian centuries[3] from the year A.D. 2000.

---

3 See Sect. 2.5.1 for a discussion of the Julian calendar.

The conversion formulae for transformations from ecliptic to equatorial coordinates are (see, for example, Lang [2.2] and Zombeck [2.3]):

$$\cos\delta\cos\alpha = \cos\beta\cos\lambda, \tag{2.9a}$$
$$\cos\delta\sin\alpha = \cos\beta\sin\lambda\cos\varepsilon - \sin\beta\sin\varepsilon, \tag{2.9b}$$
$$\sin\delta = \cos\beta\sin\lambda\sin\varepsilon + \sin\beta\cos\varepsilon. \tag{2.9c}$$

Similarly, the conversions for transformations from equatorial to ecliptic coordinates are

$$\cos\beta\sin\lambda = \cos\delta\sin\alpha\cos\varepsilon + \sin\delta\sin\varepsilon,$$
$$\sin\beta = \sin\delta\cos\varepsilon - \cos\delta\sin\alpha\sin\varepsilon. \tag{2.10}$$

Example: The position of Comet Kohoutek on 1974 January 03 at 0^h universal time[4] as given in Sect. 3.5 will be used. That calculation gave the ecliptic coordinates as

$$\lambda = 297°39\overset{m}{.}8 \quad \text{and} \quad \beta = +3°16\overset{m}{.}0.$$

These are to be converted into right ascension and declination on the assumption that $\varepsilon = 23°26\overset{m}{.}7$. Using Eq. (2.9c), we obtain directly

$$\sin\delta = \cos(+3\overset{\circ}{.}2667)\sin(297\overset{\circ}{.}6633)\sin(23\overset{\circ}{.}4450) + \sin(3\overset{\circ}{.}2667)\cos(23\overset{\circ}{.}4450)$$
$$= -0.2995368,$$

where all angles have been expressed in pure degrees and fractions thereof. Thus, the declination is

$$\delta = -17\overset{\circ}{.}4298 = -17°25'\!.8.$$

The right ascension may be obtained by solving Eq. (2.9a) for $\cos\alpha$:

$$\cos\alpha = \frac{\cos\beta\cos\lambda}{\cos\delta}$$
$$= +0.4858273.$$

Therefore,

$$\alpha = 299\overset{\circ}{.}066689 = 299\overset{\circ}{.}066689 \times \frac{1^h}{15°}$$
$$= 19^h937779.$$

Converting α into time units, the final result is

$$\alpha = 19^h 56\overset{m}{.}3, \quad \delta = -17°25'\!.8.$$

2.2.5.2 The Galactic Coordinate System. For investigations of stars and galactic structure it is convenient to use the galactic plane, or Milky Way, as the fundamental plane. The coordinates in this scheme are called *galactic coordinates* and are defined as follows (see, e.g., Mihalas and Binney [2.4]):

Galactic latitude b runs from 0° to 90°, and is positive to the north and negative to the south of the galactic plane. *Galactic longitude l* is measured from 0° in the direction of the galactic center and runs eastward along the galactic plane to 360°.

---

[4] See Sect. 2.5.2 for an explanation of universal time.

The north galactic pole has equatorial coordinates

$$\alpha = 12^h 49^m.0, \qquad \delta = +27°24' \quad \text{(equator 1950)},$$

while those of the galactic center are

$$\alpha = 17^h 42^m.4, \qquad \delta = -28°55' \quad \text{(equator 1950)},$$

The inclination of the galactic plane against the 1950 equator is $90° - 27°.4 = 62°.6$.

Prior to the 1959 definition, a different system was used and can be found in earlier literature. It counted galactic longitude from the ascending node of the ecliptic plane on the equator. In the new system this node has the coordinates

$$\alpha = 18^h 49^m.0, \qquad \delta = 0°$$

and

$$l = 33°.0, \qquad b = 0°,$$

so that apart from a slight change in the inclination, $l(\text{new}) = l(\text{old}) + 33°$. During the transition period, the coordinates l, b in the old and new systems were distinguished by superscripts I=old and II=new. Since problems involving galactic coordinates do not require high positional accuracy, auxiliary tables are adequate for most purposes [2.1].

2.3 Time and the Phenomena of Daily Motion

The simplest and most obvious periodic change that can be used to measure time is the continual passage from day to night and night to day; for longer intervals, the change of seasons is convenient to use. Thus the rotation of the Earth and its revolution about the Sun are the basis for nearly all astronomical time calculations. Since most of the important astronomical quantities are time dependent, each measurement should include the specific date and time when it was made.

Decades ago, it was recognized that the period of rotation of the Earth about its axis, that is, the length of the day, is not strictly constant. Therefore, the day consisting of 24 hours can no longer be considered an accurate measure of time when high precision is needed. Far more accurate units of time defined in terms of atomic oscillations now exist and must be employed instead. For the purpose of most astronomical observations, however, such extreme precision is not required, and the older methods of time determination, most of which were in use before the variability of the day was recognized, are still of value. The modifications of the definition of time owing to the variability of the day will be treated in Sect. 2.6.

2.3.1 True and Mean Solar Time

Originally the position of the Sun provided the interval of the day. However, when observations were made from a fixed location of the intervals between two successive upper transits of the Sun, it was noticed that these intervals are subject to conspicuous variations. Thus the hour angle of the "true" Sun, which gives the *true solar time*

(TST), is nonuniform and therefore unsuitable for measuring time with precision. There are two major reasons for this nonuniformity:

1. The Earth's orbit around the Sun is not a circle, but an ellipse. Because of Kepler's second law it is therefore impossible for the Sun to traverse equal arcs in the sky in equal times.
2. Even if the Sun's motion along the ecliptic swept out equal arcs in equal intervals of time, the projection of these arcs onto the celestial equator, along which time is measured, would lead to unequal time segments owing to the mutual inclination between the equator and the ecliptic.

It is for these reasons that the concept of a *fictitious mean Sun*, whose position coincides with that of the true Sun at the vernal equinox and which moves with uniform speed along the celestial equator, is introduced. This defines *mean solar time* (MST) as a uniform measure of time.

The hour angle of this mean Sun is called *mean local time* (MLT) and differs from place to place. Mean solar time and also the true solar time are therefore both *local times*.

The difference between the true and mean solar times, TST − MST, is called the *equation of time*. The equation of time becomes zero on four dates each year, namely on April 16, June 14, September 2, and December 25. The extreme values, on the other hand, are $-14\overset{m}{.}3$ on February 11, $+3\overset{m}{.}7$ on May 14, $-6\overset{m}{.}4$ on July 26, and $+16\overset{m}{.}4$ on November 3. The behavior of the equation of time over the course of a year is shown in Fig. 2.3 (see also Fig. 10.7 in Chap. 10).

Before 1930, the definition of the equation of time used to be opposite in sign to that given above. In current nomenclature, the term is mostly avoided and is replaced with the term *ephemeris transit of the true Sun*.

2.3.2 The Relation Between Sidereal Time and Mean Time

Sidereal time has already been defined as the hour angle of the vernal equinox (Sect. 2.2.3). Strictly speaking, it is not a uniform measure of time, since nutation (see Sect. 2.4.2) causes the vernal equinox to perform a small periodic variation, although in practice this effect can be neglected. A sidereal day, that is, 24 hours of

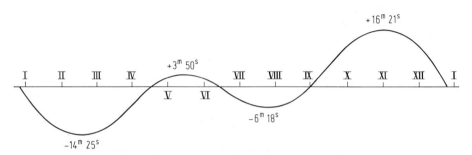

Fig. 2.3. The equation of time.

sidereal time, is the interval between two upper transits of the vernal equinox. Sidereal time is also a local time, since, from Eq. (2.5),

$$\tau = h + \alpha.$$

On the meridian, $h = 0^h$, and therefore $\tau = \alpha$. This means that the right ascension of a star at its upper transit is equal to the sidereal time. This can also be put another way: sidereal time is the right ascension of all those stars which simultaneously traverse the meridian at their upper transit. Of course, the fictitious mean Sun also has a certain right ascension at its upper transit, and this is called *sidereal time at mean noon*.

Astronomical almanacs generally give the sidereal time for 0^h universal time from day to day. This is valid only for the meridian of Greenwich, and since sidereal time is a local time, it is necessary to add a correction $\Delta\tau$ to obtain the sidereal time at mean midnight at a point on the Earth whose longitude is $\Delta\lambda$ (in hours) measured from Greenwich:

$$\Delta\tau = \pm 9\overset{s}{.}8565 \Delta\lambda, \qquad (2.11)$$

where the $+$ and $-$ signs are for longitudes west and east, respectively, of Greenwich.

A sidereal day is somewhat shorter than a mean day since the Sun moves daily some amount toward the east and thus arrives at upper transit later than a fixed star. Dividing a sidereal day into 24 hours of 60 minutes each, with each minute of 60 seconds, it is found that every unit of sidereal time must be somewhat shorter than the corresponding unit in mean time. During a *tropical year* (see Sect. 2.5), the Sun moves from vernal equinox to vernal equinox. This requires 365.2422 mean solar days. During this time the Sun has completed one revolution; a star, therefore, has one more upper transit per year than the Sun. The relation between the two days is thus

$$\text{One mean solar day} = \frac{366.2422}{365.2422} \text{ sidereal days},$$

$$\text{One sidereal day} = \frac{365.2422}{366.2422} \text{ mean solar days}.$$

From this it follows that

$$24^h 0^m 0^s \text{ mean solar time} = (24^h + 3^m 56\overset{s}{.}555) \text{ sidereal time},$$
$$24^h 0^m 0^s \text{ sidereal time} = (24^h - 3^m 55\overset{s}{.}909) \text{ mean solar time}.$$

Or, per hour,

$$1^h \text{ mean solar time} = (1^h + 9\overset{s}{.}856) \text{ sidereal time},$$
$$1^h \text{ sidereal time} = (1^h - 9\overset{s}{.}829) \text{ mean solar time}.$$

For most purposes the conversion can be achieved with sufficient accuracy using a simple approximate relation given in 1902 by Börgen:

For every hour of mean solar time, the sidereal time *gains* $(10 - 1/7)^s$.
For every hour of sidereal time the mean solar time *loses* $(10 - 1/6)^s$.

The error due to this approximation is only $0\overset{s}{.}00067$ per hour in the first case and $0\overset{s}{.}00379$ in the second.

2.3.3 Other Phenomena of Diurnal Motion

During the 24 hours in a day, the apparent motion of celestial bodies causes some particular phenomena which are to be commented on. Upper transit has already been mentioned; there is also *rising* and *setting*, *passage through the prime vertical*, and *largest digression*. A summary of formulae for the more important phenomena of diurnal motion, supplemented by additional formulae, is presented below.

The zenith angle of a star at upper and lower transit is

Upper transit:
$$z_s = (\varphi - \delta) \text{ for a star south of the zenith}$$
$$z_n = (\delta - \varphi) \text{ for a star north of the zenith} \quad (2.12)$$

Lower transit:
$$z = 180° - (\varphi + \delta) \text{ for northern latitudes}$$
$$= 180° + (\varphi + \delta) \text{ for southern latitudes} \quad (2.13)$$

Stars are circumpolar (in the northern hemisphere) if (see Eq. (2.4))

$$\delta \geqq 90° - \varphi.$$

The hour angle of a rising or setting star is (without refraction)

$$\cos h_0 = -\tan \delta \, \tan \varphi, \quad (2.14)$$

where h_0 is called the *semidiurnal arc* (see Appendix Table B.5).

The evening and morning elongation A_0 is measured from the south:

$$\cos A_0 = -\frac{\sin \delta}{\cos \varphi}, \quad (2.15)$$

where $A_0 \leqq 90°$ if $\delta \leqq 0°$.

Transit through the prime vertical $(A = \pm 90°)$
The zenith distance z is found from

$$\cos z = \frac{\sin \delta}{\sin \varphi}, \quad (2.16)$$

and the hour angle h by

$$\cos h = \tan \delta \, \cot \varphi. \quad (2.17)$$

Comment: Passage through the prime vertical takes place only if δ and φ have the same sign and if $|\delta| < |\varphi|$.

Maximum azimuth (largest digression)
The zenith distance is given by:

$$\cos z = \frac{\sin \varphi}{\sin \delta}, \quad (2.18)$$

and the hour angle found from

$$\cos h = \cot \delta \, \tan \varphi. \tag{2.19}$$

Comment: Maximum azimuth takes place only if δ and φ have the same sign and if $|\delta| > |\varphi|$. Maximum digression and transit through the prime vertical are mutually exclusive events.

2.4 Changes in the Coordinates of a Star

While the coordinates of a star measured in the horizon system usually undergo very rapid changes with time, the coordinates in other systems vary only slowly. The causes of these variations are discussed in the subsections below.

2.4.1 Proper Motion

Edmund Halley recognized in 1718 that the stars change their positions on the celestial sphere relative to one another. These continuous displacements are due to the fact that all stars, including the Sun, are actually traveling through space, and the component of the yearly displacement which is perpendicular to the line of sight (i.e., the projection of the annual displacement onto the celestial sphere) is called the *proper motion* μ of the star. Because the stars are very distant this proper motion is very tiny and therefore escaped detection until the eighteenth century. Halley was able to demonstrate the proper motions of only a very few stars; today over 500 000 proper motions are known. The largest annual proper motion is $10''\!.27$, exhibited by a faint star of magnitude $9^\mathrm{m}\!.7$. Only about 500 stars are known to have proper motions which exceed $1''\!.0$. The annual proper motion is resolved into two components, one in the direction of right ascension, the other in declination; these are denoted by μ_α and μ_δ. The observer will, apart from a few exceptions, have no need to make allowance for the annual proper motion.

2.4.2 Precession and Nutation

Precession is of the utmost importance since it causes relatively rapid changes in the coordinates α and δ. The gravitational pull of the Sun and the Moon on the equatorial bulge of the rotating Earth causes the polar axis to slowly move, eventually describing a complete revolution about the pole of the ecliptic in about 26 000 years. This period, which is more precisely stated as 25 725 years, is also called the *Platonic year*, and the movement of the polar axis about the pole of the ecliptic, in analogy to the behavior of a spinning top, is called *precession*. This phenomenon was recognized by Hipparchus as early as the second century B.C. and appears as a secular retrograde movement of the vernal equinox. The largest part of the precession, the so-called *lunisolar precession*, is caused by the inequality of the moments of inertia of the Earth and can be determined only by empirical methods. The annual lunisolar precession, according to international convention, is found to be

$$p_0 = 50\rlap{.}''3878 + 0\rlap{.}''000049\, t, \tag{2.20}$$

where t is measured in tropical years from the year 2000. To be added to this is a corresponding effect due to the planets in the solar system, the annual *planetary precession*:

$$p_1 = -0\rlap{.}''1055 + 0\rlap{.}''000189\, t. \tag{2.21}$$

The total effect is the so-called *general precession*:

$$p = p_0 + p_1 \cos \varepsilon = 50\rlap{.}''2910 + 0\rlap{.}''000222\, t. \tag{2.22}$$

In consequence of this precession the ecliptic longitude of the star increases continuously.

In order to evaluate the effect of precession on the coordinates α, δ of a star, it is helpful to introduce the two components m and n of the annual precession in the directions of right ascension and declination, respectively:

$$\begin{aligned} m &= +46\rlap{.}''124 + 0\rlap{.}''000279\, t, \\ &= 3\rlap{.}^s 0749 + 0\rlap{.}^s 0000186\, t, \\ n &= +20\rlap{.}''043 - 0\rlap{.}''000085\, t. \end{aligned} \tag{2.23}$$

For any star with coordinates α and δ, the annual variations are given by the formulae

$$\begin{aligned} p_\alpha &= m + n \sin\alpha \tan\delta, \\ p_\delta &= n \cos\alpha. \end{aligned} \tag{2.24}$$

It is evident that p_α is usually positive, while p_δ has the same sign as $\cos\alpha$. The values of p_α and p_δ calculated from Eqs. (2.24) are given in Appendix B, Table B.10.

Currently, the north celestial pole is in close proximity to the star α Ursae Minoris; it will come nearest to that star in the year A.D. 2115, when its angular distance will be only 28′. Around A.D. 14 000, the bright star Vega (α Lyrae) will be the new "pole star."

To calculate the effects of precession on star positions a few years, decades, or centuries into the future, one generally uses a series expansion of the form

$$\alpha, \delta = \alpha_0, \delta_0 + t\mathrm{I} + \frac{t^2}{200}\mathrm{II} + \frac{t^3}{100}\mathrm{III}, \tag{2.25}$$

where time t is measured in years. The quantities I, II, and III are usually given in star catalogs and have the following names:

I = variatio annua (= annual precession + proper motion),
II = variatio saecularis (= secular variation),
III = third term.

For very large time intervals and for stars near to the poles, however, a transformation requires the strict formulae of spherical trigonometry.

The revolution of the Earth's axis about the pole of the ecliptic is not entirely uniform. The term "precession" means the progressive part of this motion; the small fluctuations superimposed on this motion are known as *nutation*. In 1747, the British astronomer J. Bradley discovered an oscillation of the polar axis with a period of 18.6 years. This short period movement is the major contribution of nutation and is caused by the fact that the lunar orbit does not lie in the plane of the ecliptic, and by the retrograde motion of the nodes of the intersection of the lunar orbit and the ecliptic. It therefore has the period of one revolution of the line of nodes, namely, 18.60 tropical years. Owing to nutation alone, the celestial pole describes an ellipse on the celestial sphere. The semimajor axis of this ellipse (i.e., the amplitude of the nutation oscillation) is called the *constant of nutation* and is equal to $9''.202$.

2.4.3 Aberration

The phenomenon of *aberration of starlight* originates from the fact that the orbital velocity of the Earth is a finite fraction of the speed of light. Because of this, the stars appear displaced a little in the direction of motion of the Earth. This phenomenon was discovered accidentally by Bradley in 1728. The *annual aberration*, caused by the motion of the Earth in its orbit about the Sun, must be distinguished from the much smaller *diurnal aberration* resulting from the rotation of the Earth around its polar axis. Because of the annual aberration, a star changes its position periodically throughout the year, tracing out a small aberration ellipse on the celestial sphere.

The maximum value of the displacement, called the *aberration constant*, is

$$k = 20''.496.$$

If the ecliptic coordinates of a star affected by aberration are λ' and β', then the displacements due to aberration are given by

$$\begin{aligned}\lambda' - \lambda &= -20''.496 \cos(\lambda - \lambda_\odot) \sec \beta, \\ \beta' - \beta &= +20''.496 \sin(\lambda - \lambda_\odot) \sin \beta,\end{aligned} \quad (2.26)$$

where λ and β are the coordinates the star would have in the absence of aberration and $\lambda_\odot$ is the ecliptic longitude of the Sun.

The semiaxes of the aberration ellipse are

$$\begin{aligned} a &= 20''.496, \\ b &= 20''.496 \sin \beta. \end{aligned} \quad (2.27)$$

At the pole of the ecliptic ($\beta = 90°$), the aberration ellipse becomes a circle, and on the ecliptic ($\beta = 0°$) it degenerates to a linear segment.

The diurnal aberration originates from the rotation of the Earth. The speed of displacement of a point at the Earth's equator gives the constant of daily aberration in the amount of $0''.31$. The influence of daily aberration on right ascension and declination is found from the formulae

$$\begin{aligned} \alpha' - \alpha &= 0''.31 \cos \varphi \cos h \sec \delta, \\ \delta' - \delta &= 0''.31 \cos \varphi \sin h \sin \delta, \end{aligned} \quad (2.28)$$

where α' and δ' are the coordinates affected by diurnal aberration.

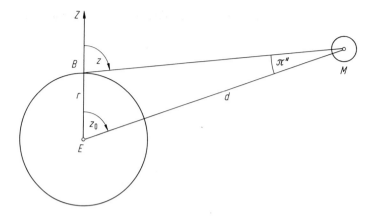

Fig. 2.4. Parallax.

2.4.4 Parallax

Diurnal parallax is due to the fact that angles to a celestial object vary somewhat depending on whether they are measured from a point on the surface of the Earth or from the center of the Earth. In Fig. 2.4 the angle z_0 indicates the zenith distance of the center of the Moon as it would appear from the center of the Earth; however, the zenith distance observed at the same time from a point B on the surface of the Earth would be z. Thus there exists a small angle π'' subtended at the center of the Moon which is the parallax. The larger the distance between the centers of the celestial bodies, the smaller the value of π''. For $z = 90°$, the maximum value of π'', the *equatorial-horizontal parallax* π_0'' is

$$\pi_0'' = \frac{r}{d \sin 1''}, \tag{2.29}$$

where d is the distance, in units of equatorial radii of the Earth, between the two bodies and r is the distance to the center of the Earth. For any given zenith distance z,

$$\pi'' = \pi_0'' \sin z. \tag{2.30}$$

The equatorial-horizontal parallaxes of the Moon and Sun are

$$\pi_{\mathcal{C}}'' = 3422.''44 = 57'2.''44,$$

$$\pi_{\odot}'' = 8.''794.$$

In exact calculations the fact that the Earth is flattened at the poles must also be taken into account. The difference between the geographic and the geocentric latitude, $\varphi - \varphi'$, and the definition of the corresponding radius r' have been given in Sect. 2.2.1. For all the bodies of the Solar System, except the Moon and other nearby objects, the following approximate formulae allow for the correction of diurnal parallax in α and δ:

$$\alpha' - \alpha = -\frac{r'\pi_0'' \cos\varphi'}{d \cos\delta} \sin h,$$

$$\delta' - \delta = -\frac{r'\pi_0'' \sin\varphi'}{d} \frac{\sin(\gamma - \delta)}{\sin\gamma},$$

(2.31)

where $\tan\gamma = \tan\varphi' \cos h$, and π_0'' is the equatorial-horizontal parallax of the star. Furthermore,

α, δ, d are the geocentric coordinates of the celestial body,
$h = \tau - \alpha, \varphi', r'$ are the geocentric coordinates of the observing site, and
α', δ', d' are the coordinates of the body referred to the observing site.

For the Moon and other nearby bodies such as artificial satellites and passing comets and asteroids, however, the more exact formulae must be used.

The *annual parallax* is defined analogously, except that the baseline is now the semimajor axis of the Earth's *orbit*. The annual parallax of even the nearest fixed star is $0\rlap{.}''772$, but most observers will seldom, if ever, need to take such minute angles into account.

2.4.5 Reduction of Mean Position to Apparent Position

The *mean position* α_0, δ_0 of a star is its position at some chosen instant of time referred to the mean coordinate system at that time from a star catalogue and with precession taken into account according to the formulae of Sect. 2.4.2; it also includes the constant term of aberration. On the other hand, the *apparent position* $\alpha_{app}, \delta_{app}$ is the position of the star at perhaps a different instant, referred to the true coordinate system at that instant. Thus the apparent position differs from the mean position in that it is affected by nutation, the variable part of aberration, and the displacement due to precession and proper motion between the two instants of time. The reduction from mean to apparent position can be effected via the following formulae:

$$\alpha_{app} = \alpha_0 + t\mu_\alpha + Aa + Bb + Cc + Dd + E,$$
$$\delta_{app} = \delta_0 + t\mu_\delta + Aa' + Bb' + Cc' + Dd'.$$

(2.32)

Here

$a = m + 1/15 \ n \sin\alpha \tan\delta,$ $a' = n\cos\alpha,$
$b = 1/15 \ \cos\alpha \tan\delta,$ $b' = -\sin\alpha,$
$c = 1/15 \ \cos\alpha \sec\delta,$ $c' = \tan\varepsilon \cos\delta - \sin\alpha \sin\delta,$
$d = 1/15 \ \sin\alpha \sec\delta,$ $d' = \cos\alpha \sin\delta.$

The quantities A, B, C, D, and E are tabulated for each day of the year in *The Astronomical Almanac* as "Besselian day numbers." Contrary to earlier use, these quantities now include the terms of short-period nutation, and are reckoned from the middle of the year; the reduction does not, of course, include the proper motion of the star.

Another form of transformation by trigonometric functions may be preferable over longer periods of time. However, the day numbers required for this transformation are no longer listed daily; page B20 in *The Astronomical Almanac* should be consulted.

2.5 Problems of the Calendar and Time Zones

It was already apparent to ancient cultures that difficulties are encountered in creating a useful calendar, since the year (i.e., the apparent revolution of the Sun) is not an exact multiple of the length of the day or the length of the month. Furthermore, the occurrence of sunrise, noon, and sunset are distinctly different depending on an observer's longitude, but this has been a problem only in modern times because of the dramatic increase in the mobility of our civilization.

2.5.1 The Calendar and the Measurement of Years

The beginnings of time reckoning date back to the fourth millennium B.C. to the Egyptians, who originally used a pure *solar year* of 360 days, starting from the annual flooding of the Nile. In later times, when the deviation from the position of the Sun became noticeable, five days were added. But even then a slow shift in time was found that amounted to one day in every four years. The Egyptians thereby deduced that the length of the year was 365.25 days.

In 46 B.C., Julius Caesar put an end to the arbitrary corrections made to the calendar by the priests, and introduced a new calendar, called the *Julian Calendar* after him, in which every three years of 365 days were followed by one year of 366 days. This calendar was in use up to 1582, by which time the vernal equinox occurred markedly before March 21. From this it could be deduced that the time between one transit of the Sun through the (mean) vernal equinox to the next, the so-called *tropical year*, must be shorter than 365.25 days. Thus a new calendar reform became necessary, and Pope Gregory XIII (1572–1585) decreed by a Bull of 1582 February 24 that:

1. The day following 1582 October 4 should have the date 1582 October 15 in order to restore the vernal equinox to March 21.
2. All years that are divisible by 4 should be leap years of 366 days, except those which coincide with the beginning of a century; the latter are leap years only if they are divisible by 400. (Thus 1600 and 2000 are leap years, while 1700, 1800, and 1900 are not.)

This calendar, called the *Gregorian Calendar*, therefore has (within each cycle of 400 years) three leap days fewer than the Julian Calendar. Thus the length of the mean Gregorian year is 365.2425 days; the length of the tropical year (see below) is 365.2422 days. The difference of 0.0003 day, which equals 26^s, accumulates to one full day only after some 3000 years.

The manner in which the years are numbered since the birth of Christ (A.D.) was proposed by the Roman Abbot Dionysius Exiguus in about A.D. 525. He erred in fixing the initial year, as it appears that Christ was born seven years earlier than the date assumed by Exiguus.

Later, the years before the beginning of the Christian era (B.C.) were also numbered continuously in this way, but allowance was not made for the fact that between the year A.D. 1 and the year 1 B.C. there should have been a year with the number "0." For this reason there are differences between historical and astronomical numbering of years. For instance:

$$1959 \text{ A.D. (historical)} = +1959 \text{ (astronomical)}$$
$$1 \text{ B.C. (historical)} = 0 \text{ (astronomical)}$$
$$300 \text{ B.C. (historical)} = -299 \text{ (astronomical)}$$

2.5.2 The Length and Beginning of the Year

Precession is not quite constant and thus the lengths of the various years change very slowly. Denoting by T the time in Julian centuries since 1900 January 0.5, each century comprising 36 525 ephemeris days, the following definitions are made:

1. The *tropical year* is the interval between two transits of the mean Sun through the mean equinox. During this time the mean longitude of the mean Sun increases by 360°. Because of the retrograde motion of the vernal equinox the mean Sun does not make a complete revolution. Therefore

$$\text{One mean tropical year} = 365\overset{d}{.}24219879 - 0\overset{d}{.}00000614T. \quad (2.33)$$

The motion of the vernal equinox is not quite regular, however, since precession slowly increases; thus the length of the tropical year decreases by 5.36 seconds every 1000 years.

2. The *sidereal year* is the time the Sun takes to return to the same star in the ecliptic. It must therefore be longer than the tropical year:

$$\text{One sidereal year} = 365\overset{d}{.}25636042 + 0\overset{d}{.}00000011T. \quad (2.34)$$

The sidereal year is *not* used to construct a calendar.

3. The *anomalistic year* is the time between successive transits of the Earth through its perihelion point. Since the perihelion advances along the Earth's orbit, this year is about 4.5 minutes longer than the sidereal year.

$$\text{One anomalistic year} = 365\overset{d}{.}25964134 + 0\overset{d}{.}00000304T. \quad (2.35)$$

4. The *Julian year* has already been defined above:

$$\text{One Julian year} = 365\overset{d}{.}25.$$

This year is at present $11^m 14^s$ *longer* than the tropical year.

5. The *Gregorian year*, also called the *civil year*, has been defined above. It contains 365.2425 days and is our normal calendar year.

The beginning of the astronomical year was defined by Friedrich Bessel as the moment at which the right ascension of the mean Sun (including the constant aberration term) is

$$RA = 18^h 40^m = 280°.$$

This moment coincides very closely with the beginning of our normal civil year; it is also called Bessel's *annus fictus*.

The length of the annus fictus is $365^d.24219879 - 0^d.00000786 T$ mean days. It therefore differs insignificantly from the tropical year. Bessel's year is independent of the observing site and therefore starts at the same moment all over the Earth. It is subdivided into decimals so that, for instance, the beginning of the Bessel year 1959 is called 1959.0.

The difference, beginning of the civil year minus beginning of the annus fictus, is equal to k, which is called the *dies reductus*.[5]

2.5.3 The Julian Date and the Beginning of the Mean Day

It is generally inconvenient to express large intervals of time in terms of days. For this reason it is the *Julian date* (JD) which is mainly used in reducing observations of variable stars. It was Joseph Justus Scaliger (1540–1609) who proposed a period of

$$19 \times 28 \times 15 \text{ years} = 7980 \text{ years},$$

which is named after his father Julius Caesar Scaliger. The beginning of this Julian period was set at January 1.0 of the year -4712.

Up until the end of 1924, the moment of the upper transit of the mean Sun, that is, the mean noon (beginning of the astronomical mean day), was taken as the beginning of the mean day. The time system thus described was referred to as *Greenwich mean time* (GMT). From 1925 onward, the beginning of the mean day was moved from the upper transit of the mean Sun to the preceding lower transit, that is, to mean midnight. Since 1925, the civil time at Greenwich has been called *universal time* (UT).

The days of the Julian period (Julian date within the Julian period) continue to begin at Greenwich Mean Noon (12^h UT), unaffected by the 1925 redefinition of the day (which begins at midnight).

Therefore:

$$1924 \text{ December } 31, \ 12^h \text{ GMT} = 1925 \text{ January } 1, \ 0^h \text{ UT};$$
$$1947 \text{ January } 17, \ 21^h 05^m \text{ EST} = \text{JD } 2432203^d.587.$$

Appendix Table B.8 gives the days of the Julian period between 1000 and 2000 A.D., and Table B.9 shows days, hours, and minutes in decimal fractions of the Julian Year. The modified Julian Date (MJD) is JD minus 2 400 000.5.

---

5 *Translators' note*: For reason of accomodating the time definitions in Sect. 2.6, the tropical year and the annus fictus are no longer "official" units of the International Astronomical Union. They are, however, still used; the Besselian year still serves to record events in decimal fractions of the year.

2.5.4 Time Zones and the Date Line

The fact that local time depends upon longitude makes it impracticable for use in practical life. For instance, the introduction within countries of "railway time" was of very little help in reducing time-related confusion while traveling. In North America more than 70 different railway times were in use up to 1882. In 1883 they were replaced by 5 different zone times which differed by whole hours from GMT, namely:

$$
\begin{aligned}
\text{Atlantic Standard Time (AST)} &= \text{GMT} - 4^h, \\
\text{Eastern Standard Time (EST)} &= \text{GMT} - 5^h, \\
\text{Central Standard Time (CST)} &= \text{GMT} - 6^h, \\
\text{Mountain Standard Time (MST)} &= \text{GMT} - 7^h, \\
\text{Pacific Standard Time (PST)} &= \text{GMT} - 8^h.
\end{aligned}
$$

Within each time zone, all points keep exactly the same time, namely, the local mean solar time of a "standard meridian" running more or less through the middle of the zone. The standard meridians are 60°, 75°, 90°, 105°, and 120° west longitude for AST, EST, CST, MST, and PST, respectively. Hawaii and Alaska both keep the time of the meridian 150° west longitude, which is two hours further behind PST.

Europe later followed suit by introducing the following time zones:

$$
\begin{aligned}
\text{Western European Time (WET)} &= \text{GMT}, \\
\text{Central European Time (CET)} &= \text{GMT}+1^h, \\
\text{Eastern European Time (EET)} &= \text{GMT}+2^h.
\end{aligned}
$$

Almost all countries have now accepted one of 24 existing time zones.

A traveler going round the Earth once from east to west loses one hour in time for every 15° of longitude traveled. Thus one whole day is lost during a complete journey of 360°. If he travels in the opposite direction, that is from west to east, one day is gained. On the line of longitude 180° from Greenwich the date will therefore differ by one day from that of Greenwich. This line is called the *international date line*, which for political reasons in some places does not exactly coincide with the 180° meridian. Thus the practical rule is: if the date line is crossed from east to west, skip one day; if crossed from west to east, repeat one day.

2.6 The Variability of the Time Systems

It has been known for several decades that the rotation of the Earth is subject to minute variations. When extreme precision is required, such variations must be taken into account. This complex subject is treated in [2.6],[2.7], and [2.8].

2.6.1 Various Causes for Changes in the Length of the Day

The time as determined by astronomical observations, referred to the mean Sun and to the mean vernal equinox, is called *empirical time*. Since, however, the Earth rotates irregularly, the empirical time is not identical with the perfectly uniform *ephemeris time* (see below), or inertial time based on Newtonian mechanics, which forms the basis for the ephemerides in the astronomical almanacs. For an exact comparison of observations within our solar system, the difference between the variable empirical time and the ephemeris time must be taken into account.

Three kinds of changes in the rotational velocity of the Earth can be distinguished:

1. *A secular deceleration of the rotational speed of the Earth by tidal friction.* This causes a continuous increase in the length of the day by 4.5×10^{-8} seconds per day, which adds up to 0.0016 seconds per century. Because the effect on the mean longitude of a star is cumulative, the secular acceleration being proportional to T^2, it is of considerable importance. The theory of tidal friction is the concern of dynamic oceanography; a number of theoretical investigations by Jeffreys and others have shown that dissipation of energy by tidal friction, particularly in the smaller oceans, corresponds to the secular acceleration observed in the longitude of the Moon.
2. *Irregular positive and negative accelerations of the rotational velocity of the Earth.* These are called *fluctuations*, and their real origin is still unknown but is probably displacement of mass in the interior of the Earth. These fluctuations have been derived from observations of the Moon, and the largest deviations of the length of the day in the past 250 years have been found to be $-0\overset{s}{.}005$ (in about 1871) and $+0\overset{s}{.}002$ (in 1907). This effect can also be cumulative and leads to noticeable time differences if the Earth's rotation is used as a clock. These fluctuations cannot be predicted in advance, but can only be deduced from observations of the Moon and made known after the fact.
3. *Seasonal variations of the rotational velocity of the Earth.* These are very small and are caused by meteorological events. These variations, which were first discovered in the years 1934–1937, when quartz clocks were introduced, can in general be neglected.

2.6.2 Astronomical Effects of the Variable Rotation of the Earth

Mean solar time is always used for observations in astronomical work. Consequently, it is necessary to transform a given mean solar time into a measure of time which is independent of the variations in the Earth's rotation. In order to implement this goal, additional terms have recently been introduced into astronomy.

The fundamental unit of uniform time is defined in atomic physics to be

1 second = 9 192 631 770 oscillations of the cesium 133 atom in the ground state.

One day consists of 86 400 seconds of this kind. The time thus defined is the argument of astronomical ephemerides, and is called *terrestrial dynamical time* or TDT. When, on the other hand, the ephemerides refer not to the Earth but rather to the *barycenter* of the solar system, *TDB* is used instead.

Experimentally, this dynamically defined time is best reproduced by atomic clocks; the time thus defined is called *international atomic time* (TAI), and its relation to

Table 2.1. Correction of Time ΔT.

Year	ΔT	Year	ΔT
1950.0	$+29^{\rm s}\!.15$	1970.0	$+40^{\rm s}\!.18$
1955.0	$+31.07$	1975.0	$+45.48$
1960.0	$+33.15$	1980.0	$+50.54$
1965.0	$+35.73$	1985.0	$+54.34$

dynamical time is defined as

$$\text{TDT} = \text{TAI} + 32^{\rm s}\!.184. \tag{2.36}$$

The term *ephemeris time* is also still encountered. It uses the length of the tropical year for 1900 January 1, $12^{\rm h}$ as unit, and thus is also independent of the irregularities of the Earth's rotation. From 1950 until 1972, it was the fundamental definition of time but was thereafter superseded by TAI.

The relation between dynamical time and the mean solar time is derived only from observation (i.e., after the fact), although preliminary values of limited precision are obtained by extrapolation of the measures from previous years. The usual Greenwich time, or universal time (UT), does contain the irregularities of the Earth's rotation, and is directly found by observations of stars. When the small effects due to latitude variations are subtracted from it the time obtained is called UT1. For purposes of highest precision, a time UT2 is used which results from UT1 by subtracting the seasonal variation in the Earth's rotation.

For most astronomical observations, the time UT1 is the one to use. The difference between dynamical time and the time UT1, which includes the variations of Earth's rotation, is

$$\Delta T = \text{TDT} - \text{UT1}. \tag{2.37}$$

Numerical values of this difference over longer intervals are listed in the astronomical almanacs. Values of ΔT for the time 1950 to 1985 are given in Table 2.1.

The time signals transmitted by broadcasting stations do not exactly represent the time UT1, but rather *coordinated universal time* (UTC), which differs from the atomic time TAI by an integer number of seconds. UTC is adjusted whenever necessary by the introduction of *leap seconds*, so that the difference between UTC and UT1 never exceeds $0^{\rm s}\!.9$. Leap seconds are always inserted at the beginning or the middle of the year. Since 1972 June 30, twelve such insertions of one second each have been made as was necessary to compensate for variations in the Earth's rotation; additional insertions will be required in future years. Once an accurate law is found from which the irregularities in the Earth's rotation could be calculated in advance, only then will it be possible to establish a firm intercalation formula for time adjustments.

2.7 Spherical Trigonometry

2.7.1 Fundamental Equations

A spherical triangle with sides a, b, c and corresponding (opposite) angles A, B, C is portrayed in Fig. 2.5. The sides and angles are interrelated via three fundamental relations known as the *law of sines*, the *law of cosines*, and the *law of sines and cosines*, which are, respectively,

$$\sin a \sin B = \sin A \sin b, \tag{2.38a}$$
$$\cos a = \cos b \cos c + \sin b \sin c \cos A, \tag{2.38b}$$
$$\sin a \cos B = \cos b \sin c - \sin b \cos c \cos A. \tag{2.38c}$$

Two additional formulae are obtained by interchanging sides and angles; they are called the *law of cosines for angles* and *law of sines and cosines for angles*:

$$\cos A = -\cos B \cos C + \sin B \sin C \cos a, \tag{2.39a}$$
$$\sin A \cos b = \cos B \sin C + \sin B \cos C \cos a. \tag{2.39b}$$

2.7.2 Derived Equations

The *rule of cotangents* can be applied when dealing with four consecutive elements of a triangle, three of which are known, the fourth needed:

$$\cos c \cos A = \sin c \cot b - \sin A \cot B. \tag{2.40}$$

In the years before the widespread use of electronic digital calculators, the following abbreviations were used for logarithmic work:

$$s = \frac{1}{2}(a+b+c) \quad \text{and} \quad \sigma = \frac{1}{2}(A+B+C). \tag{2.41}$$

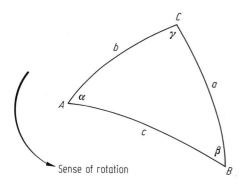

Fig. 2.5. The spherical triangle.

Then the *half-angle formulae* are:

$$\sin\frac{A}{2} = \sqrt{\frac{\sin(s-b)\sin(s-c)}{\sin b \sin c}},$$

$$\sin\frac{B}{2} = \sqrt{\frac{\sin(s-c)\sin(s-a)}{\sin c \sin a}}, \quad (2.42)$$

$$\sin\frac{C}{2} = \sqrt{\frac{\sin(s-a)\sin(s-b)}{\sin a \sin b}}.$$

And the *half-side formulae* are:

$$\sin\frac{a}{2} = \sqrt{\frac{\sin(\sigma-B)\sin(\sigma-C)}{\sin B \sin C}},$$

$$\sin\frac{b}{2} = \sqrt{\frac{\sin(\sigma-C)\sin(\sigma-A)}{\sin C \sin A}}, \quad (2.43)$$

$$\sin\frac{c}{2} = \sqrt{\frac{\sin(\sigma-A)\sin(\sigma-B)}{\sin A \sin B}}.$$

Often the *Gauss–Delambre equations* can be profitably employed:

$$\sin\frac{A}{2}\sin\frac{b+c}{2} = \sin\frac{a}{2}\cos\frac{B-C}{2},$$

$$\sin\frac{A}{2}\cos\frac{b+c}{2} = \cos\frac{a}{2}\cos\frac{B+C}{2},$$

$$\cos\frac{A}{2}\sin\frac{b-c}{2} = \sin\frac{a}{2}\sin\frac{B-C}{2}, \quad (2.44)$$

$$\cos\frac{A}{2}\cos\frac{b-c}{2} = \cos\frac{a}{2}\sin\frac{B+C}{2}.$$

Upon dividing these equations by each other, the *Napier equations* are generated:

$$\tan\frac{B+C}{2} = \cot\frac{A}{2}\frac{\cos\left[(b-c)/2\right]}{\cos\left[(b+c)/2\right]},$$

$$\tan\frac{B-C}{2} = \cot\frac{A}{2}\frac{\sin\left[(b-c)/2\right]}{\sin\left[(b+c)/2\right]},$$

$$\tan\frac{b+c}{2} = \tan\frac{a}{2}\frac{\cos\left[(B-C)/2\right]}{\cos\left[(B+C)/2\right]}, \quad (2.45)$$

$$\tan\frac{b-c}{2} = \tan\frac{a}{2}\frac{\sin\left[(B-C)/2\right]}{\sin\left[(B+C)/2\right]}.$$

Table 2.2. The Solution of Spherical Triangles

Given	Required	Formulae
a, b, c	A, B, C	Law of cosines (2.38b), or half-angle formulae (2.42).
A, B, C	a, b, c	Law of cosines for angles (2.39a), or half-side formulae Eq.(2.43).
a, b, C	A, B, c	Basic equations (2.39), or Gauss–Delambre equations (2.44), or Napier's equations (2.45).
A, B, c	a, b, C	Basic equations for angles (2.39), or Napier's equations, or Gauss–Delambre equations.
a, b, A	B, C, c	B from law of sines (2.38a); c and C from Napier's equations.
A, B, a	b, c, C	b from law of sines; c and C from Napier's equations.

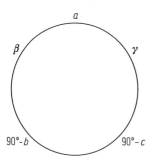

Fig. 2.6. Scheme for formulae in right-angled spherical triangles.

Procedures to calculate spherical triangles can be presented in a general scheme. Three of the six parameters must be known. Thus there are six different cases (see Fig. 2.6) which can be solved according to the recipes presented in Table 2.2.

2.7.3 The Right-Angled Spherical Triangle

If a is the hypotenuse of the spherical triangle in Fig. 2.5, then $A = 90°$.

Assembling the elements of the triangle in the form of Fig. 2.6, omitting the right angle and replacing the two other sides by their complements, then Napier's rule takes the following form:

> The cosine of one element = the product of the *sines* of the two elements shown furthest away from it = the product of the cotangents of the *adjacent* elements.

This yields the following formulae:

$$\cos a = \cos b \cos c = \cot B \cot C,$$
$$\sin b = \sin a \sin C = \tan c \cot B,$$
$$\sin c = \sin a \sin B = \tan b \cot C, \qquad (2.46)$$
$$\cos B = \cos c \sin C = \tan b \cot A,$$
$$\cos C = \cos b \sin B = \cot a \tan c.$$

References

2.1 Schütte, K.: Die Transformation beliebiger spärischer Koordinatensysteme mit einer einzigen immerwährenden Hilfstafel. *Astr. Nachr.* **270**, 75 (1940).
2.2 Lang, K.R.: *Astrophysical Formulae*, Springer, New York Berlin Heidelberg 1980, p. 504.
2.3 Zombeck, M.V.: *Handbook of Space Astronomy and Astrophysics*, Cambridge University Press, Cambridge 1982, p. 71.
2.4 Mihalas, D., and Binney, J.: *Galactic Astronomy* 2nd ed., Freeman, San Francisco CA 1981, p. 39.
2.5 Boss, B.: *General Catalogue of 33342 Stars for the Epoch 1950*, Carnegie Inst. Wash. Pub. No. 486, Washington DC 1936.
2.6 Mulholland, J.D.: *P.A.S.P.* **84**, 357 (1972).
2.7 Winkler, G.M.R., and Van Flandern, T.C.: *Astronomical Journal* **82**, 84 (1977).
2.8 Guinot, B., Seidelmann, P.K.: *Astronomy and Astrophysics* **194**, 304 (1988).
2.9 Gondolatsch, F.: *Veröff. Astr. Recheninstitut Heidelberg* **5** (1953).

3 Applied Mathematics and Error Theory

F. Schmeidler

3.1 Introduction

Practical astronomical work requires, to a large extent, the evaluation of mathematical formulae by numerical calculation. For this reason, astronomy was ranked as a subfield of mathematics until well into the nineteenth century. Today, more than ever, the treatment of many astronomical problems is impossible without the use of mathematical tools. Every observer should possess a certain minimum of mathematical expertise, and it is for this reason that the most important techniques of mathematics will be collected in this chapter. Derivations and proofs of the formulae will not, however, be given here; the interested reader can find such details in standard textbooks on mathematics.

Some knowledge of elementary mathematics, for instance, on roots, logarithms, and trigonometric functions, is assumed in this chapter. On occasion, an acquaintance with the basic notions of differential and integral calculus is also assumed.

Numerical performance of calculations was until a few decades ago in the domain of slide rules and logarithm tables. Meanwhile, pocket calculators have become widely used and are available commercially in various makes. Furthermore, most professional astronomers and a surprising number of amateurs now have access to powerful electronic computers.

3.1.1 Pocket Calculators

Some general comments on the use of pocket calculators follow. Of course all makes perform the four standard arithmetic operations of addition, subtraction, multiplication, and division, but thereafter the similarities end. Some calculators are business oriented, while others are best suited for scientists. For astronomical use, it is also very desirable that the calculator possess keys to perform logarithms, roots, and trigonometric functions, as well as the inverses of these functions. The capacity of the *memory*, where intermediate results can be stored temporarily to be recalled at a later time, is also quite important. Using a calculator with a very limited memory space not only increases the time spent doing a calculation owing to intermediate results having to be written down on paper, but can also result in needless mistakes. Pocket calculators with only one memory are readily available and usually quite inexpensive, but those with several are preferable, even though at somewhat higher cost.

3.1.2 Computers

Electronic computers employ *programs* consisting of a set of instructions in the form of mathematical equations encoded into some computer language and supplied with input data to calculate the desired results, or output. Computer programs can be run as often as desired and thus possess a distinct advantage over the pocket calculator if numerous operations of a similar nature, but with different input data, are to be executed. As a simple example, consider the addition of two quantities a and b. Two numerical values are inserted as input, the computer runs the program and performs the operation $a + b$, and then records and prints the result. Thereafter, any arbitrary values for a and b can be inserted and the program runs to yield the required sum.

Once a computer program has been constructed, it must in general be "debugged" before it can be run. Since the debugging process frequently requires considerable time and effort, the use of large computers is best justified when extensive calculations of a repetitive character are to be executed.

In brief, a computer system is comprised of three basic components, each of which in turn may have two or more subcomponents. Each component performs a specific task, and yet is interconnected with the other components to form the computer system. The components are 1. *input/output* (I/O), 2. the *central processor* (CPU), and 3. *storage*. The CPU unit is often referred to by manufacturers as the *mainframe*, while the other units—input, output, storage—are known as *peripherals*. Each of these units is described in more detail below:

1. *Input/output*. The I/O often consist of two separate units, although this need not be the case. The *input unit* takes coded data, often in the form of magnetic tape or diskette, and translates it into electrical impulses which then are fed into the CPU for processing. The *output unit* takes the processed information from the CPU and presents it in a format which is easily read by a human. A typical output device is a high-speed printer, but output can also be written to a computer terminal, perhaps the same one from which the input was entered.
2. *Central processing unit*. The CPU is the heart of the computer and the equivalent of the human brain. Contained within it are two components: the *control unit* and the *arithmetic-logic unit*. The control unit does no actual processing but instead reads and interprets the instructions contained within a program, and thereby directs other machines to execute the necessary program steps. The arithmetic-logic unit is also contained within the CPU, and performs all processing—calculations, comparisons, or decisions—made by the computer. During processing, the data is moved from temporary storage within the CPU to the arithmetic-logic unit.
3. *Storage*. The storage component is the equivalent of human memory and is, as the name suggests, responsible for storing data until it can be retrieved at high speed by the CPU. The data are stored in electronic form, usually on magnetic disks or magnetic tapes.

The computing equipment, which does the actual reading, writing, processing, and storing, is usually referred to collectively as the *hardware*, whereas the programs, which are written to direct the operation of the hardware, are known as *software*.

A good general introduction to the computer and its functions can be found in the textbook by Frates and Moldrup [3.1].

The details regarding all of these processes and the constituent parts of both the pocket calculators and the program-operated computers vary considerably from make to make, and therefore the accompanying manuals must be consulted.

3.2 The Theory of Errors

Every measurement is subject to *errors* (more properly called *uncertainties*) because man-made instruments and human senses are not capable of infinite precision. One can distinguish between two different types of errors: *systematic* and *random*. *Systematic errors* depend in a known manner on some external circumstances and tend to shift the observed values consistently toward either higher or lower values than the "true" value. Such systematic shifts can be determined, although usually not without tedious analysis. *Random errors* (also called *accidental errors*), on the other hand, tend to "scatter" the observed data so that some of the values fall above the true value and some fall below; by their very nature they cannot be predicted. It is only the latter with which the general theory of errors is concerned. Its task is to establish the law of frequency of errors and to judge the accuracy of a measurement or of a calculation.

The fundamental principle of the theory of errors was established by Gauss, and may be stated as follows: Suppose a series of measurements have been made to obtain a particular required quantity. Then the "best" estimate of the sought quantity is that value for which the sum of the squares of the individual errors is a minimum. This theory expresses not a fundamental law of nature but rather a very plausible definition. A good introduction to error theory can be found in Taylor [3.2]; another, older reference is the book by Mendenthall [3.3].

3.2.1 Direct Observations

Consider the simplest case of a single quantity l that can be measured directly. Let there be n measured values $l_1, l_2, \ldots, l_n$ which would be identical if there were no errors; obviously this is never true in practice. Here, in accordance with the *principle of least squares*, the theory of errors asserts that the "most probable value" of the required quantity is given by the *arithmetic mean*

$$\bar{l} = \frac{1}{n}(l_1 + l_2 + \cdots + l_n). \qquad (3.1)$$

The *standard deviation*, or *mean error*, is expressable in terms of the differences, also called *residuals*, $v_i = l_i - \bar{l}$ between the individual measurements and the mean value $\bar{l}$. Denoting the sum of the squares of the differences, $\sum_{i=1}^{n} v_i^2$, by $[vv]$, as is customary in the theory of errors, then the *standard deviation of a single measurement*[6] is

$$\sigma_l = \sqrt{\frac{[vv]}{n-1}}, \qquad (3.2)$$

---

6 This definition is more commonly referred to as the *standard deviation of the population*.

and the *standard deviation of the mean* is

$$\sigma_{\bar{l}} = \frac{\sigma_l}{\sqrt{n}} = \sqrt{\frac{[vv]}{n(n-1)}}. \tag{3.3}$$

Thus, by forming the mean of n individual measurements, the accuracy of the results can be improved by a factor of $\sqrt{n}$. These formulae assume that all measurements are of equal reliability, but this is often not the case. For example, an observer with a small telescope of low magnification will be able to measure the distance between the components of a double star less accurately than someone using a large instrument. Such differences in the accuracy can be allowed for, however, by assigning to each measurement a *weight*. The larger the degree of reliability, the larger the weight. The determination of these weight factors p_i depends in each case on an assessment, frequently only a very approximate one, of the quality of the observation. Once the weight factors have been assigned, the mean is no longer equal to the arithmetic mean of the single measurements but instead is given by the *weighted mean*

$$\bar{l} = \frac{l_1 p_1 + l_2 p_2 + \cdots + l_n p_n}{p_1 + p_2 + \cdots + p_n} = \frac{[lp]}{[p]}. \tag{3.4}$$

In this case, the standard deviation of one measure of unit weight is

$$\sigma_l = \sqrt{\frac{[vvp]}{n-1}}, \tag{3.5}$$

where the term $[vvp]$ denotes the sum $\sum_{i=1}^{n} v_i^2 p_i$. The standard deviation of the mean is

$$\sigma_{\bar{l}} = \sqrt{\frac{[vvp]}{[p](n-1)}}. \tag{3.6}$$

Very often the required quantity cannot be measured directly but is a known function of other quantities that can be measured. For instance, it is impossible to measure the absolute magnitude of a star directly, but it can be calculated if the apparent magnitude and the distance of the star are known. Since these two quantities can be measured only with a certain mean uncertainty (standard deviation), it is desirable to know the expected uncertainty of the computed absolute magnitude. The answer to this question is provided by the *law of the propagation of errors*.

Let the required quantity x be a known mathematical function of n other quantities $x_1, x_2, \ldots, x_n$, i.e.,

$$x = \varphi(x_1, x_2, \ldots, x_n). \tag{3.7}$$

If each of these quantities x_i is affected by a mean uncertainty σ_i, then the standard deviation of x is given by the formula

$$\sigma_x^2 = \left(\frac{\partial \varphi}{\partial x_1}\right)^2 \sigma_1^2 + \left(\frac{\partial \varphi}{\partial x_2}\right)^2 \sigma_2^2 + \cdots + \left(\frac{\partial \varphi}{\partial x_n}\right)^2 \sigma_n^2. \tag{3.8}$$

The standard deviation, or mean error, is a measure of the accuracy of the observation. However, one must be aware of the fact that the actual uncertainty exceeds the mean error. The following rule of thumb can be formulated: The discrepancy between a single measurement and the mean could, in unfavorable circumstances, be as much as 2.5 to 3 times the mean error. Therefore, if a previously known quantity is to be reexamined by new measurements and the resulting value turns out to be different, then this difference is meaningful (i.e., real) only if it exceeds the mean error of the measurement by at least a factor of 2.5; otherwise the result of the new measurement should be interpreted as a confirmation of the old value.

3.2.2 Indirect Observations

If one aims at the determination of several unknown quantities whose mathematical connection with the measured values is known, then a least-squares solution of indirect observations is made. In most cases the relation between the measured and the unknown quantities is linear, and if this is not the case the computation can be linearized by using approximations. It is assumed for the sake of simplicity that there are only three unknowns; the same principles are used for any other number of unknowns.

Let the measured quantity l be a function $\varphi(x, y, z)$ of three unknowns x, y, z. If, on the basis of some plausible hypothesis, the approximate values x_0, y_0, and z_0 are introduced for the unknowns and their true values denoted by $(x_0+\xi), (y_0+\eta), (z_0+\zeta)$, we are left with the determination of the *corrections* ξ, η, and ζ. According to *Taylor's theorem* a power series can be set up as follows:

$$f = \varphi(x_0+\xi, y_0+\eta, z_0+\zeta) = \varphi(x_0, y_0, z_0) + \xi\frac{\partial\varphi}{\partial x} + \eta\frac{\partial\varphi}{\partial y} + \zeta\frac{\partial\varphi}{\partial z} + \cdots. \quad (3.9)$$

The numerical values of the derivatives are to be calculated at the point (x_0, y_0, z_0). Using the notations

$$\frac{\partial\varphi}{\partial x} = a, \quad \frac{\partial\varphi}{\partial y} = b, \quad \frac{\partial\varphi}{\partial z} = c, \quad f - \varphi(x_0, y_0, z_0) = l, \quad (3.10)$$

a linear relation between the measured quantity l and the unknowns is obtained:

$$l = a\xi + b\eta + c\zeta. \quad (3.11)$$

Of course, at least three such equations connecting the numbers l_1, l_2, l_3 with the three unknowns ξ, η, and ζ are required. Usually, however, there are more equations than unknowns, and our task is to determine the most probable values of the unknowns ξ, η, and ζ from the measured quantities l_i, each of which carries a random error of measurement.

The various measurements must be performed in such a way that the values of the derivatives of the various functions φ_i differ from each other as much as possible (in order to separate the unknowns). If there are n different measured quantities, then n *equations of condition* are to be satisfied:

$$
\begin{aligned}
a_1\xi + b_1\eta + c_1\zeta &= l_1, \\
a_2\xi + b_2\eta + c_2\zeta &= l_2, \\
&\vdots \\
a_n\xi + b_n\eta + c_n\zeta &= l_n.
\end{aligned}
\tag{3.12}
$$

From these are formed the *normal equations*:

$$
\begin{aligned}
[aa]\xi + [ab]\eta + [ac]\zeta &= [al], \\
[ba]\xi + [bb]\eta + [bc]\zeta &= [bl], \\
[ca]\xi + [cb]\eta + [cc]\zeta &= [cl].
\end{aligned}
\tag{3.13}
$$

The algebraic solution of Eqs. (3.13) yields those values of the unknowns that are most probable according to the theory of errors.

The coefficients of the normal equations (3.13) are symmetrical with respect to the main diagonal of their determinant—e.g., $[ab] = [ba]$, $[bc] = [cb]$, etc. This property facilitates the algebraic process of the solution. First of all the first normal equation is multiplied by $-[ab]/[aa]$ and the result added to the second equation; next the first equation is multiplied by $-[ac]/[aa]$, and the result added to the third equation. Each of these two operations leads to an equation which does not contain the first unknown ξ. The same process is applied to the two resulting equations for η and ζ and this leads ultimately to one equation in ζ only. If ζ is known, η and ξ can then be determined from the previous equations.

In order to find the mean errors of the three unknowns, the *mean error of unit weight* and the *weight coefficients* must be known. The mean error m of unit weight is given by

$$
m^2 = \frac{[vv]}{n - \mu},
\tag{3.14}
$$

where μ is the number of the unknowns (in this case $\mu = 3$) and $[vv]$ the sum of the squares of the residual errors. This sum can be found either by calculating the right-hand sides of the equations of condition (3.12) and then forming the differences with the observed values l_i, or from the easily verifiable equation

$$
[vv] = [ll] - [al]\xi - [bl]\eta - [cl]\zeta.
\tag{3.15}
$$

The weight coefficients Q_{1i} follow from the normal equations by replacing the right-hand sides by $1, 0, 0$:

$$
\begin{aligned}
{[aa]}Q_{11} + [ab]Q_{12} + [ac]Q_{13} &= 1, \\
[ba]Q_{11} + [bb]Q_{12} + [bc]Q_{13} &= 0, \\
[ca]Q_{11} + [cb]Q_{12} + [cc]Q_{13} &= 0.
\end{aligned}
\tag{3.16}
$$

In an analogous manner the coefficients Q_{2i} are obtained by solving the normal equations with the right-hand sides set at $0, 1, 0$, and finally the Q_{3i} by setting the right-hand sides equal to $0, 0, 1$. Usually the greatest labor required for a complete least-squares solution is that of solving the system of normal equations three times (or μ times in the case of μ unknowns). The standard deviations σ_x, σ_y, and σ_z of the three unknowns can be found from the equations

$$\sigma_x^2 = m^2 Q_{11}, \quad \sigma_y^2 = m^2 Q_{22}, \quad \sigma_z^2 = m^2 Q_{33}. \tag{3.17}$$

Thus the task of the determination of the most probable values of the unknowns and of their standard deviations is completed. [7]

In all of these formulae, it has tacitly been assumed that all of the equations have the same weight. If the available measurements are of different precision then it can no longer be assumed that this assumption is satisfied, in which case the individual equations of condition have unequal weights. The coefficients of the normal equations are then no longer $[aa], [ab]$, etc., but must be replaced by the weighted sums $[aap], [abp]$, etc. In practice the weight factors can be very simply allowed for by multiplying each equation of condition by the factor $\sqrt{p}$, that is, by the square root of the assigned weight. The resulting set of equations can be treated as equations of equal weight by application of the previously presented procedures.

The subject of uncertainties described above is one in which familiarity with the concepts involved comes only after considerable practice with numerical cases. An example of the least-squares solution for the case of two unknowns is given below.

Example: On 1987 March 08 the zenith distances of six stars were measured in Munich, and were subsequently compared with the values which for these stars were theoretically derived using the catalogued coordinates of the star. Small differences resulted which may be attributed to three causes:

1. Each of these six measurements, needless to say, possessed random errors.
2. It was to be expected that the measuring instrument itself had a systematic deviation of constant amount.
3. It was known that the telescope had a flexure which resulted in measuring errors that, to good approximation, were proportional to the sine of the zenith distance.

Because of error sources 2 and 3, the theoretical equation of approach for the differences $z - z_0$ between the measured zenith distances z and the theoretical values z_0, the equation

$$z - z_0 = x + y \sin z, \tag{3.18}$$

was assumed; the remaining errors of the individual measurements as mentioned under 1 are then the remaining differences v.

The results are presented in the following list, giving in the first column the name of the star observed, in the second column the approximate zenith distance z, and in the third column the differences $z - z_0$:

---

[7] The normal equations are more commonly solved with high-speed computers using the method of determinants, in which case the weight factors are implicitly calculated in that process since they are the ratio of cofactor over determinant of the diagonal terms in the system of normal equations.

35 G. Columbae	$z = +75°$	$z - z_0 = -1\rlap{.}''78$
16 Orionis	$+38°$	-0.48
51 Geminorum	$+32°$	-0.40
κ Aurigae	$+19°$	-0.85
ι Aurigae	$+15°$	-0.12
o Draconis	$-72°$	$+1.19$

Introducing the respective values of $\sin z$, then, according to Eq. (3.18), the following equations of condition for the two unknowns x and y are obtained:

$$x + 0.97y = -1.78$$
$$x + 0.62y = -0.48$$
$$x + 0.53y = -0.40$$
$$x + 0.32y = -0.85$$
$$x + 0.26y = -0.12$$
$$x - 0.95y = +1.19$$

In this example, all of the coefficients a_i of the unknown x have the value 1, thereby simplifying the numerical evaluation. The sums of the products are now formed; as an example, the sum $[bl]$ is here given in detail:

$$[bl] = -0.97 \times 1.78 - 0.62 \times 0.48 - 0.53 \times 0.40 - 0.32 \times 0.85 - 0.26 \times 0.12 - 0.95 \times 1.19$$
$$= -3.6699.$$

The other sums of the products are calculated analogously, but extreme care must be taken with the algebraic signs! As a safeguard against errors, it is recommended that the sums be calculated twice: once from beginning to end, and the second time in reverse.

The next step in the present example is the formation of the two normal equations

$$6.0000x + 1.7500y = -2.4400, \qquad \| \quad -0.291667,$$
$$+ 2.6787y = -3.6699,$$

where, by convention, the first coefficient of the second equation is not shown since by definition it is identical to the second coefficient of the first equation (+1.7500). The first of the two normal equations is multiplied by the factor given on the right (−0.297667), which is calculated as the quotient −1.7500/6.0000. Adding the resulting expression to the second normal equation yields an expression without the unknown x:

$$+2.1683y = -2.9582.$$

Then, with the help of the first equation, x can also be evaluated. The result is

$$y = -1\rlap{.}''3643,$$
$$x = -0\rlap{.}''0088.$$

For the standard deviation of unit weight, and for the weight coefficient, it is found that

$$m^2 = 0.1710,$$
$$Q_{11} = +0.2060,$$
$$Q_{22} = +0.4612.$$

Therefore, the final result of the least-squares solution, rounding the result to two decimals, is

$$x = -0\rlap{.}''01 \pm 0\rlap{.}''19,$$
$$y = -1\rlap{.}''36 \pm 0\rlap{.}''28.$$

The unknown x is practically zero, whereas y is almost five times as large as its standard deviation and is, for this reason, certainly real. Of course both unknowns do not equal their "real" values; what have been found here are simply those numbers which best represent the observations of one night. A longer series of observations would have yielded more reliable values.

3.3 Interpolation and Numerical Differentiation and Integration

When a mathematical function has been numerically tabulated, one frequently wants to know values for intermediate arguments. Thus, for example, it is found that the coordinates of the celestial bodies in the annual almanacs are usually given from day to day for 0^h UT, so that the value at any required moment needs to be found by *interpolation*. If it is assumed that the graph of the function is, with reasonable accuracy, a straight line between tabulated values, then the interpolation is trivial and is called linear interpolation. If, however, the variation of the function is not sufficiently uniform, then one must resort to the formation of a *difference array*, as shown in the following table, where w is the difference between successive entries:

$$
\begin{array}{llllll}
f(a-2w) & & & & & \\
 & f'\left(a-\tfrac{3}{2}w\right) & & & & \\
f(a-w) & & f''(a-w) & & & \\
 & f'\left(a-\tfrac{1}{2}w\right) & & f'''\left(a-\tfrac{1}{2}w\right) & & \\
f(a) & & f''(a) & & & \\
 & f'\left(a+\tfrac{1}{2}w\right) & & f'''\left(a+\tfrac{1}{2}w\right) & & \text{etc.} \\
f(a+w) & & f''(a+w) & & & \\
 & f'\left(a+\tfrac{3}{2}w\right) & & f'''\left(a+\tfrac{3}{2}w\right) & & \\
f(a+2w) & & f''(a+2w) & & & \\
 & f'\left(a+\tfrac{5}{2}w\right) & & & & \\
f(a+3w) & & & & &
\end{array}
$$

The entries in this table are calculated as follows. In the left-hand column are the given values for the function at the points $(a-2w)$, $(a-w)$, etc. The other columns are calculated successively from the left by putting the difference between two successive entries in one column as the entry in the column to the right and halfway between the two entries. Differences are calculated by subtracting the upper entry from the lower entry. The columns f', f'', f''', etc., are called the first, second, third differences, etc.

If one wishes to know the value of the tabulated function f for any intermediate value $a \pm nw$ ($n < 1$), the following formulae can be employed:

$$f(a \pm nw) = f(a) \pm nf'\left(a \pm \frac{1}{2}w\right) + \frac{n(n-1)}{1\cdot 2}f''(a \pm w)$$
$$\pm \frac{n(n-1)(n-2)}{1\cdot 2\cdot 3}f'''\left(a \pm \frac{3}{2}w\right) + \cdots, \tag{3.19}$$

$$f(a \pm nw) = f(a) \pm nf'(a) + \frac{n^2}{1\cdot 2}f''(a)$$
$$\pm \frac{(n+1)n(n-1)}{1\cdot 2\cdot 3}f'''(a) + \cdots. \tag{3.20}$$

Equation (3.19), which is known as *Newton's formula*, is used if the initial value lies at the beginning or at the end of the table; otherwise *Stirling's formula* (3.20) will be more suitable. In order to calculate the $f'(a)$, or differences of odd order, evaluate the mean of $f'(a + \frac{1}{2}w)$ and $f'(a - \frac{1}{2}w)$:

$$f'(a) = \frac{1}{2}\left[f'\left(a - \frac{1}{2}w\right) + f'\left(a + \frac{1}{2}w\right)\right]. \tag{3.21}$$

And similarly, for $f'''(a)$,

$$f'''(a) = \frac{1}{2}\left[f'''\left(a - \frac{1}{2}w\right) + f'''\left(a + \frac{1}{2}w\right)\right]. \tag{3.22}$$

For the interpolation at the center of an interval (i.e., $n = 0.5$), the following simple formula applies:

$$f\left(a + \frac{1}{2}w\right) = \frac{1}{2}\left[f(a) + f(a + w)\right] - \frac{1}{8}f''\left(a + \frac{1}{2}w\right) + \cdots, \tag{3.23}$$

according to which the value of the function for $a + \frac{1}{2}w$ is equal to the arithmetic mean of the two neighboring values minus one-eighth of the second difference on the same line; the error arising is considerably smaller than the fourth difference (which can nearly always be neglected). This formula can be very advantageously employed if the differences are inconveniently large and it is desired to reduce them by switching over to the half-intervals. A numerical example is provided below.

Example: Compute the right ascension of the Moon on 1977 October 06 at $3^h 12^m$ (UT). *The Astronomical Ephemeris* for the year 1977 tabulates the coordinates of the Moon for any day from hour to hour. The right ascensions are as follows:

0^h	$7^h 20^m 42\overset{s}{.}139$		
		$+124\overset{s}{.}594$	
1^h	$22\ 46.733$		$-0\overset{s}{.}005$
		$+124.589$	
2^h	$24\ 51.322$		-0.005
		$+124.584$	
3^h	$26\ 55.906$		-0.005
		$+124.579$	
4^h	$29\ 00.485$		-0.004
		$+124.575$	
5^h	$31\ 05.060$		-0.004
		$+124.571$	
6^h	$33\ 09.631$		

Next to the values of the right ascensions those of the first and second differences are listed; the third differences are negligible. Choose as the starting point a the value 3^h because it is nearest to the value of $3^h 12^m$, which is to be calculated. With the Stirling formula (3.20), putting in $n = 0.2$ since 12^m is one-fifth of an hour,

$$f(3^h 12^m) = f(3^h) + 0.2 f'(3^h) + \frac{1}{2}(0.2)^2 f''(3^h)$$

$$= 7^h 26^m 55\overset{s}{.}906 + 0.2 \cdot 124\overset{s}{.}582 - 0.02 \cdot 0\overset{s}{.}005$$

$$= 7^h 27^m 20\overset{s}{.}822.$$

With the help of the difference array, it is also possible to find errors in calculated values of the function. Let us assume the calculated value of $f(a)$ to be incorrect by a quantity ϵ, but the neighboring values to be correct. The following difference array of the error of the function is obtained:

```
0
     0
0        +ε
    +ε      −3ε
ε       −2ε     +6ε   etc.
   −ε       +3ε
0       +ε
    0
0
```

For the higher differences the error becomes more and more conspicuous, particularly in the line which contains the faulty initial value. If, therefore, particularly large jumps in the difference array occur on a certain line, a check of that particular value of the function is advisable.

The scheme of differences also allows numerical calculation of derivatives and definite integrals of the tabulated function. The following formulae for numerical differentiation hold:

$$\frac{df(a)}{da} = \frac{1}{w}\left[f'\left(a+\frac{1}{2}w\right) - \frac{1}{2}f''(a+w) + \frac{1}{3}f'''\left(a+\frac{3}{2}w\right) + \cdots\right], \quad (3.24)$$

$$\frac{df(a)}{da} = \frac{1}{w}\left[f'(a) - \frac{1}{6}f'''(a) + \cdots\right]. \quad (3.25)$$

One easily obtains the rule of thumb that the first differences are approximately equal to the first derivatives multiplied by interval length w. Naturally, corresponding formulae apply to higher derivatives.

The integral of the tabulated function can be found by forming a summation series. This is done merely by assuming that the values in the first column play the role of the first differences and by forming an extra column to the left of the difference array. It is inherent in the nature of the difference array that any arbitrary constant can be added to the complete column without altering the differences. If a is the lower limit of the integration, then a first approximation ^1f for the series follows from the formula

$$^1f\left(a - \frac{1}{2}w\right) = -\frac{1}{2}f(a) + \frac{1}{12}f'(a) - \frac{11}{720}f'''(a) + \cdots. \quad (3.26)$$

The other values of the first summation follow from this by simple addition of the corresponding values of the function. The value of the integral of the tabulated function for any argument can be obtained from the formula

$$\int_a^{a+iw} f(x)dx = w\left[^1f(a+iw) - \frac{1}{12}f'(a+iw) + \frac{11}{720}f'''(a+iw) + \cdots\right]. \quad (3.27)$$

Methods of numerical integration are of greatest importance in those cases where an analytical expression for the integral cannot be found. Even if the analytical expression for the integral is known, its calculation may be very cumbersome; in such instances numerical methods can also be successfully employed.

Besides the formulae presented here there exist many other expressions that are partially variations of this method, and some which are independent as well. These are discussed in great detail in a publication by the Royal Observatory Greenwich: *Interpolation and Allied Tables* [3.4]. The notation given there differs from that used above, but the relations to the notations used here can be easily seen. The foundations of integration and interpolation are presented in [3.5]. Readers who own a computer or programmable calculator might want to refer to the books by Burgess [3.6] and Tattersfield [3.7], which are specifically designed for use with a microcomputer.

3.4 Photographic Astrometry

In photographic astrometry an observer uses photographic plates of a certain region of the sky to determine the coordinates of the stars. It is necessary to base this operation on a sufficient number of stars, at least three, whose coordinates on the plate are known and which can serve to define the orientation of the coordinate system. H.H. Turner has given a complete derivation of this method which can be found in the well-known book by W.M. Smart, *Textbook on Spherical Astronomy* [3.8]. Here only the most important formulae will be reproduced as needed for their application. It is assumed that all effects that change the coordinates of a star in different regions of the sky by a different amount (e.g., refraction, aberration, etc.) vary linearly within this particular region. This assumption is nearly always fulfilled and the rare deviating cases are of little importance to the amateur astronomer.

Consider now the image of a part of the celestial sphere when projected onto a plane, i.e., the plane of the photographic plate. If the center of the plate has the right ascension A and the declination D, then the rectangular coordinates X and Y of a given star with the right ascension α and the declination δ are given by the formulae

$$X = \frac{\tan(\alpha - A) \cos q}{\cos(q - D)}, \qquad Y = \tan(q - D), \tag{3.28}$$

where

$$\cot q = \cot \delta \cos(\alpha - A).$$

Here it has been assumed that the positive Y axis points toward the north celestial pole in the sky. These coordinates are called *standard coordinates* and can be calculated if the spherical coordinates of the star are known.

A comparison of the standard coordinates of the reference stars with the coordinates of the same stars measured on a linear scale yields the *plate constants*. The measurement of the plate gives the rectangular coordinates x and y of each star image, where the origin of the xy system should be very close to the center of the plate, with the direction of the positive y axis parallel to the north direction. If both X and Y, calculated from their spherical coordinates, as well as x and y, found from direct measurement, are known for a certain number of stars, then the plate constants follow from the equations

$$\begin{aligned} X &= ax + by + c, \\ Y &= dx + ey + f. \end{aligned} \tag{3.29}$$

Since each of these coordinates contains three constants, we require at least three reference stars; if there are more reference stars available, then the most probable values of the six plate constants can be found by the method of least squares (see Sect. 3.2.2).

The spherical coordinates of the other stars can be easily calculated as soon as the plate constants are determined. Since the values of x and y have been measured for each of the star images on the plate, the standard coordinates X and Y can be found directly from Eqs. (3.29). From X and Y, the spherical coordinates can be derived by inverting the equations (3.28), yielding

$$q = D + \arctan Y,$$
$$\tan(\alpha - A) = X \cos(q - D) \sec q, \qquad (3.30)$$
$$\tan \delta = \tan q \cos(\alpha - A).$$

Of course, the calculation becomes more accurate if more reference stars are used. On the other hand, experience shows that not much is gained by using more than six reference stars because only a very limited gain in accuracy is achieved as a compensation for the greatly increased tedious numerical work. It is essential that the reference stars be uniformly distributed over the plate.

3.5 Determination of the Position and Brightness of Planets and of the Planetographic Coordinates

Although the ephemerides of the planets are published in the various astronomical almanacs, it may sometimes be desirable to calculate them directly if, for instance, greater accuracy is sought, or if the calculation concerns a minor planet not yet contained in the almanacs. For this purpose six orbital elements are required:

$T =$ time of the passage through the perihelion of the orbit
$\mu =$ mean daily motion
$e =$ eccentricity
$\Omega =$ ecliptical longitude of the ascending node
$\omega =$ angular distance of the perihelion from the node in the orbit
$i =$ inclination of the orbit to the ecliptic plane

The semimajor axis a of the elliptical orbit can be determined from the mean motion μ by the formula

$$\mu a^{3/2} = k \sqrt{M_1 + M_2}, \qquad (3.31)$$

where $k = 3548\rlap{.}{''}18761$. Here, μ is given in arcseconds per day. The square root on the right-hand side contains M_1 and M_2, the masses of the central body and the orbiting body, respectively, both in solar masses. Thus if the orbit of a planet of low mass around the Sun is considered, as is normal, then the square root on the right-hand side is equal to 1. The semimajor axis of the orbit then results in units of the mean Earth–Sun distance.[8]

If it is desired to calculate the spherical position of the planet for a given instant of time t, the so-called *Kepler equation*,

$$E - e \sin E = \mu(t - T) = M, \qquad (3.32)$$

where M is the *mean anomaly*, is solved for the *eccentric anomaly E*. Equation (3.32) is what is termed a *transcendental equation*, and can be solved only by iterative

---

8 Note that this is not a *definition* of the astronomical unit, or AU.

techniques. Using an initial value $E_0 = M + e \sin M$, iteration of the two formulae

$$M_0 = E_0 - e \sin E_0,$$
$$E_1 = E_0 + \frac{M - M_0}{(1 - e \cos E_0)}$$

converges quickly.

Using the eccentric anomaly, the radius vector r and the *true anomaly* f can be calculated from the formulae

$$r \cos f = a(\cos E - e),$$
$$r \sin f = a\sqrt{1 - e^2} \sin E. \tag{3.33}$$

From these equations there follow at once the rectangular heliocentric coordinates of the planets, referred to the ecliptic:

$$x = r[\cos \Omega \cos(f + \omega) - \sin \Omega \sin(f + \omega) \cos i],$$
$$y = r[\sin \Omega \cos(f + \omega) + \cos \Omega \sin(f + \omega) \cos i], \tag{3.34}$$
$$z = r \sin(f + \omega) \sin i.$$

To convert the heliocentric coordinates into geocentric ones, the heliocentric coordinates of the Earth, which are equal to the geocentric coordinates of the Sun taken with the opposite algebraic sign, are required. Denoting the ecliptic coordinates of the planet by λ and β, the ecliptic coordinates of the Sun at time t by $\lambda_\odot$ and $\beta_\odot$, and the distances from the Earth to the Sun and from the Earth to the planet by R and d, respectively, then the following relationships hold:

$$d \cos \beta \cos \lambda = x + R \cos \beta_\odot \cos \lambda_\odot,$$
$$d \cos \beta \sin \lambda = y + R \cos \beta_\odot \sin \lambda_\odot, \tag{3.35}$$
$$d \sin \beta = z + R \sin \beta_\odot.$$

This completes the calculation. Of course, attention must be paid to the fact that the orbital elements which characterize the position of the orbit in space (Ω, ω, i) must be referred to the same equinox as the coordinates of the Sun.

The calculation of the positions of comets having parabolic orbits is in principle quite similar, except that Kepler's equation (3.32) is now replaced by another formula which enables one to determine the true anomaly as a function of time, namely,

$$\tan \frac{f}{2} + \frac{1}{3} \tan^3 \frac{f}{2} = \frac{k(t - T)}{\sqrt{2} \, q^{3/2}}. \tag{3.36}$$

Here, as in the case of the elliptical orbits, the time difference $(t - T)$ is given in days. Also, q denotes the *perihelion distance*, i.e., the distance of closest approach to the Sun achieved by the comet in its orbit. It is one of the orbital elements and must be known before the calculation of ephemerides is embarked upon. After the true anomaly has been determined, the radius vector is obtained from

$$r = q \sec^2 \frac{f}{2}, \tag{3.37}$$

and the remaining calculation is then carried out in the same way as for an elliptical orbit.

Example: The position of Comet Kohoutek for the date 1974 January 03 at 0^h universal time is to be calculated. The comet was discovered in the spring of 1973, and the following orbital elements were published in October:

$$T = 1973 \text{ December } 28.463$$
$$\omega = 37°874$$
$$\Omega = 257°715 \quad \text{(Equinox 1950.0)}$$
$$i = 14°297$$
$$q = 0.14242 \text{ AU}$$

The time interval $t - T$ between perihelion passage and the time for which the position is to be calculated is 5.537 days; since $k = 0.0172021$ radians, the result

$$\frac{k(t-T)}{2q^{3/2}} = 1.253096$$

in radians of arc is obtained. Putting this result in Eq. (3.36) yields

$$\tan \frac{f}{2} + \frac{1}{3} \tan^3 \frac{f}{2} = 1.253096,$$

from which the true anomaly f may be determined. It is immediately seen from calculating the left-hand side that the result is

too large with the hypothesis $\tan \frac{f}{2} = 1$,

too small with the hypothesis $\tan \frac{f}{2} = 0.9$.

Thus the true value of $\tan f$ is constrained to lie between 0.9 and 1.0. Further approximations give the solution

$$\tan \frac{f}{2} = 0.95905,$$

and so

$$\frac{f}{2} = 43°48'.1,$$

or

$$f = 87°36'.2 = 87°603.$$

With this value of f, the distance $r = 0.2734$ AU is found. Further calculations then give, with the above values of the orbital elements,

$$\cos \Omega \cos(f + \omega) - \sin \Omega \sin(f + \omega) \cos i = +0.89454,$$
$$\sin \Omega \cos(f + \omega) + \cos \Omega \sin(f + \omega) \cos i = +0.39916,$$
$$\sin(f + \omega) \sin i = +0.20110.$$

These three numbers by multiplication with the value of r give the rectangular heliocentric coordinates x, y, z of the comet; that is

$$x = +0.24457,$$
$$y = +0.10913,$$
$$z = +0.05498.$$

The coordinates of the Earth (with respect to the Sun) for 1974 January 03 at 0^h UT can be found in an astronomical almanac to be

$$R = 0.983267 \text{ AU},$$
$$\lambda_\odot = 281°53'27''.2,$$
$$\beta_\odot = +11'',$$

referring to the equinox of 1950.0, the same for which the orbital elements of the comet are valid. With these values, it is found that

$$d \cos \beta \cos \lambda = +0.44716,$$
$$d \cos \beta \sin \lambda = -0.85303,$$
$$d \sin \beta = +0.05498,$$

where the solar latitude $\beta_\odot$, which amounted to only a few arcseconds, was neglected (this causes an error of merely a few units in the last decimal place). From these numbers the geocentric distance d, the ecliptic coordinates (λ, β), and, upon applying the conversion formulae Eqs. (3.30), the equatorial coordinates (α, δ) are computed. The results are:

$$d = 0.9647 \text{ AU}$$
$$\lambda = 297°39'.8$$
$$\beta = +3°16^m.0$$
$$\alpha = 299°.068 = 19^h 56^m.3$$
$$\delta = -17°26^m.0.$$

This completes the computation. If it is carried out for each day over a certain period of time (or for every second or every fourth day), then an ephemeris for this time period is generated from which intermediate values can be readily found via interpolation.

The apparent brightness of a planet changes substantially with its distance from the Earth as well as from the Sun. Furthermore, because of the changing relative positions of the planet, Earth, and Sun, the degree of illumination of the planetary disk as viewed from Earth will vary; thus the *phase angle* will also influence the apparent brightness. The phase angle α_p is defined as the angle subtended at the planet by the directions to the Sun and to the Earth, respectively, and is given by the formula

$$\tan \frac{\alpha_p}{2} = \sqrt{\frac{(\sigma - r)(\sigma - d)}{\sigma(\sigma - R)}}, \qquad (3.38)$$

where

$$\sigma = \frac{1}{2}(R + r + d). \qquad (3.39)$$

The changes in the brightness of a planet because of the changing distances from the Earth and the Sun follow strictly the geometric law that the diminution in brightness is inversely proportional to the square of the distance. This leads, after conversion to the astronomical scale of magnitudes, to the formula

$$m = m_0 + 5 \log r + 5 \log d, \qquad (3.40)$$

where the constant m_0 has a different value for every planet and, moreover, is dependent on the phase angle. Photometric measurements give, for each of the planets, the following values of m_0:

Mercury	$m_0 = +1\overset{m}{.}16 + 0\overset{m}{.}0284(\alpha_p - 50°) + 0\overset{m}{.}0001023(\alpha_p - 50°)^2$,
Venus	$+4\overset{m}{.}00 + 0\overset{m}{.}0132\alpha_p + 0\overset{m}{.}000000425\alpha_p^3$,
Mars	$-1\overset{m}{.}30 + 0\overset{m}{.}0149\alpha_p$,
Jupiter	$-8\overset{m}{.}93$,
Uranus	$-6\overset{m}{.}85$,
Neptune	$-7\overset{m}{.}05$.

For planets more distant than Mars, the influence of the phase angle can be neglected. Saturn has not been included in the above list, since its apparent brightness depends in a rather complicated fashion on the orientation of the rings relative to the Earth.

Frequently the observer is faced with the task of calculating the coordinates of a point on the observed disk of a planet relative to the equatorial planetary plane. Although in most cases this problem can be solved with sufficient accuracy by graphical methods, some computation cannot be avoided. Required quantities are the right ascension A and the declination D of that point on the sphere to which the northern extension of the axis of rotation of the planet points. These quantities are:

Mercury	$A = 281\overset{\circ}{.}0 - 0\overset{\circ}{.}033T$,	$D = +61\overset{\circ}{.}4 - 0\overset{\circ}{.}005T$,
Venus	272.8,	+67.2,
Mars	$317.7 - 0.108T$,	$+52.9 - 0.061T$,
Jupiter	$268.0 - 0.009T$,	$+64.5 + 0.003T$,
Saturn	$40.7 - 0.036T$,	$+83.5 - 0.004T$,
Uranus	257.4,	-15.1,
Neptune	295.3,	+40.6,
Pluto	311.6,	+4.2.

This set of data refers to the year 2000, and the time T is to be counted in centuries from 2000 backwards.

For the further treatment of the problems, two quantities, b_0 and β, are needed. They are found from the formulae

$$\sin b_0 = -\cos \delta \cos D \cos(\alpha - A) - \sin \delta \sin D, \qquad (3.41)$$

$$\tan \beta = \frac{\sin(\alpha - A)}{\sin \delta \cos(\alpha - A) - \cos \delta \tan D}. \qquad (3.42)$$

If now the planetary disk is graphed on paper as a circle with radius s, the visible pole of the planet appears at a point P, which has the distance $s \cos b_0$ from the center, and is at the position angle β if the latter is counted from the north clockwise from $0°$ to $360°$. The quantities b_0 and β are tabulated in most astronomical ephemerides. The quantity b_0 is known as the *planetographic latitude* of the Earth with respect to the equator of the planet.

Once the position of the visible pole on the planetary disk is determined, it is possible to graph a "degree net" in steps of planetographic longitude and latitude, the

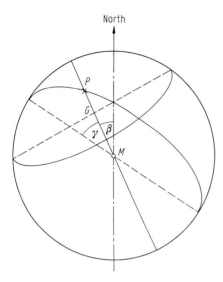

Fig. 3.1. Planetographic coordinates.

grid width of which will depend on the accuracy required. The diameter MP of the disk, which connects the center with the visible pole (see Fig. 3.1), is the projection of the central meridian at the moment of observation. All other meridians project not as straight lines, but rather as ellipses which pass through the point P. If it is desired to draw a meridian, for example, the longitude of which differs from the central meridian by the amount $(l - l_0)$, a line representing the diameter of the disk must be drawn through the center M of the disk. The direction of this line forms the angle γ with MP, which follows from

$$\tan \gamma = \sin b_0 \tan(l - l_0), \tag{3.43}$$

where l_0 is the length of the central meridian. This diameter is the major axis of an ellipse which also passes through the visible pole and which represents the required meridian.

The latitude circles also project onto the planetary disk as ellipses. If one wishes to draw a circle corresponding to the planetographic latitude b, then one must first find a point G on the straight line MP which has the distance $s \cos b_0 \sin b$ from M. The major axis of that ellipse, which represents the required latitude circle, is perpendicular to MP and has the length $s \cos b$; the minor axis has length $s \cos b \sin b_0$.

If the grid of longitude and latitude circles is sufficiently fine, then it can simply be placed on top of the sketch of the planetary surface as obtained from a telescope, and thereupon the planetographic coordinates of the various required features can be read off directly. All that is needed is the value of the planetographic longitude l_0 of the central meridian, which is published for daily intervals in ephemeris tables. The formulae for calculating it are published each year in *The Astronomical Almanac* [3.9].

3.6 The Reduction of Stellar Occultations[9]

Although as a rule it is recommended that the observers of stellar occultations send their results directly to the Nautical Almanac Office, it is conceivable that an observer would want to carry out the reduction of the observational data himself. It is for this purpose that the necessary formulae are summarized here.

Suppose that the occultation of a star is observed at the instant t (UT). The corresponding *sidereal time* τ can then be computed using the formulae in Sect. 2.3.2. If, furthermore, μ_0 is the sidereal time at 0^h, as taken from an ephemeris and corrected by the amount $9\overset{s}{.}8565 \, \Delta\lambda$ (again refer to the formulae presented in Sect. 2.3), the angle h is obtained as

$$h = \mu_0 + \tau - \lambda - \alpha, \tag{3.44}$$

where λ is the geographic longitude of the observing site; places east of Greenwich count as negative; α and δ are the right ascension and declination of the star. Next calculate the quantities

$$\xi = \frac{r' \cos\varphi' \sin h}{k}, \tag{3.45}$$

$$\eta = \frac{r' \sin\varphi' \cos\delta}{k} - \frac{r' \cos\varphi' \cos h \sin\delta}{k}, \tag{3.46}$$

where $k = 0.2724953$. Here the geocentric coordinates r' and φ' of the observing site are calculated from the formulae presented in Sect. 2.1, while δ is the declination of the occulted star. The coordinates of the Moon are

$$\begin{aligned} x &= \frac{\cos\delta' \sin(\alpha' - \alpha)}{k \sin\pi''_{\mathbb{C}}}, \\ y &= \frac{\sin\delta' \cos\delta - \cos\delta' \sin\delta \cos(\alpha' - \alpha)}{k \sin\pi''_{\mathbb{C}}}, \end{aligned} \tag{3.47}$$

where α' and δ' are the right ascension and the declination of the Moon at the moment of occultation. These quantities are taken from the hourly ephemerides of the Moon in *The Astronomical Almanac*, and are then interpolated with high precision, using the method presented in Sect. 3.3. The quantity $\pi''_{\mathbb{C}}$ is the parallax of the Moon, which is found for the instant of occultation by interpolation from the *Almanac*.

The crucial result of the observation obtained at a given site is a quantity $\Delta\sigma$, which is calculated using the formula

$$\Delta\sigma = \frac{k \sin\pi''_{\mathbb{C}}}{\sin 1''} \left[\sqrt{(x - \xi)^2 + (y - \eta)^2} - 1 \right]. \tag{3.48}$$

It is equal to the angular distance from the center of the Moon to the star minus the radius of the apparent disk of the Moon. If the ephemerides were exactly correct, then $\Delta\sigma$ would have to be zero (apart from observational errors). However, owing

---

[9] See also Chap. 17.

to incomplete knowledge of the laws of celestial mechanics (which were employed to generate the tables in the ephemerides), the values of the mean longitude $\lambda_{\mathbb{C}}$ and latitude $\beta_{\mathbb{C}}$ of the Moon taken from the ephemerides include the errors $\Delta\lambda_{\mathbb{C}}$ and $\Delta\beta_{\mathbb{C}}$. Hence the quantity $\Delta\sigma$ differs from zero, and the resulting equation is

$$\Delta\sigma = \cos(\varrho - \chi)\Delta\lambda_{\mathbb{C}} + \sin(\varrho - \chi)\Delta\beta_{\mathbb{C}}, \tag{3.49}$$

where χ is the position angle of the star with reference to the center of the Moon and ϱ is the position angle of the direction of the motion of the Moon.

Observations obtained at one station yield only one quantity $\Delta\sigma$ and not $\Delta\lambda_{\mathbb{C}}$ and $\Delta\beta_{\mathbb{C}}$ separately. If, however, observations from several sites are available, and the values of χ differ appreciably, then the most probable values of $\Delta\lambda_{\mathbb{C}}$ and $\Delta\beta_{\mathbb{C}}$, and thus also the errors in the ephemerides of the Moon, can be found with the aid of the method of least squares.

References

3.1 Frates, J., and Moldrup, W.: *An Introduction to the Computer*, Prentice-Hall, Inc., Englewood Cliffs NJ 1980.
3.2 Taylor, J.R.: *An Introduction to Error Analysis*, University Science Books, Mill Valley CA 1982.
3.3 Mendenthall, W.: *Introduction to Probability and Statistics* (2nd ed.), Wadsworth Publishing Co., Belmont CA 1967.
3.4 *Interpolation and Allied Tables*, Royal Observatory, Greenwich.
3.5 Wepner, W.: *Mathematisches Hilfsbuch für Studierende und Freunde der Astronomie*, Düsseldorf 1981.
3.6 Burgess, Eric: *Celestial Basic: Astronomy on Your Computer*, Sybex, Inc., Berkeley CA 1982.
3.7 Tattersfield, D.: *Orbits for Amateurs with a Microcomputer*, Wiley, New York 1984.
3.8 Smart, W.M.: *Textbook on Spherical Astronomy*, Cambridge University Press, Cambridge 1977.
3.9 *The Astronomical Almanac 1989*, p. E87.

4 Optical Telescopes and Instrumentation

H. Nicklas

4.1 Introduction

Most instruments employed for astronomical observations are designed primarily to increase the intensity of illumination on the surface of an image-recording device, be it a conventional photographic plate, a sophisticated electronic detector, or merely the human eye. This may be achieved by enhancing the light-collecting area or by sharper imaging, with the aims of improved angular resolution in order to separate closely adjacent objects and high light-gathering power in order to clearly pick out faint objects. There is no all-round, "universal" instrument available that satisfies all of these requirements simultaneously. Rather, an observing instrument is optimized for a specific purpose and is often named after a particular type of construction or observational mode (e.g., zenith telescope, transit circle, binoculars, rich-field telescope, coronograph, astrograph, or Schmidt camera).

Another kind of classification refers to the method by which the optical image is formed:

1. *dioptric systems*, which employ refraction of light;
2. *catoptric systems*, which employ reflection of light via mirrors;
3. *catadioptric systems*, which employ both refraction *and* reflection to image objects optically.

This chapter surveys the variety of observational instruments and telescope systems in optical astronomy which are available to the observer. Besides a description of optical systems, a discussion of different imaging errors will be relevant too. First, the basics of optical imaging are treated, including the calculation of optical systems in order to judge the imaging quality. This is of fundamental interest at this juncture insofar as it permits the determination of the value of individual imaging errors. The explanation of the various kinds of errors and their origins is followed by a description of the most widely used telescope systems and their imaging properties. There is no attempt at completeness since the possible variations of telescope types are extraordinarily diverse. Additional literature will be referred to for special optical systems. The description of telescope accessories, by means of which the telescope system supplies "results," is treated at considerable length. This includes not only the optics inserted for visual observation but also "post-focal" instruments which are used for more objective measurements of stars in quantitative as well as qualitative respects. Finally, the processing of the light collected by the telescope by means of detectors ranging from the eye to the modern semiconductors will be discussed.

The purpose of this volume is not to cover each subject in minute detail, but rather to present only the relevant relations and formulae; their derivations are listed in the "References" section at the end of each chapter. The authors have emphasized literature which is easily accessible to amateur observers. One or the other source can doubtlessly be found in a nearby college or public library. Observers specializing in the construction of telescope optics and other instruments may particularly wish to consult the section "Gleanings for ATMs" of the monthly publication *Sky and Telescope*, as well as the three volumes by A.G. Ingalls *Amateur Telescope Making*, which are a treasury of necessary information on the building, testing, and adjusting of optical systems, and for the construction of numerous instruments.

4.2 Basics of Optical Computation

4.2.1 Ray Tracing and Sign Convention

The calculation of optical elements is not at all as complex or difficult as it is generally assumed to be. To investigate the imaging process of a particular optical system, the path of each of the various rays through the system is followed. The manner in which the rays converge at the end of calculation permits a judgement on the quality of the image that the system produces. The computation of rays through an optical system yields meaningful results as long as the relevant dimensions (diameter, radii of curvature, etc.) of the imaging system are substantially larger than the wavelengths used. In this case, the use of the optics of rays, which is based upon elementary geometrical relationships, is justified and is called *geometrical optics*. Basic to geometrical optics are two physical laws which follow from *Fermat's principle* of the shortest optical path:

1. The rectilinear propagation of light in homogeneous media.
2. Snell's law of refraction, $n \sin i = n' \sin i'$, with n, n' being the refractive indices of the two media and i, i' angles between the normal to the surface and the incident and refracted ray, respectively.

The normal to the surface and the incident and refracted rays are in one plane so that the simple calculation is limited to a plane (i.e., two dimensions) for the entire system. If this plane furthermore contains the optical axis of the system, it is then called the *meridional plane*, and we may restrict the discussion to the calculation of *meridional rays* of that plane, since the majority of optical systems consist of surfaces of revolution with a common axis and significant conclusions with respect to imaging errors can already be deduced. Optical systems consisting exclusively of spherical surfaces are emphasized since, for reasons relating to manufacture and production, these are preferred for commercially available optics. The calculation is appreciably simplified for a *centered* system, where the centers of curvature of the spherical surfaces all lie along a straight line, the optical axis. A spherical surface is completely determined by the position of its center of curvature and by the radius of curvature, the latter always being normal to the spherical surface.

Before turning to calculations, it is necessary to introduce a sign convention as the formulae use specifically directed lines and angles. The direction of light propagation is always counted positive, and usually propagates on the graphs, from left to right. Segments above the optical axis, similar to the positive y axis, are counted positive, below negative. From these rules follow the sign conventions:

1. A ray coming from above intersects the axis with a positive angle.
2. The radius of a spherical surface is counted positive when the center of curvature *follows* the vertex point in the sense of light propagation. In other words, a surface of positive curvature faces incident light with its *convex* side.
3. Since reflection on a mirror reverses the direction of propagation, for the sake of simplicity it is preferable to consider the mirror reversed around its vertex after the reflection, so that conventional left-to-right light propagation is retained. Correspondingly, the radius of curvature changes sign after reflection.

The radii of curvature of a bi-convex lens are therefore $r_1 > 0$ and $r_2 < 0$. In a concave mirror, $r < 0$ before reflection and $r' > 0$ after reflection.

As with spherical surfaces, the meridional ray is characterized by exactly two coordinates: the intersection angle u with the optical axis, and the corresponding intersection distance or segment s. These, by way of refraction, are converted into the coordinates u', s'. All quantities are primed after refraction for distinction. In Fig. 4.1, in addition to the coordinates of the ray, the *height of incidence h*, the *angle of incidence i*, and the corresponding primed quantities are displayed. From these the following relations are found:

$$r \sin i = (s - r) \sin u \quad \text{and} \quad r \sin i' = (s' - r) \sin u'. \tag{4.1}$$

By means of the angular relation $\varphi = u + i = u' + i'$ and the law of refraction $n \sin i = n' \sin i'$, the coordinates u', s' of the refracted ray are found by Eq. (4.1) to be

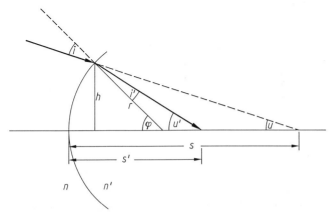

Fig. 4.1. Refraction at a spherical surface.

$$\sin i = \frac{s-r}{r} \sin u \quad \text{and} \quad \sin i' = \frac{n}{n'} \sin i, \tag{4.2}$$

$$u' = u + i - i' \quad \text{and} \quad s' = r\left(1 + \frac{\sin i'}{\sin u'}\right). \tag{4.3}$$

Equations (4.1–3) form the basic scheme by which the refraction of a ray at any individual surface of an optical system can be calculated. All relevant quantities will henceforth be subscripted with an index ν in order to distinguish the coordinates, angles, etc. before and after the refraction at the νth surface. Since the direction of the ray in homogeneous media is conserved, the angle with the optical axis is unchanged, and the exit ray from the νth surface becomes the incident ray of the $(\nu+1)$th surface:

$$u_{\nu+1} = u'_\nu. \tag{4.4}$$

The second coordinate, the new intercept $s_{\nu+1}$, changes only insofar as it diminishes by the distance $d_{\nu,\nu+1}$ between the centerpoints of the two surfaces, that is,

$$s_{\nu+1} = s'_\nu - d_{\nu,\nu+1}. \tag{4.5}$$

The scheme of calculations is then applied again using these new coordinates. The calculations are successively continued in this way until they have been carried out through the entire optical system. Equations (4.2–5) are valid for any rays except those parallel to the axis; the latter occur in the special (but certainly relevant) case of rays from distant astronomical objects on the optical axis. The object-side angle is now $u = 0$ and the intersection distance $s = \infty$; the first equation of (4.2) is then replaced by

$$\sin i = \frac{h}{r}. \tag{4.6}$$

Any path of rays can be calculated through a system by successive application of Eqs. (4.2–6). The concentration of rays in the image plane then will show the imaging quality of the optical system. Traving [4.1] gives a simple computer program in BASIC to calculate meridional rays. Also indicated are other relations which are significant for the design of optical systems. These include broad ranges in the choice of refractive indices (glass types) and of some radii of curvature which can be employed by the manufacturer to lessen or remove imaging errors. In the second part of his paper, Traving gives a computer program in BASIC for the *spatial* calculation of skew rays, which at the end leads to a *spot diagram* from which the concentration of light in the image plane can be read directly in two dimensions. A short BASIC program that traces one meridional ray through the optical system but without discussion of the system's free parameters is given by Larks [4.1a]. Readers interested in a deeper understanding of optical calculations may wish to consult references [4.2] and [4.3].

4.2.2 The Cardinal Points of an Optical System

The equations presented above permit the calculation of the characteristic quantities of an optical system. It must be emphasized that these quantities characterize the optical system only in a limited area close to the optical axis, the so-called *paraxial field*. It

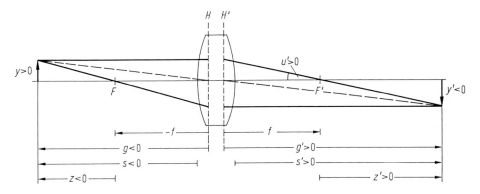

Fig. 4.2. Image construction in the paraxial range.

is only there that an ideal, error-free image can be achieved. In this area, the angles that occur are small enough that the sine of an angle can be replaced by the angle itself, that is, $\sin i = \tan i = i$, but this approximation is invalid for large apertures or large angles. The quantities obtained by paraxial-ray tracing can, however, be used for the determination of imaging errors in real systems.

One important characteristic of an optical system is its *focal length* f, which is fundamental to image construction. The focal length is calculated by tracing a paraxial ray of infinite object distance through an optical system of k surfaces by

$$f = \frac{n_1}{n'_k}\left(s'_1 \prod_{\nu=2}^{k} \frac{s'_\nu}{s_\nu}\right) = \frac{n_1}{n'_k}\left(\frac{s'_1 \times s'_2 \times s'_3 \times \ldots \times s'_k}{s_2 \times s_3 \times \ldots \times s_k}\right), \tag{4.7}$$

where $s_1 = \infty$. The focal length of a system in air ($n_1 = n'_k = 1$) is simply a product of the relative intersections s'_k / s_k. For a single lens in air, the *refractive power* F, which is the reciprocal value of the focal length, can immediately be calculated from

$$F = \frac{1}{f} = (n-1)\left(\frac{1}{r_1} - \frac{1}{r_2}\right) + \frac{(n-1)^2}{n}\frac{d}{r_1 r_2}, \tag{4.8}$$

where d is the thickness of the lens. This equation shows that a convex lens ($r_1 < r_2 \Rightarrow f > 0$) has a positive focal length, while a concave, diverging lens ($r_1 > r_2 \Rightarrow f < 0$) has a negative focal length. Knowledge of the focal length suffices to determine the position of the image using *Newton's equation*,

$$z\,z' = -f^2, \tag{4.9}$$

where z, z' are respectively the focal distances in object space and image space (see Fig. 4.2). Because of its simplicity, Eq. (4.9) is certainly preferable to the well-known lens equation

$$\frac{1}{f} = -\frac{1}{g} + \frac{1}{g'}, \tag{4.10}$$

since it contains only focal distances. The *imaging scale* or *lateral magnification* is

$$m' = \frac{y'}{y} = -\frac{z'}{f} = +\frac{f}{z}. \tag{4.11}$$

It is immediately seen that an object distant by one focal length from the focus will be imaged in the ratio $1 : -1$ (i.e., inverted). The position z' of the image on the axis is not sufficient to construct the image. Additionally, the principal planes H and H' are needed. They are the planes through the principal points between which the imaging scale is $m' = +1$ by definition. Their distances from the respective foci are given by

$$z = f \quad \text{and} \quad z' = -f. \tag{4.12}$$

By twofold calculation of the ray, once from left to right with $s_1 = \infty$, and once from right to left with $s'_k = \infty$, the axial focus in image and object space is found from the two last intersection segments $(s'_k)_{s_1=\infty}$ and $(s_1)_{s'_k=\infty}$. From these, and assuming that the focal length is known, the positions of the principal points are immediately found.

The focal and principal points constitute four of the six *cardinal points* of a system. The other two are the *nodal points*, defined as those points on the axis through which the object ray passes unrefracted (i.e, angles of incidence and refraction are equal). This property leads to the condition $m' = n_1/n'_k$, and hence the position of nodal points is

$$z = n'_k f \quad \text{and} \quad z' = -n_1 f. \tag{4.13}$$

For an optical system in air ($n_1 = n'_k = 1$), the nodal points evidently coincide with the principal points and thus play virtually no role in practical optics. As stated, all considerations in this section are valid only for the paraxial field.

4.2.3 Pupils and Stops

The components of an optical system, which include lenses, mirrors, diaphragms (stops), prisms, etc., have finite dimensions and therefore the cross-sectional size of a beam will be limited. This is of great significance in matters of image brightness and field size, and can contribute to the correction of some imaging errors. Berek [4.2] has discussed the importance of the proper positioning of a diaphragm or stop. Any consideration of imaging properties of the system as a whole must include not only the imaging rays, which are represented by the cardinal points, but also the illuminating rays, which are limited by the various stops.

The beam size is always limited by two separate stops: the *aperture stop* and the *field stop*. Aperture stops, such as the support frame of a lens or the edge of a mirror, are those stops which appear under the smallest angle when viewed from the intersection of the optical axis with the object or image plane. To determine the real aperture stop, each stop and lens frame of a system is imaged by the preceding or following element into the object or image space. Some of these images can have smaller diameters than the aperture stop, but lying closer to the object or image plane

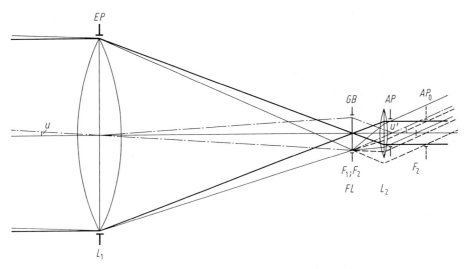

Fig. 4.3. Image beam (*solid line*) and pupil beam (*dot-dashed line*) in a telescopic system: objective L_1, field lens FL, magnifying lens L_2, field diaphragm GB, entrance pupil EP, exit pupil AP.; also pupil beam and vignetting (*dashed line*) with exit pupil AP_0 in the absence of the field lens FL. (See text for details)

they appear under larger viewing angles. The image of an aperture stop in the object space is called the *entrance pupil*, while in the image space it is the *exit pupil*. Since both pupils are images of the same diaphragm, they are also simultaneously images of each other. In telescopic systems, the aperture of the instrument generally acts as the aperture stop; i.e., the entrance pupil in refracting telescopes is the objective or the corrector plate in catadioptric systems; in reflectors, it is the primary mirror. Each optical element, such as the secondary mirror, field lenses, eyepiece, etc., images this stop until it forms the exit pupil in image space, the eye position in visual observation. This location should be at a suitable distance behind the eyepiece so that the exit pupil and eye pupil can properly coincide. If it is too close behind the lens, then visual viewing will be inconvenient and the field of view will be correspondingly narrowed, giving rise to what is termed "keyhole viewing."

The intersection of the pupils with the optical axis is of particular importance. Any ray through these points of the axis is called a *principal ray*, which is equivalent to an optical axis for the corresponding skew beam (see Fig. 4.3). Since entrance and exit pupils are images of each other, principal rays intersect in the axial points of both pupils. The inclination u of the principal ray with respect to the optical axis is transformed by the optical system into the image angle u', which is enlarged corresponding to the magnification of the telescopic system. The inclination u of the principal ray will enter the following discussion of image errors.

A diaphragm in the image plane (e.g., the format of a photographic plate or eyepiece diaphragm) acts as a field stop. Its images are called *windows*. In object space it is called the *entrance window*, in image space the *exit window*. ts position and size is found as for those of pupil images by using the imaging equations. Since the windows are always in the image plane or in planes conjugated to it, aperture stops and pupils can never be in window planes.

To recognize possible vignetting (i.e., missing of rays by one optical element) in the optical system, both the imaging beam, whose conic points are at the axis points of the windows, as well as the pupil beam, whose principal rays intersect the axis at the pupils, must be graphed (Fig. 4.3). All frames, stops, plate formats, etc. are thereby imaged with the preceding or following elements respectively into the object or image space, and the corresponding pupils and windows are graphed to scale in both position and size. The potential vignetting will then be seen. Fig. 4.3 shows the beam in an astronomical Keplerian telescope with three convex lenses: the objective lens L_1, the field lens FL, and the magnifying or eyepiece lens L_2. The entrance pupil EP coincides with the aperture stop (the objective frame), whereas the exit pupil AP_0 gives the pupil position *without* the field lens. The vignetting at the eyepiece lens is immediately seen and leads to light loss for off-axis objects with large inclinations u. If a field lens is inserted into the joint focus F'_1, F_2, then the imaging rays are unchanged, but the exit pupil $AP_0 \to AP$ moves closer to the eyepiece lens and thus the pupil beam (dashed and dotted) is also shifted. For an equal or smaller diameter of the eyepiece lens, this vignetting of off-axial objects is avoided. The exit pupil AP is accessible to the eye, which, adjusted to infinity (i.e., relaxed), will focus the parallel rays onto the retina.

4.3 Imaging Errors

4.3.1 Seidel Formulae

The *Seidel theory* provides the connection between the ideal paraxial image and the imaging errors of real, wide-aperture systems. Seidel (1821–1896) showed that the coefficients required for the determination of the errors can be found by calculating just *one* paraxial ray.

This section will not deal with the theory of imaging errors of third and higher order but will provide the formalism to calculate the partial coefficients and the individual imaging errors. These are much less complicated than they look and can easily be determined with the aid of any of the now-affordable personal computers. Theory dictates that each of the five primary imaging errors is composed of the sum of the individual errors of each surface. Thus, the five coefficients for each optical surface are calculated and, at the completion of the calculation, finally added. From these sums, the contributions of the individual imaging errors can be immediately found. First, a paraxial ray is calculated through the optical system according to Eqs. (4.2-6); all needed quantities such as incidence height h_ν and intersect width s_ν are then known. In addition to the system parameters (i.e., refractive index n_ν, radius of curvature r_ν, and surface distance d_ν), the distance of the entrance pupil from the first surface vertex d_{EP} (< 0 when pupil *precedes* vertex) enters one of the formulae since the position of the entrance pupil has a decisive influence on some of the imaging errors. The formulae are from Köhler [4.4], whereas Berek [4.2] deals with the conditions that one or more coefficients vanish. The latter author appreciates Seidel's theory less for determining aberrations and more for systematically studying the effect of the single surfaces, and in this way to influence these effects. For the sake of clarity, auxiliary quantities are introduced in calculating the coefficients themselves. These quantities are

$$Q_\nu = n_\nu \left(\frac{1}{r_\nu} - \frac{1}{s_\nu}\right) = n'_\nu \left(\frac{1}{r'_\nu} - \frac{1}{s'_\nu}\right) \quad \text{(Abbe Invariant)}$$

$$\Delta\left(\frac{1}{ns}\right)_\nu = \frac{1}{n'_\nu s'_\nu} - \frac{1}{n_\nu s_\nu}$$

$$\frac{h_\nu}{h_1} = \frac{s_2 \times s_3 \times s_4 \times \ldots \times s_\nu}{s'_1 \times s'_2 \times s'_3 \times \ldots \times s'_{\nu-1}} = \prod_{j=2}^{\nu} \frac{s_j}{s'_{j-1}}$$

$$\vartheta_\nu = \sum_{k=2}^{\nu} \frac{d_k}{n_k(h_{k-1}/h_1)(h_k/h_1)} - d_{EP}$$

$$p_\nu = \frac{1}{(h_\nu/h_1)^2 Q_\nu} + \vartheta_\nu$$

$$P_\nu = -\frac{1}{r_\nu}\left(\frac{1}{n'_\nu} - \frac{1}{n_\nu}\right) \quad \text{(Petzval Sum)}$$

(4.14)

From these auxiliary numbers, the partial coefficients of each surface ν, each of which represents an imaging error by itself, are found to be

Spherical aberration	$I_\nu = (h_\nu/h_1)^4 Q_\nu^2 \Delta(1/ns)_\nu$
Coma	$II_\nu = p_\nu I_\nu$
Meridional astigmatism	$III_\nu = 3p_\nu^2 I_\nu + P_\nu = 3IIIa_\nu + P_\nu$
Sagittal astigmatism	$IV_\nu = p_\nu^2 I_\nu + P_\nu = IIIa_\nu + P_\nu$
Distortion	$V_\nu = p_\nu^3 I_\nu + p_\nu P_\nu = p_\nu(IIIa_\nu + P_\nu)$
Astigmatic difference	$IIIa_\nu = p_\nu^2 I_\nu = (III_\nu - IV_\nu)/2$
Mean field curvature	$IVa_\nu = 2p_\nu^2 I_\nu + P_\nu = (III_\nu + IV_\nu)/2$

(4.15)

The interpretation of the different aberrations will follow in Sect. 4.3.2. These errors can be either partly or entirely removed by specific deformation of individual surfaces. A well-known deformation is the reshaping of a spherical into a *parabolic* mirror in order to remove the spherical aberration. If, in addition, the secondary mirror is deformed, other errors can be removed (cf. the Ritchey–Chrétien system). The deviation of such an aspherical surface from spherical shape is expressed by the *deformation constant k*. The rotation surface from a conic section is expressed by the relation

$$z(\rho) = \frac{1}{k+1}\left(r - \sqrt{r^2 - (k+1)\rho^2}\right)$$
$$= \frac{\rho^2}{2r} + (k+1)\frac{\rho^4}{8r^3} + (k+1)^2\frac{\rho^6}{16r^5} + \cdots. \tag{4.16}$$

The variable ρ expresses the distance to the axis of rotation (in the present case, the incident height h above the optical axis). The quantity r is the radius of curvature of that spherical surface which approximates the rotation surface at its vertex. The meridian curves correspond, according to the deformation constant k, to the following conic sections:

$$\begin{array}{ll} \text{Circle} & k = 0 \\ \text{Ellipse} & -1 < k < 0 \\ \text{Parabola} & k = -1 \\ \text{Hyperbola} & k < -1. \end{array} \tag{4.17}$$

From these constants further summation terms for the Seidel coefficients must be added for each nonspherical surface. The coefficients of every *deformed* surface are (Köhler [4.4])

$$\begin{aligned} \text{I}_\nu^* &= k\,\frac{1}{r_\nu^3}\left(\frac{h_\nu}{h_1}\right)^4 \Delta n_\nu \\ \text{II}_\nu^* &= \vartheta_\nu\, \text{I}_\nu^* \\ \text{III}_\nu^* &= 3\vartheta_\nu^2\, \text{I}_\nu^* \\ \text{IV}_\nu^* &= \vartheta_\nu^2\, \text{I}_\nu^* \\ \text{V}_\nu^* &= \vartheta_\nu^3\, \text{I}_\nu^* \\ \text{IIIa}_\nu^* &= \vartheta_\nu^2\, \text{I}_\nu^*, \\ \text{IVa}_\nu^* &= 2\vartheta_\nu^2\, \text{I}_\nu^*, \end{aligned} \tag{4.18}$$

where for reflecting surfaces $\Delta n = (n' - n) = (-n - n) = -2$. Table 4.1 lists the Seidel sums for a few widely used telescope systems. Many more are to be found in [4.4] and [4.5].

It should be pointed out that all lengths are *normalized* to the focal length f of the system (i.e., divided by f). From the Seidel sums, it is easily seen by what amount each surface contributes to the imaging errors, and thus the optical system can be improved in a specific way. The correction state of the system is already seen from the sums, although they do not define the absolute value of imaging errors. The latter depend on the focal length f of the system, the focal ratio $N = f/D$ (where $D =$ diameter of objective lens or mirror) and principal ray inclination u (object distance from the optical axis or image field radius in radians). The values of the primary aberrations are compiled below, while Köhler [4.4] lists further aberrations. All formulae are valid under the premise that the Seidel sums $\sum \text{I}$ to $\sum \text{V}$ are calculated with dimensions normalized to the focal length f of the system.

Table 4.1. Seidel sums of some telescope systems. I: spherical aberration, II: coma, IIIa: astigmatic difference, IVa: mean field curvature, P: Petzval sum. Taken from *Handbuch der Physik* [4.6].

	ν	R_ν	$e_{\nu-1,\nu}$	h_ν/h_1	p_ν ; (ϑ_ν)	I	II	IIIa	P	IVa
1. Spherical mirror (stop at vertex)	1	−2.000	−	1.000	−2 .000	+0.250	−0.500	+1.000	−1.000	+1.000
2. Parabolic mirror (stop at vertex) $b = -1.000$	1	−2.000	−	1.000	−2 .000	+0.250	−0.500	+1.000	−1.000	+1.000
	1*	−	−	−	(0 .000)	−0.250	0.000	0.000	0.000	0.000
					$\sum$	0.000	−0.500	+1.000	−1.000	+1.000
3. Cassegrain system	1	−0.500	−	−	−0 .500	+16.000	−8.000	+4.000	−4.000	+4.000
Primary mirror = aperture stop $b_1 = -1.000$	1*	−	−	−	(0 .000)	−16.000	0.000	0.000	0.000	0.000
Secondary magnification $m = 4$	2	+0.150	0.194	0.225	−0 .917	−4.219	+3.867	−3.545	+13.333	+6.224
$b_2 = -2.778$	2*	−	−	−	(+0 .861)	+4.219	+3.633	+3.128	0.000	+6.256
					$\sum$	0.000	−0.500	+3.583	+9.333	+16.480
4. Ritchey–Chrétien system	1	−0.667	−	−	−0 .667	+6.750	−4.500	+3.000	−3.000	+3.000
Primary mirror = aperture stop $b_1 = -1.081$	1*	−	−	−	(0 .000)	−7.296	0.000	0.000	0.000	0.000
Secondary magnification $m = 3$	2	+0.267	0.244	0.267	−0 .958	−2.133	+2.044	−1.959	+7.500	+3.582
$b_2 = -5.023$	2*	−	−	−	(+0 .917)	+2.679	+2.456	+2.251	0.000	+4.502
					$\sum$	0.000	0.000	+3.292	+4.500	+11.084
5. Schmidt camera	0*	−	−	−	(0 .000)	−0.250	0.000	0.000	0.000	0.000
	1	−2.000	2.000	1.000	0 .000	+0.250	0.000	0.000	−1.000	−1.000
					$\sum$	0.000	0.000	0.000	−1.000	−1.000
6. Schmidt-Cassegrain system after Baker	0*	−	−	−	(0 .000)	−0.622	0.000	0.000	0.000	0.000
Secondary magnification $m = 1.538$	1	−1.300	1.394	1.000	0 .094	+0.910	+0.085	+0.008	−1.538	−1.522
$b_1 = +0.0165$	1*	−	−	−	(1 .394)	+0.015	+0.021	+0.029	0.000	+0.058
$b_2 = 0$	2	+1.300	0.4225	0.350	0 .350	−0.304	−0.106	−0.037	+1.538	+1.464
					$\sum$	0.000	0.000	0.000	0.000	0.000

For a two-mirror system, $h_2/h_1 = a$ because of $f = 1$ (cf. Eq. 4.32)..

Lateral spherical aberration (Radius of the scattered circle)	$S = \frac{1}{16} \frac{f}{N^3} \sum \mathrm{I}$	(4.19a)
Longitudinal spherical aberration	$\Delta z_S = \frac{1}{8} \frac{f}{N^2} \sum \mathrm{I}$	(4.19b)
Lateral comatic aberration	$K = \frac{3}{8} \frac{f}{N^2} \tan u \sum \mathrm{II}$	(4.19c)
Longitudinal comatic aberration	$\Delta z_K = \frac{3}{4} \frac{f}{N} \tan u \sum \mathrm{II}$	(4.19d)
Lateral astigmatic aberration (on surface of mean field-curvature)	$A = \frac{1}{4} \frac{f}{N} \tan^2 u \sum \mathrm{IIIa}$	(4.19e)
Longitudinal astigmatic aberration (on surface of mean field-curvature)	$\Delta z_A = \frac{1}{2} f \tan^2 u \sum \mathrm{IIIa}$	(4.19f)
Mean radius of curvature of the image surface	$R_m = -f / \sum \mathrm{IVa}$	(4.19g)

The signs of the Seidel sums have the following meaning. If $\sum \mathrm{I}$ is positive, the rays from the edge of the aperture intersect the optical axis *before* the paraxial rays do, resulting in a condition known as *spherical undercorrection*. If $\sum \mathrm{II}$ is negative, then the coma shape is directed radially outward. Astigmatism is called positive ($\sum \mathrm{IIIa} > 0$) when the light ray first crosses the meridional and then the sagittal focal surface. Correspondingly, the field curvature is positive when the focal surface is concave toward the object, i.e., the radius of curvature in this case is negative ($\sum \mathrm{IVa} > 0$; $R_m < 0$). The fifth sum, when positive, expresses barrel distortion ($\sum \mathrm{V} > 0$), in contrast to a pincushion distortion. Wiedemann [4.7] explains in three examples to what extent these sums express the correction state of the optics.

4.3.2 The Primary Aberrations

After calculation of the Seidel coefficients and the determination of their values, the significance of the various types of aberrations will now be mentioned. The *primary aberrations* as obtained from the Seidel theory, including third-order polynomials, are discussed. Worthy of further study are the theoretical roots of aberrations, including diffraction analysis, for which the standard book by Born and Wolf [4.8] can be strongly recommended. Cagnet, Francon, and Thrierr [4.9] give a readable and illustrative presentation of the most important optical phenomena, such as images of the primary aberrations, diffraction images, interferograms, etc.

Spherical aberration, also called *aperture error*, occurs, as the name implies, when lenses or mirrors with spherical surfaces are used in optical systems. These spherical elements hold great importance in view of the ease with which they can be manufactured commercially. The intrinsic error of the spherical surface is that near-axial and near-limb rays do not fit into one point. Each annular *zone* of the spherical surface possesses its own distinct focal length. This difference in the focal lengths of the individual zones is called *spherical aberration* and leads to the caustic, which, like zones,

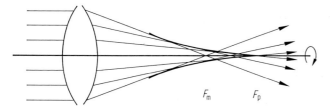

Fig. 4.4. Spherical aberration and caustic: F_p paraxial focus, F_m marginal focus (i.e., focus for light coming from the edge of the objective); the objective in this example is spherically undercorrected since $f_m < f_p$.

is rotationally symmetric about the optical axis. The dependence of this aberration on the incidence height and the focal length, and therefore on the focal ratio $1/N = D/f$, is evident from Fig. 4.4. From Eq. (4.19a), it follows that the diameter of the scattered circle increases with the cube of the aperture ($S \propto D^3/f^2$). Spherical aberration thus can be reduced in two ways: by reducing the aperture, which is obviously not very desirable for astronomical telescopes, or by choosing sufficiently long focal lengths. It can be shown that spherical aberration of an optical system is negligibly small (in the sense of the Rayleigh criterion) if the F-number N exceeds a particular value, namely

$$N \gtrsim 3.4 \sqrt[3]{D}, \qquad (4.20)$$

where D is the aperture size in centimeters. Spherical aberration can be partially removed if two lenses of a suitably chosen refractive power are combined. It is removed entirely only by a departure from the spherical shape, that is, by *aspherical surfaces*. For a single mirror this is the condition of a *rotation paraboloid*. Use is made of the defining property of the parabola, namely, that all points on that conic curve have the same distance (or optical path length) to one point and to a straight line. The straight line represents the plane wavefront of an infinitely distant object which is reflected over the entire aperture into one point, the focus. Other methods of removing spherical aberration will be described in Sect. 4.5 on telescope systems. Whether or not an objective has this type of aberration can be ascertained in various ways. For instance, the diffraction disk can be examined at various distances from the paraxial focus F_P (cf. Fig. 4.5). If the objective is free of spherical aberration, then the radius of the diffraction or Airy disk increases symmetrically on both sides of the paraxial focus; otherwise the increase is asymmetrical. This asymmetry hints at the presence of spherical aberration. Other test methods will be mentioned in Sect. 4.4.

Coma is purely an asymmetry error occurring for beams incident at an angle, i.e., in the present case for objects within the field but off the optical axis. Through the inclination of the beam, the rotational symmetry with respect to the optical axis is lost and limited to a plane of symmetry, the *meridional plane*, which contains the optical axis and the inclined principal ray. Since the aperture stop (or entrance pupil) is in this case no longer spherically symmetric, none of the rays of the meridional beam keeps a special position on the basis of symmetry properties. In general, this beam contains not only the spherical aberration which is already present in the axial rays,

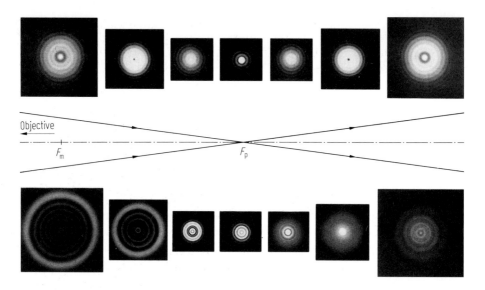

Fig. 4.5a, b. Diffraction images of **a** an ideal objective and **b** one which is affected by spherical aberration. See text for explanation. Taken from Cagnet, Francon, and Thrierr [4.9].

but also an asymmetry of this aberration. This emerges in a light pattern with some resemblance to a comet whose tail is directed radially toward or away from the optical axis (Fig. 4.6).

Coma, according to Eq. (4.19c), increases in proportion to the inclination angle u, and therefore becomes more pronounced with increasing distance from the optical axis. Figure 4.7 shows a highly magnified diffraction image of a single point source in the case of *pure* coma (see the *Atlas of Optical Phenomena* [4.9]). How coma (as an asymmetry error) can be removed by simply restoring the symmetry for beams incident at an angle will be shown in Sect. 4.5.6 on Schmidt cameras. Optical systems corrected for spherical aberration and coma are called *aplanatic* (nondeviating) by Abbe (1840–1905). Aplanatic systems enjoy a prominent role in astronomical optics owing to their relatively good correction state; see [4.4]. The treatment of the *sine condition* $f = h/\sin u'$, which indicates—if fulfilled and in absence of spherical aberration—the absence of coma by calculating only an axial beam, can be found in Berek [4.2] and in Born and Wolf [4.8].

Astigmatism and *field curvature* are interrelated and should therefore be discussed together. For skew beams, the stigmatic focus is lost and split into the two foci F_{mer} and F_{sag}, as is shown in Fig. 4.8. This effect (called *astigmatism* for *no* point imaging) occurs, in contrast to coma, even for beams of small inclination. Through the inclination of the beam, the refractive conditions in the two planes, the *meridional* and the *sagittal*, differ. Considering simply the refraction angles for the edge-rays of the meridional and sagittal section, it is seen that these rays, because of different incidence angles, lead to different foci of incident skew beams. Since the sagittal rays

Fig. 4.6. Aberrations in the image field of a camera objective of 35-mm focal length. The coma increases distinctly with increasing distance from the center, to which the coma shape is radially directed. The picture shows the Milky Way at η Carinae, the Southern Cross, and the Coalsack; exposure time 6 min ($f/2.8$) on Agfachrome 1000.

are not yet converged in the meridional focus, and the meridional rays have already diverged upon reaching the sagittal focus, the two foci F_{mer} and F_{sag} degenerate into focal *lines*. In the center between the two lines (on the surface of mean field curvature) the distribution of light is circular, but elsewhere, except at the three points named, it is more or less elliptical, and the ratio of the ellipse axis depends on its distance from the focal lines. This consideration also holds for beams of infinitesimally small aperture, i.e. the paraxial field. Thus the two foci must coincide on axis in the *paraxial focus*, while their distance off axis, called the *astigmatic difference*, increases with the square of the inclination angle (Eq. 4.19e, f). The radii of curvature for the meridional and sagittal sections are rotationally symmetric about the optical axis, so that each of the foci forms a shell tangent to the other in the ideal focus on the optical axis. The radius of curvature of the meridional focal surface is given by $\sum$ III, while that of the sagittal surface by $\sum$ IV. It is seen that the coefficient $\sum$ IIIa of the astigmatic difference, Eq. (4.15), represents *half* the distance of both focal surfaces and $\sum$ IVa the *mean* field curvature. Vanishing astigmatic difference causes the meridional and sagittal surfaces to coalesce into one focal surface, since their radii of curvature are equal, and the resulting system is called an *anastigmatic system*. The optics that will in general then be free of field curvature are called collectively an *anastigmat*.

Distortion results from the fifth coefficient of Seidel theory, and affects not the sharpness but rather the shape of the image. When present, the optics lacks a constant

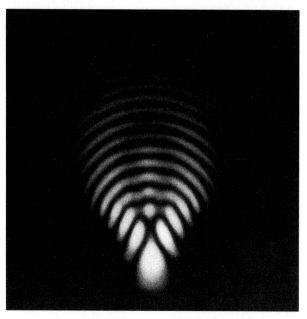

Fig. 4.7. Diffraction image of an optic with a round entrance pupil affected by coma. From *Atlas of Optical Phenomena* by Cagnet, Francon, and Thrierr [4.9].

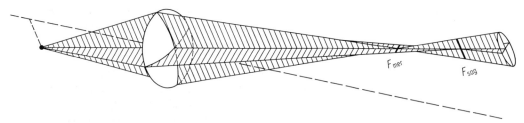

Fig. 4.8. Astigmatism of skew beams. The graph shows the two main sections, the meridional (vertical) and sagittal (horizontal) planes with the corresponding focal lines F_{mer} and F_{sag}.

imaging scale over the field of view. This simply means that the effective focal length of the system (responsible for the image scale or magnification) changes with the field angle. Distortion is revealed when straight lines not crossing the optical axis are imaged into "lines" which are curved outward or inward. The former case is termed "barrel" or negative distortion, while the latter is called "pincushion" or, as it represents an increase of scale, positive distortion. This can be readily demonstrated by observing a square mesh or lattice with the optics. In summary, three of the five primary image errors in Seidel theory (spherical aberration, coma, and astigmatism) affect the sharpness of the image, the other two (field curvature and distortion) the position of the image.

Table 4.2. Some of the most common spectral lines in technical optics.

Line	λ (nm)	Color	Atom
A′	768.2	deep red	potassium
B	686.7	red	oxygen
C	656.3	red	hydrogen (Hα)
-	632.8	red	helium-neon (laser)
D_1	589.6	yellow	sodium
D_2	589.0	yellow	sodium
d	587.6	yellow	helium
e	546.1	green	mercury
F	486.1	green-blue	hydrogen (Hβ)
g	435.8	blue	mercury
h	404.7	violet	mercury
H	396.8	violet	calcium
K	393.3	violet	calcium
-	365.0	ultraviolet	mercury

4.3.3 Chromatic Aberrations

Previous considerations were based upon strictly monochromatic light. As the refractive index n, which occurs in numerous formulae, changes with wavelength λ, the intersection width in the calculation of a ray depends upon color. Thus, in addition to the primary aberrations, lens systems are also affected by color, or *chromatic*, aberrations. In order to determine the refractive index $n(\lambda)$ and the resulting chromatic aberrations for separate wavelengths, monochromatic light of different wavelengths (which is available from the emission lines of various kinds of atoms), is required. Hence it is useful to refer all color-dependent quantities to the light of certain spectral lines as supplied by the most common spectral lamps. Table 4.2 lists the wavelengths of the most prominent spectral lines with their coding as introduced by Fraunhofer.

Color aberrations are generally divided into primary and secondary chromatic aberrations. The primary errors include the *longitudinal* and the *lateral aberrations*. The first of these causes a positional error on the optical axis; according to Eq. (4.8) the refractive power, $F = 1/f$, of a lens depends on the wavelength; i.e., a lens has a different focal length f for each color. The paraxial focus is thereby split into two foci for two different wavelengths; the focus of one color is surrounded by the "defocused" diffraction circle of the other color (and vice versa), the diameter of which depends upon the F-number N of the system and the difference in focal length. The difference in focal length between the red and blue color extremes of the brightest portion of the visual spectral range is of order

$$f_C - f_F = \frac{f_e}{\nu_e}. \qquad (4.21)$$

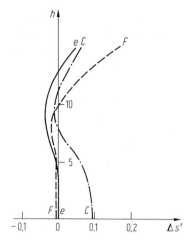

Fig. 4.9. An example of aberration curves for a simple cemented achromat.

The indices "C," "F," and "e" refer to the respective wavelengths 656.3, 486.1, and 546.1 nm (Table 4.2), and ν_e is an important quantity called *Abbe's number*,

$$\nu_e = \frac{n_e - 1}{n_F - n_C}, \qquad (4.22)$$

that gives the *reciprocal dispersive power* of the glass used. The subscript "e" in Abbe's number refers to the green e-line ($\lambda = 546.1$ nm) of mercury, which is nowadays preferred over the yellow sodium D-line as the "mean" wavelength of the visual spectral range.

For the case of an infinitely distant object, Eq. (4.21) gives the *chromatic separation* of the focus of wavelengths C and F. To remove this on-axis chromatic aberration, the optics must satisfy the *condition of achromacy*:

$$\frac{F_1}{\nu_1} + \frac{F_2}{\nu_2} + \cdots = \frac{1}{f_1 \nu_1} + \frac{1}{f_2 \nu_2} + \cdots = 0. \qquad (4.23)$$

Equation (4.23) contains only the refracting power F_i (or the focal length f_i) and the reciprocal dispersive power $\nu_i = (n_i - 1)/(n_F - n_C)$ for each lens, but not the shape of the individual lenses. Thus suitable radii of curvature and bending of the lenses can still be chosen to correct other aberrations. As the Abbe numbers are always positive, then, according to Eq. (4.23), a two-lens objective can merge two colors into one focus only when composed of a convergent and a divergent lens, since $f_1 \nu_1 = -f_2 \nu_2$. Correcting the on-axis displacement in the paraxial focus for two wavelengths does not guarantee that the edge rays of wide-open beams also pass this focus. Even if spherical aberration is corrected for one wavelength, this need not be so for other wavelengths; this leads to the chromatic error of the spherical aberration. Its correction is difficult, if not impossible, and often the color correction on-axis is abandoned in favor of a chromatic correction of the focal length of an annular zone.

Information on the correction state of an objective is often given by so-called *aberration curves*. These diagrams graph the on-axis intersections s' versus the incident height h or the incident angle w. Due to the dependence on wavelength, every color has a distinct curve from whose shape the color errors are identified. Aberration curves of a simple achromat may look like Fig. 4.9, which graphs, instead of the axis intersection, the *intersection difference* $\Delta s' = s'_\lambda - s'_0$, measured against the paraxial intersection s'_0 of a "mean" wavelength, versus the incident height h. The difference $\Delta s'$ (and also often h) are usually referred to a focal length $f = 100$ so that the deviations are read immediately as percentages of the focal length. The color correction for green and blue light near the axis is clearly seen; it decreases for larger off-axis rays because of the chromatic error of the spherical aberration. In general, an *achromat* is a lens system which corrects the color displacement on-axis for two wavelengths and the spherical aberration for a mean wavelength. This correction of a primary chromatic aberration leaves a chromatic error of second order known as the *secondary spectrum*, which is caused by the different dispersion curves of glass types whose refractive indices do not increase proportionally from the red to the blue spectral range. The dispersion curve in the visual range is measured by the *relative partial dispersion* ϑ, which has the representative value

$$\vartheta_g = \frac{n_g - n_F}{n_F - n_C}. \tag{4.24}$$

A vanishing secondary spectrum requires equal partial dispersions ϑ at simultaneously different ν-values of the glasses; a satisfactory solution is obtained only by a combination of glass and fluorite. If, for reasons of cost, the choice is restricted to glasses exclusively, then the secondary spectrum of the two-lens achromat will merely be diminished. The addition of a third lens can correct an objective for three wavelengths. An objective which corrects the chromatic focus displacement for three wavelengths and the spherical aberration for two wavelengths is called an *apochromat*. The various glass types can be graphed according to their characteristic optical properties, the refractive index and Abbe number (see Fig. 4.10). Table 4.7 lists the relevant quantities of several glass types.

Another primary chromatic error is the *lateral displacement* (chromatic magnifying difference), which, in contrast to the chromatic error along the axis, causes a displacement *across* the axis. It can also occur when the along-axis error is removed, in particular, for lens systems where the components are situated asymmetrically with respect to a stop located between them. Thus, the lateral chromatic aberration is often seen in telescope eyepieces which are composed of at least two or more lenses and with their aperture stop in front of the field lens. A detailed treatment of chromatic imaging errors with examples for calculation of objectives and error graphs has been given by Flügge [4.3].

4.4 Methods of Optical Testing

4.4.1 Manufacturing of Optical Quality Surfaces

This section concerns only the principles used to manufacture surfaces of optical quality. The procedure and special hints are described in detail in a variety of literature (e.g., *Amateur Telescope Making* Vols. I–III; see the Supplemental Reading List for

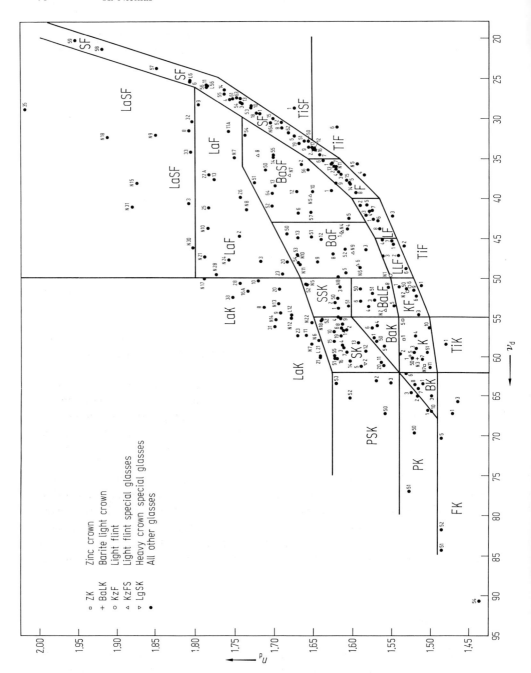

Fig. 4.10. Characteristics of optical glasses (displayed in the form of an n–ν diagram) of the Schott glassworks.

Vol. 1). For obvious reasons, amateurs nearly always prefer to grind *mirrors*: as only one surface must be manufactured for each optical element, there is no need to center both lens surfaces with respect to the optical axis during manufacture, and a large mirror diameter can be more easily engineered than a lens of similar proportions.

Grinding a mirror starts with two plane, raw disks of glass of equal diameter and thickness. The first raw disk is attached to a fixed bed, while the second disk is placed above the lower disk after a grinding powder such as carborundum or other substance has been sprinkled between the plates. The upper disk is then moved off the concentric position in a *straight direction* to about one-third of its diameter back and forth into the concentric while the optician walks around the bedding and rotates the upper disk between his hands. This motion wears off the edge of the lower disk and the center of the upper one. All prominent unevennesses are ground off first, and the two disks are beveled to fit snugly, the lower (grinding disk) with convex, and the upper (mirror disk) with concave shape. The curvature of both spherical surfaces increases as the procedure continues. Thus one should soon progress from coarse to fine grinding. The careful guiding and moving of the upper disk is of decisive importance. Should the radius of curvature become smaller than desired, then the two disks must be interchanged: the mirror disk is held fixed, and the grinding disk works the larger curvature on it. As work progresses, powders of finer and finer grain are used until, at the completion of fine grinding, an optical surface with unevennesses (errors) of fractions of a μm has been achieved. Strict attention must continually be paid to cleanliness as one coarse grinding grain can annihilate the work of hours.

Fine grinding is followed by polishing. In this process, warm pitch is placed upon the grinding disk (the lower convex part) and, after it has cooled, lines are scratched along and across in a quadratic pattern. After adding rouge paste and water, the polishing follows with the same pattern of motions as the grinding. The form of the scratched pitch pattern and the changes of motion also provide the opportunity to remove optical errors from the surface or to make spherical surfaces non-spherical. Maksutov [4.10] discusses this in detail and also describes tests and the manufacture of refraction optics. The surface shape of the mirror must be checked frequently during the entire procedure so that it does not deviate significantly from the desired shape. For an initial crude measure of the radius of curvature, a *spherometer* will suffice. Later tests require more sensitive measuring techniques. If the surface reflects light sufficiently, the "knife-edge" test of Foucault can be used to test the focal length and the surface.

4.4.2 Determination of Focal Length

The focus of a convex lens is found simply by picking up the real image of a distant (by definition, at infinity) object on a screen. The image of a faint object can be inspected using a magnifying lens with reticle; if both the image and the reticle simultaneously appear sharp, then the focus of the objective is at the position of the reticle. The focal length can also be measured in the laboratory by a procedure introduced by Bessel (see Fig. 4.11). If the distance between object and focusing screen is a constant l, then there are two intermediate positions at which the lens forms a sharp image of

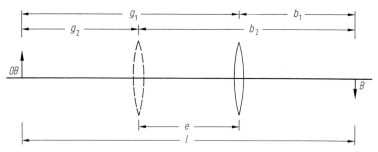

Fig. 4.11. Focal length determined by the Bessel method: $g_1 = b_2$; $g_2 = b_1$.

the object on the screen. If e is the distance between these two positions of the lens, then the focal length f is given by

$$f = \frac{l^2 - e^2}{4l}. \tag{4.25}$$

To obtain two focal positions, the distance between object and screen must exceed four times the focal length f of the lens. If it *equals* $4f$, then the image scale is 1:1 ($z = z' = f$ in Newton's equation) and the displacement e vanishes (i.e., $f = l/4$). To determine the focal length more precisely from Eq. (4.25), it is advisable to procure a series of readings from which a mean and a standard deviation can be calculated.

The focal length of a divergent or negative lens is determined using the scattered circle of a very distant point source. If D is the diameter of the lens, b the diameter of the scattered circle, and l the distance between lens and screen, then the focal length is

$$f = \frac{lD}{D - b}. \tag{4.26}$$

As a simple rule, f equals the distance l when the scattered circle on the screen has a diameter twice that of the lens. If this method is used with the solar disk, a term $0.0095\,l$ is added into the denominator of Eq. (4.26) because of the finite extension of the disk. The focal lengths of objectives and eyepieces can also be found from the telescope magnification (see Sect. 4.6.2) when one of them is known. The magnification is found by viewing a known scale (a meter stick, for instance) through the telescope with one eye and directly with the other eye, and comparing. The focal length of a concave mirror is best determined by the "knife-edge" method of Foucault (Sect. 4.4.4).

It should be noted that mirrors change shape and focal length with temperature. Therefore, an ideal mirror material should have a very high thermal conductivity in order to reach thermal equilibrium immediately after a temperature change. A metal mirror, in contrast to one made of glass, does nearly possess the desired high conductivity, and a change of temperature therefore causes a scaled increase or decrease of all dimensions without changing the shape. That is, the spherical shape is conserved, while a paraboloid changes only its focal length. This change may amount from fractions of a millimeter up to a few millimeters per degree of temperature change, and is given by Maksutov [4.10] with $\Delta f \simeq -2f^2 \alpha \Delta T/$thickness of mirror. For a glass mirror, with its high thermal inertia, the internal temperature gradients cause a deformation most pronounced in the zone at the edge of the mirror; this is known appropriately as the *edge effect*.

High thermal conductivity, however, is not the only criterion for choosing the raw material for the mirror. Elasticity or Young's modulus E, thermal expansion coefficient α, and density ρ are also important. Other significant quantities are the material hardness (for the behavior while polishing), transparency (to study internal stress), processes of aging, long-time behavior, and so on. Table 4.3 compiles the most important constants for currently used mirror materials. Representing the variety of possible quality factors is the combination of material constants as preferred by Maksutov, but it is merely a guideline. The choice of the "best" mirror material is not an intrinsic one, but depends rather upon the specific facilities of the user.

Of unquestionable importance is the *thermal expansion coefficient* α, whose diminution has been the object of longstanding research in science and industry. Now available are materials like *Zerodur* from Schott glassworks and *Cer-Vit* from Owens-Illinois, which have practically no thermal expansion and thus are currently the preferred materials for mirror blanks. These are not glasses but rather *glass ceramics*, a mixture of ceramics and glass with respectively negative and positive expansion coefficients. The production and properties of the outstanding material Zerodur are detailed in an article by Petzoldt [4.11]. Glass ceramics have now superseded borosilicate glasses such as *Duran* (Schott glassworks) or *Pyrex* (Corning glassworks) of which, for instance, the famous Palomar 5-meter telescope was made, and also fused quartz as mirror material. With the current trend to cast or weld the new lightweight mirrors into a kind of "honeycomb" structure, however, some of these materials have regained attention.

4.4.3 The Hartmann Test

To test astronomical optics by the *Hartmann method*, the Hartmann screen, a mask consisting of holes that are radially symmetric about the optical axis (or on a quadratic grid), is placed in front of the objective or mirror. This method approximates the rays of geometrical optics by small beams generated by the holes, each representing a certain *height of incidence h* (see Fig. 4.12).

Illuminated by a plane wavefront (e.g., from a distant artificial point source or a real star), two out-of-focus photographs are taken, one inside and one outside the telescope focus. Measuring the intrafocal and extrafocal distances s'_i, s'_e and the corresponding ray separations d_i, d_e from the geometry of Fig. 4.12, the focal distance follows:

$$s'_F(h) = s'_i + \frac{d_i}{(d_i + d_e)}(s'_e - s'_i). \qquad (4.27)$$

The dependence of s'_F on the incidence height h permits conclusions on imaging errors such as spherical aberration or zonal errors. A change of focal distances for holes circular to the optical axis on the Hartmann screen (i.e., the absence of rotational symmetry) reveals astigmatism. Nevertheless, a spot diagram, from which light concentration of

Table 4.3. Material constants and derived quality factors of potential mirror materials.

Material	Density ρ	Young's Modulus E	Thermal Expansion Coefficient α	Thermal Conductivity λ	Specific Heat c	Thermal Diffusivity $\frac{\lambda}{c \cdot \rho}$	Quality Factor (After Maksutov) $\frac{E}{\alpha} \cdot \frac{\lambda}{c \cdot \rho}$
Flint glass	4.4	5.0	8.5	0.0015	0.13	0.0026	0.15
Crown glass	2.5	7.5	7.5	0.0018	0.17	0.0042	0.42
Duran 50 (Pyrex)	2.23	6.3	3.2	0.0028	0.18	0.0070	1.4
Fused silica (quartz)	2.2	7.2	0.4	0.0033	0.18	0.0081	15
Zerodur (Cer-vit)	2.53	9.1	<0.1	0.0039	0.20	0.0077	>70
Reflective bronze	8.6	8.0	18.6	0.20	0.08	0.29	12
Steel	7.8	21.0	12.0	0.12	0.11	0.14	24
Aluminum	2.7	7.1	23.8	0.49	0.22	0.82	25
Invar	7.9	14.0	0.9	0.0263	0.12	0.028	43
Silver	10.5	8.1	19.5	1.0	0.06	1.6	66
Copper	8.9	12.5	16.5	0.92	0.09	1.15	87
Beryllium	1.85	29.0	12.3	0.43	0.23	1.01	>100
CFRP (carbon fibers)	1.65	23.0	−0.8	0.013	0.20	0.039	>100

The dimensions for the specific quantities are as follows: ρ (g·cm^{-3}), E (10^{8}p·cm^{-2}), α (10^{-6}·grad^{-1}), λ (cal·cm^{-1}·s^{-1}·grad^{-1}), c (cal·g^{-1}·grad^{-1}).

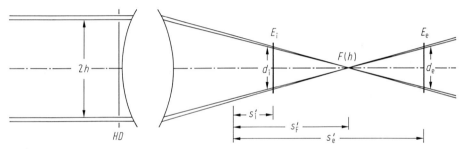

Fig. 4.12. Objective testing after Hartmann [4.12]: only one pair of holes with height h is graphed; the pairs of holes are generally arranged on several diameters of the Hartmann screen HD; intrafocal and extrafocal widths s'_i and s'_e are measured from an arbitrary reference point.

the system can be directly seen, can be calculated from the measured and computed distances. It has been suggested that the Hartmann screen be used not just for testing the optics but also, in a modified form, for "active" optics (i.e. the control of thin and flexible mirrors). For a discussion on this topic, the reader is referred to Ulrich and Kjär [4.26]. The difference in distances can also be used to construct a map of the optical surface from which the irregularities can then be removed directly by targeted polishing. The Hartmann test supplies quantitative results i.e. numerical values of the aberrations but it is limited by its nature to a pointwise checking of the optics under test.

4.4.4 Foucault's Knife-Edge Test

The knife-edge test encompasses the entire optical surface at once and is also quite sensitive to small-scale deviations which are not recorded by, for instance, the Hartmann test. Its effect is also explained by geometrical optics. All rays emanating from a point source at the center of curvature C of a spherical mirror are reflected back to that center. In the actual laboratory test (Fig. 4.13) the center of curvature replaces the normal focal point. Thus the terms *intrafocal* and *extrafocal* for knife-edge positions near the center of curvature will be used. If a sharp-edged screen (such as a razor blade) is inserted from the side into the beam, the shadow moves in the same direction as the screen in an intrafocal position but opposite to the screen in an extrafocal position (i.e., behind the center of curvature). Geometrical optics explains this by the obscuring of individual rays. If the edge is inserted exactly at the position of the center of curvature C (the focus), the obscuration of the brightly lit mirror surface occurs suddenly (or at least evenly) over the entire surface, since the image of the source is not a point but has some extension owing to diffraction and other effects. Here is thus a *highly* sensitive method for determining the center of curvature of a mirror or the precise focal position of a real star.

If the mirror deviates from its ideal form by zones or by small-scale irregularities, these deviations will become visible as bright and dark regions on the shadow and the illuminated parts respectively, as the rays bypass the center of curvature because of their different inclinations. The mirror surface then appears as a "relief" or embossed

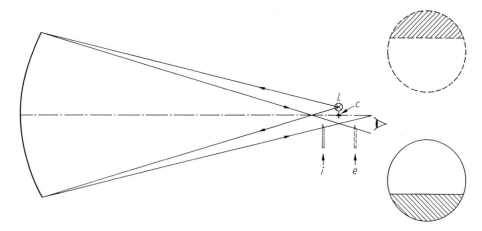

Fig. 4.13. Knife-edge test after Foucault. Intrafocal (*solid line*): edge and shadow move in the same direction; extrafocal (*dashed line*): edge and shadow move in opposite directions.

surface illuminated from the side (opposite to the edge). Foucault's test therefore is primarily a qualitative test to find minute surface errors. The method is sensitive enough to show deviations down to $\lambda/100$. In order to separate object and image, the light source L is placed slightly off the center of curvature C in the actual laboratory test (Fig. 4.13). The image point is then axisymmetric to the light source and therefore accessible. Foucault's test also shows imaging errors. When an ideal or error-free spherical mirror shows no deviations when examined in the center of curvature (and thus is brightly illuminated) it is evident that a parabolic mirror in the same setup will show shadows since the spherical wave does not converge at the center of curvature. However, if both mirrors are tested with a plane wavefront (as from a distant star), the parabolic mirror will show no focal deviations, whereas the spherical mirror distinctly displays spherical aberration (Fig. 4.14). The same holds true for zonal errors. Needless to say, this method also applies to the testing of lens objectives in *transmitted* light.

The test as described includes a point source (spherical wave) which can be replaced by a slit. This results in a strong increase in the image brightness which is very valuable in visual testing. The illuminating slit must be strictly parallel to the edge. The image brightness can be further increased by observing past the edge an extended light source which is partially covered by the edge itself. This ensures that both edges, knife and slit, are parallel. The construction of such a test device is described by Malacara [4.13], who also gives the diagnostic images of the individual imaging errors as they occur within the Foucault test.

4.4.5 Interferometric Tests

Interferometric tests are based on wavefronts orthogonal to the rays considered thus far. In this sense, the ideal wave surface has a spherical shape after passing through the optical system with its center of curvature at the image point, the focus. All departures

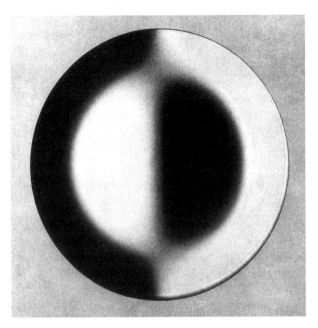

Fig. 4.14. Foucault graph of spherical aberration; the edge is inserted between the paraxial and the marginal focus into the beam. From *Atlas of Optical Phenomena* [4.9].

from this wave surface are called *wave aberrations*. In interferometric methods, these errors are made visible by the interference of two wavefronts, the wavefront being tested and a reference wavefront. Their advantage consists in permitting the qualitative *and* quantitative determination of imaging errors simultaneously.

From the wide variety of interferometric tests, many of which would be beyond the scope of this book, only a few types will be mentioned here briefly as examples. References such as Malacara [4.13] and DeVany [4.14] can be consulted for detailed instructions and interpretations of the test results obtained. The best known test device is probably the one by Twyman-Green, originating from the two-armed Michelson interferometer, but replacing one plane mirror by the optical system to be tested. Superposition of both reflected wavefronts by the beam splitter yields characteristic fringe patterns. Malacara [4.13] gives a list of the various patterns for perfect and defocused optics and such affected by spherical aberration, coma, or astigmatism. Counting the fringes gives the amount of deviation in units of half of the wavelength ($\lambda/2$). Interferometric methods are quite demanding in setup and are quite sensitive to vibrations, air turbulence in the optical path, adjustment, and so on. To overcome such difficulties, a type of *shear interferometer* has been developed. This device does not require two separate interferometer beams, since the optical system which is under test produces *both* the test and the reference wavefronts and superposes them after a displacement (i.e., shear). The possible variations are so numerous (lateral, radial, rotated, and inverted shear) that Malacara [4.13] and Bryngdahl [4.15] must be referred to. A relatively simple setup for the amateur is given by Bath [4.16] with instructions and with interferograms of optical errors.

Finally, there is the *Ronchi test*, which is similar to the Foucault test except that the knife-edge is replaced by a fine grating. In its original form, a point source in (or near to) the center of curvature was used to illuminate the mirror which was then viewed through a grating. For the purpose of better illumination, the point source has been replaced by a slit. For a further increase of the image brightness, the slit can be replaced by an extended light source which illuminates the mirror through the same grating which will be used to observe it. In this way the parallelism of "multiple slit" and grating are guaranteed, which is important for this method too. Malacara [4.13] shows the Ronchi-grams for all imaging errors and for aspherical surfaces and compares them with the interferograms of a Twyman-Green interferometer. It is then seen that the Ronchi-gram for spherical aberration agrees with the Twyman-Green coma interferogram, and the Ronchi-gram for coma with the astigmatism interferogram. Image generation in the Ronchi test can be explained by geometrical optics, but the interpretation is usually carried out interferometrically. A comparison between Ronchi-grams and Foucault-grams is given by DeVany [4.14], who also shows an illuminating device for both tests. In order to simplify the interpretation by standardizing the Ronchi test patterns, Schultz [4.17] proposes a grating whose line number depends on the aperture ratio $1/N = D/f$. His figures show for gratings with $12N$ lines/inch (e.g., $f/8 \to 100$ lines/inch) the same Ronchi-grams for mirrors of quite diverse apertures and focal lengths. The fine gratings with equidistant bright and dark lines of equal widths can also be produced photographically by amateurs. Thus, for amateurs, the Ronchi test together with the Foucault test will undoubtedly provide the most attractive method for testing astronomical optics.

4.5 Telescope Systems

4.5.1 Refractors

Perhaps the most distinctive feature of refracting optics is their chromatic error due to the wavelength dependence of the refractive index n (Sect. 4.3.3). An objective corrected for two wavelengths was first computed and ground by Joseph Fraunhofer in the early 19th century. This *achromatic* type of objective, also known as *Fraunhofertype*, consists of a convex front lens of crown glass and a concave rear lens of flint glass, with a small air space in between. The procedure for calculating and correcting a simple achromat is shown in detail, for instance, in Klein [4.19], who also addresses the fine correction of the objective. The Fraunhofer achromat corresponds to Zeiss type E (Fig. 4.15) or to the Lichtenknecker FH-objective. The secondary spectrum (see Sect. 4.3.3) of achromats can be reduced by a proper choice of glasses with suitable constants ν and θ. These somewhat more expensive special glasses lead to the two-lens *half-apochromat* of Zeiss type AS (Fig. 4.15).

The addition of a third lens of proper glass type can achieve chromatic correction for three wavelengths. This classical apochromat, corresponding to Zeiss type B or the

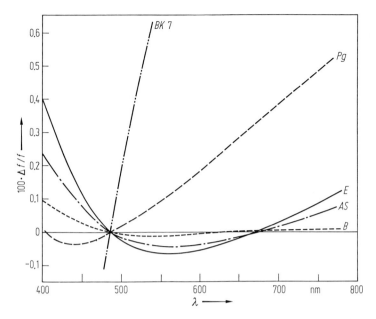

Fig. 4.15. Chromatic aberration of several different objectives: single lens of glass BK7, Fraunhofer achromat E, half-apochromat AS, apochromat B, example of an objective corrected for photography Pg. From *Handbuch der Physik* [4.6].

Lichtenknecker VA-objective, not only possesses nearly ideal chromatic correction but is also free of coma and spherical aberration. Manufacturing data and aberration curves for various wavelengths (C, d, e, F, g) of this and a variety of other objectives are given by König and Köhler [4.18]. Cemented three-lens apochromats of $f/8$ and $f/9$ (e.g., Starfire Triplet, APQ Fluorite of Zeiss) that can be used for astrophotography are also now available too. In order to reduce the tube length of the refractor, amateur observers have sometimes used plane mirrors to "fold" the beam. The telescope length is correspondingly reduced to 1/2 to 1/3 of the focal length. Needless to say, the quality of the mirrors enters into the imaging performance of the telescope. If they are not quite plane, the result will be increased astigmatism. Also very critical are the stability of the tube and of the support frame of the mirrors, conditions that can nevertheless be fulfilled by amateur telescope makers.

Robb [4.20] describes a novel method to select glasses for chromatic correction. He gives combinations of glasses which not only achieve on-axis correction of three wavelengths for a two-lens objective, but also promise on-axis correction for up to five wavelengths. Instead of the usual parameters Abbe number ν and relative dispersion ϑ (where a better color correction is achieved with ν-values as different as possible at the same dispersion curve ϑ), Robb's method uses a new description of the dispersive behavior of glass originating from Buchdahl. In place of the wavelength λ, a new parameter, the *chromatic coordinate* ω, is introduced and permits the expression of the refractive index n by a power series. Applying this model gives new dispersion coefficients characteristic for each glass. Robb lists these new dispersion constants for

813 glass types, as well as the conditions to be met in order to achieve color correction for $\geq i+1$ wavelengths when using i glasses. In addition, he shows chromatic aberration curves of two-lens objectives in comparison with two classical achromats.

Figure 4.15 indicates that the objectives of telescopes used primarily for photographic studies are in general corrected for different wavelengths than objectives used for visual observations. The vertex of the color curve lies at shorter wavelengths, and this is conditioned by the higher blue-sensitivity of earlier photographic emulsions. Along with the Schmidt telescope (Sect. 4.5.6), lens objectives are nowadays still in use as astrographs. The demands on these objectives are, of course, much higher than on those used for visual observations, at least regarding field size and flatness of the image. To be mentioned first is the venerable *Petzval objective* (cf. Berek [4.2] for construction data and Seidel sums) which, at $f/3.4$, was a fast system for its day. While its 7° field of view was quite large, it suffered from field curvature (cf. Sect. 4.5.6 for the removal of field curvature by a planing lens). Later, the *Taylor triplet* (up to $f/3$) followed from which the well-known *Ross* and *Sonnefeld four-lens objectives* were developed. These have useful fields of up to $10° \times 10°$. In contrast to the Petzval objective, which is composed of two two-lens achromats with a large air space in between, these objectives have a negative lens as the central component. More on these objectives is reported by König and Köhler [4.18]. Modern forms of light-powerful, catadioptric astrocameras with aperture ratios up to $f/1.5$ can be obtained from Wiedemann [4.39] who also gives Seidel-sums and chromatic aberration curves for these objectives.

Refractor objectives normally do not need any re-collimating of their optical components as they are well fixed in their frame. Only the adjustment of the objective's optical axis to the central axis of the eyepiece tube can be checked from time to time. If the objective is not tilted too much, so that the optical axis passes close to the eyepiece's center, a star test is sufficient. In this test one chooses a bright star and centers it in a high-magnification eyepiece. It is then defocused until the first diffraction rings become visible. If these rings appear circular, everything is fine; otherwise the adjusting screws of the objective have to be tuned until the oval diffraction rings become circular and concentric in the center of the eyepiece, assuming there is no defect in manufacturing present. In cases of large tilts, Valleli [4.71] describes two methods to align the objective before testing it on the star. A lot of care must be taken when dismounting and re-assembling a multiple-lens objective; this is a task best left to the experts since imaging and manufacturing errors are compensated for at a certain orientation of the lenses or will be introduced when a lens is stressed by its mount.

4.5.2 The Newtonian Reflector

All previously mentioned problems with chromatic aberrations of refraction optics do not, of course, occur in purely reflecting systems, as these are, owing to the law of reflection, wavelength independent and are hence *absolutely* achromatic. For these systems, only the geometrical aberrations of the Seidel theory (Sect. 4.3) need to be reduced to an acceptable value.

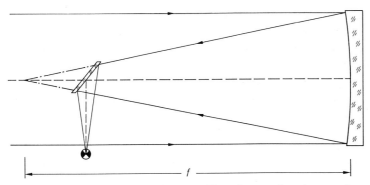

Fig. 4.16. An $f/3$ Newtonian telescope: The primary mirror is a rotation paraboloid, the secondary is a plane mirror of elliptical shape.

A simple arrangement of reflecting optics is the combination of a concave primary and a small secondary mirror to deflect the beam outward (Fig. 4.16) and to make it accessible to the eye. The simplest form of the primary is a concave spherical surface, which is easily manufactured but which contains a severe imaging error—spherical aberration. Despite this serious disadvantage, the spherical mirror can be used for astronomical observations if its f-ratio is chosen sufficiently small (Sect. 4.3.2). If the disadvantages thus caused—the low light power and the large tube length—are not deemed acceptable, a two-lens corrector can remove spherical aberration and coma almost entirely to an f-ratio of $f/2$. More on the corrector-type and the aberration curve are given by Wiedemann [4.21], who provides the construction data.

Another way of correcting the aperture error or spherical aberration is by using a mirror of *parabolic* shape. This uses the property of the parabola (the conic curve of equal distance from a point and a straight line) to concentrate light parallel to the optical axis at one focus, irrespective of aperture. A plane secondary mirror inclined by 45° against the optical axis moves the focus outside the telescope tube. This construction was conceived back in the 17th century by Isaac Newton and is named in his honor (Fig. 4.16). The parabolization of a spherical mirror can be executed by anyone with the necessary instructions (e.g., Rohr [4.22]; see also the Supplemental Reading List for Vol. 1.) and with a good deal of patience. After removing the spherical aberration, the chief remaining imaging error of the Newtonian telescope is the *comatic error*, which gives rise to a comet-shaped distortion of stellar images that increases toward the edge of the field (Fig. 4.6). The coma thus determines the useful field size of a paraboloid mirror.

Two ways of defining the useful field are possible. A maximal comatic lateral aberration of $1''$, for example, may be permitted during visual observations. For telescope apertures larger than 10 cm, this corresponds approximately to the resolution limit set by the Earth's atmosphere (see Sect. 4.6.1). Inserting $K_{max} = 1''$ and the Seidel sum of the parabolic mirror $\sum \text{II} = 0.5$ into Eq. (4.19c), the useful angular field diameter is determined to be

$$\phi \simeq 0°003 \, N^2. \tag{4.28}$$

An $f/8$ Newtonian thus gives useful images to a field diameter of $0°\!.2$ or $11'$. For photographic observations, the criterion will be guided by the resolving power (i.e., grain size) of the emulsion. Again, Eq. (4.19c) supplies the sought relation, with K as the linear extension of the coma shape:

$$\phi = f\, 2\tan u \simeq 10.7 N^2 K. \tag{4.29}$$

Assuming a resolving limit of 10 μm for a film, the coma shape of an $f/8$ Newtonian will become perceptible outside an image field diameter of about 7 mm. The corresponding angular field diameter can be calculated via the focal length; for example for $f = 1000$ mm, it is about $24'$. For emulsions with higher speed and thus larger grain size, the useful field increases proportionally. The field size increases in both cases (visual and photographic observations) with the square of the f-number N. Still, in order not to be too limited in light-gathering power, a three-lens corrector of the Wynne-type can be used, as it almost entirely removes the coma of a parabolic mirror to an f-ratio of $f/2.5$. Wiedemann [4.21] gives details and elsewhere [4.23] specifies a modification of the Baker corrector, which, largely complying with the needs of an amateur, achieves a quite satisfactory correction up to a ratio $f/5$.

The position and size of the secondary mirror depends on the focal length and f-number and can be found graphically. The plane mirror is usually of elliptical shape in order to reduce obscuration, and its two axes are chosen somewhat larger in order to deflect, free of vignetting, not only the axial rays but also the desired image field out of the tube. The path and intersection of these marginal rays with the secondary can be seen by drawing the desired field size in scale at the primary focal plane. To hold the secondary mirror in place, a four-arm support frame (preferred over a three-arm frame) is used as a spider. Since the diffraction of light is always perpendicular to the disturbing edge, the diffraction images of two arms coincide (i.e., four spikes in the case of orthogonal arms), whereas the three-arm spider leads to six radial rays in the diffraction image. The tube should not be sealed at the primary mirror end so that warm air in the tube can mix with the ambient air and flow out of the tube; warm air convecting within the tube would heavily degrade the image enlarged by the eyepiece.

Collimating a Newtonian will be briefly described here. For a comprehensive treatment of the procedure, the reader is referred to Valleli [4.71]. When starting the aligning, one should be sure that eyepiece tube and optical axis (i.e., the tube axis) are orthogonal. The plane secondary mirror is inserted and moved along the optical axis until its circular shape is centered in the eyepiece tube. Next, the three adjusting screws of the secondary are tuned until the image of the primary appears concentric with the circular rim of the secondary. Finally, moving and tilting the primary mirror can be stopped when the image of the secondary with its spider appears concentric to the primary's rim. This rough collimating, where all rims of the mirrors and their images must appear concentric to each other is carried out in daylight following the test on a star. As in the procedure on the refractor, one observes the defocused image of a bright star in an eyepiece of high magnification. The screws of the primary are fine-tuned until the oval diffraction rings of the star image appear concentric in the eyepiece. In the case of a manufacturing error of the mirror, this will not be successful. To test this one moves the eyepiece to the other side of the telescope focus. If the defect in the concentric diffraction rings turns to the opposite side of the defocused image, the mirror possesses a defect or is stressed in its cell.

4.5.3 The Cassegrain Telescope

The fact that the large tube length of the Newtonian telescope equals the focal length will be disadvantageous for some applications. Owing to its imposing size and weight, the Newtonian is normally not transportable, and requires, for occasions when a moderate wind is blowing, an extraordinary rigid mount. The telescope length can be substantially reduced, yet retaining the equivalent focal length, by replacing the plane secondary mirror with a convex one also resulting in a conveniently accessible focus location (Fig. 4.17). The curved secondary S_2 images the primary focus F_1 into the telescope focus F. The focal length of the telescope is then no longer the primary focal length f_1 but rather the distance between the system's focus F and the intersection of the narrow beam with the incident marginal rays. If $m = f/f_1$ is the magnifying factor of the primary focal length, the tube length of the Cassegrain telescope is less than $1/m$ of the telescopic focal length. The following formulae taken from the *Handbuch der Physik* [4.6] are useful in relating the Cassegrain system parameters:

$$f = mf_1 = \frac{f_1 f_2}{f_1 + f_2 - e},$$
$$a = mc = \frac{(f_1 - e)f_2}{f_1 + f_2 - e}. \tag{4.30}$$

For instance, to modify a Newtonian telescope into a Cassegrain system, and with given f_1, m, and g (see Fig. 4.17), the following equations apply:

$$e = \frac{mf_1 - g}{m + 1} \qquad f_2 = -\frac{m}{m^2 - 1}(f_1 + g). \tag{4.31}$$

The minimum diameter of the secondary follows from the *law of rays*,

$$D_2 = \frac{a}{f} D_1, \tag{4.32}$$

where an additional amount ($2e \tan u$) must be taken into account for the angle u of the image field radius. The original Cassegrain system consists of a parabolic primary (deformation constant $k_{1,\text{Cass}} = -1$) and a convex secondary. In order to transform the spherical-aberration-free prime focus into the Cassegrain focus, the secondary mirror must have a deformation which ensures the constancy of the light path for both foci F_1 and F. This constraint leads to a hyperbolic meridional cross section, and its numerical eccentricity $\varepsilon = (m+1)/(m-1)$ gives the deformation constant of the secondary (cf. Sect. 4.3.1)

$$k_{2,\text{Cass}} = -\varepsilon^2 = -\left(\frac{m+1}{m-1}\right)^2. \tag{4.33}$$

Under these conditions, the Cassegrain focus is free of spherical aberration but retains the off-axis coma of the parabolic primary (Table 4.1) whose astigmatism,

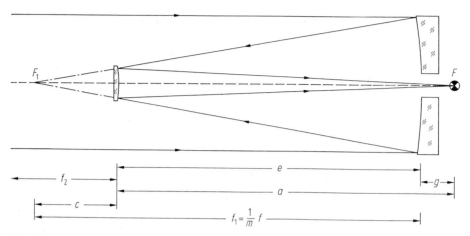

Fig. 4.17. A Cassegrain telescope ($f/12$): The primary is a rotation paraboloid, the secondary a rotation hyperboloid. The focal length is increased by the factor $m = 4$.

however, is increased by the factor m. More specifically,

$$\sum \text{III}a = \frac{f-e}{a} = \frac{mf+g}{f+mg}. \tag{4.34}$$

Insertion of the secondary mirror causes decentering as an additional source of errors. Its tolerance depends strongly on the f-number and lies in the range of 0.1 mm to several mm. A first approximate adjustment is made by fixing the center of the front end of the tube through a crosswire. If the primary mirror is viewed from this center point, then the image of the secondary mirror and its support structure should be seen exactly centered; if not, then the primary must be shifted by adjustment screws. The same holds for examination through the eyepiece tubes: all mirror contours must be seen concentrically. The easily manufactured eyepiece by Cox [4.24] may be used for fine adjustment. A sensitive method is the test with the *Ronchi grating*, which places a fine line grating at the focus in place of the eyepiece and illuminates it by means of a parallel slit. The eye to the side of it sees, at correct adjustment, parallel fringes on the secondary mirror. A test setup to examine the shape of the Cassegrain secondary has been described by Richter [4.25].

Also noteworthy is the *Gregory telescope*, which accomplishes the magnification of focal length by inserting a *concave* secondary mirror behind the prime focus F_1. This position lengthens the telescope tube by the amount $2c$ (Fig. 4.17) compared with the Cassegrain system. This may be one of the reasons why the Gregory system has not successfully competed with the Cassegrain system. In this system the spherical-aberration-free imaging of the parabolic primary is maintained by deforming the concave secondary to an elliptical curve in the meridional cross section. As in the Cassegrain system, coma and astigmatism are retained.

The Cassegrain focus is modified to the so-called *Nasmyth focus* by inserting a tertiary, plane mirror in front of the primary mirror. This third mirror deflects the beam after reflection from the secondary into the elevation (altitude) axis out of the tube. The positioning of the focus to the side of the tube, however, is preferable only in an alt-azimuthal mount (i.e., with the axes in the horizontal system; see Chap. 2) over the Cassegrain position, because in this case, it is fixed with respect to Earth's gravity and also does not burden the tube with the instrumental weight. Therefore, very heavy or bending-sensitive instrumentation (e.g., spectrographs) can be accommodated on the Nasmyth platforms, which are integral parts of the heavy azimuthal yoke on both sides of the tube close to the elevation axis. Possible disadvantages are the exchange of the tertiary mirror when switching from Cassegrain to Nasmyth, the loss of reflection, and field rotation; the latter is intrinsic to the alt-azimuth mount for all foci.

Another distinctive focus is the *Coudé focus*, whose position is absolutely fixed in the hour axis of the equatorial mount or in the azimuth axis of the alt-azimuth mount. After reflection at the Cassegrain secondary, the beam passes several plane mirrors until arriving in the rotation axis of the telescope. Normally the heaviest spectrographs are placed here; the image rotation (which for the Coudé focus also occurs in an equatorial mount) is without significance in the spectroscopy of point sources. If necessary, however, this can be compensated for by additional optical elements. The Coudé focus has gained significance nowadays for its use when combining several telescopes to form an optical interferometer. This new observing method still requires numerous technological developments, for example, for coatings of highest reflectivity.

4.5.4 The Ritchey–Chrétien System

The mirror systems considered thus far remove spherical aberration by parabolizing the primary mirror. To maintain this state of correction in the system focus (the transformed primary focus), the secondary must be correspondingly deformed into an aspherical hyperbolic mirror; i.e., the focal properties of conic sections are employed. Correction of spherical aberration can also be accomplished by deforming both mirrors in relation to each other instead of correcting each mirror individually. Thus, by insertion of a secondary mirror into the beam, this second surface adds another degree of freedom which can be used for further correction of imaging errors. This gives exactly *one* combination of aspherical mirrors which removes spherical aberration and coma simultaneously. In this system, named after Ritchey and Chrétien, who first elaborated its design, the two mirrors of the Cassegrain are even more strongly deformed according to the two deformation constants

$$k_{1,\text{RC}} = k_{1,\text{Cass}} - \frac{2a}{em^3} = -1 - \frac{2(f-em)}{em^3},$$
$$k_{2,\text{RC}} = k_{2,\text{Cass}} - \frac{2f}{e(m-1)^3} = -\left(\frac{m+1}{m-1}\right)^2 - \frac{2f}{e(m-1)^3}; \quad (4.35)$$

i.e., both mirrors are rotation hyperboloids ($k < -1$). Of course, the primary itself loses its correction for spherical aberration so that a special corrector is needed when observing at the primary focus. The system focus, however, is free of spherical aberration and coma because of the combination of the two aspherics. The remnant astigmatism and field curvature are somewhat stronger than in a comparable Cassegrain system (see for example Table 4.1). These imaging errors of the Ritchey–Chrétien (RC) system have the following quantities:

$$\sum \text{IIIa} = \frac{2f - e}{2a} = \frac{(m + 1/2)f + g/2}{f + mg} \quad \text{(Astigmatism)},$$
$$R_m = -\frac{af}{(m^2 - 1)e + f} \quad \text{(Mean field curvature)}. \tag{4.36}$$

These errors limit the size of the useful field of view. A field diameter of 1/2 degree can be considered typical, where the astigmatism of the RC system leads to a scattered circle of some $1''$ diameter at the edge of the field. Since the RC system is a further development of the Cassegrain, it has also, by virtue of the similar mirror arrangement (Fig. 4.17), a tube length of less than $1/m$ of the telescopic focal length. Because of its good optical correction state (absence of chromatic, sperical, and comatic aberrations) while having only two optical surfaces and a short telescope length, many of the existing large telescopes have been designed as RC systems, for instance, the Calar-Alto 3.5-m, the ESO 3.6-m, the Anglo-Australian 3.9-m, the Kitt Peak 4.0-m, and the ESO 3.5-m New Technology Telescope. Also, most of the large telescopes of the next generation, with effective mirror diameters of 8 to 16 meters, are being built as RC systems, for example the Californian 10-m Keck telescope and the 16-m ESO-VLT. More detailed information on these and other large telescope projects can be found, for example, in *Proceedings of the ESO Conference on Very Large Telescopes and Their Instrumentation* [4.26] or in the *Mitteilungen der Astronomischen Gesellschaft Nos. 67 and 70* [4.27].

4.5.5 The Schiefspiegler

The extreme difficulty in the accurate manufacturing of two strongly deformed aspherical surfaces is likely to keep the RC system just mentioned out of the reach of the amateur telescope maker. The situation is totally different for the *oblique mirror telescope*, or *Schiefspiegler*, which is used almost exclusively by amateurs; it can be regarded as a development of the Newtonian design. The simplest variation of the Schiefspiegler consists of a concave primary and the eyepiece at the periphery of the front end of the tube. Even apart from the disadvantage of the long, cumbersome tube, the primary imaging errors now have an enhanced effect according to Eq. (4.19) since this is a pure "off-axis" observation with inclination angle u without any correction. If a plane secondary is placed adjacent to the tube so that it reflects the light from the primary again, the result is a shortened tube length along with a larger focal length, which, via the increased f-number N, leads to a decrease in imaging errors. To further reduce the errors, the secondary mirror can be deformed or a suitable correction lens inserted (Fig. 4.18).

The theory and practice of the Schiefspiegler have been comprehensively presented by Kutter [4.28]. The manufacturing of the plano-convex correction lens may be difficult; therefore Kutter [4.29] replaced it with a third mirror, creating the Tri-Schiefspiegler (Fig. 4.19). The specialty of this instrument consists in all three mirrors

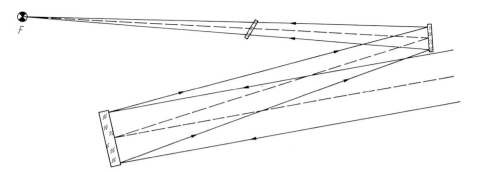

Fig. 4.18. The Kutter Schiefspiegler ($f/20$).

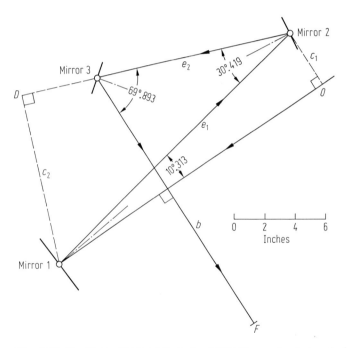

Fig. 4.19. The Kutter Tri-Schiefspiegler ($f/14.7$), constructed exclusively from spherical mirrors. See Table 4.4 for the construction data.

having spherical surfaces; they are thus of particular interest to amateur telescope makers. Table 4.4 collects the data necessary to construct a Tri-Schiefspiegler according to Fig. 4.19. The signs correspond to the convention introduced in Sect. 4.2.1. The constructional data can also be used for larger apertures. The values in Table 4.4 must then be multiplied by the desired scale factor.

Table 4.4. Construction data for the Kutter Tri-Schiefspiegler. All measurements are given in inches and in cm (the latter in parentheses).

Primary mirror			Distance e_1	21.77	(55.30)
Free aperture	4.30	(10.92)	Distance e_2	13.19	(33.50)
Radius of curvature r_1	+87.40	(+221.99)	Distance M_3-focus	18.90	(48.01)
Secondary mirror			Distance c_1	3.90	(9.91)
Diameter	2.40	(6.096)	Distance c_2	11.02	(27.99)
Radius of curvature r_2	−127.56	(−324.00)	Equivalent focal length	63.00	(151.19)
Tertiary mirror			Total length	25.00	(56.35)
Diameter	2.40	(6.096)	Aperture ratio	$f/14.7$	
Radius of curvature r_3	+704.00	(+1788.15)			

4.5.6 The Schmidt Camera

The telescopes discussed above possess a rather small field of view of less than 1°. They are usually employed to observe individual objects. The type of telescope to be described in this subsection is used for the photographic recording of large celestial areas with diameters of several degrees, but it can be modified to another focal position and then be used for other purposes. The ingenious idea of the optician Bernhard Schmidt consisted in removing the asymmetry of the known optical systems for inclined beams (and the resulting imaging errors) and removing the remaining spherical aberration by a correction lens. To eliminate the asymmetry, he used a spherical mirror and moved the entrance pupil of the system from the mirror vertex into the plane through the center of curvature of the mirror (Fig. 2.20). This eliminates the central role of the optical axis; a beam incident parallel to the optical axis is no longer distinguished from a beam incident at an angle which remains symmetric to its chief ray. Thus the asymmetry errors (Sect. 4.3.2) coma and astigmatism cannot occur.

Schmidt eliminated the spherical aberration of the mirror by inserting a correction plate at the entrance pupil. It still has the imaging errors of a lens (e.g., chromatic aberration) but these remain insignificantly small since the plate is nearly without any refractive power. This so-called Schmidt plate possesses one plane and one aspherical surface. Schmidt also showed how to make the complex aspherical surface (shown much exaggerated in Figure 2.20); with zonal differences of several tens of μm, it is strongly deformed. Schmidt's procedure (grinding a glass plate that is bent over a cylinder by evacuating it) and other methods are detailed by Ingalls [4.30]. Instructions for the do-it-yourself production of Schmidt plates using a vacuum machine are given by Cox [4.31]. The adjustment of a Schmidt camera is described, for instance, by Weigel [4.32]. The field curvature remains as the only optical error. Its radius of curvature R_F corresponds to one-half of the radius of curvature of the spherical mirror, which is equal to the focal length of the telescope, and is given by

$$\text{Field curvature} \qquad R_F = \frac{r_1}{2} = f. \qquad (4.37)$$

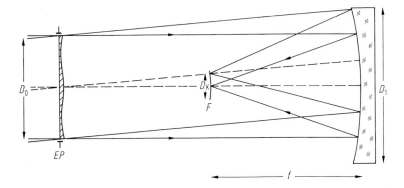

Fig. 4.20. Schmidt camera ($f/1.5$): free aperture (corrector) D_0, diameter of spherical mirror D_1, diameter of plateholder D_{PH}, field curvature $= r/2 = f$. The deformation of the Schmidt corrector is much exaggerated in the graph for illustration.

This is a rather strong curvature, but the thin photoplates used with larger instruments can be bent by a suitable plateholder and adapted to this curvature. Smaller Schmidt cameras must use emulsions on film. The problem of the strong curving of the focal plane can be avoided with the aid of a flattening lens; the image is flattened by a single plano-convex lens with radius of curvature r_P adapted to the telescope focal length f. The radius of the curved front surface of this plano-convex lens is given by

$$\text{Flattening lens curvature} \qquad r_P = \frac{n-1}{n} f. \tag{4.38}$$

The quantity n refers to the refractive index of the lens which is placed immediately in front of the image plane. Shifting the entrance pupil from the vertex to the center of curvature of the mirror requires that the mirror's diameter be larger than the correcting Schmidt plate; otherwise vignetting for inclined incident beams would occur. The amount depends on the size of the field, i.e., on the diameter D_{PH} of the plateholder used. Figure 4.20 readily shows that the mirror diameter D_1 must be

$$D_1 = D_0 + 2D_{PH}, \tag{4.39}$$

where D_0 is the size of the correction plate. The angular diameter of the imaged field ($2u$) can thus be found from

$$\tan u = \frac{D_{PH}}{2f}. \tag{4.40}$$

By virtue of its unrivaled imaging quality, Schmidt systems can be constructed as fast telescopic systems ($f/3$ and less); for example, the Schmidt telescope in Tautenburg, Germany, has $D_0/D_1/f = 134/200/400$ cm with a $3°\!.4 \times 3°\!.4$ field, and the Mount Palomar Schmidt has $122/183/307$ cm with a $6°\!.5 \times 6°\!.5$ field. The latter produced the photographs for the famous and widely used *Palomar Observatory Sky Survey* (POSS).

Drawbacks of the Schmidt system are the large tube length of twice the focal length, the field curvature (which can be compensated for), and the relatively inaccessible focus. Numerous modifications of the Schmidt camera were constructed to eliminate these problems, but only at the expense of image quality. Various forms and further developments of the Schmidt system can be found in Köhler [4.4] and Slevogt [4.5]. One such development is the *Wright–Väisälä System*, with the correction plate and entrance pupil in the plane through the focus, which is free of the asymmetry aberration coma through additional deformation of the mirror but readmits astigmatism. Despite this astigmatic aberration ($\sum \text{III}a = \frac{1}{2}$), the mean field curvature is plane but the useful field remains less than that of the original Schmidt camera. The manufacture and testing of such a system are described by Waineo [4.33]. Owing to high production costs, the use of Schmidt systems without correction plates is being contemplated. Schmadel [4.34] has listed the maximum focal length and f-ratio for which such a system may still produce useful images.

4.5.7 Schmidt–Cassegrain Systems

In order to remove the drawbacks of the original Schmidt system, namely the ponderous tube length and the inconvenient focus location, it was soon suggested that the Schmidt system be modified into a *Schmidt–Cassegrain* design by insertion of a secondary mirror. One such Schmidt–Cassegrain system which eliminates all four significant errors (spherical aberration, coma, astigmatism, and image field curvature) comes from Baker. He obtained this excellent state of correction by deforming one of the two mirrors while retaining the spherical shape of the other. Apart from removing the field curvature, the lengths could be reduced to 1.4 to 1.7 times the telescopic focal length and the focal plane placed near the primary's vertex. One disadvantage is that the increase of focal length by the secondary mirror was limited to values less than 2, which resulted in a rather strong central obscuration (15%).

Extremely short Schmidt–Cassegrain systems (tube length about 1/4 of the system focal length) have been developed for amateur use (Fig. 4.21). These systems, usually with f-ratios of $f/10$ to $f/12$, have recently become quite popular. They are a compromise between image quality, very short tube length, and easily accessible focal plane outside the optical system. The construction data are not obtainable from the manufacturers; therefore, no specific data on individual imaging errors are known. Data of one such Schmidt–Cassegrain system have been determined from optical cal-

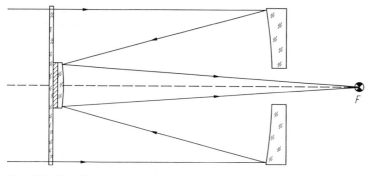

Fig. 4.21. Schmidt–Cassegrain system.

culations by Rutten and Venrooij [4.35], who also discuss imaging quality, sensitivity to adjustment, and the different focusing mechanisms of the system. It is assumed that a spherical primary mirror and a correction plate deformed on one side, lying within the primary focal length and defining the entrance pupil of the system, form the optical system. To further reduce the optical errors, the secondary mirror, which increases the focal length, is deformed into a rotation ellipsoid. By far the largest amount of error is contributed by the very strong field curvature. This can be compensated for by curving the receptor surface, so that the remaining imaging errors are largely within the size of the diffraction disk and are thus insignificant.

A Schmidt–Cassegrain system with a plane field has been described by DeVany [4.36], who also provides hints on manufacturing and testing the optical surfaces. It is particularly suited for amateur telescope makers, since it is based upon two spherical mirrors of identical curvature. The plane image results from the Petzval condition (4.14) that the radii of curvature of the primary and secondary mirrors must, apart from their opposite signs, be identical. Both mirrors thus result from grinding a single optical surface. The small secondary mirror is cut out of the convex grinding shell and glued underneath the correction plate. Thus, the diffraction spikes of the secondary's spider are removed, a feature shared by all commercially available Schmidt–Cassegrain systems. By abandoning the condition of a plane image field, for instance, by using the instrument visually with a field of $1°–2°$, a variety of possible Schmidt–Cassegrain systems with aplanatic or anastigmatic imaging can be created. Formulae for calculating the system parameters and the primary imaging errors of such systems are given with related diagrams by Sigler [4.37]. A similarly detailed treatment of Cassegrain and Gregory systems with one- or two-lens correctors (composed of purely *spherical* surfaces) in the telescope aperture, as well as of Maksutov–Cassegrain systems, is again given by Sigler [4.38]. He also gives the aberration formulae in closed form, including the imaging errors to be expected for various configurations; they are of great interest indeed to amateur astronomers.

Collimating the Schmidt–Cassegrain is done directly with a star in an eyepiece of low magnification since these commercially available telescopes do not provide any means to adjust the primary mirror itself. For rough alignment one strongly defocuses the star so that its disk fills up nearly one-third the field of view. A dark spot appears within the disk, the silhouette of the secondary mirror. The three aligning screws on the periphery of the secondary are tuned so that the secondary's shadow lies concentric within the defocused stellar disk. One has to repoint the telescope several times since by this procedure the star moves outside the center of the field of view. The fine tuning is done with the diffraction rings of a small defocused star image as described in earlier sections of this chapter. The collimating procedure (also for Maksutov systems) is treated in great detail by Valleli [4.71].

4.5.8 Maksutov Systems

Since the production of the aspherical correction plate for the Schmidt camera is quite difficult, attempts were soon made to replace it by dioptric elements with spherical surfaces. The optical principle involved was discovered independently and almost simultaneously by K. Penning, A. Bouwers, D. Gabor, and D.D. Maksutov. It is based

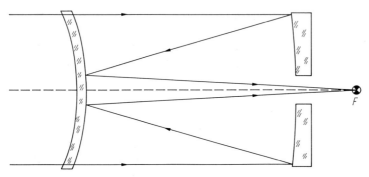

Fig. 4.22. Maksutov–Cassegrain system ($f/15$). Silver or aluminum vapored onto the back of the meniscus lens serves as the secondary mirror.

upon the use of a simple *meniscus lens*, which compensates the spherical undercorrection of the mirror by an overcorrection of the lens. The meniscus is placed concentric with the center of curvature of the spherical primary, and is located (depending on thickness and curvature) closer to the mirror so that the tube length is significantly shortened. A diaphragm (entrance pupil) in the plane through the common center of curvature results in a strictly concentric system without any distinguished axis and with a symmetrically constructed image field, which lies on a spherical surface inside the system. Despite its great similarity to the Schmidt camera, the image quality of the Maksutov camera is slightly inferior, and subsequently the former has prevailed in the design of large professional telescopes. The latter will certainly suffice, however, for high demands by amateurs, and also is simpler to make and less expensive in commercial production owing to the use of only three spherical surfaces. Many modifications of this system are known, some of which have been described by Wiedemann [4.39] as being especially attractive to amateurs.

As with the Schmidt camera, the Maksutov camera can be modified to a Cassegrain system with an improved state of corrections, resulting in the known advantages of a shorter tube length and a focal plane lying outside the system. The simplest way to introduce the convex secondary mirror into the system is by coating the rear surface of the meniscus with a reflecting layer of corresponding diameter. System data and construction hints for Maksutov–Cassegrain telescopes ($f/15$) from three spherical surfaces are given by Gregory [4.40]. The advantages include partial avoidance of light diffraction at the spider (which improves image contrast), and also relative stability against temperature fluctuations (since the tube is closed on both ends). The aberration curves by Wiedemann [4.21] show that the correction state, even at $f/15$, is not exceptionally good, so a separation of the secondary mirror from the meniscus back is advisable. The freedom thus obtained in the choice of curvature and of positioning of the secondary mirror can again be used for image correction, and Maksutov–Cassegrain systems with very good imaging properties are possible for f-ratios up to $f/7.5$. Comparisons of very different Cassegrain telescope types including spot diagrams have been given by Willey [4.41].

4.5.9 Instruments for Solar Observations

When observing the Sun, the problem is not the low light intensity which characterizes stellar astronomy. On the contrary, the radiation collected by the instrument is *too intense* for the receiver (such as the eye or a photographic plate), and causes undesirable heating of instrument parts. This is a major reason why amateurs usually prefer refractors over reflectors, since the closed refractor tube does not exchange heated air inside with the colder air outside. Mirror systems (except Schmidt–Cassegrain and Maksutov–Cassegrain) do not inhibit the heat exchange and the resulting turbulence inside the optical path can be quite detrimental to the imaging quality. Moreover, the secondary mirror (close to the primary focus) is heated, and its deformation results in a further deterioration of the image. Finally, the increased scattering of light by diffraction at the secondary mirror and its support frame diminishes the contrast. The aforementioned drawbacks are, of course, absent in a refractor. Notwithstanding the higher price and perhaps remnant chromatic errors, an observer specializing in the Sun would be well-advised to purchase a refractor.

The reduction of the intense solar radiation can be accomplished in various ways. The most obvious way to diminish the light collection power is by reducing the telescope aperture, but this, however, can cause an unacceptable loss of resolution. If the atmospheric limited resolving power, which is at best $1''\!.0$, is to be utilized, the telescope aperture cannot be less than 10 cm in diameter according to the diffraction criterion. Smaller instruments with 6- to 8-cm apertures are quite adequate for viewing sunspots and other large-scale phenomena, but for observing finer details such as granulation (see Chap. 13), which is of the order of $1''\!.0$, objective diameters of up to 20 cm are advantageous.

The most preferable (and costly!) light reduction method is by means of *objective filters*, which are plane-parallel glass plates of high surface quality (in order not to spoil the image quality) coated with a partial transparent reflecting layer. Special reflecting foils are available but they fall far short of the optical quality of glass filters; the use of the so-called "rescue foil" in reducing the radiation should be on a provisional basis only. The transmissivity of the available objective filters ranges from 10^{-3} to 10^{-5}, which corresponds to a reduction by 7.5 to 12.5 magnitudes. The reduction should not be chosen too strong, as there is the possibility of secondary filtering. Objective filters block excess radiation that otherwise could heat the instrument.

Another, less desirable technique is to use an *eyepiece filter*, which diminishes the sunlight *inside* the instrument after it has been collected by the objective. Apart from detrimental effects to the image quality due to the dissipation of heat, there is the ever-present danger that an eyepiece filter, because it is subjected to intense heating near the focal point, will suddenly crack and allow the concentrated solar radiation to reach the observer's retina unattenuated. Such exposure even for an instant can result in permanent eye damage and even blindness. A particular caution should be stated here against the use—instead of the special eyepiece filters—of soot-blackened glasses, overexposed films, or similar items, since these scarcely diminish the invisible heat radiation which can cause eye damage. More suitable than eyepiece filters are solar eyepieces, or *helioscopes*, whose light-reducing effect is based on refraction, reflection, polarization, or a combination of these properties. The *Handbuch für*

Sonnenbeobachter [4.42] compiles the most important variance of this type of eyepiece, and also treats the entire problem of light reduction in more detail than can be achieved here.

An instrument specially designed for observation of the faint solar atmosphere is the *prominence telescope*, or *coronograph*. Until a few decades ago, an astronomer would have to wait months and even years for a naturally occurring eclipse (and possibly also have to travel thousands of miles to the eclipse path) in order to observe the faint outer solar atmosphere. The coronograph, invented in 1930 by B. Lyot, is based on an optical system in which the entire solar disk is artificially eclipsed with a conic diaphragm, thus removing the blinding glare which is normally present. The main problem with this instrument is the generation of scattered light by the atmosphere and by the instrument itself. The atmospheric scatter can be reduced only by observing from a telescope site at high elevation and with optimum atmospheric conditions. As amateurs seldom have this option, they will be prevented from seeing coronal features, but they can still use the coronograph to observe the solar limb and the so-called prominences, which are much more luminous than the corona. These observations are sensitive to scattered light, so that instrumental scatter is to be avoided as much as possible. This requires that the objective O_1 (Fig. 4.23) be free of scratches, cracks, bubbles, and dust, and situated at the end of a long dewcap. In its focal plane is a conic diaphragm C masking the solar disk and reflecting the heat radiation sideways out of the tube. The scattered light arising from diffraction by the objective aperture is reduced by employing an auxiliary lens H to image the objective O_1 onto an iris diaphragm B. All parts mentioned thus far serve to remove direct or scattered sunlight. Thereafter, the solar limb, i.e., the edge of the diaphragm C and the nearby and the surrounding solar areas, are imaged with a second objective O_2 and are inspected under magnification with an eyepiece. Since the hydrogen within the prominences radiates primarily within the Balmer spectral lines, an Hα filter (656.3 nm) with a low spectral passband may be used to further increase the contrast. The eccentricity of the Earth's orbit causes the apparent solar diameter to increase from $31'.51$ in July (at aphelion) to $32'.58$ in January (at perihelion), and consequently the conic diaphragm must be changed at various times during the year in order to be adapted to the diameter of the solar disk in the focal plane of the objective O_1 (corresponding, respectively, to 0.00917 or 0.00948 times the focal length O_1 in mm). Three diaphragms of small (summer), medium (spring and fall), and large (winter) diameter will probably suffice,

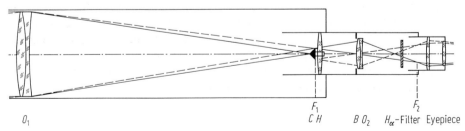

Fig. 4.23. Schematics of a prominence telescope: objective O_1, conic diaphragm C, auxiliary lens H, scattered light stop B, objective of prominence attachment O_2.

although for optimal observing a more extensive set of diaphragm diameters according to the solar diameters given in *The Astronomical Ephemeris* will be desirable. The construction of a prominence telescope and the requirements of the separate parts are detailed by O. Nögel [4.43] and a series of papers by G. Nemec [4.44].

It is possible to equip an existing refractor for observations of the solar limb using a *prominence attachment*. It has the same optical components as the prominence telescope mentioned above and the advantages of relatively short length (25–30 cm), low weight (about 500 g), and improved accessibility for changing the diaphragms; moreover, it can also be added at any time. Construction modes of the prominence attachment are described in the *Handbuch für Sonnenbeobachter* [4.42] and by Hanisch [4.45]. Additional shortening and weight reduction are obtained with a so-called *prominence eyepiece*, which is merely a Kellner-type eyepiece with the conic diaphragm on the field lens. The scattered light is reduced by two diaphragms: the field diaphragm in the focal plane of the objective O_1 (identical with the plane through the conic diaphragm) and a diaphragm at the exit pupil, i.e., just behind the eyepiece. But the removal of scattered light requires, in addition, a narrow-band Hα filter to be inserted between the field and eye lenses. Its passband and transmission critically influence the quality of the prominence eyepiece. Instructions for building this special eyepiece are again to be found in the *Handbuch für Sonnenbeobachter* [4.42] and in the paper by Richter [4.46].

Another special device for solar observations is the *spectrohelioscope*, which allows chromospheric observations to be extended from the limb to the entire solar disk. With this device, the Sun can be examined in the light of individual spectral lines. This instrument is a descendant of the spectroheliograph, in which the solar disk moves across the entrance slit of a spectrograph; an exit slit then selectively filters just one line out of the spectrum, so that as the photographic plate is slowly guided to coincide with the motion of the Sun's image, a *monochromatic* image of the solar disk is obtained. It is also possible to view the monochromatic image direct. In this case, the entrance slit is moved across the solar disk and is imaged onto a spectral grating. A single line is then extracted from the reflected spectrum by an exit slit. As the positions of the two slits must be correlated, both may be attached to a rotating disk. With a sufficiently high rotation frequency of the motor, the motion of the slit across the solar disk (1/24 s) generates a standing image of the solar surface in the light of one spectral line. Which line is filtered out depends solely on the grating angle. Simply changing the angle of incidence at the grating permits a traversal of the solar spectrum and the inspection of the solar disk in the light of each desired spectral line. The credit for making this efficient instrument accessible for amateur use belongs to F.N. Veio [4.47]. The spectrohelioscope offers a variety of modifications and refinements which will not be described here. It will only be mentioned that the use of a heliostat, which feeds the light via an equatorially-mounted plane mirror into a fixed position telescope, is advisable since the equipment behind the focal plane is rather heavy, and is also quite sensitive to bending.

4.6 Telescope Performance

4.6.1 Resolving Power

The resolving power of a good optical system is ultimately limited by the light diffraction at the entrance aperture of the telescope. Figure 4.5 shows the diffraction image of a circular aperture. A section through the center of the diffraction disk reveals an intensity distribution (Fig. 4.24) which can be calculated with the Bessel function. This gives specific values for the positions of the minima ($x = 3.83, 7.02, \ldots$) and the adjacent maxima ($x = 5.24, 8.42, \ldots$). These are connected by the expression

$$\rho = \frac{x}{\pi} \frac{\lambda}{D} \quad \text{(radians)}, \tag{4.41}$$

where ρ is the radius in radians of the corresponding ring, λ is the observed wavelength, and D is the *free aperture* of the telescope. The characteristic quantity usually quoted is the radius of the first minimum, known as the *Rayleigh criterion*,

$$\rho_0 = 1.22 \frac{\lambda}{D} \quad \text{(radians)}$$
$$= 2.52 \times 10^5 \frac{\lambda}{D} \quad \text{(arcsec)}, \tag{4.42}$$

where the conversion factor 206 264.8 arcsec/radian has been employed to obtain the second equation. Twice this angle, $2\rho_0$, is the diameter of the *Airy disk*. Two point sources are thus barely distinguished when the central maximum of the diffraction curve for one of the sources coincides with the first minimum of the diffraction curve for the second source. Since the light of a double star is *incoherent*, the two intensities add as shown in Fig. 4.24. The steep decline causes a minimum of 0.735 of total intensity to remain distinctly below the two maxima. This dip remains visible even though the diffraction images overlap; thus a double star can be resolved into its components when their separation is $\geq \rho_0$ and thus satisfies the *Rayleigh criterion*. Inserting a representative wavelength of 500 nm for visual observations, the Rayleigh criterion mandates a minimum separation of

$$\rho_0 = \frac{12.5}{D(\text{cm})} \quad \text{(arcsec)}, \tag{4.43}$$

for a telescope of aperture D. Less stringent criteria such as the *Dawes criterion*, which requires an intensity depression between the two maxima of only 3%, may be applied; by this standard, the two images can overlap a bit more before they become indistinguishable, and the constant in the above equation relating to the minimum separation is reduced from $12\rlap{.}''5$ to $10\rlap{.}''5$. The resolution limit is reached when the overlapping of diffraction patterns no longer gives a dip and the two diffraction maxima merge into one. In this case, $12\rlap{.}''5$ is replaced by the limiting value of $9\rlap{.}''77$ in Eq. (4.43).

The resolving power thus obtained holds only for ideal conditions, that is to say, for two point sources of equal brightness, no atmospheric disturbance, and no imaging errors of the optical system. Figure 4.24 indicates that the shape of the intensity distribution changes drastically for substantial brightness differences, and diminishes the separability of the components. The minimum separation for resolving unequally bright stars increases with the difference in their magnitudes. The imaging errors from

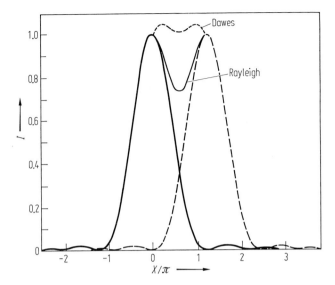

Fig. 4.24. Intensity distribution in the diffraction disk of a star from a circular aperture. Shown superposed is a second star of the same brightness and the separation between the components (Rayleigh criterion); *dashed line* = intensity sum for the Dawes criterion.

Sect. 4.3 also tend to lessen the resolving power. Furthermore, the diffraction shape of Fig. 4.24 is modified by a central obstruction such as a secondary mirror or the plateholder of a Schmidt camera. The radius of the first minimum diminishes from $1.22\lambda/D$ to the limiting value $0.73\lambda/D$, but more light is diffracted and significantly brightens the diffraction rings (Born and Wolf [4.8]). This brightening diminishes the contrast, particularly in planetary observations, and thus the refractor is the telescope of preference in this realm. Another way to influence the diffraction shape is by reducing the transmission of the outer objective zones by coating them with an absorbing layer (neutral density filter) of increasing thickness toward the periphery. This process of *apodization* causes an increase of the Airy disk ($2\rho_0$), but at the same time partly suppresses the diffraction rings (adjacent maxima). Jacquinot and Roizen-Dossier [4.49] treat this subject in great detail and display diffraction shapes for different transmission profiles.

An almost inevitable limitation of the resolving power arises in the disturbance of the light rays from celestial objects as they pass through the Earth's atmosphere. The chaotic motion of turbulent elements there causes a ray passing through them to continually change its direction (directional scintillation), resulting in the ray meandering toward the observer. The time scale of these changes in the optical range is some milliseconds, which is too rapid for the eye to discern, so that for visual observing and even for short (tenths of a second) photographic exposures, the diffraction disk will be smeared out over a diameter of several arcseconds. The amount of broadening depends on the state of the atmosphere and is called *seeing*. At exceptionally good sites, values somewhat below $0\rlap{.}''5$ have been observed, while in more ordinary locations, the seeing disk may shrink to about $1''$ on excellent nights. Taking $1''$ as the order of magnitude resolution permitted by the Earth's atmosphere, then

Table 4.5. List of 30 visual double stars which are circumpolar. The coordinates, apparent visual magnitudes of the components, and their separations in arcseconds are given.

No.	α_{2000}	δ_{2000}	m	ρ	No.	α_{2000}	δ_{2000}	m	ρ
1	12^h09^m	$+82°0$	6.5 – 8.5	67″	16	22^h48^m	$+82°9$	5.0 – 8.7	3″5
2	0213	79.7	6.5 – 7.1	54	17	1643	77.5	6.1 – 9.4	2.9
3	1500	80.6	6.9 – 7.1	31	18	1433	86.2	7.0 – 10	2.4
4	2115	78.5	7.2 – 10	26	19	2244	78.4	7.5 – 9.3	2.3
5	0312	81.3	6.0 – 10	24	20	2114	80.2	7.8 – 8.5	2.0
6	1249	83.7	5.3 – 5.8	21.6	21	0929	88.8	7.1 – 10	1.7
7	1800	80.0	5.8 – 6.2	19.2	22	1549	83.8	7.6 – 8.1	1.4
8	0221	89.0	2.1 – 9.1	18.4	23	0210	81.3	6.9 – 9.1	1.2
9	1212	80.3	7.3 – 7.8	14.4	24	1453	78.3	6.5 – 10	1.2
10	2210	82.6	7.0 – 7.5	13.7	25	1312	80.6	6.3 – 10	1.0
11	1929	78.3	7.6 – 8.3	11.3	26	0119	80.6	8.0 – 8.0	0.9
12	1255	82.5	6.9 – 11	10.5	27	0409	80.7	5.7 – 6.4	0.9
13	2008	77.7	4.4 – 8.4	7.4	28	1539	80.1	7.4 – 8.2	0.7
14	0306	79.3	5.8 – 9.0	4.6	29	0010	79.5	6.8 – 7.1	0.5
15	2015	80.5	6.8 – 11	4.0	30	0150	80.7	7.8 – 8.1	0.3

for telescopes with apertures larger than 10 cm the resolution remains about constant while for telescopes under 10 cm the resolution improves with increasing aperture according to Eq. (4.43). Nevertheless, larger telescopes are still desirable since in addition to their higher light-gathering power, their theoretical resolving power may be realized by the application of special techniques such as *speckle interferometry*, in which the atmospherical image motions are "frozen" in a rapid sequence of very short exposures. How these procedures operate—almost always at the very minimum of detectable light level—has been described, for instance, by McAlister [4.48].

In order to determine the resolving limit set by the atmosphere (seeing) or the instrument (aperture size), the diameter of the diffraction disk must be measured; this is most expediently accomplished from observations of double stars with known separations. The two star components should preferably be of equal brightness; otherwise the resolving power will be underestimated. Table 4.5 lists 30 visual double stars with apparent magnitudes m_1, m_2 in order of decreasing separation. As circumpolar stars, they are always above the horizon for most northern hemisphere observers. (A map is given by Heintz [4.50]; cf. Chap. 26.) Another source for coordinates and system parameters of double stars is the somewhat outdated *Becvar Catalogue* [4.51]; Appendix Table B.32 of this *Compendium* also contains a list of double stars.

The diffraction disk diameters of stars obtained from double star observations may be represented on a logarithmic *seeing scale* (Table 4.6) as suggested by Tombaugh and Smith [4.52]. Apart from the broadening of the stellar images by directional scintillation, there is also *intensity* scintillation, which leads to brightness fluctuations of the star images, depending on the size of the instrument used. It is usually expressed in a five-step scale (1=best,..., 5=worst)[10], but it cannot be measured by eye so

---

10 This numbering scheme is reversed in the U.S.

Table 4.6. The astronomical seeing scale as suggested by Tombaugh and Smith [4.52].

Image Quality	Image Diameter	Image Quality	Image Diameter
−4	50″	+3	2″.0
−3	32	+4	1.3
−2	20	+5	0.79
−1	12.6	+6	0.50
0	7.9	+7	0.32
+1	5.0	+8	0.20
+2	3.2	+9	0.13

objectively as the separation of double stars and is thus strongly governed by subjective impressions.

4.6.2 Magnification and Field of View

During visual observations with a telescopic system, an infinitely distant object is imaged by the objective to the focal plane of the telescope. A real image is generated and can be picked up by, for instance, a screen or photographic plate. The imaging scale is determined solely by the telescopic focal length f, which converts the object angle u into the linear image size y', given by

$$y' = f_{tel} \tan u. \tag{4.44}$$

In visual observations, this real image in the focal plane is inspected through an eyepiece which acts as a magnifying glass (Fig. 4.3). Viewed through the eyepiece, the object appears to subtend the increased angle u'. The magnification γ is defined as the ratio of the two viewing angles,

$$\gamma = \frac{\tan u'}{\tan u}. \tag{4.45}$$

The eye focuses automatically to infinity when it is totally relaxed; hence the rays from this intermediate image must leave the eyepiece in a collimated parallel beam to permit a relaxed viewing without any fatigue. To achieve this, the image is placed in the focal plane of the eyepiece. The increased viewing angle u' follows from the image size y' and the eyepiece focal length f_{eyp} as $\tan u' = y'/f_{eyp}$. The magnification is then given by the simple relation

$$\gamma = \frac{f_{tel}}{f_{eyp}}, \tag{4.46a}$$

or, according to Eq. (4.11),

$$\gamma = \frac{D}{A_P}, \tag{4.46b}$$

where D is the diameter of the entrance pupil (free aperture) and A_P is the diameter of the exit pupil. Equation (4.46b) shows that the diameter of the parallel beam emerging from the eyepiece is determined by the choice of the magnification. The telescopic system thus transforms the large entrance beam down to a much smaller "pencil" with the diameter of the eye pupil, and the conditions for the highest and the lowest useful magnification are thereby fixed.

Since all light collected by the telescope objective should reach the eye, the diameter of the exit beam or the exit pupil A_P should not exceed the diameter of the eye pupil. The limiting width results from the maximum pupil size of the eye, about 8 mm (see Sect. 4.8.1). The lowest feasible magnification, or *normal magnification*, and, correspondingly, the longest feasible eyepiece focal length then follow from Eqs. (4.46) as

$$\gamma_{\min} = \frac{D \text{ (mm)}}{8 \text{ mm}},$$
$$f_{\text{eyp}}^{\max} = 8\,N \text{ (mm)}. \tag{4.47}$$

where $N = f_{\text{tel}}/D$ is called the *aperture number* or *focal ratio*. On the other hand, the resolving power of the eye (about $2'$) determines the highest feasible magnification. Owing to adaptation of the receiving elements on the retina to the diffraction disk, this corresponds, according to Eq. (4.43), to a pupil diameter of 1 mm. The highest feasible, or *useful*, magnification, and, correspondingly, the shortest useful eyepiece focal length, thus results from

$$\gamma_{\max} = D \text{ (mm)},$$
$$f_{\text{eyp}}^{\min} = N \text{ (mm)}. \tag{4.48}$$

Pushing the magnification still higher by shorter-focus eyepieces leads to what is termed *overmagnification* (also called *empty magnification*). More specifically, magnifications far above $\gamma = 120$ should—even for large telescope apertures—be used with caution since the optimum seeing disk of only $1''$ is then spread over an apparent angular size of more than $120'' (= 2')$ and thus becomes discernible to the eye.

The *field of view* is limited by a field stop in the focal plane which is common to the telescope and the eyepiece or by the format of the photographic plate. From Eq. (4.44), the image field is

$$2u = 2 \arctan(b/f), \tag{4.49}$$

with b as either the radius of the stop or as one-half the plate diagonal, depending on the particular setup. When observing with an eyepiece, the quantity $2u$ is also called the *true diameter of the field of view*. It is determined directly at the telescope by viewing an object of known angular diameter (e.g., the Moon). The *apparent field diameter* $2u'$, as it appears to the eye of the observer at the eyepiece, is computed from the magnification γ and the true field $2u$, in good approximation by $2u' = 2u\gamma$. The apparent field of view of most eyepieces averages about $40°$, although for some wide-angle eyepieces it is about $60°$. As a rule of thumb, the following relation may be used to obtain the true diameter of the field of view:

$$\text{True field diameter} \simeq \frac{40°}{\text{magnification } \gamma}. \tag{4.50}$$

As expected, the size of the celestial field visible in the eyepiece diminishes with increasing magnification.

4.6.3 Image Brightness and Limiting Magnitude

The term *brightness* has so many different notations that its meaning in a particular context is often ambiguous. An object emitting 1 watt (=1 joule/sec) of radiation from a surface 1 cm² into an angle of 1 steradian is said to have a *luminance* of 1 stilb (sb), where 1 sb = 1 W cm^{-2} sr^{-1}. The degree of illumination on the receptor surface is described by the *illuminance*, which depends upon the luminance of the radiating surface as well as on the area and the distance between the two surfaces, and is often measured in units of *lux*, where 1 lux = 1 lumen m^{-2} = 1 W m^{-2} sr^{-1}. The illuminance is a physical measure of the radiating energy of an image as a stimulus of the retina, and is thus the relevant quantity of the perception of brightness by the observer's eye. It depends on a variety of factors in imaging systems: the apparent brightness (magnitude) and the angular extent of the object in the sky, telescope aperture D, focal length f, f-number $N = f/D$, light losses owing to absorption and reflection in all optical components, imaging quality, and scattering of light from the sky background and the instrument itself. The interested reader should refer to Chap. 8 for a more detailed treatment of some of these items.

Light losses incurred within the optical system of the telescope have several causes. First of all, some light is lost upon transmission through a glass lens, the exact amount depending on the composition, quality, and thickness of the lens. *Transmission coefficients* are usually given for glass and other materials for a thickness of 1 cm. The transmission coefficient T for any given thickness s can be calculated by exponentiation: $T = T_{1 \text{ cm}}^{s}$, where s is in cm. The untransmitted part of the light remains via absorption within the material, and thus the light loss by absorption in the lens is computed to be $A = 1 - T$. Table 4.7 compiles the transmission coefficients for some important glasses. Light losses by reflection at the glass/air boundaries are at least comparable, and occur for each dioptric element twice, once entering and once leaving the lens. As a result, considerable losses can occur for multilens systems. The mean reflection coefficient is listed in Table 4.7. Assuming perpendicular incidence of light, it can be calculated from

$$R = \left(\frac{n_1 - n_2}{n_1 + n_2}\right)^2 = \left(\frac{1-n}{1+n}\right)^2, \qquad (4.51)$$

where $n_1 = 1$ and $n_2 = n$ are the refractive indices for air and glass, respectively. Therefore, the losses by reflection amount (assuming a 4.2% loss per boundary) to as much as 30% for a four-lens objective. They can be reduced only by coating the surfaces with an antireflection layer to below 1% per surface, which, in the preceding example corresponds to a reduction of the total loss from 30% to less than 8%. Another possibility is to cement the lenses together without air spaces; this, however, results in

Table 4.7. Transmission (per 1 cm thickness) and reflective loss (per uncoated air/glass transition) of some optical glasses.

Glass	Refractive Index n_e (546.1 nm)	Abbe Number $\nu_e = \frac{n_e - 1}{n_F - n_C}$	Relative Partial Dispersion $\vartheta_g = \frac{n_g - n_F}{n_F - n_C}$	Reflective Loss $R\,(\%) = 100 \times \left(\frac{n-1}{n+1}\right)^2$	Transmission/cm λ (nm)						
					360	380	400	440	500	580	660
Crown (K3)	1.5203	58.71	0.5443	4.3	0.971	0.986	0.992	0.995	0.997	0.997	0.997
Flint (F2)	1.6241	36.11	0.5826	5.7	0.910	0.976	0.988	0.993	0.995	0.996	0.997
Heavy flint (SF6)	1.8127	25.24	0.6097	8.3	0.067	0.515	0.793	0.971	0.997	0.998	0.998
Light flint (KzF2)	1.5319	51.45	0.5505	4.4	0.887	0.967	0.980	0.992	0.996	0.998	0.998
Borocrown (BK7)	1.5187	63.96	0.5350	4.2	0.967	0.990	0.994	0.996	0.996	0.998	0.998
UV-crown (UBK7)	1.5187	64.08	0.5351	4.2	0.985	0.991	0.994	0.996	0.997	0.998	0.998
Quartz glass	1.4601	67.86	0.5162	3.5							
Fluorspar	1.4355	96.77	0.6222	3.2							
Calcite	1.6634	49.10	0.6074	6.2							

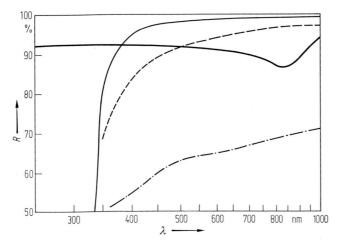

Fig. 4.25. Reflectivity of the most common mirror coatings. *Heavy line*: aluminum, freshly vapored-on; *thin line*: silver, freshly vapored-on; *dashed line*: silver, chemically deposited; *dot-dashed line*: polished mirror bronze. From *Handbuch der Physik* [4.6].

the loss of freedom in the choice of the radii of curvature and the distances between lenses in order to correct imaging errors. An expensive coating is therefore virtually mandatory for high-performance lens optics. The total transmission coefficient T of an n-lens system is then composed of the partial transmissions T_i^s and reflections R_i of the lenses (with thicknesses s_i cm) according to the relation

$$T = T_1^{s_1} \times T_2^{s_2} \cdots (1 - R_1)^2 \times (1 - R_2)^2 \cdots = \prod_{i=1}^{n} T_i^{s_i}(1 - R_i)^2. \qquad (4.52)$$

Mirror surfaces can also cause losses by absorption owing only to imperfect reflectivity. The *reflection coefficient* of a mirror surface depends strongly on the material with which it is coated and on wavelength (Fig. 4.25). Coating materials other than silver and aluminum are available but have not gained great acceptance for astronomical mirrors. A surface upon which silver has been freshly deposited has the highest reflectivity of the common materials, but also two glaring drawbacks. First of all, when exposed to air silver slowly tarnishes and thus loses an appreciable degree of reflectivity; this can be prevented only by the use of special protecting layers (e.g., MgF_2 or Al_2O_3) which in turn give rise to other disadvantages. In addition, there is a strong decline in reflectivity for wavelengths shorter than 400 nm. This can be tolerated when the spectral range between 400 and 330 nm, where Earth's atmosphere begins to lose transparency, is foregone for the benefit of higher reflectivity. Usually, however, the long-proven aluminum coating, whose reflectivity over the entire optical spectral range reaches about 90%, is employed instead of silver. The total reflectivity of an optical system of n mirrors is calculated analogously to Eq. (4.52), namely

$$R = R_1 \times R_2 \times R_3 \cdots = \prod_{i=1}^{n} R_i. \qquad (4.53)$$

The following comments refer only to visual observations, photographic quantities being deferred to Sect. 4.9.1. The brightness of a telescopic image given by the illuminance at the position of the image first depends on the size of the light collecting surface, i.e., the free aperture D. The gain of light results from the ratio of objective area to eye pupil area to be D^2/P^2. This quantity is to be multiplied with the total transmission or reflection coefficient of the optical system. Neglecting losses, a 6-cm telescope already collects over 50 times more light than the unaided eye with the pupil at maximum dilation (8 mm), corresponding to a gain of about 4.5 magnitudes ($= 2.5 \log(I_1/I_2)$). Thus the limiting magnitude of the eye is raised from 6 to 10.5. This gain of light corresponds to an increase in the telescope aperture from 6 cm to about 0.5 m, the latter being an instrument of quite reasonable size. The brightness of the telescopic image depends on the diameter A_P of the exit pupil. When A_P is smaller than or equal to the eye pupil P, the entire light beam flows into and stimulates the retina. If the exit pupil A_P is larger than the eye pupil P (i.e., below the minimal magnification), then the eye pupil acts as a stop and the increased flux of light is to be multiplied by the factor P^2/A_P^2 which is in this case less than one.

The situation is rather different for extended objects such as nebulae or comets. The light collected by the instrument is spread over a larger area by the magnification of the eyepiece. It is easily shown that the light gain in the case of extended objects is proportional to the ratios of the areas of exit pupil and eye pupil (A_P^2/P^2). If a high magnification is chosen, so that the exit pupil is smaller than the eye pupil, then the object will appear less bright in the eyepiece than with the naked eye. If both pupils are of equal size (normal magnification), then the image will appear as bright through the instrument as with the naked eye (if instrumental losses via absorption and reflection are neglected). This is also the case if the exit pupil is larger than the eye pupil, because only the area of the eye pupil is utilized while magnification is sacrificed. As a rule, the area of the exit pupil significantly determines the *relative* brightness of the extended object; this also holds for terrestrial observations during the daytime.

The *limiting magnitude* of stars just barely perceived by the eye depends on the telescope aperture D and on the brightness of the sky background at night. In the following discussion, the light losses in the optical system will be neglected. The following formula is mostly used to estimate the limiting magnitude $m_{\lim}$ for a telescope with aperture D:

$$m_{\lim} = m_0 + 2.5 \log D^2 = m_0 + 5 \log D, \tag{4.54}$$

where D is in cm, and the value for m_0 is usually given in the range 6.5–6.8 mag corresponding to limiting magnitudes of 6.0–6.3 mag, respectively, for the naked eye depending on which maximum pupil diameter is assumed. The visibility of a point source, however, depends critically upon the background brightness, or more specifically, on the square root of the brightness, as Langmuir and Westdorp showed as early as 1931. This is merely a way of saying that the limiting magnitude depends on the telescopic magnification. As stated, the brightness per unit area of the sky background diminishes with the square of the exit pupil diameter A_P, which in turn depends on the magnification γ according to Eq. (4.46). This gives the following new

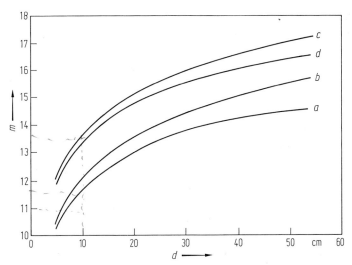

Fig. 4.26. Limiting stellar magnitudes in visual observation. (a) in dark field (H. Siedentopf); (b) with eye pupil of 7.6 mm (J.B. Sidgwick); (c) after W.H. Steevenson; (d) including light losses in the optics (after J.B. Sidgwick).

relations for the limiting magnitude $m_{\lim}$ which replaces Eq. (4.54):

$$\begin{aligned} m_{\lim} &= m_0 + 5\ \log D - 2.5\ \log A_P, \\ &= m_0 + 2.5\ \log D + 2.5\ \log \gamma, \end{aligned} \qquad (4.55)$$

where the value of m_0 is 5.5 mag, and D is in cm. The validity of Eqs. (4.55) was studied and confirmed by Bowen [4.53] at telescopes with apertures ranging from 0.8 to 150 cm. He also showed that the optimum power equals the highest applicable magnification $\gamma_{\max}$ (Eq. (4.48)), and that overmagnification does not improve the limiting magnitude. It follows that smaller telescopes can reach a limiting magnitude 1 mag fainter than expected from Eq. (4.54) by suitable choice of magnifying power, while for the normal magnification (i.e. $A_P = 8$ mm) the magnitude limit is 1 mag less than that given by Eq. (4.54). Figure 4.26 graphs limiting magnitude as a function of telescope aperture D according to data from various authors.

4.7 Accessories

4.7.1 Eyepieces

The real image in the focal plane of the collecting objective lens or mirror can be viewed direct with the eye, but this is quite disadvantageous since the eye cannot focus the image to be viewed closer than some fixed distance. In order to distinguish fine details, an auxiliary optical element, the *eyepiece*, serves as a magnifying glass for the eye to "get within" the convenient viewing distance. In the simplest case of

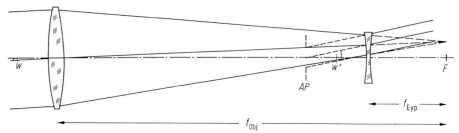

Fig. 4.27. Rays in the Galilean telescope with a diverging (concave) lens as eyepiece. The telescope produces upright images. The eyepiece lens is in front of the common focus F and leads to a virtual, inaccessible exit pupil A_P.

the Keplerian telescope, the eyepiece is merely a single collecting (convex) lens, or in the Galilean telescope a diverging (concave) lens (Fig. 4.27). Single-lens eyepieces (Fig. 4.28 a), however, carry severe chromatic errors, and image curvature and distortion are also noticeable. In order to remove these errors, Huygens introduced the *two*-lens eyepiece. A two-lens system with the individual focal lengths f_1 and f_2 and lens separation e achieves a total (effective) focal length of

$$f = \frac{f_1 f_2}{f_1 + f_2 - e}. \tag{4.56}$$

The insufficient correction of chromatic errors leads mainly to a lateral chromatic aberration so that the inclination of the chief ray becomes wavelength dependent, i.e., the magnification of the eyepiece is somewhat different for each color. This form of aberration is named *chromatic magnification difference*. The longitudinal chromatic aberration is less apparent for usual f-numbers. The (lateral) chromatic magnification difference is eliminated by adjusting the eyepiece focal lengths at different wavelengths to a common mean value via the so-called achromacy condition. This condition determines the lens separation of a two-lens eyepiece of the same glass types to

$$e = \frac{f_1 + f_2}{2}. \tag{4.57}$$

It is satisfied by the *Huygens eyepiece* (Fig. 4.28 b), which is free of distortion, chromatic magnification differences, and reflections. For this reason and owing to its simple composition of two plano-convex lenses (usually in the ratio $f_1/f_2 = 2$), this eyepiece is still in use. The image field is strongly curved, but about 50° of the apparent field of view can be used. A modification of this is the *Mittenzwey eyepiece*, which replaces the plano-convex field lens by a meniscus lens. Neither type is suited for use with a micrometer since the focal plane of the objective lies in the inaccessible region between the eyepiece lenses, and reticles (crosswires) placed there show colored edges which can lead to erroneous readings. For accurate micrometer readings, the focal plane needs to be located at an accessible point outside the eyepiece.

A more suitable type of eyepiece for this purpose is the *Ramsden* eyepiece (Fig. 4.28 c) for which the two lenses have equal focal lengths, and with $f_1 = f_2 = f = e$ also satisfying condition (4.57). The focal plane lies in front of the plano-convex field lens, which images the entrance pupil onto the eye lens in the eyepiece. This creates two serious drawbacks. First, all scratches and dust grains on the field lens become visible, and second, the eye cannot be brought to the position of the exit pupil, hence giving rise to a condition known as "keyhole observing." For these reasons, the two lenses are brought closer together, retaining the correction for distortion, but a chromatic magnifying difference occurs as the strict distance condition of Eq. (4.57) is not fulfilled. The three-lens *Kellner eyepiece* (Fig. 4.28 d) achieves the correction of the chromatic difference even without obeying the Huygens condition by incorporating a lens composed of two different glass types on the viewing side of the eyepiece. This eyepiece is nearly free of distortion but retains image curvature; therefore this type is usually arranged to give an apparent field of view of 40°. Another three-lens eyepiece (made from two different glasses) is the *monocentric* eyepiece (Fig. 4.28 e). As the name implies, all six surfaces have the same center of curvature, and the positive radii preceding the center correspond to the following negative ones. Satisfactory correction, however, is limited to field diameters of less than 30°.

Further developments include the *orthoscopic* eyepieces of Abbe (Fig. 4.28 f) and Plössl (Fig. 4.28 g). They owe their name to the impressive achromatic and distortion-free images they provide. The orthoscopic Abbe eyepiece was for many years considered by astronomers to be the one with the least distortion. It should also be mentioned that the exit pupil is rather distant from the eye-side lens and thus is conveniently accessible; its apparent field of view is between 40° and 50°. There exist today more advanced eyepiece types, where the continued diminution of the air space between the eyepiece components is appreciable and results in a decrease in the Petzval sum corresponding to larger radii of the field curvature. One such eyepiece which has both excellent corrections and a large plane field of 50° is the *astro-planocular* of Zeiss. The aberration curves (Fig. 4.28 h) show the excellent correction of this eyepiece. The so-called "wide-angle" eyepieces are usually of the *Erfle* type (Fig. 4.28 i) with a single middle lens between two cemented doublets. Fields with apparent diameters of up to 70° are viewed with these eyepieces at the expense of diminished corrections. Readers interested in the details of the construction of these and other kinds of eyepieces may refer to König and Köhler [4.18]. These authors present other aberration curves, which are generally computed against the direction of light, i.e., the rays are calculated from the exit pupil backward through the eyepiece to the focal plane.

4.7.2 The Barlow Lens

The Barlow lens is a negative (divergent) lens system inserted for the purpose of increasing the telescopic focal length. It consists of an achromatic doublet placed in front of the focus (Fig. 4.29). The off-axis chromatic errors can be kept to a bare minimum if the Barlow lens is adapted to the objective. This is the more easily achieved the smaller the focal ratio D/f and the smaller the lengthening factor B. The latter corresponds to the ratio $f'/f_{BL} = s_2/s_1$, so that from Eq. (4.56) the new effective length f' of the telescope follows from

$$f' = \frac{f_1 \, f_{BL}}{f_{BL} - s_1}. \tag{4.58}$$

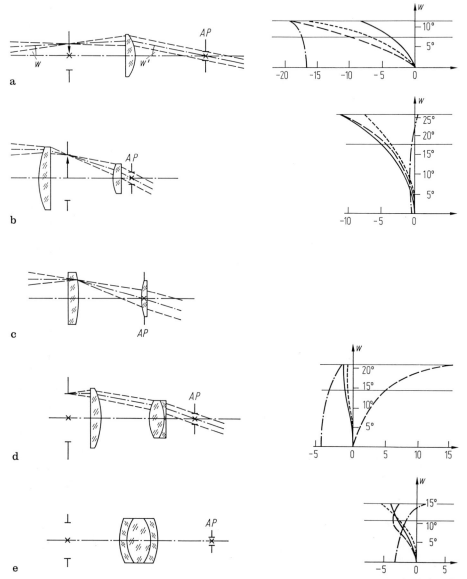

Fig. 4.28a–i. The most common eyepiece types and their aberration curves. The field stop in the focal plane and the position *AP* of the exit pupil are shown. The aberration curves graph the sagittal (*solid line*) and meridional (*dashed line*) deviations, referred to an eyepiece focal length $f = 100$ mm, versus the angle w' of the primary ray. Also given are distortion (*dotted line*) and chromatic magnification difference (*dot-dashed line*), both in %. (After König and Köhler [4.18]). **a** plano-convex lens; **b** Huygens eyepiece; **c** Ramsden eyepiece; **d** Kellner eyepiece; **e** monocentric eyepiece of Steinheil

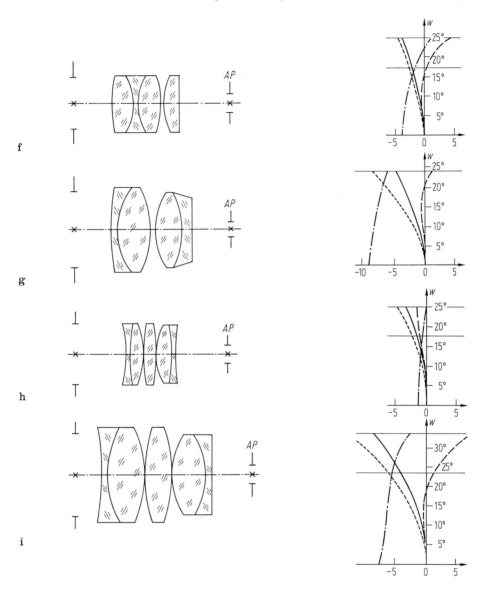

Fig. 4.28 (Continued). **f** orthoscopic eyepiece of Abbe; **g** Plössl eyepiece; **h** astro-planocular of Zeiss; **i** wide-angle eyepiece of Erfle type

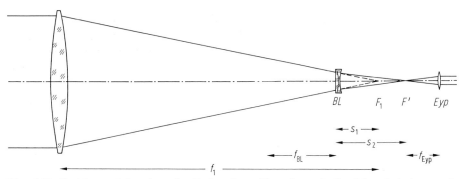

Fig. 4.29. Position and function of a Barlow lens. The telescopic focal length is increased, resulting in an increased image scale for visual and photographic observation.

The distances s_1, s_2 are obtained from the following relations:

$$s_1 = \frac{f_{BL}(B-1)}{B}, \qquad s_2 = f_{BL}(B-1). \tag{4.59}$$

Thus, if the focal length is doubled, the system focus lies at half the Barlow's focal length f_{BL} behind the objective focus. The increase in telescopic focal length is therefore achieved without a significant increase in the tube length. The Barlow lens should *not* be used as an element to shorten the tube length (with a fixed telescopic focal length) as the image quality is inferior to that of a telescope with equal focal length but without a Barlow lens. Also, the position of the Barlow lens should not be shifted in order to reach another effective focal length, because then the correction will be lost. For optical data on corrected Barlow lenses, Hartshorn [4.54] should be consulted. If an increased focal length is desired without regard to image quality (e.g., for guiding telescopes) then a single divergent lens can be used instead of a Barlow lens.

4.7.3 Tube and Dewcap

The tube serves as a structure which fixes the optical parts and also protects against rapid temperature fluctuations, humidity, and scattered light. The tube dimensions are given by the focal length and the free aperture of the objective and are found by simply constructing a true-to-scale graph of the entire optical system. The increase in diameter of the tube and of the secondary and other mirrors for the "true" field angle (corresponding to the desired field of view) according to Eq. (4.49) have to be taken into account.

A general statement regarding preference for a closed or an open tube cannot be made. For telescopes that are closed at *both* ends (e.g., refractors, Schmidt or Maksutov systems, astrocameras) the closed tube is preferred, in contrast to open-ended systems (reflectors), for which heat accumulation inside a tube has to be avoided since here a direct exchange with ambient air can take place. Either the tube diameter is chosen a bit larger than the mirror so that surrounding air can pass along the rim of the mirror and pass through the tube outside the beam, or else the tube is constructed entirely as an open framework, in which case the heat-exchange with the surrounding air proceeds much more rapidly. Upwards of a certain telescope size, an open tube is advisable anyway, as the large cross-section increases the force of the wind on the telescope mount and drive, and the prevention of vibrations caused by wind gusts requires even more attention.

Irrespective of which construction is preferred, the tube must guarantee stability for the adjusted state of the optical system. For small- and middle-sized telescopes this rigidity against flexure can be achieved. The compensation for flexure of large telescopes by so-called *Serrurier struts* is beyond the scope of this chapter. Since the mass of a telescope increases approximately with the third power of the mirror diameter, the weight-caused loads are quite small for small-size telescopes, so that compensation for flexure can be omitted in such cases. The choice of the tube material depends on personal preference and the available facilities. Suitable materials range from metals to wood and plastic, for which Schiffhauer [4.55] gives useful hints for construction.

All optical components should be attached to the telescope tube in such a way that they may be adjusted. No stress or shift of the components should occur when either the pointing of the telescope is changed or the temperature varies (this is particularly important for primary mirrors). Lens objectives are framed at the edge while small mirrors can be supported on their back by an elastic base (e.g., a felt plate). Adjustment of the mirror is achieved by supporting its back on three fixed points normally located on a circle of radius $r = D/\sqrt{2}$. For large mirrors, the three-point support is insufficient since the mirrors deform under their own weight. In such cases, an *astatic* support, which adjusts itself according to the tilt of the mirror during telescope tracking, must supply an equal counter weight to the mirror at numerous points. Possible support systems such as those proposed by Lassell and by Grubb are described, for example, by Maksutov [4.10].

Suitable diaphragms may be added to the optical components as stops in order to prevent scattered light from reaching the detector (eye, photographic plate, etc.), and should be inserted into a closed tube which has been blackened on the inside. Where and how large these diaphragms should be can also be found from a true-to-scale graphing of the optical path, or from the calculation after Wepner [4.56]; the latter promises to be somewhat more accurate. *Irrespective of the procedure, extreme care should be taken that the diaphragms prevent scattered light only and do not cut directly into the optical path.* (Take into account the maximum field angle of the image.) To check this with a refractor, the telescope is pointed with the eyepiece toward the daytime sky; the eyepiece pupil is then viewed through the objective, and it should be visible from any point on the edge of the objective. Telescopes with open-tube construction use other diaphragms, usually conic tubes attached to the central bore of the primary and the support frame of the secondary mirror. These *baffles* are especially important for telescopes of the Cassegrain-type, where light can pass the secondary mirror and fall directly onto the receiver. Owing to the conical shape of the baffle, the central obscuration here is somewhat stronger than that produced solely by the secondary mirror.

The *dewcap* can also help to suppress scattered light. It is chosen to be a rather long cylinder but should not vignette the field of view. This cylinder also prevents the formation of dew on the objective. This function can be enhanced by employing a tiny electric resistance heater as described in Chap. 5.13. If dew forms on the surface of the unheated objective, it should never be wiped dry; it should instead be evaporated by gentle fanning, or dried the next day by exposing it to natural air motion in the open dome. Moisturized optics should not be closed but rather covered by a cloth which is porous to air. In spite of even meticulous care, some dirt will inevitably deposit on the objective so that regular cleaning will be needed from time to time. Specific instructions regarding these procedures have been layed out by, for example, MacRobert [4.57].

4.7.4 Finding and Guiding Telescopes

The *finder scope* is a simple refractor with a relatively small (a few centimeters) aperture. It enables the observer to locate faint objects that are invisible to the naked eye and to point the telescope quickly, especially if it is one which lacks accurate setting circles or the amenity of computer-driven coordinate settings. The limiting magnitude of visible stars is dealt with in Sect. 4.6.3. The true field of view of the finder should not be less than 3° in order to facilitate finding the object after a rough pointing. In order to avoid unnecessary gymnastics at the telescope, it is useful to have the finder located close to the eyepiece. For exact positioning, the finder is usually equipped with a crosswire eyepiece. The finder must be pointed parallel to the optical axis of the telescope, a task best accomplished with a twin holder rings with adjustable collimating screws, which ensures parallel adjustment also after transportation. The various methods of crosswire illumination (in the dark field) and field illumination are described in detail in Chap. 5.15.4. A single half of a binocular can also serve as a finder, but there is no crosswire for centering, and the upright image may lead to some confusion with the upside-down image in the eyepiece as to which direction the pointing must be corrected. One of the simplest arrangements is a pair of sighting rings as used with some rifles. In this scheme, the observer's line of sight is framed by a small ring next to the eye and a crosswire within a larger ring close to the end of the tube at the entrance aperture. The precision increases with the distance between the rings.

Guide telescopes serve for guiding *during* photographic exposures. There are three requirements: (a) a moderately large aperture in order to see even faint guide stars, (b) a long focal length, and (c) high mechanical stiffness. If the aperture is too small, it may be difficult to find a suitable guide star in sparsely populated sky regions at high galactic latitudes. The focal length of the guide scope must be larger than that of the photographic telescope, but this depends in detail on the "pixel size" of the detector (i.e., the resolution in line pairs/mm depending on the grain size of the photographic emulsion) and on the effective focal length of the system. It is essential that the drift which can just be noticed in the crosswire eyepiece of the guiding scope remains below the angular resolution of the detector. This can be accomplished by means of a Barlow lens and the subsequent magnification by an eyepiece. Lacking any kind of crosswire illumination, the observer can resort to makeshift guiding: the guide star is defocused into a disk which is then used to illuminate the crosswire, the guiding being achieved by centering equal areas of the disk around the four sectors in the field of view. This method achieves acceptable accuracy of guiding but it reduces drastically the limiting magnitude of the guide star.

4.7.5 Eyepiece Micrometers

The *eyepiece micrometer* measures the position angle and relative separation of two objects (e.g., double-star components in the apparent field of view). It uses a crosshair eyepiece with an additional third wire, perpendicular to one of the crosshairs and movable across it by means of a micrometer screw. The entire crosshair unit is rotatable, and its position can be read at a *degree circle* outside the eyepiece. Centering one star on the fixed crosshair and the other on the movable crosshair, their position angle can be read directly at the circle, while the micrometer position—each unit corresponds to a certain displacement in degrees on the celestial sphere—after conversion gives the separation between the components.

One preferred way of making the crosshair visible is *field illumination*. To achieve this, a faint light bulb, preferably with a rheostat to regulate its brightness, is placed in front of the objective lens or primary mirror and screened from the eyepiece. The wires therefore appear dark against the brightened field of view. The illumination of the crosshairs in a dark field requires that, because of chromatic errors of eyepieces, the light source should have a similar energy distribution as the object to be measured; the use of a dark red light, for example, may lead to measuring errors if e.g. blue stars are observed.

The measurements described above can also be made without a micrometer using a simple crosshair eyepiece. One of the wires is oriented precisely east–west by switching off the telescope drive and rotating the eyepiece until a star, following its diurnal motion, moves exactly parallel to the wire. The time difference Δt between the passage of both stars through the crossed (north–south) wire is then timed with a stopwatch. This time difference is the angular separation of the stars in right ascension. Subsequently, Δt is converted from mean solar time into *sidereal time* via $\Delta \tau = 1.002738 \, \Delta t$ (see Sect. 4.7.9) and the declination δ of the double star. The separation $\Delta \alpha$ in right ascension is

$$\Delta \alpha = 15 \, \Delta \tau \, \cos \delta \quad \text{(in arcsec)}, \tag{4.60}$$

since an equatorial star moves $15''$ per second of sidereal time in the sky. The position angle is obtained from the angular difference between the just-determined "null position" and the rotated position of the eyepiece, which, with the telescope drive switched on, aligns both stars on the former east–west wire. The required degree circle at the eyepiece frame can be homemade, as can the crosshair of the eyepiece. Useful for such wires are threads made from spider web, quartz, or other materials with sufficiently small cross section (but not human hair) since these appear very much magnified when viewed through the eyepiece. The declination difference and total separation of the double star components can be calculated from the position angle and from $\Delta \alpha$ using trigonometric formulae. The error in estimating the separation increases rapidly for position angles near 0° and 180°. A precision of 1/10 of a second of time ($\approx 1'' - 2''$) can be obtained in right ascension.

4.7.6 The Photometer

Brightness estimates of stars made with the eye are naturally somewhat subjective, but attempts have in fact been made to increase the precision by certain techniques (e.g., the "step-method" of Argelander (see Chap. 8) or the "fraction method" of Pickering). To measure stellar magnitudes objectively, a (usually photoelectric) photometer is utilized. The collected photons are converted into a measurable electrical current which supplies a quantitative linear measure for the illuminance at the telescope focus. To reduce the background brightness of the night sky, suitable diaphragms (with diameters adjusted according to the current atmospheric seeing) are used in the focal plane of the telescope. The diaphragm is imaged by a lens (for either a lens or a mirror system) to the photocathode of a photomultiplier, which executes the conversion of light quanta into an electric current. Extreme caution is advised regarding the power supply of photomultipliers (which possess quite respectable voltages of around 1000 V); their functions are described in Sect. 4.9.2 and Chap. 8. Stellar brightnesses are generally measured not in integrated light but rather in precisely defined wide or narrow spectral bands, from which the colors and other physical data for stars are derived; a movable *filter slide* is therefore an essential component in the photometer.

The most commonly used color system is the *UBV*-system of Johnson. It is defined by the combination of a reflecting telescope (spectral reflectivity of an aluminum coating) with a photomultiplier tube of type RCA 1P21 and employing the following filters:

U: 2 mm Schott/UG2; maximum transmission at 360 nm; bandwidth about 55 nm.
B: 1 mm Schott/BG12 + 2 mm GG13; maximum transmission at 440 nm; bandwidth about 100 nm.
V: 2 mm Schott/GG11; maximum tranmission 550 nm; bandwith about 80 nm.

Each photometric system is associated with a characteristic set of filters, which, depending on their maximum transmission and passband, enable astronomers to determine a multitude of stellar properties. Those astronomical problems which are particularly well suited for photometric investigations will be treated in Chap. 8.

In addition to the rotating stop and filter wheels, the photometer contains additional components such as finding and centering eyepieces. The former locates the object in the field of view of the telescope, while the latter centers it on the stop. Each is brought into the optical path by movable mirrors. Details of observing with a photometer are discussed in Chap. 8, which also reports on the astronomical magnitude scale, the execution of suitable measurement sequences, and the subsequent reduction of data.

4.7.7 The Spectrograph and the Spectroscope

The spectrograph allows a *qualitative* analysis of starlight by investigating characteristic spectral lines. It has three basic components: the collimator, the dispersing element, and the camera. The collimator transforms the convergent rays from the telescope into a parallel beam which falls at a certain angle onto the dispersing element. Having passed this element, the light beam exits at some other wavelength-dependent angle, again in a parallel beam, and is imaged by the spectrograph camera onto the receiver (i.e., photographic plate or photoelectric detector). The collimator ensures that all rays fall at the same angle onto the dispersing element since angle of refraction, resolving power, and image quality all depend on that angle (cf. Chap. 7). Lens as well as mirror optics are used for the collimator, which can be thought of as an "inverted" telescope. The two dispersing media used are prisms and diffraction gratings. The prism spectrograph uses the dispersing effect of glass (which leads to the unwanted chromatic aberrations in refracting optics) while in the diffraction grating interference of the individual rays leads to spectral dispersion. The operation of these elements will not be treated in detail here, except for the comment that the grating produces a so-called *normal spectrum* since the diffraction angle depends linearly on the wavelength, while the prism response is distinctly nonlinear and disperses the blue spectral region more strongly than the long-wavelength red region. For the imaging cameras in spectrographs, usually lens optics or fast Schmidt systems are used.

The three aforementioned elements also appear in objective prism photographs. The stellar light, by virtue of its infinite distant source, arrives as a plane wavefront already collimated at the telescope aperture and passes through a wedge-shaped prism in front of the aperture. The telescope optical system then functions as a spectrographic camera which images the starlight into small spectra, thus permitting a simple stellar classification. To obtain higher spectral resolution, the spectrograph possesses a slit which lies in the focal plane of the telescope. Chapter 7 will deal with the dependency of the resolving power on slit width, grating angle, and other quantities. Spectroscopy requires high light-gathering power since the starlight, owing to dispersion and the widening of the spectrum, is distributed over a rather large receiving area and normally needs integration on the detector. One exception is the Sun. Its radiation, even after spectral dispersion, is still strong enough to stimulate the color-sensitive elements in the eye.

To view the solar spectrum and its numerous absorption lines direct with the eye, a *spectroscope* is employed. A *direct-vision prism* (Fig. 4.30) introduced by Amici (1786–1864) usually serves as the dispersing medium. It consists of up to five individual prisms of two different glasses cemented into one prism. A ray with a typical wavelength, such as that of the Fraunhofer e-line (546.1 nm), exits this prism in a direct line, while rays of shorter (e.g., the blue g-line of mercury) or longer (e.g., the red C-line, or Hα, of hydrogen) wavelengths are deflected, respectively, right or left. The wavelength-dependent inclination of the parallel beams is then magnified by an eyepiece to be viewed by the eye as a spectrum, or it is imaged onto plane film. Instructions for building such a spectroscope with a direct-vision prism are given by Gebhardt and Helms [4.58]. The spectrum need not be widened perpendicular to the dispersion since the extended disk of the Sun automatically illuminates the full

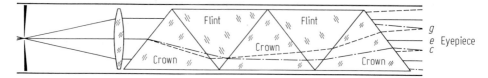

Fig. 4.30. Spectroscope with direct vision after Amici; the slit is in the focus of the telescope objective and of the collimator lens.

length of the slit. When viewing a point source, such as a distant star, on the other hand, the widening is provided by a cylindrical lens, while photographic widening is best achieved by physically moving the telescope in right ascension or declination depending on which motion is parallel to the slit.

4.7.8 Sun Projection Screens

Instruments for direct solar observation were described in Sect. 4.5.9. Lacking those elements (such as solar filters, absorbing glasses, helioscope, etc.) to reduce the intense solar radiation, the Sun can still be observed by magnified projection onto a white screen. The screen should lie in the shadow of a larger stop surrounding the telescope tube. It is convenient to connect the projection screen with the telescope tube so that it follows the telescope motion. A rigid connection is needed especially when the solar phenomena (Chap. 13) are to be not only viewed but also graphed on paper. Owing to their proximity to the telescope's focus and the heat concentration there, cemented eyepieces must *not* be used for solar projection, but only those of the Huygens, Mittenzwey, or Ramsden type. The danger of heating and damaging during continued solar observing also exists, but to a lesser degree, for a cemented objective. The diameter of the projected solar disk $P_\odot$ depends on the focal length f_{eyp} of the eyepiece and the distance d to the projection screen. An approximate formula is

$$P_\odot = D_\odot d(f_{\text{obj}}/f_{\text{eyp}}) = D_\odot d\gamma,$$

where $D_\odot$ is is the angular diameter of the solar disk. Owing to the ellipticity of the Earth's orbit, $P_\odot$ varies between 0.00917 and 0.00948 radians at aphelion and perihelion, respectively. Thus, for graphs of high precision, the distance of the screen should be adjustable. A projected diameter of 15 cm is adequate to sketch the solar disk; it is also a recognized "norm" for which graph paper or stencils with division in degrees are available that also take into account the variation in heliographic latitude of the center of the disk during the year due to the tilt of the solar rotation axis against the ecliptic plane. Distortion of the image can be avoided by attaching the screen perpendicular to the optical axis. Of course, solar images can be projected in any size, but as the light intensity decreases quite rapidly with projection scale, the end of the eyepiece and the projection screen must be located in a dark room for large-scale images to be viewed satisfactorily.

Table 4.8. Some radio stations which broadcast time signals on the shortwave band.

Location	Station	Frequency	Power	Location	Station	Frequency	Power
Great Britain	MSF	2.5 MHz	50 kW	Czech Rep.	OMA	2.5 MHz	1.0 kW
		5.0	50	USA	WWV	2.5	2.5
		10.0	50			5.0	10.0
Hawaii	WWVH	2.5	2.0			10.0	10.0
		5.0	2.0			15.0	10.0
		10.0	2.0			20.0	10.0
		15.0	2.0			25.0	10.0

4.7.9 The Clock

The time of an observation, whether visual or photographic, should always be recorded. This applies to predictable phenomena like eclipses or lunar occultations as well as unforeseen events such as fireballs or novae. It should be emphasized that every photographic picture is a document (the most recent example being Supernova 1987a in the Large Magellanic Cloud), so recording the exact observation time—beginning, or perhaps end, or total exposure time—should be routine. The demands upon the precision of the clock vary widely, depending on the task of observation. For most purposes, a pocket or quartz watch will suffice; it can be compared before (and perhaps after) an observation with a time signal which permits a much higher degree of accuracy than the times available from radio and telephone service. A number of time signal broadcasts are obtained on the shortwave band (Table 4.8); they can be readily obtained using simple radios which have the designated band. Thus, the instant of observation and the time signal can be recorded simultaneously on a tape (e.g., via microphone, recorder, and mixer) and evaluated after the observation. An instant can also be recorded photographically; see Boulet [4.59] for specific instructions.

In the U.S., station WWV provides shortwave broadcasts at 5, 10, 15, and 20 MHz and time signals over the telephone: Nos. 900-410-TIME, and, for the Washington D.C. area, 202-653-1800. In addition to shortwave signals, there are also time signals broadcast on longwave; for instance, a station in Prague on 50 kHz, a British station on 60 kHz, a Swiss station on 75 kHz, and also the German station Mainflingen (DCF77) on 77.5 kHz.

Phenomena observed in the sky are almost always recorded in *Universal Time* (UT) as obtained from the time signals, while for the telescope drive, however, *sidereal time* (ST) is needed (see Chap. 2). The conversion of a time signal to sidereal time is described in detail by Scheucher [4.60], who discusses telescope tracking and the precision obtainable. Electronic circuit diagrams for building a simple sidereal time clock can be found, for instance, in Newton [4.61].

4.8 Visual Observations

4.8.1 The Eye

The properties of the eye, especially where they relate to and limit the performance of telescopes, have been repeatedly addressed. The diameter of the eye pupil is of unquestionable importance in determining resolving power and image brightness. The pupil width adapts automatically to existing light levels, and since most astronomical observations are performed at low light levels, the pupils will usually be fully dilated at such times. The maximum attainable pupil width is often given in the range 6–8 mm; one can, without significant error, assume a linear decline from 8 mm for a person of age 20 to approximately 2.5 mm at the age of 80. Thus the minimum or normal magnification should be arranged so that the beam from the exit pupil is not vignetted by a too small an eye pupil. The influence on the brightness of telescopic images has been discussed in Sect. 4.6.3. In order to coincide the eye pupil with the exit pupil, persons who wear glasses should remove them when observing; the focal error of the eye can then be compensated by simply refocussing the eyepiece.

Another important quantity is the resolving power of the eye, given as about $120''(= 2')$ for high image contrast such as exists in double stars. This value agrees surprisingly well with the minimum pupil diameter of 1–2 mm. This means that magnifications with exit pupils under 1 mm do not gain any more angular resolution, since then the smallest diffraction element is spread over several receiving elements on the eye's retina. Likewise, the use of magnifications much in excess of $100\times$ is not practical since then the maximum resolution of about $1''$ permitted by the Earth's atmosphere reaches the threshold of perception.

The sensitivity of the eye is wavelength dependent. The retina of the eye consists of a network of nerve cells that send impulses along the optic nerve to the brain. There are two principal kinds of retinal cells, namely *rods* and *cones*, which behave according to two different (internationally standardized) spectral sensitivity curves (Fig. 4.31). The rods provide vision in dim light and reach their maximum sensitivity at 507 nm, while the cones respond best to bright light and reach their maximum sensitivity at 555 nm. Color vision is provided by the cones. The rods, on the other hand, are responsible for so-called "night vision," and are stimulated at a threshold value which, under ideal conditions, corresponds to a star of 8th magnitude. In reality, however, the background brightness of the night sky diminishes the limiting magnitude for the naked eye to about $6^{\mathrm{m}}\!.0 \pm 0^{\mathrm{m}}\!.5$. Sect. 4.6.3 has dealt with the dependence of limiting magnitude on telescope aperture and magnification. In this connection, the adaptation of the eye to the dark is of some interest and proceeds roughly as follows: the retinal sensitivity increases very little during the first ten minutes, then rises steeply from 25% to around 80% of the maximum sensitivity during minutes 15 to 25, and ultimately approaches the maximum asymptotically, so that the maximum sensitivity of the retina is essentially reached after about 1 hour.

The distribution of rods and cones in the retina is of particular interest for visual observations. The color-sensitive cones are concentrated in the center, or *fovea* of the retina. The more sensitive but color-neutral rods, on the other hand, are more numerous toward the outer periphery of the retina. This explains why some very faint light sources (e.g., nebulae) are "invisible" by direct vision, but may be glimpsed by looking slightly to one side of the sought object; this is termed *averted vision*. A comprehensive description of the human eye can be found in Fry [4.62].

4.8.2 Binoculars

Binoculars are certainly the most widespread instruments for visual observations. Depending on their entrance- and exit-pupil size, they may be better or less well suited for astronomical observations. Binoculars are of the refracting telescope type, which, since they are frequently used for terrestrial observations (birdwatching, sports, opera, etc.), are designed to provide an *erect* image. Erecting the image with the correct left-to-right orientation can be achieved with various kinds of prisms, among which the combination of two rectangular prisms—the *prism-inverter* according to Porro—is most widespread and achieves a substantial reduction of the tube length by "folding" the beam. The original cemented achromats have meanwhile been replaced by two-lens achromats with a large air space in between and which contain a divergent lens as a detached second element. The result is known as a *telephoto lens* (similar to the combination objective + Barlow lens) which again shortens the tube length. The objectives with f-ratios $f/3$ are corrected for errors (spherical and chromatic aberration) on- and near-axis. The off-axis field curvature is largely compensated for by the automatic accomodation of the eye.

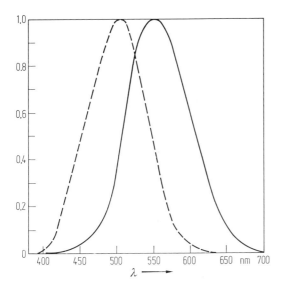

Fig. 4.31. Spectral sensitivity curves of the eye (normalized so that maximum = 1): *dashed line* = nighttime, *solid line* = daytime vision.

Binoculars are always labeled with designations such as 8×21, 7×50, or 10×50. The first number gives the magnifying factor γ, and the second the free objective diameter D in mm; together these provide all the characteristic data for the binoculars. The size of the exit pupil, which is critically relevant for the brightness of extended objects, is found by division as $A_P = D/\gamma$. Binoculars with exit-pupil diameters of about 2 mm possess the maximal useful magnification, which means that the resolving power of the eye of 2 arcminutes is attained and coincides with the resolution by the objective D. A further increase in magnification by reducing the size of the exit pupil to 1 mm or even 0.5 mm, however, is feasible and can in some instances actually enhance details on the observed object. In astronomy, these magnifications serve to achieve the resolution of close double stars, of clusters, and perhaps of the surface features of planets, but also to lower the limiting magnitude by reducing the brightness of the sky background (Sect. 4.6.3). Because of the latter effect, for observations of extended objects (e.g., comets, gaseous nebulae, galaxies) binoculars with minimum magnification should be chosen, i.e., the diameter of the exit pupil is some 8 mm (Eq. (4.47)). Binoculars with this magnification ($D/\gamma = 8$) are often called *nightglasses*. Of interest is also the *twilight number*

$$Z_D = \sqrt{\gamma D} \qquad (4.61)$$

which provides a reliable value for the attainable sharpness of vision in twilight ($L_D \approx 0.3 \, Z_D$). Gera [4.63], for instance, has subjected two binoculars (7×50 and 15×80) to various astronomical tests and has discussed their suitability with respect to the characteristic values D and γ.

The *sharpness S* of a viewed object may be put into quantitative terms as the inverse of the barely perceivable object size u' in arcminutes. The *sharpness factor L* is then defined as the ratio of sharpness of an object as seen with the naked eye S_{eye} to that viewed with binoculars S_{bnc},

$$L = \frac{S_{bnc}}{S_{eye}} = \frac{1/u'_{bnc}}{1/u'_{eye}}, \qquad (4.62)$$

where u'_{eye} and u'_{bnc} denote the just-perceivable object size in arcminutes for the naked eye and viewed with the binoculars, respectively. Ideally, the sharpness factor L should be identical with magnification γ, but because of certain physiological effects in the eye, L is always less than γ, reaching at best 90% thereof. Drastic differences in L are found between "handheld" and "mounted" binoculars. An investigation of this problem may be found in König and Köhler [4.18], which deals with the entire subject of sharpness factors for various telescope types. According to their results, an 8× (or at most 10×) magnification is feasible for handheld observations, and reaches 80% (or 60%) of the value of L for fixed-mounted glasses. Except for observing suddenly occurring phenomena such as meteors and satellites, the binoculars should be as firmly mounted as possible for astronomical observations. Homemade mountings may suffice, but it is preferable to use the readily available camera tripod on which the binoculars can be mounted with a special support cradle (Fig. 4.32) that can be purchased from most camera shops. With this device, the binoculars can be easily and quickly pointed to any region of the sky, and the vibration-free image will show a variety of celestial objects in all their splendor. Many observers have constructed special mounts with

Fig. 4.32. Binoculars mounted on a phototripod, attached at the central hinge.

a deck chair (which can be tilted and rotated) permitting comfortable body posture for all observing positions, particularly when observing near the zenith. These chairs have been repeatedly described in various amateur magazines.

A general recommendation to anyone whose interest in observing the heavens is just beginning to blossom would be to postpone purchasing a telescope for a little while, and instead to mount a pair of binoculars on a vibration-free tripod. When turned skyward, the latter instrument alone will surely delight the neophyte as it draws the firmament substantially closer to his or her eyes. The gain from naked eye to binoculars is about the same as from binoculars to a 35-cm telescope. Binoculars, owing to their high light-gathering power and wide field of view of several degrees, are particularly suited for observing extended celestial objects. A bonus is the unique visual impression which can be experienced only by viewing with both eyes. A book full of suggestions for observing with binoculars has been written by R. Brandt [4.64].

4.9 Photographic Plates and Photoelectric Detectors

4.9.1 Astrophotographs and Their Limitations

The photographic plate is an eminent tool for rendering measuring procedures more objective. Because the sensitivity of the plate is cumulative (i.e., integrating) and thus permits the detection of extremely faint light sources via long exposures, it has played a particularly valuable role in astronomy ever since its first use in the mid-19th-century by John William Draper (see Chapt. 11.5.3). Its disadvantages, however, should not be overlooked. For instance, the registration and storage of the photographic emulsion take place at the same location and can be done only *once*; therefore a direct calibration over its entire area is therefore not possible (Chap. 6 will deal more fully with this matter). Moreover, while improved emulsions and modern techniques (e.g., hypersensitization) can increase the efficiency, the quantum efficiencies of even the high-sensitivity astro-emulsions are less than 1%. In addition to the existence of a certain energy threshold value which must be exceeded to obtain a registration, the photographic plate also has a rather small dynamic range. It also lacks the clear linearity between incident light and photographic effect; the latter also depends on other factors such as temperature, developer chemicals, procedures, and so on. These comments should alert the reader that caution must be used in calibrated measurements.

Despite its drawbacks, the photographic plate nevertheless enjoys a wide range of uses in astronomy. The enormous number of grains embedded in the emulsion layer imbue a moderate-sized photographic plate with considerable storage capacity, thus making it indespensible for large-area direct photographs, such as those obtained with Schmidt cameras. It is also still used in some spectrograph cameras. While, compared with other detectors, the plate is only of limited use for obtaining images in various areas of scientific research, it is still quite popular among amateur astronomers and also reasonable to use. Virtually every issue of the many popular astronomy magazines such as *Astronomy* and *Sky and Telescope* has a section devoted to the outstanding astrophotographs, in both black-and-white and color, that have been obtained (sometimes with very simple equipment) by amateurs. The special problems of color astrophotography (direct onto chromatic films or by means of the three-color composite method) are addressed in minute detail by Malin and Murdin [4.65] in their visually stunning book, *Colours of the Stars*. Other techniques, such as "unsharp masking" or contrast-enhancement methods of black-and-white photography, are also discussed there. Since the subject is addressed in detail in a separate chapter (6) in this volume, it will not be treated further here; instead, the connection between the practical limits of photographic observations and the parameters of the telescope (or camera) will be briefly discussed. The relevant description of these factors and their confirmation are given by Knapp [4.66], from whom the most relevant relations have been extracted.

A photographic system is described by the following parameters: free aperture D, system focal length f, aperture ratio $D/f = 1/N$, linear diameter b of the stellar disk at the plate, film sensitivity in the logarithmic DIN-system[11], and the Schwarzschild exponent $p \approx 0.8$ for long exposures. The background brightness of the sky m_{sky} enters critically into the limiting magnitude m_{lim} to be reached. The background brightness in square degrees (denoted by $\beta°$) is inserted into the relations which follow. An optimal value of 4th magnitude per square degree ($4^m/\beta°$), corresponding to $22^m/\beta''$, is usually assumed. In practice, however, one must be satisfied with about $3^m/\beta°$ (corresponding to $21^m/\beta''$), while in urban areas the observer must contend with $2^m/\beta°$ and less; the astronomical magnitude scale is treated in Chap. 8. In all considerations, the relevant quantity is the size b of the stellar disk, which depends on four parameters:

---

11 The more commonly used system in the U.S. is the arithmetic ASA-system; cf. Chap. 6.

(a) Resolving power of the emulsion: $b = 1/A$ (lines/mm),
(b) Resolving power of the camera: $b = 2.44\,(\lambda/D)f$,
(c) Atmospheric-seeing disk: $1'' \leq 2.1 \times 10^5\,(b/f) \leq 5''$,
(d) Optical imaging errors (see Sect. 4.3)

Which of these four components dominates and therefore causes the maximum value of b must be determined for each photographic system. One popular technique of finding the relevant value of b is to produce "star-trail" exposures; the diameters of the trails are then measured with a microscope. The attainable limiting magnitude $m_{\lim}$ of a star is also dependent upon the density of the the sky background. Thus if m_{sky} is the magnitude per square degree of the background brightness, then $m_{\lim}$ is given by

$$m_{\lim} = m_{sky} + 5\,\log\left(\frac{f}{b}\right) - 8.5. \tag{4.63}$$

An increase in focal length influences the limiting magnitude for photographic work in the same way that an increase in magnifying power affects the limiting magnitude for visual observations. The exposure time needed to reach $m_{\lim}$ is found from Eq. (4.66) upon inserting the relevant background brightness m_{sky} for surface brightness. In general, the minimum exposure time t necessary to reach a certain stellar magnitude m_* on the photographic plate is calculated as

$$t = \left(4 \times 10^6 \frac{2.512^{m_*}}{10^{\frac{DIN}{10}}\,(D/b)^2}\right)^{1/p},$$

or

$$m_* = 2.5p\,\log t + 5\,\log\left(\frac{D}{b}\right) + \frac{DIN}{4} - 16.5, \tag{4.64}$$

where t is in seconds. The situation is different when photographing extended objects: in place of the imaging quality D/b, the aperture ratio D/f enters critically. The surface brightness m_β (in mag/$\beta°$) to be reached within t seconds of exposure time follows from

$$m_\beta = 2.5\,p\log t + 5\,\log\left(\frac{D}{f}\right) + \frac{DIN}{4} - 8 \quad (\text{mag}/\beta°). \tag{4.65}$$

The limit is again set by the sky background m_{sky}. In turn, the exposure time t needed to record the surface brightness m_β on the plate is

$$t = \left(1585\,\frac{2.512^{m_\beta}}{10^{\frac{DIN}{10}}\,(D/f)^2}\right)^{1/p}, \tag{4.66}$$

where t is again expressed in seconds. Note that m_β is inserted in mag/$\beta°$, which is to be derived from total magnitude and total area, the latter usually being given in β' for nebulous objects. In addition to diagrams for the graphical determination of the desired quantities, Knapp [4.66] also gives the characteristic parameters b, f/b, D/b, and $m_{\lim}$ for a variety of photo-objectives and photographic telescopes.

The Earth's diurnal rotation causes an apparent east–west motion of the stars which is fastest in the equatorial zone. Thus a photograph of the sky taken with a nontracking camera will display streaking of the exposed stellar images unless the exposure time is very short. It is therefore important to establish the maximum admissible exposure time t_{stat} and the corresponding attainable magnitude m_{stat} for a stationary camera. One possible criterion for a nonstreaking exposure is the time interval needed by a star to pass its own diameter b on the plate. For a star with declination δ, t_{stat} is given in seconds by

$$t_{\text{stat}} = 13{,}800 \frac{b}{f} \frac{1}{\cos \delta}, \qquad (4.67)$$

This value therefore increases with increasing declination, so that stars near the polar regions can be exposed considerably longer than equatorial ones for streakless images. The magnitude reached in this time for a stationary camera is found from

$$m_{\text{stat}} = 2.5 \, p \log \left(\frac{b}{f}\right) + 5 \, \log \left(\frac{D}{b}\right) + \frac{\text{DIN}}{4} + 10.4p - 16.5. \qquad (4.68)$$

4.9.2 Photomultipliers

The photomultiplier consists basically of a *photocathode*, a series of multiplying electrodes (called *dynodes*), and an *anode*, all arranged in a vacuum tube. When the photocathode (usually composed of an alkali metal or compound) is struck by light, electrons are ejected from it by the *photoelectric effect*. These so-called liberated photoelectrons, now subjected to the applied electric field, migrate toward and reach the first dynode. Each dynode is coated with a special layer so that the incident electrons release other, *secondary electrons* which in turn hit the next dynode and release more electrons. In order to keep the electrons on a specific path, each successive dynode is kept at a slightly higher electrical potential than the previous one by feeding the supplied voltage via a resistor chain to the dynodes. With about a 100-volt potential difference between adjacent dynodes, the total voltage at the photomultiplier tube (PMT) reaches a respectable 1000–2000 volts. The types of PMTs differ, for example, in the geometric arrangement and shape of the dynodes, depending on their application (e.g., for measuring short-period oscillations in nuclear physics). Especially significant for astronomical applications is the high quantum yield of up to 20% with present-day S-20 photocathodes. The connected dynode system supplies, via an "avalanche" increase of the electrons in the photomultiplier, multiplying factors of 10^6–10^8 of the electrons originating at the photocathode. This electron current reaches the anode and, upon exiting the PMT, delivers a voltage impulse which is then processed to yield a quantitative measure of the flux of the original light rays.

4.9.3 Charge-Coupled Devices—CCDs

The new semiconductor detectors called *charge-coupled devices*, or *CCDs* were developed in the early 1970s and are based on a silicon chip which contains a number of miniature photodiodes. Each photodiode defines a picture element, or *pixel*, with side lengths typically between 15 and 30 μm (with the expectation of even tinier dimensions in the coming decade). These pixels are produced in the manner of the masking technique of highly integrated circuits, in which chips with up to 2000 × 2000 elements are today available. The problems facing the scientist in storing the huge amount of data and in the time consumed by image processing are obvious. But new storage techniques (optical disks) and sophisticated image processors are on the way and will hopefully improve the situation.

The imaging mode of semiconductors is based on the conversion of the incident light energy into electrons. The photoelectrons released by the photons are collected in the pixels, which are separated from each other by insulating potential walls consisting of silicon dioxide. After the exposure, the chip contains an electron-encoded image which must be stored for later retrieval. To read out the electronic image, the charges collected by each pixel are transferred from one element to the next, by means of electrical potentials (similar to a serial register), until they enter a sensitive preamplifier at the edge of the chip. There, an enhancement of the charge accumulated on each pixel is achieved. The charge is then sampled and digitized via an analog–digital converter, and the data is stored for further processing by a small computer. For a more detailed description of CCD properties and techniques relating to their use, the article by Kristian [4.67] and the comprehensive review by Mackay [4.68] should be consulted.

The main characteristics of CCDs are: (a) the high quantum efficiency, which, depending on spectral range, may reach up to 80%; (b) the linearity between incident light and the measured output (the registered charge is, as in the PMT, directly proportional to the incident light intensity); (c) the high dynamic range of 10^4 and more (compared with 100 of the photographic plate); (d) the photometric precision, since their sensitivity can be calibrated over the entire area; (e) the low detector noise originating in the read-out process by the electronics; and (f) the high geometric stability, since the pixels are structurally fixed on the chip (in contrast to possible emulsion shifts which can occur during the development of photographic plates).

Despite all of these advantages, the development of semiconductor detectors is still in its infancy. CCDs attain their maximum sensitivity toward the red wavelength region of the spectrum. The steep decline in sensitivity in the blue and UV regions can at present be compensated for only by the application of special coatings which transform the radiation from short into long wavelengths, but the resulting sensitivity is still far below what would be considered optimum. The problem of the charging of the pixel by cosmic rays can be especially troublesome on long exposures, although attempts are being made to solve this by measures taken during manufacture and observation. Finally, the size of the detector surface of only a few square centimeters is still relatively small and even then considerable time is spent in processing the image. Nevertheless, the articles by Buil [4.69] and Harris [4.70] report on how these highly sensitive modern detectors have already entered amateur astronomy.

4.10 Services for Telescopes and Accessories

Most astronomical magazines such as *Sky and Telescope* and *Astronomy* print suppliers' advertisements on state-of-the-art equipment and instrumentation. In addition, information on used telescopes and accessories can usually be found in a small "classified ads" section (e.g., "Sky-Gazers Exchange" in *Sky and Telescope* and "Reader

Exchange" in *Astronomy*). Astronomical periodicals also print reports on tests of and experiences with certain designs and new developments, including do-it-yourself constructions.

Some magazines publish "specials" compiling addresses of regional suppliers. For example, the **Astronomy Equipment Directory 1989** appeared in *Astronomy* Vol. 17, No. 10 (1989) and the **Astronomy Resource Guide** in *Sky and Telescope*, Vol. 78, No. 3 (1989).

Supply centers and working groups within the frame of amateur observatories, societies, and planetaria offer details on do-it-yourself constructions. Information on the numerous national and regional amateur conventions (such as Stellafane in the USA) including meeting dates are announced in magazines; see Appendix A.

References

4.1 Traving, G.: Einiges über optisches Rechnen I u. II. *Sterne und Weltraum* **24**, 661 (1985) and **25**, 274 (1986).
4.1a Larks, L.: Optical Ray-Tracing on a Microcomputer. *Sky & Telescope* **61**, 356 (1981).
4.2 Berek, M.: *Grundlagen der praktischen Optik*, W. de Gruyter, Berlin 1970.
4.3 Flügge, J.: *Leitfaden der geometrischen Optik und des Optikrechnens*, Vandenhoeck & Ruprecht, Göttingen 1956.
4.4 Köhler, H.: Die Entwicklung der aplanatischen Spiegelsysteme. *Astron. Nachrichten* **278**, 1 (1949).
4.5 Slevogt, H.: Über eine Gruppe von aplanatischen Spiegelsystemen. *Zeitschrift für Instrumentenkunde* **62**, 312 (1942).
4.6 Bahner, K.: Teleskope. *Handbuch der Physik, Band XXIX*, Springer, Berlin Heidelberg New York 1967.
4.7 Wiedemann, E.: Über die Interpretation der Korrektionsdaten von Astro-Amateur-Optiken. *Sterne u. Weltraum* **18**, 268 (1979).
4.8 Born M., and Wolf, E.: *Principles of Optics* (6th edn.), Pergamon, New York 1980.
4.9 Cagnet, M., Francon, M., and Thrierr, J.C.: *An Atlas of Optical Phenomena*, Springer, Berlin Heidelberg 1962.
4.10 Maksutov, D.D.: *Technologie der astronomischen Optik*, Verlag Technik, Berlin 1954.
4.11 Petzoldt, J.: 'Zerodur' – ein neuer glaskeramischer Werkstoff für die reflektierende Optik. *Sterne u. Weltraum* **15**, 156 (1976).
4.12 Hartmann, J.: Objektivuntersuchungen. *Zeitschrift für Instrumentenkunde* **24**, 1, 33, 97 (1904).
4.13 Malacara, D.: *Optical Shop Testing*, Wiley, New York 1978.
4.14 De Vany, A.S.: *Master Optical Techniques*, Wiley, New York 1981.
4.15 Bryngdahl, O.: Applications of Shearing Interferometry, in *Progress in Optics IV, 37* (E. Wolf, ed.), North Holland, Amsterdam 1965.
4.16 Bath, K.L.: Ein einfaches Interferometer zur Prüfung astronomisher Optik. *Sterne u. Weltraum* **12**, 177 (1973).
4.17 Schultz, S.W.: Standardizing the Ronchi Test Pattern. *Sky and Telescope* **67**, 272 (1984).
4.18 König, A., Köhler, H.: *Die Fernrohre und Entfernungsmesser*, Springer, Berlin 1959.
4.19 Klein, W.: Bemerkungen zur Korrektion und Durchrechnung von optischen Systemen. *Jahrbuch für Optik und Feinmechanik 1983*, Fachverlag Schiele & Schön, Berlin 1983.
4.20 Robb, P.N.: Selection of Optical Glasses. *Applied Optics* **24**, 1864 (1985).
4.21 Wiedemann, E.: Verfeinerte Optiken für Astroamateure. *Sterne u. Weltraum* **15**, 366 (1976).
4.22 Rohr, H.: *Das Fernrohr für Jedermann*, 5. Aufl., Rascher Verlag, Zürich 1972.
4.23 Wiedemann, E.: Verfeinerte Optiken für Astroamateure II. *Sterne u. Weltraum*, **17**, 374 (1978).

4.24 Cox, R.E.: Notes on Telescope Making from Here and There. *Sky and Telescope*, **35**, 319 (1968).
4.25 Richter, J.L.: A Test for Figuring Cassegrain Secondary Mirrors. *Sky and Telescope* **39**, 49 (1970).
4.26 Ulrich, M.H., and Kjär, K. (eds.): Proceedings of the ESO Conference on Very Large Telescopes and their Instrumentation, ESO, Garching 1988 and 1992.
4.27 Fricke, K.J.: Workshop on Large telescopes, *Mitteilungen der Astron. Gesellschaft* **67**, 193 (1986) and **70**, 250 (1987).
4.28 Kutter, A.: *Der Schiefspiegler*, F. Weichardt, Biberach an der Riß1953.
4.29 Kutter, A.: A New Three-Mirror Unobstructed Reflector. *Sky and Telescope* **49**, 46, 115 (1975).
4.30 Ingalls, A.G.: *Amateur Telescope Making III*, Scientific American Inc., New York 1953.
4.31 Cox, R.E.: The Vacuum Method of Making Corrector Plates. *Sky and Telescope* **43**, 388 (1972).
4.32 Weigel, W.: Justieren einer Schmidt-Kamera, *Sterne u. Weltraum* **18**, 272 (1979).
4.33 Waineo, T.: Fabrication of a Wright Telescope, *Sky and Telescope* **38**, 112 (1969).
4.34 Schmadel, L.D.: Schmidt-Systeme ohne Korrektionsplatte, *Sterne u. Weltraum* **16**, 214 (1977).
4.35 Rutten, and Van Venrooij, M.: Die optischen Eigenschaften eines 200-mm Schmidt–Cassegrain-Teleskops. *Sterne u. Weltraum* **23**, 274 (1984).
4.36 De Vany, A.S.: A Schmidt–Cassegrain Optical System with a Flat Field. *Sky and Telescope* **29**, 318, 380 (1965).
4.37 Sigler, R.D.: Compound Schmidt Telescope Designs with Nonzero Petzval Curvatures. *Applied Optics* **14**, 2302 (1975).
4.38 Sigler, R.D.: Compound Catadioptric Telescopes with all Spherical Surfaces. *Applied Optics* **17**, 1519 (1978).
4.39 Wiedemann, E.: Optiken für die Amateur-Astronomie. *Sterne u. Weltraum* **19**, 411 (1980).
4.40 Gregory, J.: A Cassegrainian-Maksutov Telescope Design for the Amateur. *Sky and Telescope* **16**, 236 (1957).
4.41 Willey, R.: Cassegrain-Type Telescope. *Sky and Telescope* **23**, 191, 226 (1962).
4.42 Völker, P. et al.: *Handbuch für Sonnenbeobachter*, Veröffentlichung der Vereinigung der Sternfreunde e.V., Berlin 1982.
4.43 Nögel, O.: Ein Fernrohr zur Beobachtung der Protuberanzen für den Amateur. *Die Sterne* **28**, 135 (1952) and **31**, 1 (1955).
4.44 Nemec, G.: Das Protuberanzenfernrohr als Hochleistungsinstrument. *Sterne und Weltraum* **10** (1970) and **11** (1971).
4.45 Hanisch, H.D.: Protuberanzenansatz für kleine Refraktoren. *Sterne und Weltraum* **14**, 370 (1975).
4.46 Richter, G.: Ein vereinfachtes Protuberanzenfernrohr. *Die Sterne* **50**, 105 (1974).
4.47 Veio, F.N.: An Inexpensive Spectrohelioscope by a Californian Amateur. *Sky and Telescope* **37**, 45 (1969).
4.48 McAlister, H.A.: High Angular Resolution Measurements of Stellar Properties. *Annual Reviews of Astronomy and Astrophysics* **23**, 59 (1984).
4.49 Jacquinot, P., Roizen-Dossier, B.: Apodisation. In: *Progress in Optics III*, 31, E. Wolf (ed.), North Holland Publishing Co., Amsterdam 1964.
4.50 Heintz, W.D.: Doppelsterne in der Polumgebung. *Sterne und Weltraum* **4**, 118 (1965).
4.51 Becvar, A.: *Atlas of the Heavens–II, Catalogue 1950*, Sky Publishing Corp., Cambridge MA 1964.
4.52 Tombaugh, C.W., Smith, B.A.: A Seeing Scale for Observers. *Sky and Telescope* **17**, 449 (1958).
4.53 Bowen, I.S.: Limiting Visual Magnitude. *Publications of the Astronomical Society of the Pacific* **59**, 253 (1947).
4.54 Hartshorn, C.R.: Amateur Telescope Making III, p. 277. A.G. Ingalls (ed.), Scientific American, New York 1953.

4.55 Schiffhauer, H.: Die Verwendung von Kunststoffrohren im Fernrohrbau. *Sterne und Weltraum* **1**, 158 (1962).
4.56 Wepner, W.: Berechnung der Blenden eines Refraktors. *Sterne und Weltraum* **15**, 289 (1976).
4.57 MacRobert, A.: Caring for Optics. *Sky and Telescope* **73**, 380 (1987); **74**, 573 (1987) and **76**, 5 (1988).
4.58 Gebhardt, W., Helms, B.: Ein Selbstbau-Prismenspektrograph zum Gebrauch am C8. *Sterne und Weltraum* **15**, 58 (1976).
4.59 Boulet, D.L.: A Simple Photochronograph. *Sky and Telecope* **68**, 76 (1984).
4.60 Scheucher, E.: Sternzeituhr hoher Genauigkeit. *Sterne und Weltraum* **23**, 473 (1984).
4.61 Newton, R.J.: An Easy, Inexpensive Sidereal Clock. *Sky and Telescope* **66**, 453 (1983).
4.62 Fry, G.A.: The Eye and Vision. *Applied Optics and Optical Engineering Vol. II*, Kingslake (ed.), Academic, New York 1965.
4.63 Gera, H.D.: Welches Nachtglas? *Sterne und Weltraum* **26**, 167 (1987).
4.64 Brandt, R.: *Himmelswunder im Feldstecher, 7. Aufl.*, J.A. Barth, Leipzig 1964.
4.65 Malin, D., Murdin, P.: *Colours of the Stars*, Cambridge University Press, Cambridge 1984.
4.66 Knapp, H.: Über die Reichweite von Objektiven bei Astroaufnahmen mit kleinen Montierungen. *Sterne und Weltraum* **3**, 262 (1964).
4.67 Kristian, J., Blouke, M.: Charge-Coupled Devices in Astronomy. *Scientific American* **247**, 48 (1982).
4.68 Mackay, C.D.: Charge-Coupled Devices in Astronomy. *Annual Reviews of Astronomy and Astrophysics* **24**, 255 (1986).
4.69 Buil, C.: A Charge-Coupled device for Amateurs. *Sky and Telescope* **69**, 71 (1985).
4.70 Harris, C.: Silicon Eye: A CCD Imaging System. *Sky & Telescope* **71**, 407 (1986).
4.71 Valleli, P.: Collimating Your Telescope. *Sky & Telescope* **75**, 259 and 363 (1988).

5 Telescope Mountings, Drives, and Electrical Equipment

H.G. Ziegler

5.1 Introduction

The current trend toward larger and more powerful instruments is normally limited by the size, type, and quality of not only the optics but also the mounting. The telescope mounting is a complex structure of many parts and is far less easily comprehended than the optical path in the telescope; some parts are even more difficult for the amateur telescope maker to manufacture than a primary mirror. Thus, the major problem has always been the manufacture of the mechanical components. Regrettably, no direct help can be provided regarding this matter, since the telescope maker must depend on his or her own workshop facilities and inventiveness.

The emphasis of this chapter is placed, with good reason, on the design principles. Practice-oriented publications – *Amateur Telescope Making*, *Sky and Telescope*, *Telescope Making* – and many other periodicals and books are available. Innumerable mounting types have been presented over the years in these publications, and in fact there is no type, no concept, no detail of manufacture which has gone undescribed. Completely lacking, however, has been a systematic treatment of the criteria for the design and layout of a telescope mounting. This may be the primary reason why even now the amateur builds instruments in a pragmatic and intuitive fashion. At best, he is aware of the goal that a mounting should be "stable," and that this can be attained by making the the mounting and all its essential parts "massive." As they cannot be mathematically formulated, the terms "stable" and "massive" are virtually useless in a well-founded design. Amateur ideas regarding telescope vibrations and kinematic principles are similarly vague. But technology needs measurable and calculable quantities. In this chapter, the telescope maker will be shown systematically and for the first time the strict rules by which a mounting can be treated. The chapter begins with a discussion of basic criteria for statics, kinetics (vibrations), and the kinematics of the structure. From these, the technical principles which are relevant for determining design and layout will be derived. Extensive calculations in the design of an instrument will certainly not appeal to everyone. In order also to help the practice-oriented telescope maker with some systematic guidance, the primary consequences of the theory have been condensed into the *fundamental principles*. A similar treatment of the basics of electrical equipment is included. A substantial difference between mechanical and electrical telescope problems is that the latter are much more easily dealt with. In an electrical circuit, components may be replaced simply and at low cost, or even the entire circuit changed, whereas a mechanical structure with serious faults and weaknesses can be corrected only with prodigious effort and expense. It is for this reason that careful designing and layout are supremely important for telescope mountings.

Note for 1993 edition: Since the printing of the 1981 (German) edition of this *Compendium*, considerable progress has been made in the area of computational methods. Fast computers using the *method of finite elements* can now analyze the stiffness of complex parts and of entire structures so as to design a telescope with low demands on materials and yet be sufficiently precise for the stringent requirements of astronomical work. The previous edition has not been changed in principle, but rather partly supplemented and systematized. The sections on electrical equipment have been revised in order to present the basics of telescope drives and controls in a clear and systematic fashion. These sections present predominantly *block diagrams* without circuit details. As modern electronics also works with integrated blocks and modules, such diagrams should also suffice for practical realization. The external wiring of power and IC modules is generally very simple, using the data sheets and instructions supplied by the manufacturer.

5.2 Basic Types of Telescope Mountings

The basic parts of a telescope are the optical system and a mechanical support structure with two rotating shafts arranged at right angles to one another. With these shafts, the telescope optical system can be directed to any point on the celestial sphere above the horizon. Two types of axis orientation systems are distinguished:

5.2.1 The Alt-Azimuthal or Horizontal Mounting

This system has one axis oriented in the vertical direction, normal to the plane of the horizon. In the horizontal plane is the *azimuth circle* with the azimuth A. The other axis is horizontal and its rotation generates the *altitude circle*, with altitude or elevation a. This type is found in theodolites, large radio telescopes, and also in the "new-generation" large optical telescopes, since it is mechanically advantageous for such instruments. A typical amateur telescope with an alt-azimuth mounting is the well-known *Dobsonian mounting* (Fig. 5.8).

One disadvantage of the alt-azimuth mounting is that the tracking of the telescope to compensate for the Earth's rotation requires three complex, non-linear motions: the rotations around the azimuth and the altitude axes, and a rotation of the field of view around the optical axis. For large telescopes, these three motions are computer controled.

5.2.2 The Parallactic or Equatorial Mount

In this case, one axis, called the *polar axis* (or *hour axis*), is parallel to the rotation axis of the Earth pointing toward the celestial pole and carrying the *hour circle* with the *hour angle* h. At right angles to the polar axis is the *declination axis*, with the

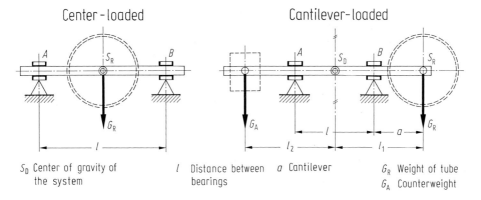

S_D Center of gravity of the system
l Distance between bearings
a Cantilever
G_R Weight of tube
G_A Counterweight

Fig. 5.1. Arrangement of load on a two-bearing shaft. A load G_R, such as the tube supporting the optical system, may be placed between the bearings A, B (*left*) or outside of them (*right*). These two cases classify the various types of mountings. The former case is *center-loading*; the latter, *cantilever-(end) loading*, requires a counterweight G_A when the system must be balanced with respect to the symmetry axis (dot-dashed line). This condition must be met with respect to the declination axis. The balance condition is given by

$$4G_R \cdot l_1 = G_A \cdot l_2. \tag{5.1}$$

Table 5.1. Types of axis systems for telescope mountings.

Type of Mounting	Polar Axis	Declination Axis	Counterweight
German mounting	Cantilever-loaded	Cantilever-loaded	Required
Fork mounting	Cantilever-loaded	Cantilever-loaded	Not required
English mounting	Center-loaded	Cantilever-loaded	Required
(Beam mounting)			
Horseshoe mounting	Center-loaded	Center-loaded	Not required

declination circle showing the angle of *declination* δ. The tracking with this arrangement needs only the polar axis to rotate with the constant angular velocity ω_p of the Earth's rotation.

A movable shaft in two bearings may support the weight either between the two bearings or external to them, as schematically shown in Fig. 5.1. The first case is known as *center loading* and the second as *cantilever (end)-loading*. The combination of these two kinds of loadings leads to the four basic types of telescope mounts, as shown in Table 5.1. Mountings with cantilever loading of the declination axis have the system rotatable around the polar axis unbalanced as the center of gravity of the tube is not in the intersection of polar and declination axes. The German and the English mountings thus require a counterweight G_A, which further loads the axes and structure, and also is disadvantageous for vibrational behavior. However, this should not be construed to mean that these two axial arrangements are unfavorable under all circumstances. A detailed comparison of all four systems shows certain advantages and disadvantages everywhere, although this may not be immediately apparent. The quality of a mounting is ultimately determined not by these pluses and minuses but rather solely upon the carefully considered and solidly founded design. Which type of mounting to choose is thus not terribly relevant, because, according to the basic rule of the designer:

"There are many ways to Rome. There is no royal way, and on every way, many false steps can be made."

The following sections aim to prevent such false steps.

Figures 5.2 to 5.5 graph the basic types; they can be modified in various ways. Pier, fork, and frame mountings can also be modified in many different ways which lead to interesting ideas of layout. Two known variations of the German mounting are the *knee mounting* after Repsold (Fig. 5.6) and the *Springfield mounting* after W. Porter (Fig. 5.7).

Especially worthy of mention are the *Folly mounting* and the *horseshoe mounting*, also designed by W. Porter. The latter has become known for its role in the design of the the famous Hale telescope on Mount Palomar in California. It is particularly suited for large telescopes but less so for amateur instruments. A very simple *fork mounting* was designed by J. Dobson, and has of late become quite popular among amateurs (Fig. 5.8). It is especially geared toward working with wood as its construction material. Generally speaking, axis arrangements which differ appreciably from the basic types are more prone to problems, and hence are more suitable for experienced designers and advanced telescope makers. The amateur who is not well versed in matters of mountings may advantageously use the *German-type mounting*. It is well known, widely used, not too difficult to build in a compact and rigid way, and is suited to optical systems with short and long tubes alike. Less well known is that the fork mounting, much used in large telescopes, is in some respects very problematic and in fact does not insure a more rigid, vibration-free instrument than do other types.

5.3 General Design Criteria

5.3.1 The General Mechanical Context and the Design Catechism

Consideration of the mounting design begins primarily with the discussion of the optical system and the loads to be supported. In addition, conditions relating to the intended application of the instrument, its placement, the housing, etc. must all be taken into account. Important items in this design manual are the limitations given by

1. available finances,
2. available sources of raw materials and components, and
3. facilities for manufacture and machining and the technical skills.

All three of these may confront amateurs with difficult problems. When drawing the layout, one should make certain that these items are concretely fixed and numerically founded. They should also be realistic and realizable within the available means. It is certainly no waste of time to fully explore such limitations before beginning the layout and starting with the construction.

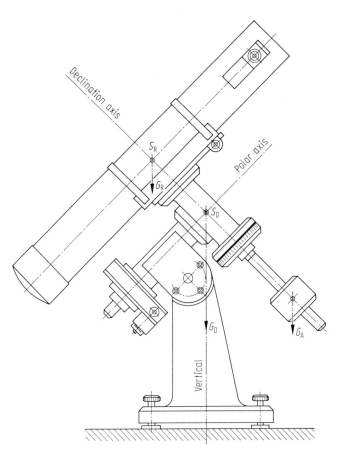

Fig. 5.2. German mounting. Classical design with a broad range of applications and suitable for long-focus refractors and Newtonian systems, as well as short-tube systems such as Cassegrain reflectors, Maksutov systems, astrographs, etc. covering the entire size range of amateur instruments. The configuration is statically good for medium geographic latitudes, less so for low latitudes. The axis system can be constructed in a compact and rigid way, thus compensating for the disadvantage of the counterweight. The axes and their casings are also simple and in manufacture are unproblematic structures that can be machined to any desired accuracy. The disadvantage is that a long, lower part of the tube is impeded near the zenith by the pier. The telescope must therefore be moved over from the west to the east position.

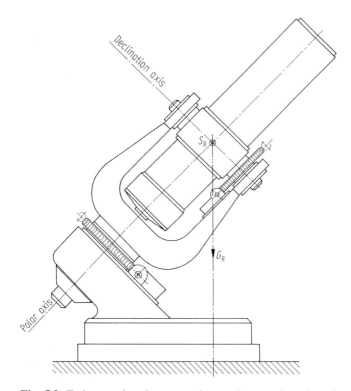

Fig. 5.3. Fork mounting. Its use on large telescopes has also triggered its increased use by amateurs, despite its drawbacks and problems. It has a favorable static configuration only at high geographic latitudes or as an azimuthal system. Even at middle latitudes, the rather unfavorable load conditions on the polar axis and fork, the low static stability (important for portable instruments), and parts that are problematic in machining are unfavorable forces outweighing the absence of a counterweight. Making the fork mounting with reasonable accuracy and bendinging-stiffness is exceedingly difficult, particularly in the declination axis bearings. At hour angles $\pm 6^h$, the force component $F_D = G\cos\varphi$ is loaded solely on one arm of the fork. In addition, a truly aligned and stiff design and construction of the two pivots of the axis at the tube is hard to reproduce. Fork mountings may be considered for small, transportable instruments with short tubes and without high demands concerning rigidity and mechanical accuracy. The faultless layout of large fork mountings requires a thorough knowledge of the theory of elasticity and extensive manufacturing facilities.

Figures 5.4 and 5.5. Cross-axis mounting and Yoke mounting. Both of these configurations are designed for use at relatively low geographic latitudes and for large reflecting optics of modest focal length. The frame and beam must be assembled from individual parts, which are likely to compromise stiffness at the joints and thus the precision. The beam or frame on two piers requires much more space and thus much more housing than a comparable German mounting. The total effort is considerable, machining not without problems, and applications are limited. Such drawbacks make these mounting types and their variations (horseshoe mountings) less attractive for use with amateur instruments.

Telescope Mountings, Drives, and Electrical Equipment 143

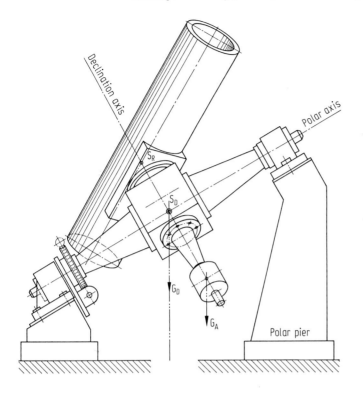

Fig. 5.4. Cross-axis mounting (for caption see page 142)

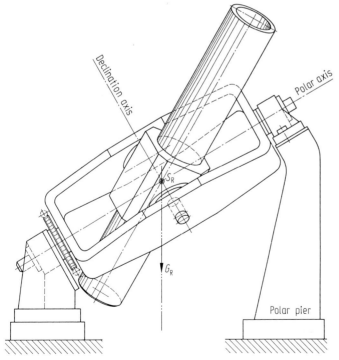

Fig. 5.5. Yoke mounting (for caption see page 142)

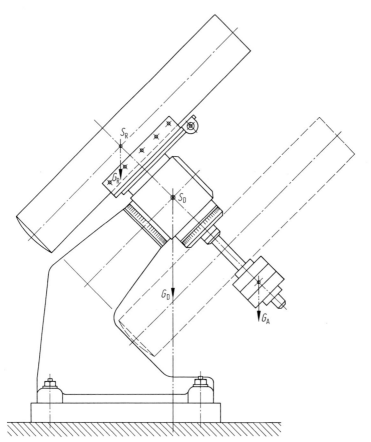

Fig. 5.6. German knee-mounting. This variant was first conceived by Repsold. Kinematic impediment of the tube's lower end near the zenith is avoided by a bent or "knee-shaped" pier. The arrangement is indicated primarily for photographic work on short-focus Newtonian systems and Maksutov and Schmidt reflectors. It is less useful with optics where the viewing is from the lower end of the tube (refractors and Cassegrains) and for systems with a greatly extended lower tube. The bent pier is more problematic with respect to stiffness, static stability, and construction than the straight pier of classical design. It needs very careful design and manufacture. Besides, anchoring of the pier to the foundation is generally required owing to the position of the center of gravity, so that this type of mounting is less feasible for transportable and unanchored instruments. Its static configuration improves with increasing geographic latitude.

Note 1 to the captions of Figs. 5.2–5.7: These figures refer primarily to amateur instruments and take into account the machining facilities generally accessible to amateurs. The explanations are not meant for very simple designs and constructions, for instance, of wood.

Note 2: The amateur is usually familiar with professional large telescopes only from their external, finished forms, but he has no access to blueprints and construction specifications. Thus merely copying the *form* which has been used in a large telescope is not terribly practical since the most important data are lacking. Such copying only rarely results in the amateur instrument fulfilling the builder's expectations. This is especially true for the cases of fork, frame, and horseshoe mountings.

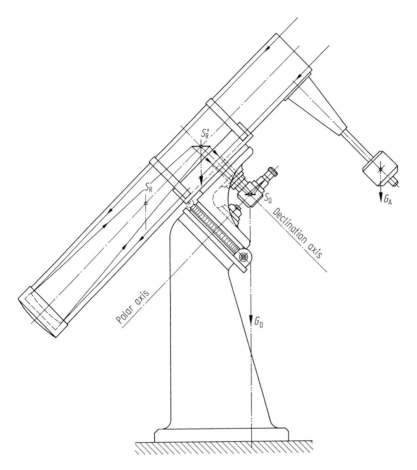

Fig. 5.7. Springfield mounting. Named after the town of Springfield, Vermont, where it was invented by W. Porter, this is a modified German mounting applicable to long-focus Newtonian systems (over f/10). The optical axis is deflected into the declination axis and is mirrored a second time at the intersection of declination and polar axes (the Coudé principle). The eyepiece thus remains in the same position and makes observing more physically convenient. This axis system is based on a disk arrangement according to Figs. 5.10 e,f or 5.10 h and can be manufactured quite compactly and rigidly by cast-, weld-, or epoxy-bonding techniques. The peculiarly placed counterweight not only compensates for the weight of the tube but also shifts its center of mass from S_R to S'_R. The double reflection and increased vignetting are optical disadvantages of this interesting mounting for amateur telescopes whose design and construction requires some experience and technical skill.

5.3.2 Basic Criteria for Telescope Mountings Regarding Statics, Kinetics, and Kinematics

In amateur terms, a mounting may be referred to as *stable* or *massive*, but these terms lack quantitative interpretation and measurable values, and thus are not useful as design criteria. Their place is taken by well-defined mechanical quantities from

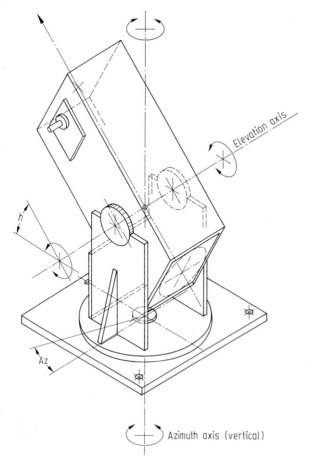

Fig. 5.8. Dobsonian mounting. The Dobsonian mounting is a fork mounting in an alt-azimuth arrangement and is, in static respects, the most favorable orientation of this configuration. J. Dobson gives a fine example for a suitable construction from wood. As such, it can be made with simple tools and at low cost, even for comparatively large Newtonian systems (> 300 mm). On the other hand, the use of wood as the construction material and the alt-azimuth positioning limit the application. Its use is predominantly for visual observations. The *altitude* or *elevation axis* is supported in a simple cradle bearing, and the azimuth circle rotates about a pinion. The bearings are coated with teflon, which permits smooth and stick-free motions. The axis system demands no particular orientation. Setting and guiding on astronomical objects is done manually.

which conclusions on the admissible deviation of images formed in the image plane can be derived.

An image remains fixed when the object does not move and when no forces act on the telescope. The instrument, however, is always subject to its own weight and various other external forces; in addition, the telescope optical system must follow the celestial motion precisely. Apart from kinematic drift effects, this also changes the orientation of components with respect to gravity. All of these factors can in various ways cause images in the focal plane to drift or vibrate.

When designing a telescope for a particular application, the first requirement is to have a firm idea of how much displacement is tolerable. One should not, however, forget that design problems, mechanical demands, and costs increase exponentially with desired precision. Calling the admissible image displacement x_0, the basic criteria for a telescope mounting can be formulated straightforwardly.

> **The Static Criterion.** Static forces shall not cause displacements exceeding a certain value x_0 in the focal plane.

With respect to kinetics, only vibrations are considered since a mounting does not admit to rotary or translatory motions other than the tracking. It is improper to regard a mounting as a *freely movable mass* in order to deduce from Newton's principle of action or the theorem of momentum that it must be "heavy." Even a mounting without anchoring is considered to be a *system at rest*. Its orientation in space must be conserved. A minimum requirement is that after a displacement it returns exactly to its original position.

> **The Kinetic Criterion.** Vibrations resulting from any source shall not excite amplitudes over $x_0/2$ in the focal plane and should be attenuated as quickly as possible.

> **The Kinematic Criterion.** Image drift in the focal plane caused by kinematic inaccuracies shall not exceed the amount x_0 for a given observing time t_b.

The following sections will show which quantities cause these displacements and how the design and layout of the parts can keep them within the desired limits.

5.4 Static Criteria of Telescope Mountings

5.4.1 Stiffness as the Static Characteristic of Telescope Mountings

Criteria I and II already imply the quantity *stiffness* (or *rigidity*) c, which is all-important for telescope mountings. A force acting on an elastic body or structure[1] always causes a displacement x at the point of contact, or more generally, a *deformation*. Stiffness is the resistance[2] of the body against elastic deformation. The force F, the displacement x, and the stiffness c are connected by the simple relation known as

---

1 Elasticity, like mass, is an elementary property of matter. All bodies possess some finite degree of elasticity.

2 To be precise, stiffness is a *tensor* represented by a three-dimensional matrix. In the theory of elasticity, stiffness is connected with the stress tensor via the elastic material constants, *elasticity modulus E* and *Poisson number ν*. The inverse quantity $1/c$ is called *compliance*.

the *force law for springs*, or *Hooke's law*,

$$F = c \cdot x, \qquad (5.2)$$

where F is the force in Newtons (N), x the displacement in meters (m), and c the stiffness in $N \cdot m^{-1}$.

Stiffness is paramount in the design of mountings for two main reasons:

1. Stiffness determines the layout of every part so as to satisfy criterion 1.
2. It also plays an important role in vibration, and permits the vibrational behavior of the telescope to be controlled in a specified way; cf. Sect. 5.9.

A convenient unit for stiffness is the megaNewton per meter (MN/m). Thus, a mounting is said to have a stiffness of 1 MN/m when a force of 10 N causes a displacement of 0.01 mm, as, for instance, for stellar images on a photographic plate.

A telescope mounting is a structure, that is, it consists of a large number of different elements which are connected by discrete joints such as bolted joints, welded joints, slip-joints, bearing, etc. Each element and each joint can be assigned an individual stiffness c_{ki}, where the quantity i in the subscript denotes the ith element. If the elements are connected consecutively in series, the system stiffness c_{ks} can be calculated by the addition theorem of stiffness[3]:

$$\frac{1}{c_{ks}} = \frac{1}{c_{k1}} + \frac{1}{c_{k2}} + \frac{1}{c_{k3}} + \cdots + \frac{1}{c_{kn}}, \qquad (5.3)$$

where n denotes the total number of elements. Note that the individual stiffnesses c_{ki} are defined in the same state k of load. Consider a force which affects the instrument at the focal plane in an assumed direction. The effect of the force would then propagate from element to element down to the foundation, and thus load each element in a characteristic manner. Depending on the direction of force, different system stiffnesses will result. For a mounting, the important quantity is merely the direction of the *minimum* stiffness, for the following reasons:

1. While the instrument is in use, every manipulation gives rise to forces. These forces and impulses are not in any specific directions, and it would be quite coincidental if they occurred in directions of high stiffness. The same holds for the force of wind on unhoused instruments.
2. Vibrations are not limited only to certain directions. They occur predominantly in the direction of lowest stiffness, and thus cause the greatest deflections in the focal plane.

The addition theorem of stiffnesses, Eq. (5.3), predicts two important consequences for the layout of telescope mountings.

> **Criterion 1.** The individual stiffnesses of n series-connected parts and joints must be n times larger than the required system stiffness.

---

[3] To be precise, the stiffness *matrices* of the element should be added. Equation (5.3) below neglects the lateral stiffness, whose influence is generally small enough to be ignored. Exceptions are bearings where the lateral effect may be substantial.

Table 5.2. Guidelines for the system stiffnesses (in MN/m) of amateur telescope mountings. The stiffness is measured at the tube near the position of the focal plane. After design and calculations by the author.

Field of Application	Mirror Diameter in mm		
	120–150	150–200	200–300
Demanding projects, photoelectric photometry, long-exposure photography, sturdy instruments for public education, mountings anchored on heavy foundations.	2–6	5–15	15–45
General work, less critical concerning image placement, requires more careful operation, mounting anchored on foundation.	0.5–2	2–6	5–15
Light and portable instruments, on tripod, operation subtle or tricky, sensitive to wind and vibrations	< 0.5	0.5–2	2–6

Criterion 2. The stiffnesses of parts should be approximately equal. Equation (5.3) demonstrates that just one weak element with low stiffness in the structure will suffice to adversely affect the system stiffness, whereas oversized other parts will not improve it appreciably.

The following example will readily demonstrate these principles:

Consider a mounting consisting of 10 parts and 10 joints. To neglect the stiffness of the links would be unrealistic. They play an important role especially in the total structure.

Case A. All parts and joints have a stiffness of 10 MN/m. Equation (5.3) results in a system stiffness of $c_{ks} = 0.5$ MN/m which may have been desired.

Case B. Similar to Case A, except that of the 10 parts, the two axes and the pier are massively oversized with 100 MN/m. The result is $c_{ks} = 0.58$ MN/m. The three overproportioned parts thus yield only an insubstantial improvement in system stiffness.

Case C. Similar to Case B with three oversized elements, but two screw joints among the 10 links are assumed to have a stiffness of 0.5 MN/m. The system stiffness is this time found to be $c_{ks} = 0.18$ MN/m. That is, the two weak joints have diminished the system stiffness by 79% and the three oversized parts could not nearly compensate for the two weak screws.

The above example is analogous to the proverb that a chain is only as strong as its weakest link.

Table 5.2 compiles guidelines for the system stiffnesses of amateur telescope mountings.

Table 5.3. Stiffness of a one-end fixed beam.

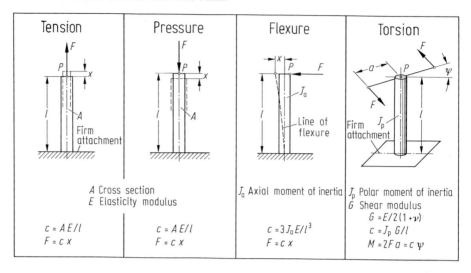

Tension	Pressure	Flexure	Torsion
A Cross section E Elasticity modulus		J_a Axial moment of inertia	J_p Polar moment of inertia G Shear modulus $G = E/2(1+\nu)$
$c = AE/l$ $F = cx$	$c = AE/l$ $F = cx$	$c = 3J_a E/l^3$ $F = cx$	$c = J_p G/l$ $M = 2Fa = c\psi$

5.4.2 Stiffness as a Design and Layout Quantity

The stiffness of an element depends on its geometric form, the material properties (elasticity modulus E, Poisson number ν), and the conditions at its fixation point. It does not depend on the acting force. Fast computers permit the calculation of the stiffness matrix via the *method of finite elements*, even for complex shapes and structures. This procedure, however, is quite demanding, not very instructive, and of little application to amateur work. The simple classical deformation formulae for "beam-like" elements[4] show instructively what is relevant for the rigid design of a telescope mounting. They also permit the calculation of the more important parts so as to satisfy the requirements. Table 5.3 lists the stiffness formulae for the four basic types of stress on one-end fastened beams. Certain engineering handbooks[5] such as the *Handbook of Chemistry and Physics* [5.2] give the corresponding formulae of stiffnesses for numerous other types of loading and fixation conditions.

The formulae of Table 5.3 show that for a given length and cross section of the beam, the flexural stiffness is smallest. Hence, this stiffness of the parts is primarily significant for the telescope mounting, and should determine their layout. If all elements have sufficient flexural stiffness, then they will not become weak parts under tension, compression, or torsion either. This gives the following important principle.

---

[4] Essential parts of mountings such as tube, axis, pier, etc. fall naturally into the category of "beam-like" elements. As a rule, even more complex parts can be subdivided into beam-like or plate segments so that some of the stiffnesses there can be calculated by the simple formulae, and then added up by the addition theorem.

[5] Note that some sources list the flexures f which are inversely proportional to the stiffness c.

Table 5.4. Flexural stiffness in some important cases of stress for telescope mountings.

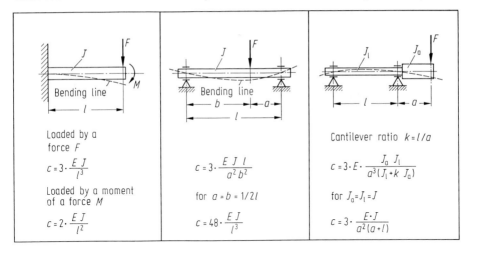

Loaded by a force F $$c = 3 \cdot \frac{EJ}{l^3}$$ Loaded by a moment of a force M $$c = 2 \cdot \frac{EJ}{l^2}$$	$$c = 3 \cdot \frac{EJl}{a^2 b^2}$$ for $a = b = 1/2\,l$ $$c = 48 \cdot \frac{EJ}{l^3}$$	Cantilever ratio $k = l/a$ $$c = 3 \cdot E \cdot \frac{J_a J_1}{a^3 (J_1 + k\, J_a)}$$ for $J_a = J_1 = J$ $$c = 3 \cdot \frac{E \cdot J}{a^2 (a + l)}$$

Criterion 3. The supreme guideline in the design and layout of telescope mountings is high flexural stiffness of the structure.

It remains now to be shown how this may be achieved.

The formulae for flexural stiffness (Table 5.4) include three variables of layout:

1. The flexing length l (a geometrical property).
2. The area moment of inertia J (a geometrical property).
3. The elastic modulus E (a material property).

The quantities E and J appear in the numerators of the formulae for c, and the design goal should be to maximize them. The flexure length l appears in the denominator and is cubed! Special attention must therefore be paid to this quantity. Any attainable shortening of a particular part during layout will lead to a substantial increase in its stiffness. A structure consisting of elements designed as short as possible will appear in its entirety to be compact and sturdy. The words "compact" and "sturdy" in this context are not meant to be exact physical terms, but they are illustrative words which describe the overall impression of a well-designed telescope mounting. (This fact can also be expressed in the following way: In a flexure-resistant structure the centers of mass of the individual parts are always relatively close together.) This gives rise to another principle.

Criterion 4. All cradles and cantilever arms in the support structure of a telescope mounting should be made as short as possible.

Table 5.5 gives the area moments of inertia J for various cross sections.
The defining relation
$$J = \int y^2 \, dA$$

Table 5.5. The area moment of inertia.

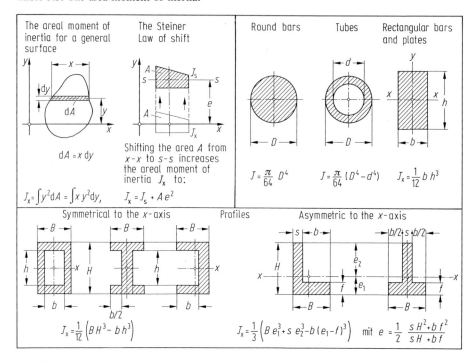

reveals that what is important regarding the area moment of inertia is not the total cross section A but the rather large distances y of the area elements dA from the bending axis x–x, because they enter in the square. The interpretation is that material near the axis and in the center does not contribute substantially to the moment of inertia and hence to the flexural stiffness. All "solid sections" (see Table 5.5) contain much superfluous material in their centers. Omitting the material around the axis gives the "hollow sections" (box and tube). Reduction of the material near the axis to narrow webs leads to the known "I," "Tee," and "channel sections." These examples show how relatively small amounts (and low masses) of material can be used to make cross sections with large area moments of inertia.

Criterion 5. All parts of the support structure of a telescope mounting should be designed with the area moments of inertia as large as possible.

Criterion 6. Statically speaking, a telescope mounting need not be massive but should be rigid. The amount of weight needed for satisfactory stiffness depends only on the skill of the designer in effecting short cradles and cantilever arms with high moments of inertia.

5.4.3 The Modulus of Elasticity

The *modulus of elasticity*, or *E-modulus*, should not be confused with the *tensile strength*, the *yield point*, or the *hardness* of a material. In numerous technically important metals such as iron and aluminum, alloying constituents increase the strength and hardness, but have practically no influence on E. When it comes to elastic properties, it may be surprising that soft iron and very hard chromium-nickel steel, or soft aluminum and the very hard alloys of light metals, have the same E and therefore identical deflections. It is important to distinguish between, on the one hand, objects for which strength and fracture resistance are the relevant quantities (bridges, cranes, wheel axles, etc.), and on the other hand, where it is the stiffness that is critical (telescope mountings, precision measuring instruments, etc.). As far as stiffness is concerned, materials of high strength and hardness are not needed, but the use of very soft materials (e.g., pure aluminum) is not feasible either because of their low resistance to scratches, dents, and wear, and also because such materials are often difficult to machine. As a rule, harder and stronger materials can be more easily machined than can soft ones. Table 5.6 compiles some of the relevant properties of some materials of interest for telescope mountings. It is seen from the E-moduli that for equal stiffnesses, a light-metal construction must have an area moment of inertia three times larger than a steel construction; for wood, it must be 15 times larger. The steel version thus requires the smallest thickness of the walls of the component parts. Also of note is the very low E-modulus of bakelite (for telescope tubes) and of plywood plates, which, because of their resistance to warping, are preferred in wooden telescope mountings.

It is hardly possible to single out one material as the best for amateur instruments as too many technical and individual conditions will affect the choice. But Table 5.7 lists a few factors to which some thought should be given.

5.5 Shafts and Bearings

5.5.1 Stiffness of Bearings

In every mounting, shafts and their bearings are significant and critical parts. In addition to stiffness, precision, and freedom from play, a few other important design criteria will be presented here. These are consequences of the Criterion III on kinematics, but they also enter as parameters into the inner stiffness of bearings.

The overall stiffness of the shaft is composed of, according to the $1/c$ rules, the stiffness of the shaft, the inner stiffness of the bearings and that of the bearing casing. Figure 5.9 shows in schematic form the individual stiffnesses of the shaft on bearings. Again, it is seen how important short cradles and cantilever arms are since they not only reduce the bending stiffness of the shaft with the third power but also enter into the formulae for the bearing stiffness. Consider for the bearing casing the bending stiffness and the buckling stiffness. The inner bearing stiffness c_L is a rather complex quantity depending on the following factors:

1. The type of bearing (ball bearing, friction bearing, hydrostatic bearing, etc.).

Table 5.6. Properties of some important mounting-construction materials.

(European) Trade Name	Kind	Semi-finished form	Density ρ kg m^{-3}	Elastic modulus E 10^3 N mm^{-2}
St 37	standard structural steel	rolled bars all profiles	7850	206
C 35, CK 35	standard steel rounds	rolled bars	7850	206
ST 34 BK	cold-rolled drawn bars	all profiles	7850	211
St 35 BK	cold-rolled drawn bars	precision steel tubes	7850	211
9 S Mn Pb 28 K	drawn steel (machining alloy)	bars	7850	211
X 12 Cr Mo S 17	stainless drawn steel	bars	7750	206
X 10 Cr Ni Ti 189	Stainless steel (St S 18/8)	bars, sheets	7900	196
GG 20	grey cast iron	diverse shapes and parts	7200	110
Cu Zn 40 Pb 3h	cold drawn brass	all profiles and tubes	8500	93
Al Mg Si 1 hbd	anticorodal	bars, tubes	2700	69
G Al 12 Si heat treated	aluminum cast alloy	Silafont 1, Silumin	2650	73
G Al 10 Si Mg heat treated	aluminum cast alloy	Silafont 3, Silumin hypo-eutectic	2650	73
	Non-metallic materials			
Hp 2065	Bakelite, Pertinax Delit, Resocel 81	Fiber-reinforced sheets, tubes phenolic	1100–1200	5–8
Hp 2066, Hp 2067	Bakelite, Pertinax Delit, Resocel 82	Fiber-reinforced sheets, tubes phenolic (dense)	1100–1300	7–9
Hgw 2375	Vetresit 52 Delit, Resocel 82	glass-filled epoxy sheets, tubes, shaped parts	1700–1900	13–15
Fir wood	loaded parallel to grain at 12% moisture content (dry air)	boards, brackets	470 330–680	10.8
Birch wood	loaded parallel to grain at 12% moisture content (dry air)	boards, brackets	650 510–830	16.2
Red beech wood	loaded parallel to grain at 12% moisture content (dry air)	boards, brackets	720 540–910	15.7

Table 5.6. (continued)

(European) Trade Name	Yield Strength R_e N mm^{-2}	Tensile strength R_n N mm^{-2}	E/ρ 10^4 N g^{-1} mm	Brinnell hardness	Machining cutting, lathing, boring, scraping,
St 37	215	365–440	26.2	140	good
C 35, CK 35	295	490–635	26.2	160	good
ST 34 BK	390	490	26.9	140	very good
St 35 BK	295	440	26.9	150	very good
9 S Mn Pb 28 K	225	370	26.9	140	excellent
X 12 Cr Mo S 17	440	690–835	26.6	210	very good
X 10 Cr Ni Ti 189	205	490–735	24.8	180	poor
GG 20	—	196–735	15.3	160	good
Cu Zn 40 Pb 3h	290	390–460	10.9	100–125	very good
Al Mg Si 1 hbd	95	195–265	25.6	60	very good
G Al 12 Si unanodized	75–90	165–215	27.6	50	fair
G Al 10 Si Mg anodized	175–245	235–295	27.6	75	very good
Non-metallic materials			Bending stiffness		
Hp 2065	100	50	5.7	—	good
Hp 2066, Hp 2067	120–140	40–60	6.7	—	good
Hgw 2375	400	180–220	7.8	—	poor texture!
Fir wood	76 48–133	88 22–240	23.0	—	good
Birch wood	144 75–152	134 34–265	25.0	—	very good
Red beech wood	120 73–206	132 56–176	22.0	—	very good

Table 5.7. Guide to the selection of materials

Considerations	Aspects	Comments, examples, materials, properties limiting use
Primary demands	Rigidity, E-modulus	High demands (cf. Table 5.6); heavy accessories: high E-modulus, hardness, no wear; precise machinability: steel metals
	Accuracy	Lesser demands: Dobson and similar telescopes: wood, plastic
	Robustness	Rough handling (public; portable instruments): stability needed; steel, metal Careful handling: permits use of softer materials and their favorable properties
	Weight, portability	Higher weight permitted (larger, permanently place instruments): steel, metal, concrete pier Low weight indispensable: wood, light metals, plastics
	Price	Possibly expensive: stainless steel, aluminum alloys, brass, bronze, hardwood, high-quality plastic Inexpensive: standard steel, softer wood, plywood (low E-modulus!), cheaper plastics, cardboard, concrete
	Durability	Prototypes, improvisations: wood, cardboard, plastics Long life: corrosion-proof metals and all alloys
Sources	Broad spectrum of supplies	Cities, wholesalers, connection to manufacturer: free to select best material
	Limited spectrum	Rural, no connections to supplier: limits materials and feasibilities of design and construction
Available machines	for wood, plastic	Access to tools and woodworking machines; choose material and design accordingly;
	For metals	Access to metal machining (lathe, milling machines and tools); choose material and design accordingly
	For welding joints	Access to autogeneous/electro-welding; consider materials that are weldable
Knowledge	Of engineering Of materials Of handicract	In principle choose materials with whose constructional properties and machining you are best acquainted
Weather/corrosion	Climate, humidity, pollutants	Placement in the open/in shelter; quality of shelter corrosion, rust; swelling/shrinkage of wood and many plastics increased corrosion by pollution (seaside); sweat at handles and button-knobs

Table 5.7. (Continued)

Considerations	Aspects	Comments, examples, materials, properties limiting use
General compatibility of materials	Thermal expansion	Widely different expansion may cause distoriton, misadjustment, optical decollimation (mirror cells), loose seats
	Intermetal corrosion	By combining metals of different electro-chemical potentials. The part *more* exposed to corrosion should be electro-negative, or to the *left* of the other part in the galvanic sequence: −Ag,Bronze,Brass,Cu,Ni,AlCuMg/neutral/V2A,steel,AlMgSi,Al,An,Pb,steel,Zn +.
	Slide/friction properties, wear	To watch for bearings and for sliding parts. Slip conditions, friction, and wear properties are very different: Good: steel/bronze,steel/brass, metals, and some plastics (teflon, delrin) Poor: steel/light metal, lightmetal/lightmetal, stainless steel, Ni coating
	Creep properties	High in plastics, light metal alloys, Cu; low in steel and cast alloys
	Adhesion of coatings	Some light metal alloys are poor for galvanoplastic coatings, many plastics poor for paints; Wood needs special impregnation and paint when exposed to weather and moisture
Compatibility in links (joints)	Rigidity different E-moduli, different hardness	Metal/wood connections are problematical Screws in wood, plastic, or soft metal (pure Al) are not rigid enough and "tear out" Tight fittings of different materials problematical
	Welded links	Not all steel alloys are weldable. Some cast alloys are not weldable, some light metals and stainless steels are only with special methods. Tempered alloys soften at the seam
	Soldered links	Some materials are difficult to solder, or need special solders and fluxing agents (bronze, stainless steel, grey cast). Fluxing agents increases corrodibility of steel, brass and alloys
	Bonded links	Some combinations are poor (some plastics; wood on metal). Avoid too different hardness and expansion coefficients. Chemical compatibility of glue and glued part (plastic).

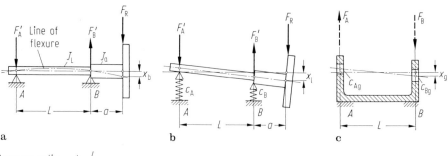

a

Laverage ratio $\quad k = \dfrac{L}{a}$

Reactive forces
of the bearing $\quad F_A = F_R \dfrac{a}{L} \; , \; F_B = F_R \dfrac{(L+a)}{L}$ (5.4)

Inflexibility of shaft c_b $\qquad$ Rigidity of bearing c_l $\qquad$ Rigidity of casing c_g

$$c_b = \dfrac{3E}{a^3} \dfrac{J_L J_a}{(J_L + k J_a)} \qquad \dfrac{1}{c_l} = \dfrac{1}{c_B}\left(1+\dfrac{1}{k}\right)^2 + \dfrac{1}{c_A}\left(\dfrac{1}{k}\right)^2 \qquad \dfrac{1}{c_g} = \dfrac{1}{c_{Bg}}\left(1+\dfrac{1}{k}\right)^2 + \dfrac{1}{c_{Ag}}\left(\dfrac{1}{k}\right)^2 \quad (5.5)$$

Overall rigidity $\quad \dfrac{1}{c_s} = \dfrac{1}{c_b} + \dfrac{1}{c_l} + \dfrac{1}{c_g}$ (5.6)

Fig. 5.9a–c. The stiffness of a pivoted shaft. The total stiffness is given by the flexural stiffness c_b, the inner bearing stiffnesses c_A, c_B, and the stiffnesses of the bearing casing frame c_{Ag}, c_{Bg}. In **b**, the bearing stiffnesses are symbolically represented as springs. The formulae show the significance of the cantilever arm A not only for the flexure of the shaft but also for the stiffness of bearing and casing. Amateur-made mountings frequently show ample and even much oversized shafts, whereas the stiffnesses of bearings and housings are often ignored.

2. The dimensions and geometry of the bearings.
3. The internal fit and the amount of bearing clearance.
4. The external precision of the shaft fits and the case bores.
5. The axial and radial preload (of antifriction bearings).

Figures 5.10a–h show the various possibilities of the bearing arrangement of the polar axis. The variation schematically shown with friction bearings (a, b, d, and e) may also be executed with corresponding roller bearings as shown under g and h. The overall stiffness of the design is determined from Fig. 5.9 by the stiffnesses of the shaft (disks in e and f), the casing frame, and the inner stiffness of the bearings. These individual quantities vary considerably over the designs shown. For instance, in the Porter's Folly mounting (c) the stiffness of the polar axis and that of the supporting structure is very high, but the rollers upon which the polar axis cone rests can hardly be made with adequate stiffness (except when – as in large telescopes – hydrostatic bearings are used in place of the rollers). The disk design with a four-point ball bearing (h) is a rigid structure in every respect. For any arrangement it is important to recognize the points that are critical for the stiffness and to work out the design accordingly.

5.5.2 Stresses on the Declination Axis

In a balanced system, the weight G of the moving, rotating parts is a spatially fixed, constant vector composed of the force components

$$F_P = G \sin \varphi \quad \text{and} \quad F_D = G \cos \varphi, \tag{5.7}$$

where φ = latitude, F_P = component of force in the direction of the polar axis, and F_D = component of force in the equatorial plane. Evidently, the position of the declination axis relative to F_D changes with the hour angle h. When the declination axis is horizontal (at hour angles of 0^h and 12^h), F_D causes radial stresses on the declination bearings, and the axis is stressed in flexure. In the positions $\pm 6^h$, the twisting moment of the component F_D vanishes (but not so for F_P), thus creating an axial load on the bearings. These variable effects cause varying deformations.

Older telescope designs compensated for this effect by the use of elaborate weight/lever "relieving arrangements."[6] The modern design philosophy is to keep deformations as small as possible by a very rigid structure, and to correct the remaining deformation errors automatically with a control computer. Position-dependent deformations are usually quite pronounced in fork-type mountings unless preventive measures are taken. Some hints follow:

At the horizontal position of the declination axis, each fork arm is transversely loaded with $1/2\ F_D$. At $\pm 6^h$, the entire bending load is carried by only one arm. In principle, the axial load component cannot be equally distributed on both fork arms (the statically indeterminate case). Hence, in this position, one arm of the fork can be omitted without harm, thereby resulting in the so-called "half-fork mounting." This mounting, which is statically determinate, can, with adequate stiffness, be realized. Also, a fork arm can be designed such that flexure stiffnesses are equal in all orientations of the tube (*isotropy of stiffness*).

There exist design possibilities in which both fork arms are equally loaded or in which a high degree of stiffness is achieved in all orientations of the tube. Briefly, they are:

- A powerful prestressing of the fork arms ($F_\text{stress} > 5$ to $10 \cdot F_D$), which requires a specially designed bearing arrangement at the axles.
- Forming a rigid frame structure to achieve high integral stiffness. This also requires specially made bearings and a rigid central yoke (crossbar).

As far as is known, these designs have not been applied in amateur fork mountings or in commercially available amateur instruments.

---

6 Such weight/lever relieving systems, such as the refined and elaborate mechanics of F. Meyer of the Zeiss company, were extensively used in older telescope mountings and transit circles. A well-known example was the celebrated *Treptow refractor* on a fork mounting with relieving systems on both axes.

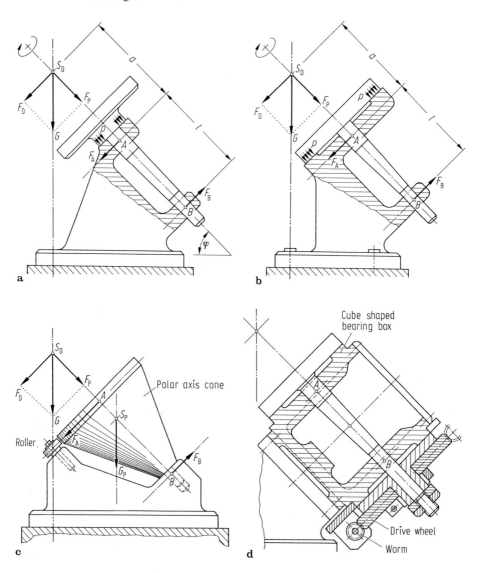

Fig. 5.10a–d. Various forms of polar axis design. The arrangement of bearings is loaded by the axial force $F_P = G \sin\varphi$ and by the moment $M = a\,G\,\cos\varphi$. Primarily M is relevant for the displacement. In **a**, F_P is transferred through the shaft shoulder onto the support. The bending moment must be absorbed by the shaft. With peripheral support of the flange **b**, the mounting also absorbs part of the moment. This increases the stiffness. *Porter's Folly mounting* **c** possesses very high stiffness of the polar axis cone. Weak parts of this concept are the support rollers. The stiffness of ball bearings in this arrangement is small. Also occurring are *Herz deformations*, bending, and kinematic displacement. The total stiffness is thus not higher than with other optimally designed shaft and bearing arrangements. **d** and **g** show cube-shaped casings (the *Badener cube mounting*). The hollow cube has high stiffness in all directions and its favorable shape with respect to machining processes accomodates the rectangular axis system ideally.

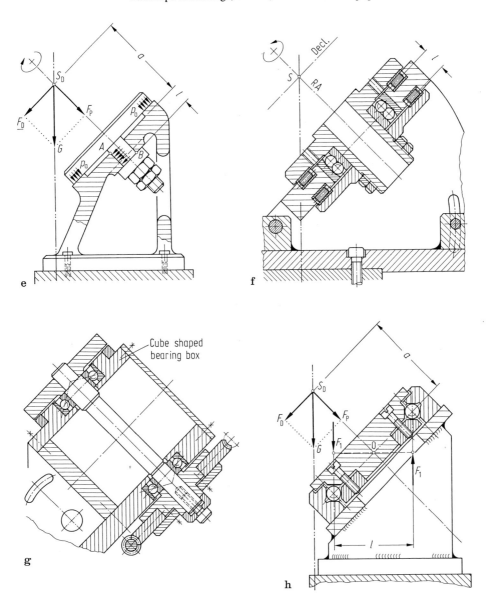

Fig. 5.10e–h. Design arrangements for the polar axis. For reasons of stiffness the distance l between bearings is short. However, it can be reduced further, leading to the disk designs **e**, **f**, and **h**. These transfer the moment M directly from the disk to the support structure. Since the disks can be made very rigid, such designs can reach very high overall stiffnesses. **f** is a design with two axial needle thrust bearings and one ball bearing; **g** is an epoxy-bonded cube construction with preloaded angular-contact ball bearings. **h** shows a do-it-yourself, preloaded four-point ball bearing design which has a highly isotropic stiffness, and can support forces and moments in any direction. Where minimum weight and space are desired, this is the most rigid and vibration-free bearing arrangement which, in addition, can be made to very high precision.

5.5.3 Friction Bearings

The advantage of friction bearings compared with antifriction bearings is their somewhat better ability to damp vibrations. However, it is more difficult to achieve high bearing stiffness and freedom from play, both of which require a very small bearing clearance and a very high accuracy of the sliding surfaces. In order that closely and accurately fitted bearings do not become pitted, suitable materials for shaft and bushings and special grease are required. Steel on bronze slides well, whereas stainless steel, most light metal alloys, and nickel-coated surfaces tend to become pitted. A good grease for all sliding surfaces in mountings, including worm gears, is Corning's *Molycoat Longterm 2*, a lubricating grease with a molycoat base.

5.5.4 Antifriction Bearings

Antifriction bearings are very attractive for use as elements in amateur telescope mountings. There are various types and versions with different properties. The following criteria may be used when choosing them:

1. Requisite stiffness in radial and axial directions.
2. Freedom from play.
3. Requisite precision (e.g., the concentricity error).
4. Constraints of price and of design and constructional demands.

Telescope shafts can be placed on bearings such that:

– the radial and axial forces are supported by separate bearing units,[7] or
– roller bearings supporting both radial and axial forces are used. These are normal ball bearings if the axial forces are not too large, but *angular-contact ball bearings* and *tapered roller bearings* otherwise.

Bearing designs of the first kind are rather expensive and thus not feasible for amateur mountings. The second variant is much simpler and raises the question of which kind of bearing is preferred. In order to properly answer that question, the following hints are provided:

1. The bearing size d is determined by the shaft diameter and not by radial and axial loads (weights and axial initial stressing forces). The bending stiffness of the shaft is the quantity which determines the leverage. The bearings thus determined are generally not fully utilized with respect to their *static load rating* (see suppliers' catalogues) nor with respect to their stiffness. A normal axially prestressed ball bearing has sufficient stiffness for use in amateur instruments. Even the light load series with serial numbers 160..., 60..., and 62... can be used; heavy ones are

---

[7] The first variant is at present used only in the very precise, rigid, and low-vibration machine-tool spindles (e.g., lathes, drilling machines, and milling machines) while the angular-contact-bearing arrangement is preferred in high-precision grinding spindles. The latter are at present among the most precise made.

not necessary. From the viewpoint of stiffness, tapered roller bearings (special heavy-duty bearings) are not required.

2. Precision and price: All types and variants of antifriction bearings are available in the precision grade *normal* (ISO-0), which will certainly suffice for most amateur mountings since the bearing precision is much better than the accuracy with which the average amateur constructor can machine the other parts of the axis system. Ball bearings of the normal accuracy grade are the least expensive elements overall, and the tapered roller bearings the most costly. Indeed, should high precision be demanded, the angular-contact ball bearings of the precision grades P6, P5, and P4A can be used.[8] Buying high-precision bearings does not in itself complete the task. Bearings require a very careful overall design (shaft, bearing, bearing housing). The shaft and case fits must be made with the same accuracy and tolerances as the bearings and be exactly aligned in the bearing frames. The latter condition is in general the most difficult to realize.

5.5.5 Stiffness of Antifriction Bearings

The stiffness of a mechanical structure is, in the *elastic range*, independent of the force, although for the antifriction bearings it is a nonlinear quantity depending on the radial bearing load F_r and the initial stressing force. The radial displacement δ of standard ball bearings can be calculated with sufficient accuracy from the *Lundberg–Stribeck formula*:

$$\delta = \frac{1.28 \times 10^{-3}}{\beta} \cdot \sqrt[3]{\frac{F_r^2}{D_0 z^2}}, \tag{5.8}$$

where δ is the radial displacement (in mm) of the shaft center, z the number of balls, F_r the radial bearing load, D_0 the ball diameter (in mm), 1.28×10^{-3} a dimension factor dependent on bearing shape, and β a dimensionless "spring" coefficient ($\beta = 0.42$ for an axially optimally prestressed ball bearing). The bearing stiffness c_l is then F_r/δ.

The axially not prestressed ball bearing has some *bearing clearance* (positive clearance). An axial initial stressing force compresses the balls slightly and elastically (negative clearance). Thus, all balls come into load contact, while in the not prestressed case only a few balls carry in the direction of the force. The spring coefficient β implicitly contains the negative clearance caused by the the prestress. This coefficient is strongly nonlinear, so that a prestress beyond a certain value only slightly increases the stiffness but drastically increases the friction, diminishes the accuracy of motion, and can even damage the bearings. Ball bearings should never be violently prestressed by forces and impacts. The axial stress for the annular ball bearings should not exceed the value $0.5 \, C_0$, where C_0 is the static load rating for the bearing from the suppliers' data sheets.

---

8 These grades follow the SKF nomenclature.

5.5.6 Stiffness and Distance Between Bearings

The opinion is widespread that in the German mounting (Fig. 5.2), the bearing distance l should be made very large. This is the case in many amateur mountings. A large distance between bearings, however, is disadvantageous with respect to stiffness. Taking the derivative of Eq. (5.6) with respect to dl and setting it equal to zero enables the optimum bearing distance to be calculated. Also in this case an optimum stiffness layout leads to surprisingly short bearing distances and thus to quite compact forms of bearing casings, as realized, for instance, in the cube mountings of Fig. 5.10 d, g.

A larger distance between bearings may be feasible at most on the grounds of the concentricity error of the shaft/bearing arrangement. Every ball bearing has some of this error Δr which causes the polar axis to describe a cone of aperture angle ξ around the celestial pole. In the least favorable case

$$\tan(\xi/2) = \frac{\Delta r_A + \Delta r_B}{l}. \qquad (5.9)$$

It makes little sense to try to reduce this error at the expense of stiffness by using a large distance between the bearings. It is so small already for bearings of the standard accuracy grade that it hardly affects amateur mountings, especially compared with other errors. If high demands are made on accuracy, then the use of precision angular-contact ball bearings may be advisable, but it should again be pointed out that this is warranted only if all other parts affecting the precision can be produced with the same accuracy.

5.6 The Foundation and Stability Against Tilting

The joint between mounting and foundation is the last part in the chain of stiffnesses. This joint should never be the weak one in the system, but in amateur designs it often is. Sometimes the instrument rests on three scantily dimensioned footscrews. In such cases one has only to compare the cross sections of the shafts with those of the screws in order to see what little sense this scheme makes. Consider also that these joining elements suffer the largest stresses. To achieve a rigid coupling between the mounting and foundation requires amply proportioned footscrews and accurate, even surfaces at the supporting points. The foundation should meet the following requirements:

1. It should go sufficiently deep into the ground, at least below the frost limit.
2. It must have a sufficiently high mass.
3. It should be placed where the ground will be free of vibrations.

Small instruments often cannot be fastened to a foundation if they are to be used on a balcony, a rooftop terrace, or in a portable manner outside on the ground. Such instruments need sufficient static stability, which is achieved when the amount of work W_k needed to topple the system over is large. A structure has a high W_k when its center of mass S is low (h_s is small) and at the same time the distance l_k to the tilting edge is large (see Fig. 5.11). In principle, a mounting should have its center of

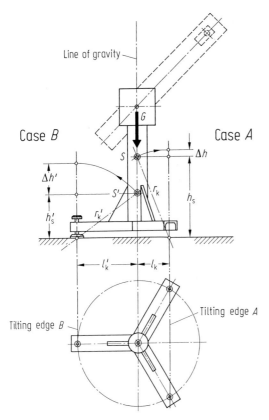

Fig. 5.11. Stability of a free-standing structure. A measure of the stability of a non-anchored structure is the work W_k required to topple it over. It is given by

$$W_k = \Delta h \cdot G, \tag{5.10}$$

where

$$\Delta h = r_k - h_s = \sqrt{h_s^2 + l_k^2} - h_s. \tag{5.11}$$

The two cases shown illustrate the situation. Evidently, a "nose-heavy" mounting has a low stability. Large masses are admitted only near the ground (low mass center S, small h_s). It is seen that a three-point base (tripod) has one tilting edge near the line of gravity.

mass as low as possible.[9] Therefore, highly elevated parts such as the tube, optics and accessories, and the upper part of the pier should always be designed and constructed to be of low weight. It has been shown that one need not counteract the stiffness criteria in order to satisfy the stability requirements. Larger masses in a mounting are

---

9 It makes little sense to place an instrument so high above the ground that steps and ladders are required for observations with it. The lowest eyepiece position should be at eye-height when the observer is in the seated or slightly bent-over position.

admitted only near the ground, that is, in the lower part of the pier (cf. Sect. 5.3.9 on vibrations).

A typical counterexample is the case of an instrument on a high tripod, since such tripods always have a high center of gravity. Besides, it is nearly impossible to produce a tripod mounting with adequate stiffness; such instruments often display unstable, wiggly images. The unsatisfactory vibrational performance can be improved by attaching braces between the tripod legs and by attaching heavy weights to them. This moves the center of gravity downward while the braces increase the stiffness.

5.7 Joining Elements

The parts of a mounting become a unit by joining the elements, such as boltings, welding joints, adhesive bondings, press fits, and bearings. One should always be aware of the fact that joints are potential weaknesses, because each joint between parts is a perturbation in the flow of force which can be distorted, deflected, stretched, or compressed. This is equivalent to increasing cantilever arms, or reducing the cross section. Calculating the stiffness of joints is not simple, and their optimal design requires much experience. It is thus advisable to avoid joints as much as possible. The general rule is:

> **Criterion 7.** The supporting structure of a mounting should consist of as few integral parts as possible with few joints. All joining elements which are unavoidable must be deliberately designed and very carefully manufactured.

The following items are important for rigid joints:
1. The parts to be joined must be sufficiently stiff at the contacting surroundings (i.e., wall thickness).
2. The contact surfaces of both parts must have accurately machined surfaces and fits.
3. The area of contact between surfaces over which the force acts must be sufficiently large. The joint should not create a bottleneck in the flow of force.
4. The force should be equally distributed over the contacting surfaces.

Particularly critical are joints by screws and bolts, which in general have very low stiffness. Even a forcefully tightened screw will not improve matters appreciably. On the contrary, if the bolt is stretched beyond its elastic range, its stiffness diminishes. On the other hand, the stiffness of bolted connection parts can be substantially improved by adhesive bonding of the surfaces. In general, adhesive bonded joints are practicable for mounting designs; they can be easily made and, when carefully done, provide a quite rigid structure. An appreciable degree of stiffness cannot be expected, however, from the common welded joints. For these, carefully welded V and X seams are needed. The strong *thermal* deformation can also present a problem in welded joints. The most critical joints in a mounting are the following:
- Tube to tube cradle and tube cradle to declination axis.
- Polar axis to bearing frame of declination axis.

- Bearing frame of polar axis with pier.
- Pier with foundation.
- All of the connections introduced by the adjusting elements for latitude and azimuth.

None of these, the key parts in the unit, should be allowed to become the dominant weak part of the instrument.

5.8 Measuring the Stiffness

The idea of "stability" as applied to telescope mountings is at about the same stage as telescope mirrors were in the time before Foucault. That is to say, the geometric form which a mirror has to have was known, but it could not be measured until Foucault made that possible with his ingeniously simple knife-edge method. "Stability," on the other hand, does not provide a way to layout and design a mounting, nor can it be quantitatively measured. The "stiffness" does both. It unambiguously determines the telescope structure and it can be readily measured.[10]

The stiffness equation (5.2) contains the known force F loaded onto a structure and the displacement x to be measured, and that by the telescope itself. The optics may be used with a reticle eyepiece aimed at stars of known positions, or a bright star can be photographed, and the displacement measured on the film. The known force is given by a weight, for instance, 1 kg = 9.81 N, connected via a cord and a pulley to the point to be tested. The stiffness can thus be determined in various directions, which is important because it is the directional variation of c (a tensor) which gives clues to the weak parts of the mounting. With f (in mm) as the focal length of the optical system, the measured angular displacement ε gives the linear displacement x (in mm) by

$$x = f \tan \varepsilon. \tag{5.12}$$

Any measurement should be made within a conventional measuring procedure. Since such conventions are not defined with respect to telescope mounting stiffness, some elementary conditions will be suggested here.

1) The stiffness must be measured in, or reduced to, the imaging plane of the telescope.
2) The stiffness is measured preferably in the zenith position of the tube, but supplementary measurements in other positions may complete the picture of stiffness behavior.
3) Since the stiffness of the connection between telescope and foundation is an important component of the overall stiffness, this measurement should be made with the telescope positioned as it will be when in use. A tripod should be placed on level and rigid ground.
4) The stiffness should be measured over 360° at a minimum of 12 points.
5) According to definition, the displacement must be measured at the position where the force acts. It is not feasible, however, to allow it to act at the crosswire eyepiece or the camera. Both of these are "touchy" instruments of low stiffness. Thus, the calibrating force is applied to the end of the tube and then recomputed for the image plane (according to the law of leverage, with the foundation or ground level as the fixed point).

These points should cover the more important general conditions.

---

10 For more details on stiffness measures, see Ziegler [5.1].

5.9 Telescope Vibrations

5.9.1 Principles of Mechanical Oscillations

The image stability of amateur telescope mountings is interfered not so much by static effects but rather by vibrations, and almost always flexural vibrations. Oscillations are rather difficult for the amateur to treat mathematically, as each statement leads to a boundary-value problem of differential equations.

A physical oscillator necessarily contains three elements that determine its behavior. These are:

1. At least two independent energy storage elements.
2. A coupling of the storage elements permitting energy to flow in both directions.
3. An excitation source which supplies (at least once) energy to the system.

In a mounting, the mass m carrying kinetic energy is one of the energy stores, the elasticity of materials represented by the stiffness c, the other. A simple mass-spring oscillator is shown in Figure 5.12. Such an oscillator is mathematically characterized by two terms. One term characterizes the vibrational properties of the structure (masses, stiffnesses, and damping properties of the parts), the other the excitation source given by the excitation function $E_F = f(t)$, as shown in the schematic diagram. The structural term supplies an important system characteristic, namely the natural frequency (or frequencies) ω_0:

$$\omega_0 = \sqrt{\frac{c}{m} - D^2}, \tag{5.13}$$

where c is the stiffness of the spring (in N/m), m is the mass (in kg), and D is *Lehr's damping number*. This frequency formula provides two important conclusions, generally valid for structures with elasticity and mass:

> 1. The mass and the stiffness of a mechanical structure are complementary with respect to the natural frequency.

> 2. Increasing the stiffness increases the natural frequency of the system, while increasing the mass diminishes it.

The natural frequency of a flexure oscillator, e.g., a lever with a mass at one end, is obtained by inserting the flexural stiffness from Table 5.5 into Eq. (5.13); the result is

$$\omega_0 = \sqrt{\frac{3EJ}{l^3} \cdot \frac{1}{m} - D^2} \quad \text{for} \quad c_b = \frac{3EJ}{l^3}. \tag{5.14}$$

This equation shows the dependency of the natural frequency on the various system parameters. In order to obtain information on the oscillation amplitude at any point (e.g., the focal plane) of a structure, the type of excitation must be considered. In other words, the mounting can be correctly designed with respect to oscillations only once the location and type of the excitation are defined.

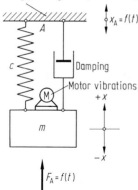

Fig. 5.12. Schematic graph of the simple spring-mass oscillator with one degree of freedom. The diagram shows schematically the three most important excitation mechanisms which act on a telescope mounting:

Case A: Excitation on the fixing point "A" by a deflection $x_a = f(t)$ caused by soil or foundation vibrations.

Case B: Excitation by external forces $F = f(t)$ acting (preferably) on the tube or other parts of the structure, e.g., manual manipulation forces and wind forces.

Case C: Internal excitation sources, e.g., excitation by imbalance forces are magnetically generated forces in the drive motors.

5.9.2 The Mounting as Oscillator Chain and Mechanical Low-Pass Filter

A mounting is a complex, three-dimensional, oscillating structure with many degrees of freedom, and for which calculations are difficult at best. For the design, however, relevant results can be also derived from simplified oscillator models in which the mounting is understood as a chain of sequential elements consisting of flexing shafts and masses. Some general statements can be made regarding such oscillator chains, thus allowing us to deduce guidelines for the layout without the necessity to calculate the system numerically.

The present discussion will begin from another interpretation, which rests on the strict analogy between arbitrary mechanical chains and electrical circuits. This analogy allows us to consider a mounting as a *mechanical quadripole* with low-pass properties (Fig. 5.13c). The points of excitation and of foundation are the entrance and exit clamps of the quadripole, whose frequency behavior is determined by the mechanical impedance Z_{mech}. The impedance and the excitation force F_a are interrelated via[11]

$$F_a = Z_{mech} \cdot \dot{x} = Z_{mech} \cdot v, \tag{5.15}$$

where Z_{mech} is in N·s·m^{-1}, F_a is in N, and v is the instantaneous speed of oscillation in m/s. The electrical analogue is related via

$$Z_{elec} = k^2 \cdot \frac{1}{Z_{mech}}, \tag{5.16}$$

---

[11] In the notation of Newtonian mathematics, differentiation with respect to time is denoted by superscripted dots, so that $\dot{x}$ means the same as the derivative dx/dt, and $\ddot{x}$ means d^2x/dt^2.

where k is the analogy constant.

The mechanical impedance is a complicated, direction- and frequency-dependent quantity[12], whose electrical analogue is the *admittance*, or the inverse of the electrical impedance. The characteristic of a low-pass filter is that oscillations below a threshold frequency ω_s pass the chain fairly easily, whereas oscillations above this frequency are not transmitted through the structure. This analogy demonstrates that a telescope mounting excited by soil vibrations must be designed differently from a mounting which is stimulated at the tube by manipulation or wind forces.

Case 1. Ground vibrations may have a very low frequency. To stop their transmission to the focal plane, the low-pass filter is operated in the *suppression range*. The limiting frequency of the mounting must therefore be well below the lowest stimulating frequency ω_a. This means that all parts have to be designed with large mass and low stiffness so that the structure will have a low mechanical impedance. While such a mounting would remain relatively unaffected by ground vibrations, even the slightest force applied at the tube would excite long-term, low-frequency vibrations. In addition, the low stiffness would conflict with the static requirements. It would be virtually useless.

Case 2. The normal manipulation forces and pushes incurred during operation and wind-caused oscillations have a frequency spectrum ranging from 0 Hz to 25 Hz or more. In order that these forces, which act primarily on the tube, create only minor oscillatory amplitudes, they must be short-circuited from the structure onto the foundation. The low-pass filter therefore operates in the transmission range $\omega_a < \omega_s$. For this to occur, the structure must have a high mechanical impedance, which is obtained by designing each part with the highest possible stiffness and smallest mass. This also satisfies the static requirement, and a high-frequency structure has the advantage that any vibrations damp out rapidly. With given damping properties of the material, there is a logarithmic relation between the frequency and the damping time.

Since the structure can never be designed optimally to satisfy cases 1 and 2 simultaneously, the following important design criteria are required:

> **Criterion 8.** The design of a telescope mounting should aim at high natural frequencies of the system and a high mechanical impedance. High natural frequencies lead also to better damping properties of the structure.

> **Criterion 9.** These properties are obtained by designing every part with stiffness as high and the mass as low as possible and by avoiding all "dead weight."

However, a mounting designed in this way requires placement of the instrument on an oscillation-free foundation. The statement on dead weight suggests another worthwhile hint: telescope mountings are occasionally seen which are burdened with

---

[12] While the amateur can measure the stiffness by simple means and for any mounting, measuring the mechanical impedance requires more sophisticated equipment.

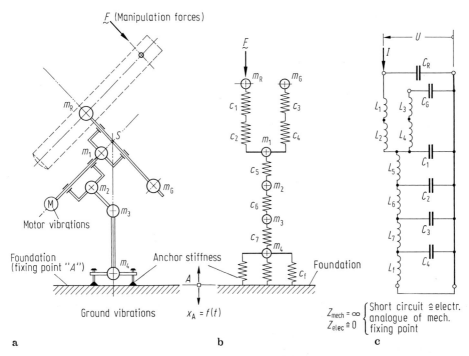

Fig. 5.13a–c. The telescope mounting as a flexural oscillator and oscillating chain, and the electrical analogue. **a** A telescope mounting can be essentially reduced to flexing shafts with stiffness c and point masses m. **b** This flexural oscillator can be interpreted as a branching oscillator chain with an attachment. **c** The electrical analogue is an LC-circuit with low-pass filter characteristics. The relations of analogy follow:

$$I \triangleq \frac{F}{k},\ U \triangleq k\dot{x},\ C \triangleq \frac{m}{k^2},\ L \triangleq \frac{k^2}{c},\ R \triangleq \frac{k^2}{D},\ Z_{elec} \triangleq k^2 \cdot \frac{\dot{x}}{F} = \frac{k^2}{Z_{mech}} \ldots, \quad (5.17)$$

where k is the analogy constant and the symbol "$\triangleq$" means "corresponds to."

a mélange of all conceivable accessories. This is not a very good idea in light of what has been said regarding vibrations.

> **Criterion 10.** A mounting should in principle be burdened only with the optical system and only those accessories which are essential for the specific observing task to be executed at that time.

5.10 Kinematic Aspects of Telescope Mountings

5.10.1 General Criteria and Instrument Errors

The subject of kinematics deals with all processes involving motion at telescope mountings. The mere presence of rotatable shafts refers to kinematics. It therefore covers pointing and tracking equipment in declination and right ascension, as well as all adjustment parts. These are for the collimation of the optical components as well as for pointing the axis system toward the celestial pole. Depending on the area of telescope use, Criterion 3 (The Kinematic Criterion; page 147) fixes the kinematical accuracy for setting on and tracking the objects of observation. Note that this accuracy implicitly contains errors of the axis system (concentricity errors, lack of perpendicularity) and adjustment errors in latitude and azimuth.

Errors of various drive elements determine how precise the tracking will be. In analogy to the treatment of stiffness, one can speak of an "error chain." In this regard, *systematic errors* and *random errors* must be distinguished, as they require different treatment. Systematic errors can be uncovered by a careful investigation and precise measuring of the system in question. They can in principle always be diminished by direct corrections of the system (although often only with much effort and cost). Not so the random errors, which are treated by the *theory of errors*. This theory makes a statement on the mean error of the chain, and on which error term contributes most to the total error.

The following are examples of random and systematic errors:

1. Parts which are to be combined into a structure are measured with a vernier caliper and a micrometer. Each measurement is affected by random errors (divisions of the scale, reading errors, and parallax errors). This error becomes associated with the length measured. The mean error of the composite structure is calculated by error theory. In this case the total error could be measured directly with a more precise measuring machine. But such high-precision instruments are hardly accessible to amateurs. The theory of errors purports to say something plausible regarding the accuracy of measurements in the absence of direct verification.
2. In the focal plane, a star shows a periodic oscillation with the rotation period of the worm wheel. This indicates a wobble or tumbling error of the worm. A correction of this or of the bearings will be needed. On the other hand, had the displacements been caused by faulty pitch and dividing errors in the worm wheel, and if these errors were random, then the wobble would be subject to the theory of errors.

What is interesting is the end result y of an error chain, which is as a rule a function φ of terms $x_1, x_2, x_3, \ldots$, and written $y = \varphi(x_1, x_2, x_3, \ldots)$. Let the terms be affected with the mean errors $\sigma_1, \sigma_2, \sigma_3, \ldots$ (see Chap. 3). The law of error propagation then gives the standard error σ_y of the system

$$\sigma_y = \sqrt{\left(\frac{\partial \varphi}{\partial x_1}\sigma_1\right)^2 + \left(\frac{\partial \varphi}{\partial x_2}\sigma_2\right)^2 + \left(\frac{\partial \varphi}{\partial x_3}\sigma_3\right)^2 + \cdots}. \tag{5.18}$$

Two examples will show how to treat and evaluate errors in telescope mountings.

1. Random errors, according to the statement above.

Example: Suppose that the four lengths l_1, l_2, l_3, l_4 are associated with a particular structure, and are measured partly with a vernier caliper and partly with a micrometer. The values are found to be

$$l_1 = 12.0 \text{ mm}, \quad l_2 = 172.4 \text{ mm}, \quad l_3 = 20.0 \text{ mm}, \quad l_4 = 80.6 \text{ mm},$$

and the corresponding errors (uncertainties) are

$$\sigma_1 = \pm 0.010 \text{ mm}, \quad \sigma_2 = \pm 0.09 \text{ mm}, \quad \sigma_3 = \pm 0.012 \text{ mm}, \quad \sigma_4 = \pm 0.07 \text{ mm}.$$

Then,

$$y = \varphi(l_1, l_2, l_3, l_4) = l_1 + l_2 + l_3 + l_4 = 285.0 \text{ mm},$$

$$\frac{\partial \varphi}{\partial l_1} \cdots \frac{\partial \varphi}{\partial l_4} = 1,$$

and

$$\sigma_y = \sqrt{\sigma_1^2 + \sigma_2^2 + \sigma_3^2 + \sigma_4^2}$$
$$= \sqrt{1 \times 10^{-4} + 8.1 \times 10^{-3} + 1.44 \times 10^{-4} + 4.9 \times 10^{-3}}$$
$$= \pm 0.115 \text{ mm}.$$

It is evident that neglecting the *errors* of the two measurements by micrometer leaves the sum practically unchanged (± 0.114 mm instead of ± 0.115 mm), since the result is dominated by the vernier reading errors σ_2 and σ_4.

2. Systematic error in a worm drive.

The *tumble*, a precession-like motion of the worm, is an error frequently encountered in worm-wheel drives. It is caused by a tilt of the axis of the worm which may occur in machining or by concentricity error in the bearings. If the errors in the bearings amount to Δx_1 and Δx_2 and the distance between bearings is l, then they contribute to the tumble, maximally $\tan \psi = (\Delta x_1 + \Delta x_2)/l$.[13] The tumble ψ superposes onto the uniform angular motion of the worm ω_s an oscillation $\psi \omega_s \cos \omega_s t$. The perturbed angular velocity ω_p at the polar axis then is

$$\omega_p = \frac{\omega_s}{u_1}(1 + \psi \cos \omega_s t), \tag{5.19}$$

where u_1 refers to the reducing gear ratio at the worm wheel and ω_s to the angular velocity of the worm.

Although the tumbling error is divided by the reducing gear ratio of the worm wheel it can severely interfere with the tracking if the worms are very inaccurate or have poor bearings. Dividing (indexing) errors of the worm wheel, however, affect the polar axis in undiminished amounts.

Error theory shows that in a structure or chain affected by errors, the part with the greatest error has the most influence on the entire system. There is thus very little

---

[13] For this reason, precision worms must be very carefully supported by bearings. The high-precision angular contact ball bearings mentioned earlier are best for this task.

purpose in machining *one* part with exceptional accuracy or in measuring only *one* quantity with particular care. In analogy to Criterion 2 on stiffness, the following holds for an error chain:

> **Criterion 11.** The relative accuracies of the parts of an error-afflicted structure or chain should be about equally large. Those parts which contribute the largest errors are those which should be targeted for improvement.

Example: The following illustration will introduce some interesting terms. Let the total mechanical error of a drive be $\pm 30''$, which is not bad for an amateur instrument. The remaining errors need to be corrected when tracking by manual guiding. The permissible frequency drift Δf at the driving motor is desired to give an equivalent precision over a tracking time of, say, 1 hour. The angular velocity ω_p of the polar axis is

$$\omega_p = \frac{360 \times 60 \times 60\,''}{86164.09\text{ s}} = 15.041\,''/\text{s}. \tag{5.20}$$

The frequency f of an AC voltage is expressed as the *angular frequency*:

$$\omega_e = 2\pi f \quad (\text{rad/s}), \tag{5.21}$$

where

$$2\pi \text{ rad} = 360° = 1.296 \times 10^6\,''.$$

The mechanical angular velocity of ω_m of a synchronous motor is

$$\omega_m = \frac{\omega_e}{p} = \frac{2\pi f}{p} \quad (\text{rad/s}), \tag{5.22}$$

with p equal to the number of pole pairs of the motor. The supply frequency is reduced by a factor

$$\Omega = \frac{1.296 \times 10^6}{u \cdot p} \quad (\text{arcsec})$$

to match the angular speed of the polar axis, with u being the reducing gear ratio of the mechanical drive stages. Thus,

$$\omega_p = \Omega f \quad (''/\text{s}), \tag{5.23}$$

where

$$\begin{aligned}\Omega &= 0''\!.30082 \quad \text{for} \quad f = 50 \text{ Hz},\\ \Omega &= 0''\!.25068 \quad \text{for} \quad f = 60 \text{ Hz}.\end{aligned} \tag{5.24}$$

The angular rotation α over a time t is $\alpha = \omega t$. For the example mentioned,

$$\Delta\alpha = \pm 30'' = \Omega_{50} t \Delta f = 0.30082 \times 3600 \times \Delta f,$$
$$\Delta f = \pm 2.77 \times 10^{-2} \text{ Hz}.$$

This corresponds to about the frequency error of the power supply system at night.

The frequency accuracy is reached or maintained by a simple *Wien oscillator*. For amateur mountings, it is not necessary that the power supply of the motor be via a high-precision quartz oscillator.[14] Considerations are more important on those parts which contribute most to the errors in the structure. It is there where improvements must begin. As a rule in amateur instruments, these are the axis system, the adjustments, and parts of the drive mechanics, but neither the drive motor nor the electronics supplying it. In essence,

"The amateur is limited by mechanics."

5.10.2 The Accuracy of Manufacture of Mechanical Parts

The manufacture and machining of a part of requisite accuracy involves a number of factors and external conditions which determine the precision. To look merely for a well-equipped workshop is to consider only one of these factors. Those factors which determine the precision are:

A. Conceptual factors.

1. A design which admits of accuracy.
2. A comprehensive knowledge of the working technologies, and skill and much practical experience in handling tools and machines.
3. Careful scheduling of the steps involved in machining and manufacture. The sequence of working operations is vitally linked with the precision of a part.
4. A measure of creativity guided by precision considerations. In other words, one devises means and appliances which allow the achievement of maximum precision from the available equipment and means. An example is the tremendous increase in the precision of mirrors through the simple but ingenious measuring device by L. Foucault.
5. Ample time. High precision is never reached in a hurry and certainly not without substantial reflection.

B. Material factors and external conditions.

1. Availability of or access to the tools, machines and measuring instruments needed. Accurate manufacture of a part requires accurate measuring.
2. Several external conditions influence the accuracy, including the absence of high-temperature heating during the machining, absence of deformations reducing the precision when the parts are chucked and clamped, a geometric form adaptable to the material, sharp cutting tool, etc.
3. Money. High precision is invariably a matter of substantial expense.

Short of discussing these items in detail, some practical hints on precision-matching design are given below. The geometrical shape and execution of the part greatly influences the precision to be realized, and this leads to the consideration of the following rules:

---

[14] Amateur activity is a hobby, and a hobby means that one need not follow strictly rational considerations slavishly. Anyone who feels challenged by the high-precision quartz regulator should contemplate building one. This chapter will restrict itself to the technical basics.

1. Design the parts so that all surfaces and fits are made in a simple and precise way. Not every shape can be made with an equally high accuracy!
2. Aim at a minimum of fits that limit the precision. Successive or superimposed fits should be avoided in view of the *law of propagation of errors*.
3. Shape and design the parts such that, if possible, they can be machined completely in one chucking on the lathe. Every re-clamping impairs the precision.
4. Design the parts so that they may be easily clamped and chucked onto a machine tool. The deformations suffered by the clamping must not result in diminished precision.

5.11 Drives in Right Ascension and Declination

5.11.1 General Considerations

Basically, telescope drives can be categorized into three construction parts:

1. Drive mechanics (gears).
2. Drive motor.
3. Power source and control unit.

Parts 2 and 3 together comprise the drive system. Drives and tracking train the optical system onto the objects of observation and follow the celestial motion. The functions of motions are:

(a) Coarse (ω_g) and fine (ω_f) settings in declination and right ascension of the objects.
(b) Tracking (ω_p) which acts on the polar axis in parallactic-mounted instruments. As mentioned before, alt-azimuthal mounting telescope require tracking in three complex, nonlinear operations which can be effectively executed only via a control computer.
(c) Fine corrections (ω_{cor}) on both axes, needed, for instance, in long-exposure photographs.

Table 5.8 compiles values for these angular velocities. These velocities determine the tracking motor, the mechanical gear ratios u, Ω of the drives, and the range of electronic controls.

Most amateur instruments still employ the operator's hand to work the slow motion control as a drive system for moving, setting, and correcting the instrument. From the manually pointed comet finder or Dobsonian telescope to the integrally computer-controled instrument, the choice of a drive will depend on the intended application, on the amount of technical effort, and on costs. Criterion 11 suggests that precision and thus the technical effort for the control electronics should be about the same as that for the mechanical parts.

The almost unlimited possibilities of modern electronics may lead to an undue preference to group type 3 and to neglect the mechanics. It is important to coordinate the three groups well, and not to draft them independently. For the layout of the gearing, the following procedure is advisable:

Table 5.8. Angular velocities and rotational speeds.

Operation	Angular velocity ("/s)	Speed ratio relative to ω_p
Stellar tracking speed	$\omega_p = 360 \cdot 60 \cdot 60 \div 86164.09 = 15.04$	—
Correction operations	$\omega_{cor} = \pm 20 \div 30\% \omega_p = \pm 4.5$	1:0.2÷1:0.3
Fine setting	$\omega_f = 120 \div 300$	1:8÷1:20
Coarse setting, slewing	$\omega_g = 1 \div 2°/s = 3600 \div 7200$	1:240÷1:480

Rotational speed (rpm) $n = \omega \cdot \frac{30}{\pi} = 9.55 \cdot \omega$.

1. Choose the driving system, by synchronous, stepping, or DC-motor; determine which operations are executed manually.
2. Explore which makes and types of motors are available, and compile the mechanical and electrical data. For instance, the *starting* and *nominal operation torques*, characteristics of the torques, upper and lower limits of the rotational speed in rpm, current, regulating dynamics, requirements of the power source, and control.
3. Explore which gears are available for the chosen type of motor. Note that the gears of such small motors are often only weakly constructed and therefore are not suitable to drive the main worm directly (e.g., for clocks and small instruments). The gears should be chosen with a torque of 0.1 Nm or more. Commercially available gears of DC and stepping motors often have maximum admissible revolutional speeds which are much lower than those of the motors they are coupled with.
4. Synchronous and stepping motors cannot be operated at very low angular speeds (because of saturation and jerky motion). The angular speed of a shaft rotating with n revolutions per minute is

$$\omega = \frac{\pi n}{30} \quad (°/s). \tag{5.25}$$

The lowest angular speed $\omega_{m,min}$ and ω_p determine the required reduction ratio $u = \omega_{m,min}/\omega_p$ of the gear. The fastest attainable slew speed follows from the maximum rpm number of the motor or drive. Or else, a separate motion for the slew (fast) speed is designed with a summation drive (a differential or planetary gear).
5. Power supply and control must fit the motor characteristic and the drive mechanics (the mass moments of inertia) well – particularly so for stepping motors.

5.11.2 The Mechanics of Drives

The drive is usually frictionally connected with the polar or declination axis through a clamp. Instead, it is advisable to insert a finely tuned *friction clutch* between drive and axis. This has two functions: to permit a fast manual motion of the tube, much faster than by any simple motor-drive design, and to protect the sensitive gear parts from damagingly high forces and momenta.

Operating the telescope always exerts forces and pushes (jerks) onto the tube. Much amplified by the high leverage ratio, they will be transmitted to the worm wheel and other parts, possibly resulting in impaired precision and even damage.

A simple clutch design is the axial disk clutch, where the worm wheel is constructed in disk form, and is moved along by two axially pressing clutch disks. Important for friction clutches are carefully machined surfaces and a suitable combination of materials. Good sliding properties and freedom from the "slip-stick effect" are afforded by wheel bronze (often used for worm wheels) against clutch plates of hardened steel.

For the mechanical drive itself there are numerous design variants, among them these well-known devices:

(a) *Worm wheel drives* – the most widely occurring in telescopes.
(b) *Tangential spindle drives*, where a lever or a sector is moved by a tangentially placed worm-gear spindle. It permits to realize high gear ratios, but the range of motion is limited. At the end the nut on the spindle is to be returned to the initial position.
(c) *Band drives*. A steel band is wound around a disk, which can also be moved by a threaded spindle. Hydraulic band drives have also been devised. In this type, the friction clutch is already given by the drive principle (i.e., band-disk).
(d) *Friction wheel drives*. A large and precisely machined disk is moved by rollers.

The following general rules apply to designing and building these drive types:

1. The first drive step connected with the shaft should have a gear reduction ratio as large as possible, $u_1 > 150$. For this to be realized, the worm wheel, the lever (in tangential spindle designs), or the band or friction disk must be made large. Guideline values for the worm wheel diameters are given in Table 5.9.
2. The first stage largely determines the precision of the drive mechanics. Division errors of the teeth or cogs, rotatory errors, irregularities, and backlash of the subsequent stages are divided by the gear ratio of the preceding drive stages. If high precision is expected from the drive, then attention should be paid to the first stage.
3. When designing the gear mechanics, the stiffness aspect should not be overlooked. No undesired lateral deflections or vibrations should occur in the motion direction of the drive.

Table 5.9. Guideline values for the main worm wheel diameters.

Refractor D (mm)	Reflector D (mm)	Worm wheel pitch circle (mm)	Module of wheel
60–100	150	130–170	0.70–0.90
100–120	200	180–220	0.80–1.25
120–150	250	220–270	1.00–1.50
150–200	300	270–350	1.50–1.75

Wheel pitch circle = Module × Number of Teeth

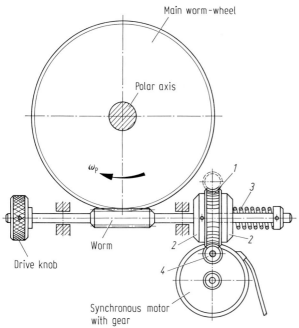

Fig. 5.14. Simple drive design with two worm wheel stages (Badener mounting after H.G. Ziegler). The essential element of this design is the axial friction clutch at the second worm wheel stage. The small worm wheel *(1)* driven by the synchronous motor drives the main worm wheel via the two disk clutches *(2)*. The friction torque can be fine-adjusted by the spring *(3)*. With this clutch type, manual setting operations ω_f can be performed while the tracking motor runs. Fine corrections ω_{cor} are, however, performed via the variable frequency.

In worm drives, the precision is given primarily by the accuracy of indexing (pitch) of the worm-wheel, by the concentricity of the worm and worm wheel, and by a precisely adjusted mesh. The gear mesh between worm and wheel should be adjusted carefully. Here, the worm is adjusted centrically relative to the gear teeth, precise in angle so that it is not tilted, and radially so that there is no play. The worm bearings have to provide the adjustment devices needed here. Band- and friction-wheel drives require precisely worked disks. A recurring problem in friction wheels is dirt grains getting between the driving rollers and the disk. A boxed arrangement with wipers helps alleviate this.

Stiffness in worm drives is conditioned by the *module* of the worm wheel, which should not be too small, and by whether or not the worm is in a rigid bearing housing. For this reason, a worm pressed against the wheel with springs is only a stopgap if the worm sticks and does not revolve smoothly. The stiffness is thereby considerably reduced.

In manually operated drives, it should be noted that the human hand is not capable of arbitrarily small motions. The angle of operation σ_{man}, by which the knob is turned in operation, should not be smaller than 1–2°. When operated over a flexible shaft,

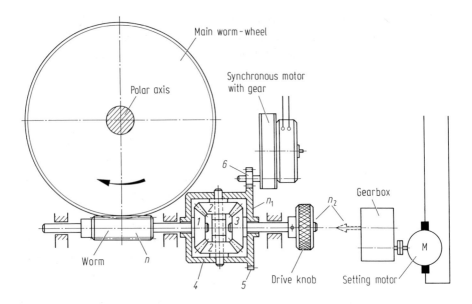

Fig. 5.15. Drive design with differential gear. Differential and planetary gears are real summing gears in which two input speeds n_1 and n_2 are summed to an output speed n_0. n_1, n_2, and n_0 can be functionally arbitrarily interchanged. For the differential gear, $n_0 = (n_1 \pm n_2)$. On the main worm shaft is the bevel gear *(1)*, which is driven by these two star bevel gears *(2)*. The rotational speeds n_1 and n_2 to be summed are added via the bevel gear *(3)* and the differential star *(4)*. The drive of this differential star comes from the cogwheel *(5)* and the pinion *(6)*. In place of the manually operated drive knob, setting motions may also be executed via a servo motor, a DC motor being particularly suitable; see illustration to the right.

this value should be even larger in order to allow for torsion elasticity. If $\Delta\sigma$ is the requisite precision of setting an angle on the axis, then the gear ratio u_f between the operating knob and axis is

$$u_f = \frac{\sigma_{\text{man}}}{\Delta\sigma}. \tag{5.26}$$

Figure 5.14 presents a simple drive design with two worm-wheel stages and Fig. 5.15 a variation with a *differential summing gearing*, both shown schematically. In the gearing, the setting knob can be replaced by a servo motor.

5.11.3 Drive Motors

The primary types of motors to be considered for telescope drives are *synchronous*, *stepping*, and small *DC motors*. They are available from many manufacturers and have a wide range of characteristics. Many suppliers also have gears fitting these motors, which simplify the design and construction of the drive mechanics considerably.

5.11.3.1 Synchronous Motors and Stepping Motors. These are motors which employ a rotating magnetic field. The stator contains windings which are fed by AC or by pulses. This induces a magnetic field rotating with the angular velocity ω_m given by

$$\omega_m = \frac{\omega_e}{p} = 2\pi \frac{f}{p}, \tag{5.27}$$

where p is the number of *pole pairs* of the motor. A two-pole motor has $p = 1$, a 16-pole motor has $p = 8$, and so on. Note that in stepping motors a two-pole stator winding is often combined with a multipole rotor. The rotating field "drags" along with it a soft magnet rotor or a permanent magnet rotor. In the first case, this is called a *reluctance motor*, in the second case, a *permanent magnet motor*. In these motors, the rotational speed remains fixed; it is synchronized with the frequency. Thus, the speed can be regulated via the frequency. There is more to the stepping motor: each step corresponds to a precisely defined rotation angle. By simply counting the step pulses, the angular position can be digitally displayed and processed. Thus in a certain sense stepping motors are drive elements and encoders simultaneously. Small synchronous motors have the following operation characteristics:

- The torque due to the load has no influence on the rotational speed as long as it does not exceed the *pull-out torque*. At that limit, the motor simply stops.
- The motor torque depends on the rotary magnetic field and thus upon the current. If, in a coil of inductivity L, the frequency is changed, then the voltage must also change proportional to frequency for the same current. Almost all frequency-regulated tracking systems are supplied from constant voltage AC sources. If, in a synchronous motor, the frequency is diminished at constant voltage, then the magnetic circuit quickly reaches saturation; when the frequency is increased, the strength of the rotating field, and thus the torque, drops.
- In these inexpensive, small synchronous motors, the magnetic circuits are not designed for higher frequencies, nor are the mechanical rotor components made for higher speeds.
- Often the power is supplied not by a sinusoidal voltage, but rather by a *square-wave voltage*. The harmonics of these pulses generate counterrotational magnetic fields which brake the rotor, develop pulsating torsions and noise, and cause high energy losses.

If small synchronous motors are not operated with sinusoidal current, then their speeds cannot be varied by more than ±25% from the nominal value. For pure tracking gears this is sufficient. Stepping motors normally have a two-pole stator winding supplied alternatingly with square-wave pulses of 90° electrical phase shift and of changing polarity (bipolar operation). In addition, there are *half-step* and *microstep* devices. Stepping motors are very complex structures. Whenever a system composed of magnetic fields, mass moment of inertia, and friction suffers impulses and thrusts, it can be expected that highly nonlinear behavior, resonances, and a complicated frequency response will ensue. Besides, a stepping motor cannot be considered by itself as its performance depends strongly on the power supply and the mechanics to be driven (moments of inertia and of friction). For this reason, the data sheets of such motors hardly permit any conclusions regarding their actual performance in operation.

Table 5.10. Properties of small motors.[1,2]

	Synchronous Motor	Stepping Motor[3]	DC Motor[4]
Principle	Rotating field, $\omega_m = 2\pi f/p$	Rotating field; field vector advancing in steps.	Internal commutation.
Revolution range, useful adjustable range	Used with constant rpm, adjustable about ±20%, much larger for sinusoidal, f-proportional voltage. ω_{max} defined by magnetic and mechanical design of motor.	Large range, 0 to several 10^3 rpm. Low range not useful owing to momentum pulsations and vibrations, highest range not useful as rotational momentum diminishes to 0.	Very large range, 0–5 × 10^3 (sometimes over 10^4) rpm. Entire range useful. Continuous operation at high rpm and high momentum increases wear on brushes and commutator.
rpm constancy	Rotation synchronous with frequency, which determines constancy and precision; no influence by the load momentum.	Rotation synchronous with step frequency, the latter determining constancy and precision	Rotation depends on voltage, momentum, and temperature. Low constancy of rpm without control circuit.
Behavior with respect to the load momentum	Constant for sinusoidal and f-proportional voltage; otherwise much diminished at highest and lowest frequencies; can be used to breakdown torque, then stops.	Strongly nonlinear: diminishes to 0 at high step rates, pulsates at low rates. Distinct resonances and momentum instabilities, depending much on feed electronics and gear properties.	Constant and pulsation-free over entire range; high peak momenta (accelerations) possible. Momentum depends linearly on rotor current.
System behavior	Linear in operating range, 1st-order system.	Highly nonlinear, 2nd-order system, mathematical relation complex.	Strictly linear 1st-order system.
Behavior in control circuit	Operated in open circuit with given frequency f.	Operated in open circuit; behavior depends on feeding electronics and mechanical properties of driven system.	Linear characteristic with very good dynamics

Table 5.10. (Continued)

	Synchronous Motor	Stepping Motor[3]	DC Motor[4]
Positioning properties	Not a typical motor for positioning.	Typical positioning motor, needs no encoder as rotational motion is in counted angular steps. May jump steps in ranges of momentum instabilities.	Not a typical positioning motor, requires encoder and position control circuit. The latter entirely determine the positioning accuracy
Area of use	Very good tracking motor corrections $\pm\omega_{cor}$ are within operating range. ω_f and ω_g can be realized with some more mechanics via a summing gear.	Good motor for tracking and positioning; operation in nonpulsating rpm range or in microstep mode. Fast motion ω_g possible if speed is not too high.	Covers entire range ω_p to ω_g. rpm-constant tracking and positioning require tachodynamo, encoder, and control circuits.
Costs for electronics	Rather low, not problematic.	Medium or high, especially for use of full rpm range and for optimum operation. Optimization problematic because of complexity of system.	Costs for tachometer, encoder, etc. very high; design, optimization, and adjustment of control circuits is rather demanding.

Notes: [1] Based on common commercial types (no special makes). [2] Evaluated with special scope on telescope drives. [3] Stepping motors with permanent magnet; their properties may vary much between different makes, and also depend on the supply electronics. Data given apply to full-step and half-step operation. [4] Modern DC motors with ironless drum or disk rotor (not cheap makes used in toys).

Table 5.10 compiles some properties which are important for telescope drives; the reader should consult the Supplemental Reading List for Vol. 1 for additional literature on this subject.

5.11.3.2 DC Motors. Modern small DC motors usually have an iron-free drum or disk rotor of small mass moment of inertia and very low inductivity for the purpose of controling and driving. Their properties differ strongly from those of stepping motors. Their torques are high, are constant over a very large range of speeds, and are free of pulsating torques. For the tracking motion, they can be operated at very low speeds ($n < 20$ rpm). Most of them have speeds ranging from 0 to 5000 rpm, some even as high as 10 000. They can start up and stop very quickly and have no natural frequencies or tendency toward oscillations. The rotational speed is proportional to the voltage at the armature winding, and is also controled that way. The disadvantages compared with stepping motors are:

- The commutator (collector) and the brushes suffer wear, especially when continuously operated in the upper range of angular speed and torque.
- The speed depends on the load. If the motor is loaded, then the rpm drops somewhat. Also the temperature has, via the armature resistance, some influence on the speed. If a highly constant speed is required, then the motor should be operated by a feedback loop controling the rotational speed.
- DC motors are not directly positioning units as stepping motors are. For the positioning task, another control loop with an encoder is needed.

Owing to the large range in angular speed and the excellent controling dynamics, large telescopes are now often equipped with DC-disc armature and pancake motors. The somewhat larger demand on control technique is insignificant.

5.11.4 The Control Circuit of Telescope Drives

The control circuit consists of the following basic units:

- A unit supplying the desired setpoint of the frequency or the desired value of the angular speed.
- An amplifier unit supplying the control voltages for the output power stage or, in a control loop, providing the required loop amplification.
- The output or driver circuit feeding the motor.

Add to the above the necessary power-supply units or voltage sources. These are the basic parts of open control circuits common in synchronous and stepping motor drives. For closed control loops, other functional units need to be added:

- Actual selsyn (angle encoder, tachometer), supplying the actual value of position and angular speed.
- A *comparator* unit, which compares the actual value with the nominal value and transfers the difference as a deviation to the amplifying and driving stages.
- Perhaps units in which control quantities are mathematically processed, such as integrators, filters, limiters, and so on.

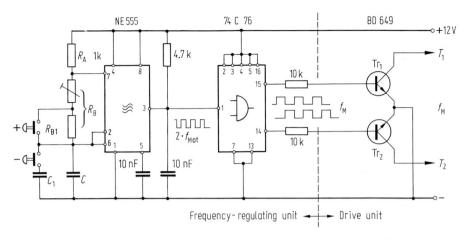

Fig. 5.16. Frequency-control unit for synchronous motor drives (after S. Witzigmann and K. Güssow). The primary elements of a frequency-control drive are an integrated circuit generating the frequency and a power stage or driver. The driver is represented here only by the two transistors $Tr1, Tr2$. The frequency is generated in the *Timer-IC 555* (newer type ICM 7555). The frequency is determined by the resistors R_A and R_B and the capacitors C_1 and C_2. The frequency can also be set by a voltage applied at terminal 5 of the IC, as is, for instance, needed for photoelectric tracking. The push-pull power stage operates on two input signals, phase-shifted by 180°, and which are obtained by inverting from the logical block IC 74C76.

Certain functions can be digitized and performed in microprocessors. These are the functions where signals are generated, processed, logically connected, and displayed. Analog elements, however, are the amplifiers and the output stage. For these digital and analog units, ready-made IC (integrated circuit) modules are available which need only a few connections with external circuit elements. They are offered by numerous IC manufacturers and much simplify the layout and wiring of the control electronics. For most IC modules, explicit directions for applications are available containing detailed data, operating data, and instructions on the external wiring and on their use. A detailed treatment of telescope drives is given by Güssow [5.3].

5.11.5 Control Electronics for Synchronous Motor Drives

The synchronous motor is operated in an open control circuit for the tracking speed ω_p. Owing to the absolute synchronism between frequency and rotational speed, feedback of the actual value and a control loop are not needed. The frequency generated by an oscillator directly feeds the power stage. Figure 5.16 shows the basic circuit arrangements using the oscillator IC-NE 555. The parts determining the frequency for the sequence of pulses at the IC exit are R_a, R_b, and C. By using high-quality components such as good metal-film resistors and precision polystyrene capacitors, the error caused by frequency drift will be smaller than the mechanical errors of the drive and the corrections required for atmospheric refraction.

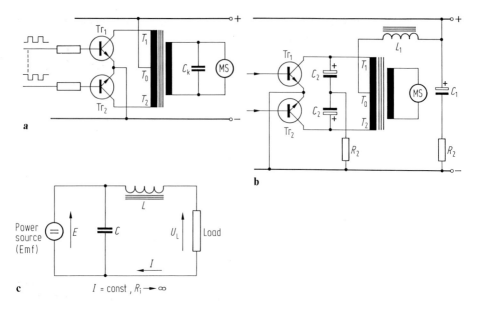

Fig. 5.17a–c. Push-pull motor drives. In the simplest case **a**, the output stage consists of the two transistors $Tr1, Tr2$ switched in antiphase, the transformer T with center tap, and the compensating capacitor C_k. A *Darlington pair* can also be used as transistors. The invariant, nonsinusoidal voltage is disadvantageous for the motor and permits only small frequency variations. Substantially better is the drive stage **b**. The original voltage has to be proportional to frequency, and better sine-shaped. The series resonant circuit shown in **c** serves as a source of constant current and thus forces the voltage proportional to frequency. The two electrolytic capacitors C_2 absorb the harmonics and relieve the original circuit from the reactive load.

For R in kΩ and C in μF, the pulse frequency at the NE 555 exit is

$$f = \frac{10^3}{\ln 2} \cdot \frac{1}{R \cdot C} = \frac{1443}{R \cdot C}, \tag{5.28}$$

where

$$R = (R_a + 2R_b).$$

Example: The gear ratio is chosen for a 60 Hz synchronous motor. Owing to the frequency division 1:2 in the inverter unit, the oscillator frequency must be 120 Hz. Choosing $R_a = 1$ kΩ and $C = 0.5$ μF, Eq. (5.28) then gives $R_b = 11.53$ kΩ. When for tracking corrections ω_{cor} the frequency is to be variable by $\pm 20\%$, about 10% of R_b is short-circuited by a push button, and a second switch adds a capacitance of $C_1 = 0.125$ μF to C. C_1 can be an inexpensive capacitor, since for ω_{cor} constancy of frequency is an irrelevant condition. For fine setting of f, R_b is split into a fixed resistance R_{b1}, for instance 10 kΩ, and into a 2.5 kΩ, 10 turn potentiometer.

The power transistors of the push-pull output stage are circuited in ON/OFF mode. This requires two inverted signals generated by the IC-74C76 from the input sequence of pulses, the frequency being halved. The ON/OFF mode causes only small losses in the transistors. Therefore special cooling elements are in general not needed. The simplest layout of such a push-pull stage (Fig. 5.17a) has distinct drawbacks for the motor, and gives a poor overall efficiency:

- The motor voltage is constant. What should be constant, however, is the motor *current*.
- The motor is fed with a square-wave voltage instead of a sinusoidal one.

The consequences of these disadvantages have already been mentioned (Sect. 5.3.11.3). Considerably better is the circuitry in Fig. 5.17b.

- The final stage is powered from a series resonance circuit consisting of L_1 and C_1 as shown schematically in Fig. 5.17c.
- The reactive current component is compensated by the capacitor C_2. This relieves the switching transistors and the feeding circuit with respect to current.
- Harmonics[15] caused by the circuitry are not led through the motor (where they would cause inverse torques) but instead are absorbed by C_2.

A series resonance circuit has at the point of resonance the property of a source of constant current ($R_i \rightarrow \infty$). The motor circuit therefore adjusts automatically to the requisite value. A resonance circuit has

$$\omega_0 = 2\pi f_0 = \frac{1}{\sqrt{LC}}, \qquad (5.29)$$

where f_0 is the operating frequency of the motor for ω_p. The compensating capacitor C_2 is placed at the primary side. Electrolytic capacitors of small dimensions may be used. The capacitance C_2 must be determined from experimental trials. For commercial synchronous motors, C_2 will be in the range of 100–200 μF. R_2 can be assumed to be roughly 200–300 Ω. The optimizing is with respect to the current mininum and a motor voltage which is as sinusoidal as possible. For ω_{cor} frequencies this compensation is not optimal, but even then, the operating conditions are markedly better for the motor than without reactive current compensation.

If higher constancy of frequency is required, the feeding frequency can be obtained from a quartz oscillator. One should always consider, however, the elementary rules of error theory within the frame of the entire design in order to avoid chasing after exaggerated precision demands at those points where they are rather unimportant. There are also ready-made IC modules for oscillators on quartz bases, which normally contain integrated frequency dividers. Figure 5.18 shows such circuitry. The switches S_1 through S_7 adjust the required frequency-division ratio from the quartz frequency to the operating frequency. Closed switches give logic level H, open switches logic level L. These levels are set by the microprocessor corresponding to ω_p, ω_{cor}, and also to tracking speeds of nonstellar objects.

---

[15] The *Fourier series* for a square wave is

$$U_\sqcap = U_0 \sin \omega t + \frac{U_0}{3} \sin 3\omega t + \frac{U_0}{5} \sin 5\omega t + \cdots, \qquad (5.30)$$

which has the harmonics $\sin 3\omega t$, $\sin 5\omega t$, $\sin 7\omega t$, $\sin 9\omega t$, ...

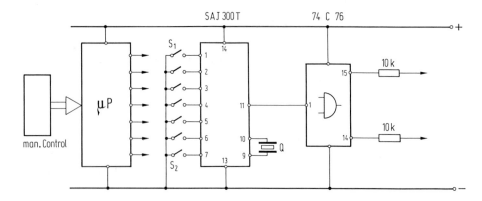

Fig. 5.18. Pulse generator with quartz oscillator. A tracking drive of highly constant frequency can be constructed with a clock or pulse-generator integrated circuit. Such modules are offered by several IC manufacturers, and contain the oscillator for the external quartz and the frequency-dividing cascades. With the switches S_1–S_7 the frequency-dividing ratio, and thus the required input frequency, is adjusted. The switch positions correspond to the logic levels H (high) and L (low). These levels can be given by a microprocessor or a computer. The inverting and the output stage are the same as in Figs. 5.16 and 5.17.

5.11.6 Control Electronics for Stepping Motor Drives

Stepping motor gears are designed primarily for instruments which are to be positioned digitally. In such drives, however, considerable costs must be reckoned with. The stepping motor is substantially more expensive than a small synchronous motor. The large span of revolutions ($\omega_p \to \omega_g$) and the characteristic properties of stepping motors require well-harmonized gear and circuit components. To ensure optimal adjustment of stepping motor and control electronics, it is advisable to contact the motor manufacturer, or even to use indexers and gears from the same producer. Many manufacturers of stepping motors supply complete drive systems. The operation of the stepping motor (full-step, half-step, microstep), the limiting speeds for ω_p, ω_g, and the mechanical gear ratio u are correlated quantities. This may be shown for two variants:

Variant 1. Stepping motor in full-step or half-step mode. This is the most common operation. Stepping motors with step angles of $1°\!.8$, $3°\!.6$, and $6°$ are commercially available. The number n_p, which describes the revolution speed ω_p for the motor, should not be below 100 rpm in this operation. Below this level, most stepping motor makes show distinct resonance phenomena and "eigen frequencies." Moreover, they cause pulsating torsional moments on the motor shaft and annoying telescope vibrations.

$$\omega_{mp} = \frac{\pi n_p}{30} \quad \text{for } n_p = 100 \text{ rpm,}$$

$$\omega_{mp} = 10.5 \,°/s = 3.77 \times 10^4 \,''/s,$$

$$u = \frac{\omega_{mp}}{\omega_p} = \frac{3.77 \times 10^4}{15.04} = 2500 : 1$$

Assume a fast motion speed of $\omega_g = 0.5°/s$, which loads to a stepping motor rotational speed n_g of

$$n_g = u\omega_g \cdot \frac{30}{\pi} = 2500 \cdot 0.5 \cdot \frac{30}{\pi} = 11\,936 \text{ rpm.}$$

At such speeds, most stepping motors no longer have any substantial torque. Besides, gears with reduction ratios for such high speeds are rarely available. The consequence is that ω_g must be reduced by a factor 5. The less-convenient operation is compensated by a lesser demand on the control electronics.

Variant 2. Stepping motor in microstep mode. In this operation, each full step is divided in a stepped or terraced fashion. This provides, as is to be shown, several advantages. For this example, it is important that in this mode, the motor can for ω_p be run at 10–20 rpm.

Example: Assume a 3.6° stepping motor with 12 microsteps and a tracking speed of 15 rpm. With the same relations as above, it is found that

$$\omega_{pm} = 5.66 \times 10^3 \text{ ''}/s, \quad u = 376 : 1, \quad n \approx 1800 \text{ rpm.}$$

This case provides an acceptable rpm value for gear ratios, and each microstep corresponds at the polar axis to a rotation by $2\rlap{.}''8$. The gear design could consist of a main worm wheel with 180 teeth and a cogwheel stage 2:1 in front of it. This advantageous gear concept requires more costly electronics. For ω_g, 36 000 step pulses per second must be produced with exactly defined amplitudes.

5.11.7 Circuitry for Microstep Operation

The telescope motions ω_p and ω_f can be adequately operated in the full-step or half-step mode. Over the entire range of speeds and for correct positioning, however, it is advisable to operate in microsteps. This is by far the best operating mode for telescope drives. The advantages are:

- Low pulsation of the torque.
- Strongly reduced tendency toward vibrations and eigen-resonances. The motor is practically a system of first order.
- A constant torque over a large range of rotational speeds.
- Increased angular resolution when positioning.
- A better motor efficiency.

However, not all motors are suitable for the microstep mode. The motor torque M, which depends on the turning angle φ, must strictly obey the formula

$$M = k_T I \sin\varphi, \tag{5.31}$$

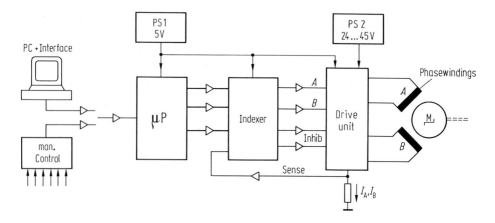

Fig. 5.19. Block diagram of a stepping-motor control: motor phases *A* and *B* are sequentially excited by the driver by step pulses. Signals are supplied by the indexer, which has other functions as well. It limits the motor current and blocks via the inhibit logic the bridge transistor during the phase of declining current. The current-proportional signal arrives at the resistor R_s and is fed via "sense" to the indexer. The control function proper is in a microprocessor (μP). To enforce rapid current increase in motor phases requires a supply voltage much higher than the motor's nominal voltage. The driver thus has to be fed from a separate network power supply unit *PS2*.

a condition which is not true for all makes and types. Here *I* is the current and k_T is a constant.

Figure 5.19 shows the block diagram of a stepping motor control. It consists of the following units:

- The *driver*, generating the two motor phases sequentially.
- The *indexer* or *translator*, which makes the control pulses for the driver inputs, and has other functions.
- A *microprocessor* (μP), generating the necessary functions and level values, and determining and coordinating the series of operations. It receives commands such as "faster," "slower," "go right," "go left," "go to position *yy* with ω_g," etc. from a manual control unit (man. control) or from a personal computer (PC).

Also available are the special indexer IC modules or complete stepping motor controls containing all units. Even an electronics expert would probably not find building one of these devices by himself very worthwhile.

5.11.7.1 The Driver (Fig. 5.20). For each motor phase, there is a transistor bridge with free-wheeling diodes *D*1 to *D*4. In full-step mode, the bridges are switched in alternating and sequential fashion, as in Figure 5.21a. Owing to the winding inductivity, the motor current does not follow this scheme. It rises and declines exponentially (Fig. 5.21b). As the torque needs a strong magnetic field, the motor winding inductivity cannot be made arbitrarily small. For this reason, such circuiting admits only rather limited step frequencies. A high supply voltage with a current chopper forces a

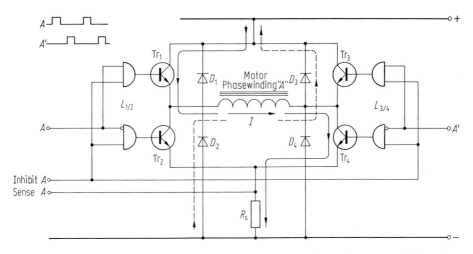

Fig. 5.20. Bridge driver for stepping motors. Each motor phase is supplied over a full transistor bridge with the corresponding free-wheel diodes. For a positive pulse, transistors $Tr1$, $Tr4$ are connected (the current follows the solid line), for a negative pulse $Tr2$, $Tr3$ are connected. Blocking all four transistors by the *inhibit logic* causes a sudden decline in the current. The current then goes via the diodes $D2$, $D3$ back to the source (*dashed line*) which acts as a *current sink*.

sudden increase of current and, by a so-called *inhibit circuit* of the bridge transistors and a current sink, causes a rapid decline (Figs. 5.20 and 5.21c). In the phase of rising current, for instance, the transistors $Tr1$ and $Tr4$ are switched, which puts a voltage of 45 V or more on the 6 V motor windings. The quickly rising phase current is sensed via a sense resistor R_s and compared in a comparator with the nominal level (corresponding to the nominal motor current). When this level is reached the chopper starts to keep the current at that level. At the end of the step, the inhibit logic blocks all four bridge transistors. The phase current then flows rapidly over the free-wheeling diodes $D2$ and $D3$ and back into the voltage source, which acts as a sink. The inhibit function is also activated with the motor at rest, so that no unnecessary current flows through it.

The indexer supplies to the driver circuit the step pulses in the correct sequence and phase; it activates the inhibit-logic at the proper times, deblocks the choppers and provides their frequency cycle, it contains the comparators with their logic, and has additional functions such as *Reset, Go right, Go left, Synchronize*, etc. In microstep mode, every full step (90° electrical phase) is cascaded into N microsteps, for instance, 9 microsteps in Fig. 5.22. The two current comparators are not given a fixed nominal level, but values for each microstep which relate to the rotary field equation:

$$\sin^2 \varphi_i + \cos^2 \varphi_i = \text{const.}, \tag{5.32}$$

where the sine term on the left corresponds to phase A, the cosine term to phase B; $i = 1 \ldots N$, and N is the number of microsteps. As is seen from Fig. 5.22a, the phase current approaches the ideal sinusoidal form by the step curve. The larger the number

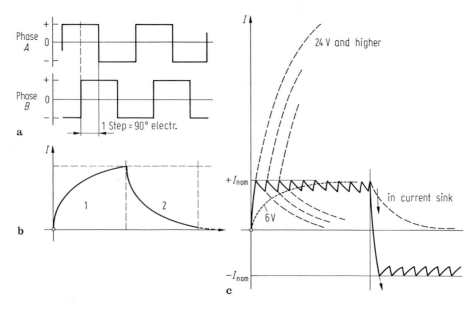

Fig. 5.21a–c. Driver, pulse shapes, and chopper mode. In full-step mode, the driver bridge is controled by a sequence of pulses, as in **a**. The inductivity makes the current rise and fall exponentially **b**. An adequately high voltage at the windings yields a fast increase in the current. When the nominal value is reached, the chopper is triggered via "sense" **c**. Decline of the field is reached through a current sink switching the windings via "inhibit."

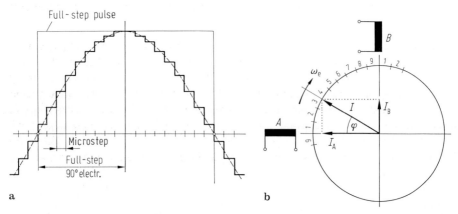

Fig. 5.22a, b. Microstep operation, step curve, and vector diagram. Anomalies of torque of stepping motors in full-step mode can be avoided by microstep operation. The field vector I in the figure revolves in microsteps; the graph is drawn for 9 microsteps. The motor phases A and B are excited by the current components I_A and I_B. The current in the windings then follows a step-curve **a** approximating the ideal sine curve.

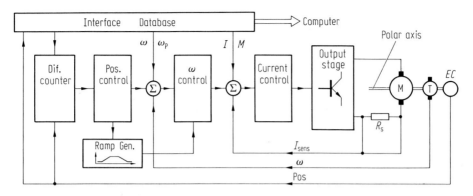

Fig. 5.23. Professional telescope control with disc armature DC motor. Large telescopes are often operated by the disc armature DC motor and a control cascade. The block diagram shows this drive circuit. Feedback quantities are the current or the torque of the motor, the angular speed and position of the instrument. A computer provides coordination and control.

of steps, the better the approximation. For a telescope drive, even 12–20 microsteps afford a sufficiently pulsation-free rotor torque.

There are two ways to provide the current levels:

1. For each step, the quantities $\sin \varphi_i$ and $\cos \varphi_i$ are calculated and stored in an EPROM (= *erasible, programmable, read-only memory*). From here they are called cyclically and fed into the comparators.
2. Continuous sine/cosine functions are generated and supplied to the comparator input.

The microprocessor generates these functions or stores the corresponding numerical values. It also provides the cycle frequency for the requisite rotational speed and the "ramps" for the motor's accelerating phases. The motor cannot be suddenly run up from ω_p to ω_g; it would get out of phase. This must be done over the acceleration ramps. The ramp functions are to be adjusted in such a way that in the phases of acceleration or deceleration, about 40% of the motor pull-out torque is not exceeded. Only this guarantees that no losses of steps and thus erroneous positionings will occur. The details of circuiting and programming the indexer and the microprocessor will not be discussed here.

5.11.8 Control Electronics for DC Operation

Many professional telescopes now have all operations (positioning, tracking, guiding, scanning) performed by a single disc armature DC motor. The control is through a control cascade, which is a chain of sequentially connected circuits (Fig. 5.23). The governing functions are performed by a computer. Such expenditure for control is seldom justified for amateur telescopes. Less rigorous demands can be met with a simple DC telescope drive. Only ω_p needs to be run with some precision, and for this a closed circuit with tachometer is needed. All other operations ω_g, ω_f can be performed

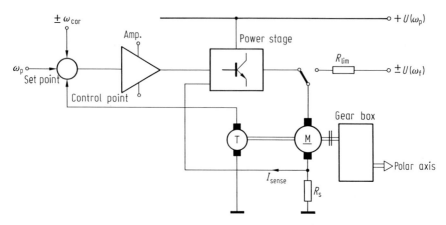

Fig. 5.24. DC telescope tracking with tachometer. Precise control is needed only for ω_p. The motions ω_f and perhaps also ω_g can be performed uncontroled through corresponding voltages $U = n \cdot k_e$. The figure shows the n-control circuit with tachometer. A double-throw switch feeds the unregulated voltages to the motor. Current limitation is needed to protect the motor. It is provided, as already noted, by R_s and the "sense."

over uncontroled circuits. For these operations, the motor is directly supplied with the corresponding voltages:

$$U \approx n \cdot k_e,$$

where k_e is the EMF (*electro-motive force*) constant for DC motors and $n = 9.549\omega$. Note that DC motors have a very low internal resistance R_i. Thus, if the voltage changes suddenly, a very large current will flow through the armature winding (rotor winding) and may damage the commutator and brushes. The control circuit thus needs a current limiter (I_{sense}), and the open circuit a current-limiting resistance $R_{\lim}$. These are to be arranged so as not to exceed the nominal current. The schematic of such circuitry is shown in Fig. 5.24; IC modules for this purpose are commercially available.

An even simpler DC tracking without tachometer and with only one OPAMP (= *op*erational *amp*lifier) and serial transistor comes from the following reasoning (Fig. 5.25): The largest speed fluctuations in a DC motor are caused by variations of the driving torque. Every change ΔM of torque causes a change of current $\Delta I = \Delta M/k_T$ and hence, via R_i, a change in the rotational speed. This action by R_i may be compensated by an additional voltage ΔU. The current-dependent potential difference at the resistor R_s controls the serial transistor Tr over the differential and loop amplifier A so as to satisfy the condition $\Delta U = \Delta I \cdot R_i$. This current-dependent circuitry reaches a tracking precision adequate for many applications. As a hint, the efficiency is primarily determined by the energy losses in Tr. The voltage from a battery or power supply and the motor voltage should be adjusted so that, in the given range, these losses are small.

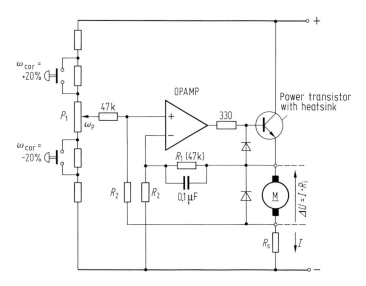

Fig. 5.25. Simplified DC telescope tracking. In disc armature DC or drum motors, the largest variations in speed are caused by the nonconstant driving torque of the gear. Compensating the potential difference $I \cdot R_i$ can remove these speed fluctuations. The current signal from R_s is fed to the + input of the OPAMP. The setting of the nominal value of ω_p is at $P1$, and the keys perform the corrections ω_{cor}. For the compensation condition $\Delta U = I \cdot R_i$, the following OPAMP relation is valid:

$$\frac{R_1}{R_2} = \frac{R_i + R_s}{R_s}. \tag{5.33}$$

5.12 Photoelectric Guiding Systems

5.12.1 Direct and Off-Axis Guiding

Long-exposure photographs in cold weather can make guiding at the eyepiece a bit unappealing. Modern electronics now make it possible to carry out this task automatically and with greater precision. The eye is replaced by a photoelectric sensor and corrections are effected by control circuits. Photoelectric tracking has two basic designs:

(a) The guide star is sensed with the guide scope which has a *tracking head* supplying the signals for the control circuits.
(b) The photographic telescope is used in off-axis guiding (OAG). That is, at the edge of the photographic field, the light of the guide star is deflected by a small prism and fed through a transfer optics into the tracking head.

Advantages and disadvantages of these two modes are summarized in Table 5.11. It is often a problem to find a sufficiently bright guide star because even for fairly bright stars the signals given by the photoelectric sensors are quite small. For this reason,

Table 5.11. Photoelectric guiding systems.

Separate guide scope	Off-axis guiding
Advantages	
Guider and photographic (observing) system are independent units and can be designed best for their tasks.	Only one optical system for which the mounting has to be designed
Useful when off-axis is not feasible: Schmidt, Maksutov cameras, apertures under 300 mm, etc.	Equal aperture and light-gathering, reaches fainter guide stars.
No interference in focal plane of telescope; simple attachment of guiding head at guide scope.	Optical path the same, avoids guiding error caused by axis flexure
Disadvantages	
Requires two optical systems; guider must not be too small (over 100 mm)	Feasible only for larger telescopes
More costs for optics and mounting, heavier counterweight	Not for Schmidt and similar systems
Deformative displacements between systems may be troublesome; stiff coupling required	Limitations and complex mechanics near focal plane (prism, special plateholders, etc.)
	Field of telescope not seen while guiding.

the aperture of the guide scope should be as large as possible: about 100–120 mm is a lower limit well worth the expenditure of photoelectric tracking.

The deformative displacement between the two optical axes may require attention in long-exposure photographs. The connection and joints between the two systems should be very rigid. Moreover, the optical components in their cells should be free of displacement (or tilt) and the guiding head rigidly fitted to the guide scope.

In OAG mode, the photographic and guide systems use the same optical light path; thus, the hard-to-avoid flexure displacement is not harmful. Of course, flexures in the light ray behind the deflecting prism and in the tracking head should not occur here either. In OAG mode, the mechanics for the plate holder, the prism, and the guiding head are more complex and expensive than with a separate guide scope. On the other hand, only one optical system is needed in OAG mode.

5.12.2 The Guiding Head

The displacement Δ of a star from the nominal position (at the center of the cross-wires) is a vector with the components $\Delta\alpha$ and $\Delta\delta$ in right ascension and declination,

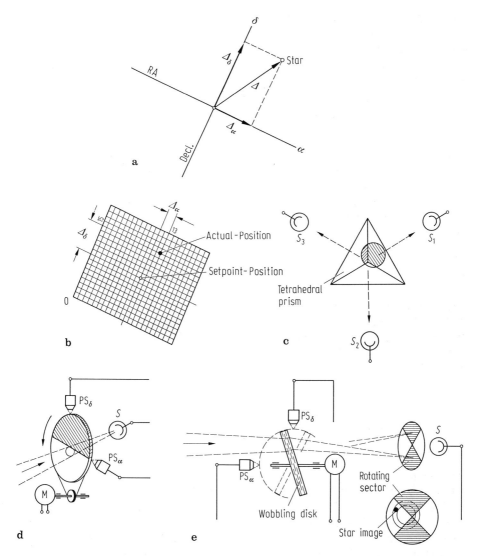

Fig. 5.26a–e. Basic methods of photoelectric guiding (after P. Höbel and H. Ziegler). The displacement of a star is a vector Δ with components $\Delta\alpha$ and $\Delta\delta$ **a**. The basic designs differ in the method by which control signals are generated for these components. **b** A *position-sensitive detector* (PSD) supplies direct matrix values (in this case 15/13) for the displaced image point which are digitally processed. **c** The position-selective beam splitting divides the light with a tetrahedral prism into three beams. A vector algorithm computes the components from these beams. **d,e** The *phase-selective light-variation method* modulates the light by an optical chopper. The chopper phase and the modulated light signal permit the securing of control quantities for α and δ.

respectively, as shown in Fig. 5.26a. Three basic methods to obtain the control signals exist:

(a) *The direct position method.* The coordinates for $\Delta\alpha$ and $\Delta\delta$ are supplied by a photoelectric position sensor (PSD) as analog or digital values (image-point values, pixels; see Fig 5.26b).
(b) *Constant-light method with position-selective beam splitting.* The optical path contains a small tetrahedral prism (Fig. 5.26c). The orientation of the prism fixes the angular orientation. It divides the light from the guide star into three partial beams, which are received by sensors S_1–S_3 and processed by the control electronics. The three sensors read equal intensities when the star is centered.
(c) *Phase-selective alternating light (AL) method (chopper method).* Here, the beam is periodically interrupted (modulated) by an optical chopper. Figures 5.26d and e show how this may be achieved with a rotating disk, or with a stationary sector and a wobbling plane plate. Other arrangements are possible. In the graphed design, phase detectors PS_α and PS_δ get the phase signals at the rotating elements. They are simply light barriers.

Each of these methods has advantages and drawbacks, which are discussed by P. Höbel [5.4]. At the present state of technology, the alternating light devices probably offer the most reasonable investment for the amateur. Their features are:

- They require only one sensitive light detector, which gets the entire current from the guide star. Thus fainter guide stars can be used than with method (b).
- Only one *electrometer-amplifier* (an amplifier to enhance very small currents of $< 10^{-9}$ A) is needed. This is less delicate and not affected by drift effects because AC signals are amplified.
- The signal-processing electronics are, however, somewhat more complex, and the guidehead with the chopper contains some precision mechanical components, in particular the wobbling chopper elements, which should be made quite accurately in order not to transfer vibrations to the telescope.

It should be generally noted that photoelectric guiders cannot be assembled by "cookbook" methods. Some mechanical and electronic knowledge and expertise are called for, as well as the necessary tools and parts.

5.12.3 Light Receivers

The light power received from a star is very weak, particularly when limited by the optical window of the Earth's atmosphere. The signals displayed by a light receiver will be correspondingly small. To reach faint stars, the light sensors must efficiently convert the available light energy and supply an initial signal which is as strong as possible. Therefore, the following criteria need to be used to assess and select light receivers:

(a) The sensitivity S should be high in the wavelength range of the optical window (400–700 nm with center at 560 nm). Some receivers, such as photodiodes, have their maximum sensitivity outside this range, but for ground-based instruments a sensitivity around 560 nm is most valuable.

(b) Even in the absence of incident light, a radiation receiver will always display a faint signal, the so-called *dark current* caused by noise effects. A true light signal is detected only when its intensity lies substantially above this noise level. A receiver designed to pick up weak signals must therefore possess, besides high sensitivity, a very low dark current. A quality criterion for this is the *NEP-value*, where NEP = *n*oise *e*quivalent *p*ower, which is the light power equivalent to the dark current present. The following relation holds:

$$\text{NEP} = \frac{\sqrt{2qI_d GB}}{S} \quad \text{W} \cdot \text{Hz}^{-0.5}, \tag{5.34}$$

where
q = electron charge (1.6×10^{-19} C).
I_d = dark current (A).
G = current multiplying factor for photomultipliers ($G = 1$ for photodiodes).
B = frequency bandwidth (Hz).
S = radiation sensitivity (A/W), in the present case referred to as $\lambda = 560$ nm.

Suitable receivers for guide systems and for photoelectric photometers are photomultiplier tubes and semiconductor photodiodes. The favorite for astronomical work is the classical photomultiplier 1P21. It has a radiation sensitivity $S_{560} \approx 5 \times 10^4$ A/W and an NEP-value of about 6×10^{-16} W Hz$^{-0.5}$. The highest sensitivity photodiodes are the S-2386-18K from Hammamatsu [16] and the BPW32 from Siemens with sensitivities of 0.3 A/W and NEP-values of $1-5 \times 10^{-15}$ W Hz$^{-0.5}$. A good photomultiplier reaches about 1.5–2.0 magnitudes fainter than a diode. The choice of receiver, however, should be made with several other important properties and considerations in mind:

Photodiodes:
– Photodiodes are very small. The guide head can therefore be designed and constructed compactly and with little weight. This is significant as a space and weight criterion for small instruments.
– Photodiodes are operated at low voltages. They are not dangerous in operation and are compatible with semiconductor circuits.
– They are mechanically sturdy, and also simple to operate.
– They are inexpensive compared with photomultipliers.

Photomultipliers:
– They are comparatively large glass tubes. The guide head is large and heavy.
– They require high voltages (1000 V) and the requisite high-voltage sources. Working with such high voltages can be very risky for the nonexpert.
– Photomultipliers are mechanically fragile and also, in electrical respects, quite demanding. Their smooth operation requires some prior experience.
– Good photomultipliers with a high-voltage source are expensive.

---

[16] Hammamatsu is a company specializing and leading in the production of radiation receivers. Other reputable manufacturers of photomultiplier tubes are RCA, EMI, and Philips.

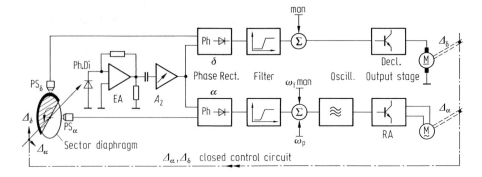

Fig. 5.27. Photoelectric tracking – schematic (after H. Blikisdorf [5.5]). The circuitry shown operates on the phase-selective alternating light method from Fig. 5.26d. It requires only one light sensor whose signal is amplified in EA and A. The amplified signal is fed to the phase-selective rectifiers (*Ph. Rect.*) These, together with the chopper-phase signals, form the components in α and δ. After further conditioning by scintillation and noise suppression filters, the signals are fed to the output stages for α and δ.

In conclusion, photodiodes are preferred for use by the amateur. The advantages noted fully compensate the somewhat lower limiting brightness.

5.12.4 Signal Processing and Control Electronics

Processing the light and phase signals depends on the modulating property of the photo-chopper. Usual modulation procedures are:

- *Amplitude modulation.* With the rotating disk, the signal amplitude is proportional to the star displacement for small amounts. This is the actual control range. Outside this range the control element gives full output pulse.
- *Modulation of the keying ratios.* In the wobbling disk and the quadrant sector of Fig. 5.26e, the *keying ratio*, which is the ratio of pulse width to pulse period, is changed by the star displacement.

Other modulating devices exist but cannot be discussed here.

A complete guiding system is shown in Fig. 5.27, where the schematics and suggestions are from the article by H. Blikisdorf [5.5], which should be consulted for more details. The disk rotates at about 5 rpm. A higher chopper frequency could increase the speed of correction. The control circuit would then, however, be more strongly affected by stellar scintillation. The light receiver is a high-sensitivity photodiode placed directly on the print of the electrometer amplifier (EA) behind the rotating sector. The EA preamplifies the signal current into the nA-range to an easily processable value. This preamplifier is carefully designed and wired and well-shielded. Another capacitor-coupled amplifier (V_2) then follows. The important elements are the two *phase amplifiers* (*Ph. Rect.*). These derive, from the light signal and the two-phase signals, the actual signals for the corrections, which are proportional to the two components $\Delta\alpha$ and $\Delta\delta$ of the displacement. A filter follows each to smooth the signals.

The declination is corrected through a DC motor. In right ascension, the control signal changes the oscillator frequency delivered to the AC synchronous motor by ±10%. The correction speed in declination is ±2 ″/s.

5.13 Adjust Elements and Telescope Alignment on the Celestial Pole

5.13.1 Adjustment Elements

To align the equatorial axis system with respect to the celestial pole requires adjusting elements in the meridional and horizontal planes of the mounting; in the former, the latitude φ and, in the latter, the azimuth A are adjusted. These important telescope parts are often rather poorly made in amateur telescopes and constitute weak points. Adjusting the telescope precisely with such deficient components can be a tedious task indeed.

The adjusting parts can be placed between foundation and pier (base adjustment), or between pier and bearing case of the polar axis. Instruments on tripods adequately provide an azimuth range over 360°. Here are some basic hints for the design of the adjusting elements:

1. The elements should respond precisely to the operation of adjustment and should not suffer from backlash. The use of the telescope determines the precision $\Delta\sigma_j$ for adjusting it. As mentioned in Sect. 5.3.11.2, the human hand has a limited accuracy grade of motion σ_m. The adjusting elements should therefore have a gear ratio u of at least $u \geq \sigma_m/\Delta\sigma_j$ This ratio is usually realized by a combination of screw and lever. Figure 5.28 shows schematically the function of such adjustment.
2. The elements for adjusting azimuth and latitude must be kinematically decoupled (i.e., independent). Kinematic uncoupling means that when adjusting one coordinate, the other is not also changed. This is a trivial requirement, but one which is not always satisfied in all amateur mountings.
3. The elements should not be in the line of main force, that is, they should not be part of the chain of stiffness. Figure 5.28 shows that in this widely used basis adjustment, the line of force continues via the lever and the screw. These two elements are thus a coupling link in the chain and can disadvantageously influence the overall stiffness. Adjustment precision and stiffness can make demands on the design which are largely mutually exclusive. A fine adjustment needs fine-pitch screws and long levers; both are elements which are quite unfavorable from the view of stiffness. Therefore, the main flow of force should be in a parallel path past the adjustment parts; otherwise, these should be relieved with respect to forces.
4. The elements should be lockable so that unintentional offsetting is prevented.

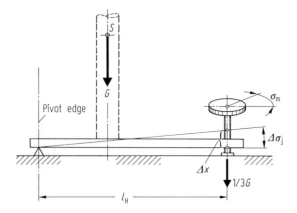

Fig. 5.28. Base adjustment for latitude on three-point contact with the ground. A fine adjustment needs a long lever l_H and a fine-pitch screw, which are related via

$$\tan \Delta \sigma_j = \frac{\Delta x}{l_H} = \frac{\tau \sigma_m}{360 l_H}, \tag{5.35}$$

where τ is the pitch of the screw. It is evident that here the adjusting elements are in the main line of the whole telescope weight and are part of the chain of rigidity. The requirements of adjustment conflict with those of stiffness (short leverage l_H, thick screw with large pitch τ). Elements in the line of force need special consideration.

5.13.2 Adjustment of the Axis System by the Scheiner Method

Before the adjustment, the placement of the optical axis at right angles to the declination axis should be checked and, if necessary, corrected by aiming the telescope tube alternately at both sides of the pier to a fixed sighting point. The adjustment of the instrument after Scheiner is in four steps, with steps 3 and 4 repeated alternately until the desired precision is reached. A reticle eyepiece with a sufficiently large field of view is needed. The mounting should be adjusted using the primary optics or the guide scope, but *not* the finder.

Step 1. Align the mounting very approximately in the north–south direction. This can be done during the daytime using a compass, or the Sun, or familiar terrestrial marks; at night, the Pole Star can be used. Note that ferrous material in the mounting can deflect the compass needle.

Step 2. Set the declination wire. This wire in the eyepiece serves as the reference line for the instrumental coordinate system. The declination wire is set as follows: set a star on the wire, clamp the declination axis, move the telescope by its RA drive slightly back and forth and observe the star. The star should run exactly along the wire. If not, rotate the eyepiece until this condition is met. Thereafter keep the eyepiece fixed.

Step 3. Adjust the instrument in azimuth. Point the instrument at a star near the zenith, set the star on the declination wire, clamp the declination axis. With tracking on, the star will gradually move to above or below the declination wire. Apply a correction in azimuth until this deviation is eliminated.

Step 4. Adjust the instrument in latitude. Point the telescope at a star with a declination of about 70° near its eastern elongation [5.6]. Set the star on the declination wire, then clamp the declination axis. With tracking on, the star will gradually drift off the wire. Correct the latitude until this drift disappears, then repeat Step 3. When the instrument is perfectly adjusted, lock the adjusting elements.

The mounting is considered correctly oriented when the star, after a long tracking period, does not drift off the declination wire. A slow drift along this wire is caused by the tracking speed and not by the adjustment. It should be mentioned that the instrumental axis system is in reality tilted against the true latitude by an angle corresponding to the refraction at the pole, as the instrument has been adjusted via an incident beam refracted in the Earth's atmosphere.

5.14 The Setting Circles and Their Adjustment

5.14.1 Setting Circles

Setting or divided circles are a valuable tool for setting the telescope on celestial objects, and should therefore be a component of any large instrument. Since their function on an equatorial mounting is merely to set on a star, and not to ascertain its coordinates (as is the case with a theodolite or transit circle), the divisions need not be extremely fine or accurate. The circles should be readable at very low light levels. The following hints will serve to divide a circle.

1. Recommended angular distances Δ between divisions are, in declination (360° division): $\Delta = 1°$ or $2°$, labeled: $+90° \ldots 0° \ldots -90°$. In hour angle ($24^h$ division): $\Delta = 2$ or 5 min, labeled: $\ldots 1^h \ldots 12^h \ldots 24^h$. When labeling, be careful of the direction of the polar axis (northern and southern hemisphere), and the sign of the declination, and the arrangement of the circles at the instrument. To use a moving circle with fixed index, simply reverse the labeling compared with a fixed circle with moving index.
2. The minimum required circle diameter D_K is given by the recommendation that the divisions $\Delta = 0.5 D_K \cdot \mathrm{arc}\Delta$ should not be less than 1 mm apart, or better, 2–3 mm.
3. Use thick markings (broader than 0.25 mm). The divisions can also be made as bars 0.5Δ wide. Such a division is readable even in poor illumination and permits an estimate for intermediate values of quarters of Δ.
4. White lines and numbers against a black background correspond better with the accomodation of the eye to the star-studded night sky.

The divisions can be made radially on a flat disk, or on the surface of a cylinder or cone. Ready-made division circles can be obtained from astro-supply centers and in specialty stores. There are also numerous possibilities for making them oneself in an inexpensive and useful way.

5.14.2 Adjustment of the Circles

To set the divided circles to the *instrumental coordinate system* requires that the circles can be finely rotated relative to the pointers, or vice versa. Amateur mountings often have the circles clamped with bolts directly on the shafts. Such an attachment nearly always causes damaged seats on the shafts, and it is hardly conducive to precise adjustments. Tangential and radial clutches (clamps), as mentioned in the discussion of slow motion, are also the best joining elements between circles and shafts.

5.14.2.1 Adjusting the Circles by the Kolbow Method. In this method, the instrument is first adjusted with respect to the polar axis (Sect. 5.13.2). Next, the instrument is pointed to the zenith and the declination axis made horizontal: the declination circle should then indicate the latitude φ of the geographical position. As the instrument is also in the meridional plane, the hour circle should read 12^h on one side of the pier and 24^h on the other. The adjustment procedure is as follows: Point the instrument in both the east and west positions – or, for fork mountings, in both positions of the fork – to the zenith. Read the circles in both positions and repeat a few times. Find the differences of readings from the nominal φ and 12^h or 24^h for averages, and correct the circles accordingly.

The problem with this method is that the zenith is not a visible reference point, and the zenith position has to be found with mechanical aids of limited precision, such as a bubble level or plumb bob. Also, not every telescope optical system has straight and plane reference surfaces perpendicular and parallel to the optical axis, or is suitable for attaching a level or a plumb bob. Thus this method affords only limited precision.

5.14.2.2 Setting the Circles from Known Star Positions. This method can, in principle, be made as accurate as desired. Point to several stars near the zenith and find their positions from the circles and the clock. Compare them with the correct positions from a star catalogue, allowing for precession if necessary, and determine from the difference the circle corrections. A least-squares calculation (see Chap. 3.2) may be used if it is thought that this will improve the precision.

5.14.3 Digital Position Displays

Professional telescopes are now usually equipped with digital position indicators and computer-operated positioning devices. The signals needed for this are generated by angle encoders coupled with the polar and declination axes. The angle encoders or increment encoders replace the divided circles. Their precision and resolution determine the precision of the positioning and display they control. There are two encoder systems:

1. *Coded-angle encoders (GRAY-code encoders).* Each angle of position is represented by a code number. The numbers are displayed as *bar codes* on a glass circle and are read by reading optics transferred into binary digits or "bits," and the angle can be directly displayed on a simple indicator, or interfaced into a computer. GRAY-code encoders are quite expensive, especially for higher angular resolution.

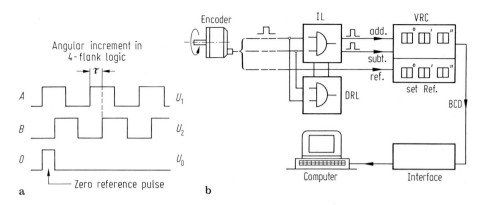

Fig. 5.29a, b. Digital position indication with incremental encoder. **a** graphs the pulse sequences given by the encoder. The phase-shifted signal B discerns the direction of motion. **b** The pulse logic IL/DRL instructs the counter whether to count forward or backward. If the counter VRC is equipped with a BCD-exit, the position signal can be fed into a computer.

2. *Incremental encoders.* These devices do not provide absolute angles but rather a defined number of pulses for a change of angle $\Delta\beta$. The original position β_0 of the pulse sequence is therefore entered manually. It is erased when the power is switched off and so always has to be entered anew.

Incremental encoders have a glass circle with ruled grating of division τ which is scanned by photodiodes. Generally, these code tracks have three outputs. One track generates a signal U_1, another an electrically phase-shifted (by 90°) signal U_2, and the third supplies the signal of a reference mark U_0. Signal U_2 is needed for the *direction discriminator* (DRL) which instructs the counter to add or to subtract the angle increments to or from the reference angle β_0. This needs a forward–backward counter VRC. The phase-shifted signal also increases the angular resolution by a factor of 2. Incremental encoders are comparatively inexpensive. Numerous makes and a variety of encoder types are on the market and cover the full range of applications. For reasons of costs, only these should be considered by amateurs.

Figure 5.29 shows the signals and schematics of digital processing of position. If the circular grating has Z lines, the angle increment is $\Gamma = 360/Z$. The input logic IL can register from the two signals either the front slope, or back slope, or both, and forward them as pulses to the counter. Correspondingly, an angular resolution of $\Gamma/2$ or $\Gamma/4$ is obtained. An encoder with an 1800-line grating therefore has an angular resolution of 3′.

Digital positioning should also include realistic accuracy considerations of the entire instrument. It makes little sense to use precise and expensive encoders of arcsecond resolution if the adjustment and mechanical precision of the telescope are in the range of arc*minutes*. The circuit schematics are simple and need no further explanation. The incremental encoder can be linked in two ways to the astronomical coordinate system:

1. With the electronics switched on, the telescope is pointed to a star whose coordinates are then input into the counters as reference angles.
2. Another method uses the reference mark U_0. In the first step encoders are mechanically adjusted just as circles are. This assigns certain angle positions to the reference marks. Before use, the telescope is set on the reference position, and the corresponding angular values are then entered into the counters.

5.15 Electrical Equipment

5.15.1 Power Sources and Safety Aspects

Power for the electrical telescope accessories can be obtained from a standard electrical outlet connected to the local power line (220 V/50 Hz or 110 V/60 Hz) or from a battery. Batteries are needed for portable instruments or when observing in places where a power line is not available. In such cases, the car battery can be tapped. The type of power supply to be employed is an important factor in planning for all telescope applications. The properties of the intended power source should be considered as early as when planning the telescope installation and when acquiring the electrical accessories. Thus, battery operation should be planned with battery-matching voltages and a low current consumption by all supplied instruments.

The current from a power line, as well as from batteries, presents potential hazards. Nonexperts may not be aware of all these risks or may underestimate them. It is thus necessary to comment on them here.

Dangers from the Power Line Voltage. It is common in amateur circles to build one's own electrical equipment and to connect them direct (via an electrical outlet) to the power line, or even to build electrical installations. The risks involved in using amateur telescopes are especially enhanced. They are operated in the dark and often in the presence of dew and humidity. The observer is rarely on well-insulated ground, and the electrical parts are often improvised or concocted from cheap components. The risk of contacting the line voltage is thus disturbingly high. The physiological effects of a current on the body are described in the following paragraph:

Currents over 10 mA cause such strong muscular spasms that it may be impossible for the victim to release the current-carrying parts on his own. Currents over 40–50 mA are lethal. For a given voltage, the current flowing through the body depends in practice only on the skin resistance at the points of contact. This in turn depends on several factors and ranges between a few hundred and 2000 Ω. As a consequence, for the standard line voltages of 220 or 110 V, the condition for a fatal dose of current is always present.

Thus every amateur constructing and operating electrical tools at the telescope should take the following principle to heart:

> **Criterion 12.** It is highly irresponsible to work directly with power-line voltages of 110 or 220 V at an amateur instrument. Only when the voltage is below 48 V can the operation of equipment be considered safe and admissible. The supply for the telescope's electrical power system must be fed via a faultless *low-voltage insulated transformer*.

Unfortunately, even commercial manufacturers market instruments with direct line-voltage connections to the instruments and which fail to match safety conditions and regulations. The author knows of cases where instruments connected in this manner via a cable reel are operated freely outdoors. Furthermore, from a technical viewpoint, there is simply no convincing argument which can be made in favor of using the standard line voltage.

A useful voltage for the telescope is 12 V-AC. This is a widely used standard in automobile electrical systems, and thus many components are commercially available. Also IC modules and digital circuitry can be powered with it after rectification. The electrical installation is not critical even when not entirely expertly made. Even in the event of a short-circuit, the consequences are not serious. But the individual branches and consuming devices should be protected by correctly dimensioned fuses. The safety-insulated transformer should be a tested, high-quality product, and should have a certain power reserve. A low-voltage power supply is graphed in Fig. 5.30. Low-voltage installations should have amply dimensioned wire cross sections so that a voltage drop has little impact. The voltage drop ΔU of a two-wire copper line (conductivity $\rho = 0.017 \ \Omega \ mm^2/m$) carrying a current I (A) is

$$\Delta U = 0.043 \cdot \frac{I \cdot L}{D^2}, \tag{5.36}$$

where L (m) is the length of the wiring and D (mm) is the wire diameter. ΔU should be below 5%.

5.15.2 Batteries: Properties and Hazards

Amateur instruments employ primarily rechargeable lead or nickel-cadmium (NiCd) batteries. Both types are available in two makes: in open or half-open cells with liquid electrolyte (e.g., a car battery) or in a gastight casing. Gastight batteries are now contained in many pieces of electronic equipment. They contain only small amounts of electrolyte taken by an absorbing separator between the electrodes. They are useful for nearly every purpose, do not leak, and are thus particularly suitable for portable instruments. Lead batteries contain sulfuric acid ($H_2SO_4 + H_2O$) as the electrolyte and the alkaline NiCd batteries potash lye ($KOH + H_2O$). Handling batteries also has its hazards:

– *Danger from the electrolyte.* H_2SO_4 and KOH are strongly caustic compounds. In contact with the skin, they cause painful burns which are slow to heal. Particularly dangerous is any contact with the eyes. Leaking sulfuric acid also attacks many metals, textiles, and materials. Potash lye affects textiles, colors, paints, wood, and plastics. Lead compounds and cadmium are toxic and are strong environmental

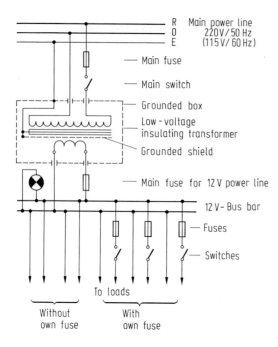

Fig. 5.30. Low-voltage power supply for an amateur observing platform. All electrical circuits of the stand or telescope are powered from an insulating transformer with 6 or 12 V. Only this guarantees adequate safety of the observer. The circuits can be built at home without risk. Parts without their own fuse protection are provided with fuses at the distributing bus (seen in the graph) from which the various branches depart.

hazards. Batteries must therefore be handled very carefully. When transporting them, do not damage or tip over open cells. Rubber gloves, protective glasses, and an apron should always be worn when working with them (e.g., changing the electrolyte). The electrolyte should never get into the sewage system or into groundwater. Old batteries should be dropped off at official collecting points. Inquire at municipal offices or in drugstores. If contact is accidentally made with the electrolyte, the affected area should immediately be flushed with water and then treated with one of the following depending on the battery type:
 – soapy water to neutralize the sulfuric acid from lead batteries, or
 – vinegar diluted with water to counteract the potash lye from nickel-cadmium batteries.
– *Risk of explosion by overcharge and oxy-hydrogen formation.* Charging the battery dissociates the electrolyte. The highly explosive oxy-hydrogen develops and quickly escapes from open cells. For gastight cells, the forming oxygen is chemically bound but only to a limited extent. If there are high charging currents and extended overcharging, the cell may, despite overpressure valves, explode. Open-cell batteries should be charged in well-ventilated places (because of toxic fumes)

with no open flames nearby. For gastight cells, avoid overcharging for long periods, which, incidentally, also diminishes the lifetime of the battery.
- *Risk of burns from short circuits at the clamps.* Batteries have very low internal resistance. If the outer circuit is also of low resistance, it can, for a short time, carry a 100 A current or more. Short circuits can thus lead to severe burns. Hence, never provoke short circuits at the clamp of the battery, for instance, for the purpose of checking with a wire to see if it is still carrying a charge.

The vital data and properties of lead and NiCd accumulators with liquid electrolyte are compiled in Table 5.12. Least expensive are lead car batteries, but they are rather problematic and have substantial drawbacks when compared with NiCd batteries. When not constantly charged their lifetime is short. The properties of gastight lead batteries are in some respects better, but they are more expensive. The reverse holds for gastight NiCd batteries: they have a shorter lifetime than cells with a liquid electrolyte and are in some respects quite delicate.[17] Considering the long life, the sturdiness, and other advantageous features of NiCd batteries with a liquid electrolyte, they provide by far the most preferable and least expensive solution. Their only disadvantage is the requirement that they be used in an upright position. Modern makes have plugs which prevent the electrolyte from immediately leaking out should the cell tip over. The only maintenance required is a change of the electrolyte every four or five years. Only the electrolytes prescribed by the maker of the accumulator should be used, and the instructions for changing the electrolyte should be followed to the letter. If drives and other electrical equipment are designed so as to conserve current, then the power supply from dry batteries (coal/zinc or lithium batteries) is feasible and should be considered.

5.15.3 Power Supplies for Electronic Circuits

Electronic parts are powered with defined and stabilized DC voltages, for instance +5 V for digital circuits with TTL-logic. Certain IC modules contain an integrated voltage regulator supplying the circuit with the required voltage. These modules can be operated in a wider voltage range, for example 5–18 V, as can be learned from the operating instructions. Operational amplifiers (OPAMPs) need a centrally symmetric to common voltage of ± 15 V and, for the driving stage of stepping motors, even as much as 45 V may be required. All of these voltages can be operated without any risk. The only exceptions are photomultiplier tubes which need stabilized voltages of 1000 V or more. For these, high-voltage modules, integrated in the tube base and hermetically sealed, are now obtainable. Nevertheless, the handling of a photomultiplier is a risky and very demanding job which should be left to an electronics expert.

Building a stabilized power supply today is not worth the required effort, as ready-to-use power supply modules and complete power packs are available at reasonable

---

17 Included on this count is the *memory effect*, the reduced lifetime, and the sensitivity to overcharging. The author has already, despite careful handling, witnessed the explosion of some cells!

Table 5.12. Characteristic data of batteries (open cells with liquid electrolyte).

Properties, characteristics		Lead accumulator	NiCd accumulator
Electrodes	Anode +	PbO_2	$Ni(OH)_3$ (multivalent nickel hydroxide)
	Cathode −	Pb (metallic)	Cd (metallic)
Electrolyte		$H_2SO_4 + H_2O$	$KOH + H_2O$ (potash lye)
		participates in elect.-chem. reaction	chemically inactive, solely current conductor
Charge–discharge reactions		PbO_2–$H_2SO_4 + H_2O$–Pb charged ↑↓ $PbSO_4$–H_2O–$PbSO_4$ discharged	$2Ni(OH)_3$ Cd charged ($KOH + H_2O$) ↑↓ $2Ni(OH)_2$ $Cd(OH)_2$ discharged
Cell voltage at rest (V)		2.05	1.36
Mean discharge voltage		1.9–2.0	1.1–1.2
Mean charge voltage		2.3	1.45–1.55
Final charge voltage		2.7	1.65–1.75
Efficiency (%)		About 85	About 70–85

Table 5.12. (Continued)

Properties, characteristics	Lead accumulator	NiCd accumulator
Self-discharge	High: about 1% per day	Low: 0.05 to 0.1% per day (at 25°C)
Capacity/load dependence	Strong capacity decline at high load	Capacity depends little on drawn current
Sensitivity to shorts	High: diminishes lifetime	Low; special shockproof cells commercially available
Discharge when low: overcharge	High: diminishes lifetime	Not sensitive; discharge down to 0 possible overcharge is even needed for cell regeneration (except gas-tight cells; avoid overcharging)
Temperature dependence	Quite pronounced, strong decline of capacity at low temp., much reduced lifetime at high temperature (cell corrosion)	Not substantial; a good low-temperature battery; lifetime not affected by higher temperatures
Storage properties	Poor; is quickly damaged by sulfates when stored uncharged	Stores well (charged or discharged); recommended for long-time storage; discharge down to 0.6 V/cell
Servicing demands	High; needs charge all time, needs check on electrolyte density	Less demanding, more rugged; electrolyte change at about 4–5 yr intervals
Corrosion of other items	High; charging releases aggressive acid fumes and some oxy-hydrogen gas	None; charging releases dry oxy-hydrogen gas
Lifetime	Short; cells self-corrode	Very long
Weight	High	Medium
Costs	Low; car batteries are inexpensive mass production	Quite expensive, but a good buy, considering Durability, ruggedness, other advantages
Method of charging	By constant-voltage charger; fast charging not advisable, reduces lifetime	By constant-current charger; fast charging Possible with multiple nominal current

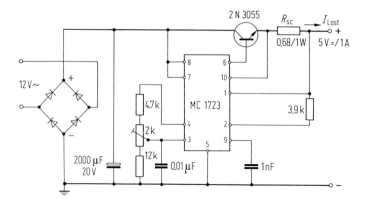

Fig. 5.31. Stabilized power-supply unit to power TTL circuits using a power module MC 1723 and a series transistor 2N 3055 by Motorola. The module is powered directly from the 12 V AC power supply over a diode bridge. The 2 kΩ trimmer adjusts the output voltage. The current limitation is via the resistor R_{SC} and prevents short circuits. With cooling of the 2N 3055 ($I_c = 15$ A) the circuitry can be designed for higher currents.

prices. The connection of the modules is quite simple and can be read from the instruction sheets provided by the manufacturer (e.g. *Voltage Regulator Handbook: Theory and Practice* by Motorola). Figure 5.31 (reprinted with kind permission from the Motorola brochure) shows such a power pack. An external power transistor permits an increase in the current range of such modules and power packs with variable voltage and current limitation for general laboratory tasks. Automatic recharging units for NiCd batteries can also be built. Power modules are also available for ± 15 V. To perform the regulating function, the input voltage U_{in} of the modules must be about 25% higher than the required output voltage U_{out}. The secondary voltage (effective no-load voltage) at the network transformer should then be higher still than U_{in} by a factor 1.15. This takes care of the conversion $U_{DC} \to U_{eff}$ and the voltage losses.

5.15.4 Illumination

The general illumination of the observing platform can be made from the 12 V AC network with small bulbs such as the ones used for interior illumination in automobiles. The 110 or 220 V line is not needed. Simple installations can be built by anyone without risk and with considerable cost savings. Low-voltage incandescent lamps are, by virtue of their higher efficiency, preferable to normal light bulbs. Even better are *low-voltage halogen lamps*. Some 6–10 W small bulbs should suffice to illuminate the platform. With a *dimmer module*, a thyristor with a phase-gated control, the brightness can be adjusted to what is needed. Arrange the lamps so that they provide an indirect, not blinding, illumination.

Crosswire and circle illumination, signal lamps on electronic parts and installations, and reading lamps need only low light levels. The brightness of these lamps should be

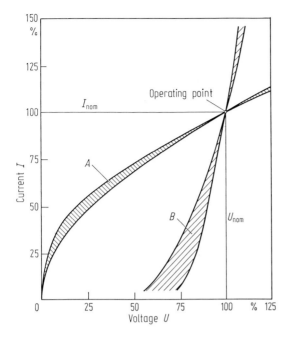

Fig. 5.32. Current/voltage characteristics of small light bulbs and light diodes. Small bulbs *(A)* and diodes *(B)* have very different characteristics. This is to be considered when dimensioning the resistors. The steep characteristic curve especially apparent in LEDs, which makes LEDs sensitive to excessive voltages, is to be noted. The generalized characteristics cover practically all makes and applications.

compatible with the dark-adapted eye. For the purposes of illumination, light diodes, which use less current than standard light bulbs, can be effectively used. *Light emitting diodes* (LEDs) are available in the colors red, yellow, and green. Red is especially preferred for the dark-adapted eye. A typical LED has a luminous intensity in the range 2–12 mcd[18]. Also available are high-power LEDs with luminous intensities of 40–1000 mcd. This covers the range from crosswire to reading lamps. The operation current of LEDs is between 5 and 30 mA at forward voltages of 1.8–2.8 V. Figure 5.32 graphs the typical current/voltage characteristics of small bulbs and light diodes. The steep characteristic curve makes LEDs sensitive to voltage changes. They are thus always operated over a properly dimensioned resistor. A second regulating resistor switched in series permits the brightness to be made comfortable for the eye.

A crosswire or reticle used for nighttime work can be made visible either by *wire illumination*, in which case it is illuminated so that the wire marks appear bright in the dark field, or by *field illumination*, where the entire field is brightened slightly so that the wires appear slightly dark before it. Wire illumination requires special eyepieces with built-in illumination. Such eyepieces are expensive. It is less costly to build a crosswire eyepiece for field illumination. This can be done with ordinary Kellner-

---

[18] mcd = millicandela, where the *candela* is the photometric unit for *luminous intensity*.

or Ramsden-type eyepieces. A frame with fine wires (hairs, webs) is placed into the eyepiece tube so that the wires are in the focus of the eyepiece. The field illumination is a separate unit in the beam of the telescope and is located as far as possible from the eyepiece. In a refractor this may be at the dewcap, while in a reflector it is at the secondary mirror or at the top of the tube. Field illumination gives favorable contrast since the wire is easily distinguished from the bright stellar disk, but about 1/2 magnitude is lost by this method owing to the brightened field of view.

5.15.5 The Dewcap and Its Heating

Summer nights with high humidity provide favorable conditions for the formation of dew, which can be especially annoying if it occurs on optical surfaces. It primarily occurs on the front lenses of photo-objectives and refractors, on meniscus lenses of Maksutov systems, and on Schmidt corrector plates when these surfaces radiatively cool at high air humidity to below the dew point. This cooling below the temperature of the surrounding air is purely a radiation effect[19], and it seldom occurs on Newtonian and Cassegrainian mirrors. The reasons for this are the small emission value of the mirror-coated surface and the tube's acting as a radiation screen. To avoid dewing of the optical surfaces, the dewcap should have two functions:

1. *To reduce to a minimum the radiation transmitted to the cold sky background.* The front surface of glass faces not only the small field of view but, with regard to radiation, a large solid angle. As glass in the given temperature range is an excellent radiator (emission coefficient $\varepsilon = 0.93$), the radiative loss into this angle is quite large (about 5–10 mW per cm^2 of glass surface). The dewcap serves to reduce as much as possible the extent of the solid angle into which the glass surface radiates. This is done with a radiation screen placed well ahead of the objective, whose aperture is not larger than the optical aperture and which is covered on the outside with a material (usually aluminum foil) which emits very little radiation toward the sky, while the inside is covered with material which is a good thermal insulator (styrofoam).
2. *To compensate the remaining radiative loss by a suitable energy (heating) supply.* The aperture in the screen still allows energy to radiate away from the glass. This has to be compensated by a targeted energy supply. Glass is an excellent heat radiator but a very poor heat conductor. Thus it is not feasible to add the energy by the conductive mechanism in the glass as is the case for many heated dewcaps. The requisite large temperature gradient generates strong turbulence. Figure 5.33 shows how to replace the radiative losses by radiating in energy. The heating power needed is very small, and the heated surface need be only slightly above the surrounding air temperature.

---

19 Measurements by the author indicate that when the optical element is dewed over, the surrounding air is still 2–3° above the dewpoint. If this were not the case, then thick fog would surround the observer, putting him or her out of work.

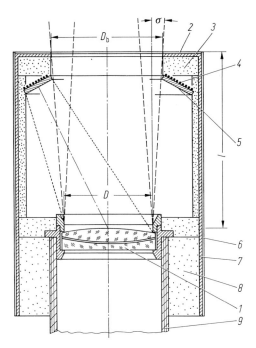

Fig. 5.33. The heated dewcap. The radiation screen with aperture D_b and visual field angle σ should expose the objective *(1)* to a small sky sector, where

$$D_b = D + 2l \tan \sigma \quad \text{with} \quad l \geqq 2D. \tag{5.37}$$

The front side of the screen *(2)* and the outside of the dewcap *(6)* are covered with high-reflecting aluminum foil. The inside of dewcap *(3)* and the objective are thermally insulated with styrofoam *(8)*. Other features include *(7)* cardboard tube for dewcap, *(9)* optical tube, *(4)* filament winding, and *(5)* radiating surface of heating.

5.16 Comments on Literature Cited Here and in the Supplemental Reading List

The introductory part of this chapter has already mentioned several (although certainly not all) sources which provide information on numerous telescope mountings. A large part of the designs presented have a substantial creative level. A critical analysis, however, uncovers violations of elementary principles of mechanics in many of these constructions. Designs should stimulate fresh designs and ideas, but should not be uncritically copied without searching for technical discordances and weak points.

Where can the reader find literature beyond the basics presented? As already mentioned, there is no work that treats the mechanics of telescope mountings in an elementary way. The interested reader will therefore have to resort to a general text on mechanics, vibrations, machine elements, and so on. Data and numerical values are also to be found in engineering handbooks. The problem for the technically less-experienced amateur is how to make use of this general literature with respect to his special objective. There is one area of engineering in which the same or very similar basic criteria apply: machine tool design. Here, the emphasis is on small deformations, the high freedom from vibrations, precisely pivoted shafts, and high kinematic accuracy. On this subject there are various publications which are relevant and of interest in the design of mountings.

References

5.1. Ziegler, H.G.: *Orion* **36**, (1978).
5.2. *Handbook of Chemistry and Physics* (56th ed.), CRC Press, Cleveland 1975.
5.3. Güssow, K.: Elektrische Fernrohrantriebe. *Sterne und Weltraum* **26** (1987).
5.4. Höbel, P.: Photoelektrische Nachführsysteme. *Sterne und Weltraum* **7/8**, 225, and **12**, 372 (1973).
5.5. Blikisdorf, H.: Eine optoelektrische Nachführung für die Langzeitfotografie. *Orion* **202**, 121 (1984).
5.6. Schürer, M.: The Influence of Refraction on Adjustment and Tracking of Equatorial Mountings. *Orion* **38** (1980).

6 Astrophotography

B. Koch and N. Sommer

6.1 Introduction

Compared with direct visual observing, photography has the distinct advantage, owing to the light-integrating properties of the photographic emulsion, of being able to record the images of faint objects not within reach of the eye. Once developed, photographs also provide a permanent document which can be unbiasedly inspected and analyzed at any convenient time.

Since the last edition of this book[1], the field of astrophotography has developed in a manner which could not have been foreseen. The availability of increasingly faster emulsions and more powerful lenses has dramatically improved the situation regarding astrophotography within the amateur realm, and thus has stimulated many more observers to turn to this sector of astronomy. It is not surprising that a large number of amateurs are now obtaining results of "professional" quality. The techniques employed are essentially the same as those of professionals and are evidently limited not by instrumental, but rather by financial constraints. One purpose of this chapter will be to refute the notion that astrophotography is not worth pursuing unless elaborate equipment, the operation of which is rather complex and difficult, is available. Fortunately, this is not true. In fact, whatever equipment is employed, successful astrophotography depends on a combination of personal tenacity, well-thought-out use of the available means, interest in experimenting, and, not least, experience.

6.2 Cameras and Lenses

6.2.1 Single-Lens Reflex Cameras

In principle, astrophotography differs from visual observations only in that the eye is replaced by another sensor, the photographic emulsion. In this sense, all astronomical optics can, given the mechanical requirements, also be used as photographic lenses. For certain purposes they are indispensible. The common wide-angle and telephoto lenses have the advantage that they are widely available and can be used for ordinary

---

[1] The authors refer to the 3rd German edition (1981).

photography. They can also record very large celestial areas—otherwise reserved for expensive special optics. The available range of focal lengths for single-lens reflex (SLR) 35-mm cameras begins with the 8–16 mm "fisheye" lenses and extends to the 500-mm telephoto objective. These correspond respectively to angular extents from nearly 180° down to 5° measured diagonally on the rectangular 24 × 36-mm format.

Aspherical lenses are the products of optical artistry, as are those made from special materials like calcium fluoride (CaF_2) or from low-dispersion glasses, and provide high imaging quality with enormous light-gathering power. They are not, however, likely to become popular as tools for astrophotography. The latter are rather the "normal objectives" with focal lengths of around 50 mm, wide-angle objectives with focal lengths of 28–35 mm, and telephoto objectives with 200-mm focal length. Unfortunately, wide-angle objectives at full aperture suffer from strong vignetting, which appears as a diminishing brightness toward the edges of the field. This is caused by the incidence of a beam at an oblique angle with the optical axis, and thus onto a perspectively reduced entry pupil. This effect increases with $\cos^4 \alpha$, where α is the angle of incidence with respect to the optical axis. Moreover, the front and rear lenses are often underdimensioned, and thus, especially in fast lenses, contribute to increased light loss at the edge of the field of view. The amount of this vignetting can be diminished by reducing the aperture, which is obviously at the expense of light-gathering power, but which does improve the image quality at the film's edges for all lenses [6.1,3]. The usefulness of lenses for larger format cameras combined with 35 mm camera bodies has been mentioned; they utilize the most often well-corrected central part of the field of view.

6.2.2 Medium and Large-Sized Cameras

While the 35-mm SLR cameras are the most widely used cameras for amateur astrophotography, there do exist larger-format cameras which can also be profitably employed. Medium-format astrocameras (6 × 4.5, 6 × 6, 6 × 7, 9 × 12 cm) and large-format cameras have advantages and drawbacks which the various users will weigh quite differently. Fewer emulsions are available, and their processing has to be done by oneself or by sending the exposed film to a professional lab. Usually the lenses are slower and much more expensive. Furthermore, telescopes—apart from rare exceptions constructed with the specific purpose in mind—do not illuminate these formats without vignetting. On the other hand, there are certain tasks where the advantages dominate, such as detailed photography of very large celestial areas, or—in cameras which use glass plates—the conservation of image scale on the emulsion-coated glass, a paramount consideration if the images on the plates are to be subsequently measured and analyzed.

While do-it-yourself construction of large- and medium-format cameras presents no insoluble problems, it is crucial to keep the film plane or glass plate perpendicular to the optical axis. Thus the correct positioning is locked securely with setscrews (see Fig. 6.1). Since most materials expand or contract with changing temperature (e.g., metals) or "warp" owing to aging and humidity (e.g., wood), corrections for these deformations must be possible.

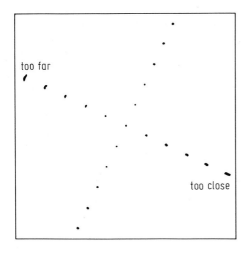

Fig. 6.1. The plane of the film is perpendicular to the optical axis. The sketch shows schematic stellar images in the case of misalignment. Radially distorted stellar images indicate that the plane of the film or plate in this part is too close to the lens; tangentially distorted images mean that it is too distant.

6.2.3 Other Cameras

Simple box cameras, old-fashioned folding cameras, or even the popular pocket cameras are also useful. When using a folding camera, the mechanics of the folding or switching mechanism should be checked so that it operates without play. A simple viewfinder is sufficient when pointing the camera but this practically eliminates using the camera for photography through the telescope.

6.2.4 Refractors

Anyone who has attempted to photograph the moon with a 135-mm telephoto lens was undoubtedly surprised by the disappointingly small image size. A refractor with a focal length of 1500 mm forms a lunar image with a 13.6 mm diameter. As a rule of thumb, each 1000 mm of focal length corresponds to roughly 10 mm of diameter of the Moon or Sun on the film. As has already been described elsewhere (Chap. 4), very good refractors offer very high contrast. To eliminate the secondary spectrum of fast Fraunhofer optics, a weak yellow filter should be used, because otherwise photographs on color film would exhibit color rims around highlights or simply blurred images on B/W film. This is avoided in apochromatic systems and in those refractors with fluorite lenses (see Chap. 4.5.1).

6.2.5 Reflectors

Compared to equal aperture refractors, fast reflectors are better-suited for deep-sky photography, i.e., detecting faint nebulae outside the solar system as well as faint planetary moons. The often extremely compact Maksutov–Cassegrain and Schmidt–Cassegrain telescopes are, in contrast to refractors of comparable aperture, especially well-suited for being carried to remote locations far from urban light pollution where serious astrophotography can be undertaken.

6.2.6 Special Optics for Astrophotography

When hunting extremely faint and extended nebulae—particularly with filters—the fastest lenses are required. At nearly all major observatories around the world the standard instrument for this purpose is the *Schmidt camera*, introduced in 1931 by Bernhard Schmidt in Hamburg. Nowadays, telescopes originating from the Schmidt design (i.e., for photography only) can be obtained at reasonable prices from various suppliers in Belgium, Japan, and the U.S.A. Of of the numerous variants of the Schmidt design, such as the *Wright–Schmidt*, *Baker–Schmidt*, *Schmidt–Cassegrain*, and the *Wright–Väisälä* telescopes, some can also be used for direct viewing. Their f/ratio approximates $f/0.7$ for meteor and satellite monitoring, whereas more commonly the range is from $f/1.5$ to $f/4$ at focal lengths of about 200–1000 mm. A property of many systems is their tube length of about twice the focal length. The corrector plates are sensitive to dew formation, but can be protected by a dew cap and, if necessary, additional heating. Their extreme speed and exceptionally good imaging quality render the tolerance of focusing very narrow (about 1/100 mm). Therefore the film holders are often mounted on invar rods so as not to be affected by temperature changes.

To cover even larger fields than can be reached by superwide-angle objectives ($f \approx 21$ mm), a fairly inexpensive alternative to popular "fisheye" optics is the *wide-angle adaptor*. This device is supplied in various widening factors, and can reach, when combined with a lens of at most 28 mm focal length, an image field diameter of nearly 180°. Unfortunately, the imaging quality is often poor and must be improved as much as possible by reducing the aperture. The illuminated field reaches 20–22 mm in diameter in the focal plane. Yet another way is to photograph the image of the sky seen in a highly reflective spherical surface. In this technique, a camera with telephoto or normal lens is suspended above a convex mirror (even a polished hubcap from an automobile will do!). The instrument should preferably be fork-mounted and guided equatorially [6.4,5,6].

6.2.7 Video Cameras, Image Intensifiers, CCDs

Recently electronic devices have become more and more popular for the recording of astronomical objects. The first encouraging results by means of *charge-coupled devices* (CCDs) and the increasingly light-sensitive video cameras were reported in the mid-1980s. The clever use of image-processing techniques combined with powerful personal computers will result in a wealth of information even from short exposures. CCDs possess several distinct advantages over photographic emulsions, the first being their very sensitive reaction to incident photons of greatly different wavelengths from the near UV of about 350 nm to the near infrared [6.60,193]. While normal astro-emulsions process merely about 0.6% of all light quanta (about 4% for hypered films), this value can reach nearly 80% for CCDs [6.194]. Second, intensity ratios of about 5000:1 can be recorded with CCDs, while films achieve only 300:1 at best. Finally, the image information can be obtained directly in computer-processible form. But the size of the imaging elements (pixels) renders photographic emulsions with their immense information content per unit area superior to CCDs as long as, especially for short focal lengths, the lens is not seeing-limited but optically limited [6.191]. *Image converters* and *intensifiers* increase the number of the electrons generated by photons striking a light-sensitive layer. Their maximum of sensitivity may reach out to the near infrared region of the spectrum, while the information from the image delivered comes into the visual range; hence the name "image converters." The image obtained with these devices often differs markedly from those obtained in the visible wavelength range, and this sometimes leads to problems of correlating the observed area with finder charts. Their price is rather high and has remained unchanged for many years [6.192].

6.3 General Considerations

6.3.1 Focusing

As far as the camera is concerned, all the objects of celestial photography lie at essentially the same distance (i.e, infinity), and therefore no special controlling mechanism is needed for the focus setting. One can usually assume that the infinity marking or the end of range of the lens have been correctly adjusted by the manufacturer. If in doubt, a "control" photograph will verify this. It can consist, for instance, of a star trail composed of several parts, each with a slightly different focus setting. The segment with the least width is closest to the ideal setting (Fig. 6.2). Or, one can make some exposures of lamps several miles away so as to serve as point-like sources (at a distance of more than about 1000 times the focal length!), again with slightly different focus settings for each exposure, and finally take the best focus position.

The need for focusing tests becomes particularly necessary when filters are used. Lenses consisting of a large number of optical elements are often well-corrected for chromatic aberration which, however, applies primarily to a limited visual spectral range. When using band-pass or narrow-band filters beyond this spectral range in front of the lens, a correction of the altered focal position must be applied according to the procedure described above. This may also hold for built-in filters which can be interchanged only as parts of the optical system. Most modern SLR cameras are equipped with an interchangable *focusing screen* containing a central split-image focusing spot which, for normal photography, provides fast and adequate focusing. For photographing astronomical objects through the telescope, however, very fine-grained matte-type focusing screens are advantageous. Clear-glass focusing screens are best suited for projection photography of the Sun, Moon, and planets. The procedure for making a screen is described by Röhr [6.36].

Indispensible is a right-angle finder, or a *focusing ocular,* or, best of all, a combination of the two. The focusing ocular, preferably with diopter adjustment, should afford 3–5× magnification. To use it, the camera body without lens is pointed toward a light source, and the focusing ocular is adjusted for best visibility of the surface structure

Fig. 6.2. Photographing a star field to find the best focus. With the lens successively covered and uncovered at 30-second intervals, exposures have been made with different accurately-recorded focus settings. The last exposure was 1 minute in order to unmistakably ascertain the correct arrangement of the sequence. The sharpest focus over the entire field is given by exposure no. 2.

of the focusing screen or acuity of the crosshairs. Marking this focus position permits the observer to check it before use. The best focus with matte-type focusing screens is reached when, in repeated slow passings through it, first, the stars look subjectively smallest, and second, the faintest stars in the field are visible. Using a clear-glass focusing screen with edged crosshairs, the exact focus is achieved when the object in the field does not shift against the crosshairs upon slight sideways motion of the eye (i.e., the absence of optical parallax). If the focusing screen cannot be replaced, then the surrounding microprism, or better still, the large, matte-style screen part should be used for focusing. The mircroprisms of the focusing screen darken from f-ratios around f/3.5 to f/4 and as such are not very useful.

When using verniers at the focusing knob, the observer should always approach the setting from the same side (direction) in order to avoid backlash. For extremely precise work, it may be necessary to prepare a table on the change in the focus as a function of temperature. Liller and Mayer [6.9] describe a successful method of averaging the vernier reading after repeated focusing and subsequent setting on the computed average.

Another method of focusing which is unquestionably the most accurate but also the least convenient is to place a *knife-edge* directly in the plane of the film. To employ this technique, the instrument with attached camera body (but without film and with

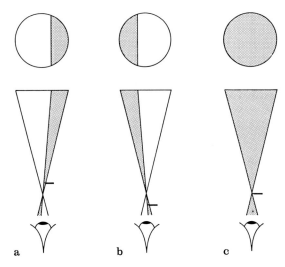

Fig. 6.3a–c. A knife-edge traverses the light beam from a star in or near the focal plane. Only in the precise focus is the light from the star suddenly blocked, whereas on either side of this point the beam is gradually narrowed.

opened camera back!) is pointed toward a star. For those optical systems which are focused by moving the primary mirror, this star should be close to the object which is later to be photographed in order that the focus setting as found does not change again when the telescope is moved. Unfortunately, the observer, who must look through the telescope past the knife-edge, can sometimes end up in an uncomfortable position. A thin metal strip sharpened on one side (the knife-edge) and which must not flex, is placed on the "rails" of the film support. Then by gradually varying the focus setting, a position is found when the star, upon shifting the edge, disappears *suddenly* behind it. Thus, the objective, which appears otherwise homogeneously illuminated to the eye, becomes *suddenly* dark. When the objective is gradually shadowed from the same side from which the knife-edge is moved, the focal plane is *between* the eye and the edge. If the directions of movement of edge and shadow are opposite, then the focal plane as seen by the observer is *behind* the edge (Fig. 6.3); see [6.2,3]. The film is loaded after this procedure, or else a separate camera body can be used for the focusing. A variation of this method uses, in place of the camera body, an adaptor with a built-in knife-edge and with the same camera body depth (i.e., distance between the lens mount and the plane of the film). In this case, the telescope is moved in one coordinate [6.170]. A ronchi grating in a special adaptor may be used instead of a knife-edge.

It may be added that—at least when photographing through a telescope—careful focusing at or near the image center is just as important as perfect guiding thereafter; thus in those cases it is always a good idea to focus anew before taking each photograph.

6.3.2 Modification of the Primary Focal Length

Methods of extending the primary focal length of the optical system are feasible only for bright objects. Every photographer is familiar with teleconverters with extension factors of 2–3 times. Even multilens systems are often of poor quality, but experimenting with them may prove worthwhile. The illumination of the film is reduced to 1/4–1/9. Increasing the distance between teleconverter and camera body also raises this factor but at the expense of image quality. The visual variant of a teleconverter is called a *Barlow lens*. With a suitable adaptor, the lens system can be used in a manner similar to the teleconverter. To obtain an even greater increase of up to 20 times the focal length, positive projection can be employed. The extension factor here should be not less than about 5 in order to achieve high resolution to the edge of the 35-mm film. Among high-quality eyepieces, there does not seem to be much of a preference. Good results have been reported from users of Erfle, orthoscopic, Plössl, and other eyepiece designs, and even microscope objectives. It should be mentioned that small extension factors tend to enhance optical errors such as field curvature [6.51].

What teleconverters or Barlow lenses do to increase the focal length, the *Shapley lens* (or telecompressor) does to reduce it. The compression is accompanied by a decrease in the image scale and an increase in the f-ratio. The disadvantage of the decreased image scale is somewhat compensated for by a shortened exposure time, and thus lessens guiding errors or other unforseen interference. With a reducing factor of 0.6, for instance, the exposure time is decreased to 36% ($0.6 \times 0.6 = 0.36$), not considering the low-intensity reciprocity failure of the film. For Shapley lenses, this advantage comes more or less at the expense of strong vignetting, caused by too small lenses or by a too small T-mount (see [6.52,53]).

6.3.3 Guiding

A basic requirement of perfectly guided photographs is to stay within the guiding tolerance as determined by various factors. The *total* system—Earth's atmosphere, the optical system, photographic film, and observer—determines a limit of resolution which can be limited by any one of these factors. The seeing disk is almost always larger than the diffraction disk of the optical system, but for focal lengths of up to about 1000 mm, it is usually smaller than the resolution afforded by the emulsion. So the photographic resolution is limited by the graininess of the film. For larger focal lengths, the seeing [6.33,34] is the limiting factor (Fig. 6.4). On those very rare occasions of steady air, a long-focus system with an aperture diameter not exceeding 300 mm may be optically limited, because then the seeing and resolution of the atmosphere are better than the resolving power of the optical system. The astrophotographer has to guide the telescope with such an accuracy that its guiding error is below the resolution-limiting factors. While periodic errors can be compensated for by electronic devices, aperiodic errors cannot.

In addition to insufficient polar alignment, virtually all telescope drives used by amateurs show periodic and aperiodic errors requiring guiding of the telescope. Sources of aperiodic errors may be, for instance, dust grains in the worm gear drive causing

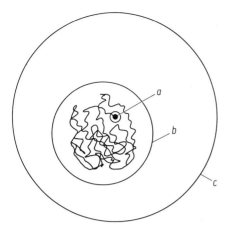

Fig. 6.4. The stellar image (*a*), whose size is determined by the lens aperture, is distorted and smeared out by random motions caused by atmospheric turbulence, thus generating the "seeing disk" (*b*). The tracking irregularities should move this disk by not more than the diameter of the scattered light disk on the film (*c*).

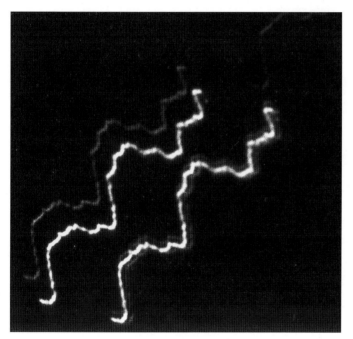

Fig. 6.5. Misalignment of the polar axis of the telescope causes the stars to drift in declination. Superimposed are the periodic and aperiodic errors of the drives, the amount of which can be determined by comparison with stars of known separation.

sudden deviations. Ground-in worms and worm-wheels may minimize the periodic error but do not entirely eliminate it. Finding its amount and period photographically has been described by di Cicco [6.11] (see Fig. 6.5).

Guiding is needed before the tracking error causes an enlargement of the stellar disk on the film. The amount Δ'' of tolerable deviation is found using the following formula based on the assumptions that an extension of the stellar image by 20% in one direction is permissible and that the smallest disks have a diameter of 0.02 mm:

$$\Delta'' = 7200 \arctan\left(\frac{0.002}{f}\right), \tag{6.1}$$

where Δ'' is in arcseconds, f is the effective focal length of the photographic optics in millimeters, and 0.002 represents the 20% increase of the smallest disks.

Example: The focal length is $f = 225$ mm. Then $\Delta'' = 3\rlap{.}''6$, the permitted deviation being to within $\pm 1\rlap{.}''8$.

Only a few special optical systems yield such minimal stellar disks of about 0.02 mm on the film. Also, the films are usually not that fine grained.

The assumptions underlying Eq. (6.1) are rigorous and, for focal lengths exceeding about 1000 mm, are almost always made insignificant by the prevailing scintillation. Since nothing can be done regarding the atmospheric conditions, the observer can only strive to maintain a *mean* deviation of the guide star from the crosshairs of the eyepiece that is within the calculated amount, and to correct deviations of the guide star as soon as possible. Any large deviations will, except for very bright stars, shown up on the photographic image as a neighboring point or line, whereas this will be barely noticeable for faint images. For stars of medium brightness, most conspicuous is an increase in the size of their images owing to poor seeing. Bright stars are—at least at short focal lengths—more affected by light scatter in the emulsion, whereas faint-star images remain comparatively small, as long as only the central part of the "seeing disk" is imaged and the outer parts remain below the sensitivity threshold.

In order to maintain some control over the tracking accuracy, the following conditions must be satisfied. The major requirement is an eyepiece which has some kind of reference markings in its field of view. These may be, for instance, a single or double crosshair or reticle, rings of various diameters, the edge of the field of view, or simply two bright points, which the guide star is kept positioned between [6.32]. Variants permitting a direct estimate of the allowable deviation are preferable. The kind of marking used ultimately depends on the preference of the observer. The simplest technique of all is to throw a star out of focus until its disk fills almost the entire field of view of a high-power eyepiece. Of course, this appreciably reduces the limiting magnitude for useable guide stars, but in the field of a short-focus objective, a suitable star can always be found in the guidescope (see Sect. 6.3.4). Markings or crosshairs should be thin enough not to cover a very faint guide star. In this respect, projected rings are quite advantageous, as they do not obstruct the guide star in the field. The technique for projecting markings onto the field of view can be found in the article by Cox [6.35]. Instructions for building a reticle eyepiece may be found in the book by Sidgwick [6.15]. The distance between crosshairs of a double reticle, the diameter of a ring, or the field of view of the eyepiece, may be determined from double stars of known separations, or perhaps by measuring the time needed for a star to travel a distance s with the drive off. For the latter,

$$s = 15\rlap{.}''04 t \cos\delta, \tag{6.2}$$

with s in arcseconds, the stopped time t in seconds, and δ the declination of the star in degrees.

To see the reference markings at night requires some form of illumination. Field illumination, which is achieved by simply placing a small light bulb into the dewcap of the guiding telescope, has the disadvantage that it noticeably reduces the number of usable guide stars. The alternative is to illuminate the crosswires (or other reference markings). This works well for lines etched on glass, or for wires made of quartz, silk, or even real spider web. The illuminator control should allow for readjusting the brightness of the crosshairs during the exposure, preferably without the need to touch the telescope. For extremely faint guide stars, or for diffuse objects (e.g., the inner coma of a comet), a circuit which continuously switches the illumination on and off for either selectable periods of time (and thus makes crosshairs and the guide object alternately visible) is recommended.

The photographing telescope itself can be used simultaneously for guiding by means of the *off-axis guiding* technique. Here, a small fraction of the light from the outer light cone is deflected into the guiding eyepiece. Photographing and guiding at the same focal length is to be done only when necessary; otherwise a ratio of 1:2 of the focal lengths is preferable. In this case, a Barlow lens should be used to avoid very short-focus guiding eyepieces. Compared to using a guidescope, the advantage of preventing any slight, but noticeable, flexure between camera and guide star which would cause guide errors is paid for by the reduced sky area in which the guide star is to be found. This point is less significant for larger telescopes, for there is a large number of fainter guide stars available.

Guidescopes have the advantage that, within the considerations of the following section, a much larger area may be scanned to find a suitable guide star. This is true so long as it is possible to easily point the guidescope off the optical axis of the main telescope tube. The found orientation between the two telescope tubes must be fixed properly without any flexure. As a test, insert reticle eyepieces in both guide and photographic scopes and adjust them parallel toward a star on the equator in the southeast. Then, move the telescopes across the meridian to the southwest and check that the star is still located accurately on both crosshairs. The guidescope together with the main telescope should form a compact unit (Fig. 6.6) and should not, as might be first thought, be placed on the declination axis to serve as a counterweight.

As a result, the disadvantages of guide scopes are also quite evident. Further, heavy guidescopes will need to be considered from the very beginning when designing or choosing the mounting. And finally, at least for larger telescopes, guidescopes are expensive (Fig. 6.6). The question arises regarding the proper magnification for guiding, and, related to it, the focal length of the guiding eyepiece, as well as the aperture of the guidescope. The magnification should be chosen high enough that the maximum permissible deviation of the guide star can be safely recognized. In the previous example ($\pm 1''\!.8$), this would be $180''/1''\!.8 = 100\times$, as $180''$ is easily resolved visually. A telescope with an aperture ≥ 70 mm resolves the $1''\!.8$; at $f/15$ (1050 mm) a 10 mm eyepiece, or a 20 mm eyepiece plus Barlow lens, will suffice. A magnification of $300\times$ is evidently a reasonable upper limit [6.169,170]. It is easily seen that, for normal eyepiece focal lengths, the guide scope should have a longer focal length than that of the photographic tube. This is a requirement easily satisfied in off-axis systems,

Fig. 6.6. The guide scope and main tube must be connected in an exceedingly stable and torsion-free fashion so that no differential flexure occurs between the two telescopes. This should be guaranteed in all observing directions. However, the option of off-setting the guide scope should also be available.

but much more difficult with a separate guidescope. In this connection, shortening the focal length of a photographic tube by a *Shapley lens* is worth considering.

6.3.4 Polar Alignment

The literature describes various methods of fast [6.11] or precise [6.12] polar alignment. The fastest and most expeditious technique for the astrophotographer with a portable instrument is to use a polar axis finder built into the polar axis or mounted parallel to it [6.15,19]. The accuracy is $\approx 2'-3'$. The markings etched on the glass plate can sometimes be used in both the northern and southern hemispheres. Otherwise, the *star-drift method* [6.20], developed by the German astronomer Christian Scheiner, may be used (see also Sect. 5.13.2). Concerning this latter method, time can be saved by immediately utilizing the entire azimuthal range of adjustment to see if it suffices; if this is not done, the telescope will eventually have to be completely repositioned, thus forcing the observer to repeat the alignment from the very beginning.

The accuracy required for precisely guided photographs is the subject of the following. Assume that the observer has expended much effort in guiding his stably mounted, 300-mm lens to an accuracy of $\pm 3''$ and yet is surprised to find trails stars around the guide star. The cause is inadequate polar alignment. This guiding error increases with growing deviation of the telescope polar axis from the Earth's axis with increasing declination, with larger film format, and with finer-grained emulsions. The resulting field rotation around the guide star is independent of focal length. For a guide star near the center of the photographic field, the following inequality given

Table 6.1. Minimum accuracy of alignment for KODAK 103a and TP2415 emulsions.

Format	103a-E/F/O	TP 2415
24 × 36 mm	9′	3′
6 × 9 cm	3′.6	1′.2

Table 6.2. Maximum allowable angle ρ between the Earth's axis and the telescope's polar axis as a function of declination and exposure time.

Declination	Exposure 30 min	Exposure 90 min
20°	$\rho = 13'7$	$\rho = 4'5$
40°	11′4	3′8
60°	7′4	2′5
80°	2′6	0′9
90°	0	0

by Riepe [6.13] is valid:

$$b \leq \frac{r}{2} \sin \rho. \tag{6.3}$$

For an exposure time $t \leq 2^h$, b is the arc described by any star on the photograph, r the radius of arc of this star around the guide star, and ρ the angle of polar axis misalignment. Table 6.1 compares the minimum accuracy of alignment for two different emulsions [6.13].

It is thus apparent that Kodak TP2415 needs about three times more accurate polar alignment than 103a-E/F/O, so that even in the least favorable case (i.e., the guide star in one corner of the picture, the target star in the opposite corner), no field rotation occurs. Table 6.1 takes into account the larger field diagonal in the 6 × 9-cm format.

The maximum permissible angle between the telescope's polar axis and the true rotation axis of the Earth, assuming an admissible trail of 0.02 mm of a stellar image 40 mm distant from the guide star, is shown in Table 6.2.

Clearly, it is best to have the guide star as close to the photographic field center as possible, and always to remember that the polar alignment for photographs of stars at higher declinations requires very high precision [6.14,15]. It should be pointed out that these requirements are independent of the guiding precision (Sect. 6.3.3); errors resulting from guiding must be added to those resulting from misalignment of the polar axis.

6.3.5 Miscellaneous

Any camera is suited for astrophotography if its shutter can be kept open with a cable release over an extended period of time. The shutter must be set on "B" or "T." On electronically operated cameras, this function is not always available and may require increased electric capacity, which—particularly in winter—may lead to failure of the battery and hence of the camera. As a remedy, an external battery pack kept warm in a coat pocket is connected with the camera using a plug-in extension cord. The cord should include an easily-separated connection to prevent damage caused by thoughtless movements of the observer.

Photographs taken of star fields at low altitude often show the stars deformed into short vertical streaks owing to differential atmospheric refraction. At some large telescopes, this is minimized by modifying the elevation of the polar axis. Because of atmospheric refraction, the two ends of the small streak are blue–green on the top and red on the bottom. This effect is easily observed in bright stars and at the edges of bright planetary disks, and can be reduced by filters or by special optical techniques [6.16,17,18].

A photographer operating the telescope on a balcony or terrace, or on a tiled or wooden floor, may wish to consider how to improve the stability of the mounting. Even a change in position of the observer may bend the floor sufficiently to shift the polar axis and thus result—especially if the guide star is somewhat distant from the photographed object—in field rotation not apparent in the guidescope.

6.4 Stationary Cameras

6.4.1 Trails, Constellations, Planetary Conjunctions

The simplest technique of astrophotography requires no astronomical mounting at all. The camera is fixed on a stable tripod, and the shutter kept open with a cable release. Depending on exposure time, shorter or longer star trails are obtained. The ever-popular photographs taken of celestial polar regions and displaying circular trails, as well as many other aesthetically appealing photographs, have been made in this manner. Again, the exposure time should be kept short enough so as to avoid fogging by the sky background. The length l of a star trail is easily computed using the following formula:

$$l = \frac{tf \cos \delta}{13\,714}, \tag{6.4}$$

where t is the exposure time in seconds, f the objective focal length in millimeters, and δ the declination of the star. The value 13 714 is the conversion factor for the length of arc at $t = 1$ s. Choosing t short enough to keep l below the resolution of the film produces point-like images of stars. The permissible exposure time t in seconds is obtained from the relation

$$t < \frac{450}{f \cos \delta}. \tag{6.5}$$

Table 6.3. Heating power and lens diameter.

Objective diameter	10 cm	20 cm	30 cm
Heating power	4 W	10 W	16 W

To image all stars point-like on the film, insert for δ the smallest declination value within the field. Therefore, with a high-speed film and a 50-mm lens, stars at the celestial equator are point-like images at exposure times shorter than 9 s; however, at $\delta = 45°$, $t < 13$ s, and for $\delta = 70°$, $t < 26$ s. For fine-grain films, replace the value 450 by about 140, which correspondingly diminishes t.

High-speed films permit, when used in conjunction with a powerful lens, photography of stars and galaxies down to about magnitude 10^m; one can then prepare one's own sky atlas. Color changes of the rising and setting Moon, planetary conjunctions, and even the rare but impressive bright comets, can thus easily be photographed.

Sources of failure in this mode of observing arise from vibrations of the tripod and dew formation on the lens. The former may be prevented by (a) shielding the tripod from the wind or by (b) the so-called "hat trick," in which the covered lens is uncovered only after the vibration caused by opening the shutter has vanished. The small "hooks" which sometimes decorate the beginning of star trails are thereby eliminated. Dew formation is almost inevitable at long exposure times, and can be avoided by a weak heating (e.g., using a car battery) of the lens mount or the dewcap, or by using a fan. A dewcap heater is built from a Nichrome wire available in most hardware stores. It is glued into the dewcap as a spiral so that the windings do not touch each other. The length d of the wire is calculated using

$$d = \frac{U^2}{P\rho}, \qquad (6.6)$$

where U is the applied voltage in volts (V), P the desired power output in watts (W), and ρ the specific resistance of the wire in Ω m^{-1}.

Example: $U = 12$ V, $P = 5$ W, $Q = 25$ Ω m^{-1}

$$d = \frac{12 \times 12}{5 \times 25} = \frac{144}{125} = 1.15 \text{ m}$$

For 5 W of heating by a resistance of 25 Ω m^{-1}, a wire of length 1.15 m is required.

Table 6.3 gives an idea of how much heating power is needed for a particular lens diameter.

Applying voltages substantially above that of the standard 12-volt car battery, especially the 110 volts or 220/240 volts supplied in most homes, is a matter that should be left to the experts!

6.4.2 Atmospheric Phenomena

The term "atmospheric phenomena" includes various effects which, while not directly connected with astronomical objects, are often visible on or near them. The "blue Sun," the "green flash," color changes of objects near the horizon, halo phenomena around the Sun and Moon [6.37,38,172], and noctilucent clouds [6.171] are some of them. For instance, the atmospheric spectrum, as well as the positional color and intensity scintillations, can be made visible on trail photographs at long focal lengths. The oval distortion of the solar and lunar disks by differential atmospheric refraction near the horizon is just as impressive as the partial or total detachment of parts of these disks by air layers of different refractive indices. Note that the Sun and Moon fit easily into a 35 mm frame for focal lengths up to 2000 mm.

6.4.3 Meteors

Meteor photography is an area of astrophotography which requires no tracking. The extremely rapid motion of the object requires high-speed emulsions together with fast, wide-angle lenses [6.22,27]. Since considerable film is consumed, it will be advantageous to purchase the black-and-white film in bulk quantities. Exposure times will depend on the brightness of the sky background and will rarely exceed 10 minutes (for an aperture ratio of $f/1.8$ and ISO $400/27°$). The times of the beginning and end of the exposure and of the appearance of meteors in the field of view should be recorded for later analysis. The sky area photographed should be about $30°$ to $60°$ away from the radiant. At distances much closer than this, the meteor trails become shorter and ultimately point-like. Even in the latter situation, they still permit the determination of the position of the radiant, and then place the greatest demands on photographic limiting magnitude. The speed of a meteor and its change while traversing the Earth's atmosphere may be determined by placing a rotating shutter in front of the lens; the astronomer can then find how the meteoroid moved with respect to the Sun [6.23].

Meteor photography is particularly suited for team work. Simultaneous photographs are obtained from at least two stations about 30–50 km apart to determine the radiant and path of the meteor. Accurate timing is absolutely essential [6.24,25,27,50]. The chance that a meteor bright enough to be recorded on the film will appear somewhere in the sky is, in the case of the Perseid shower, about 5 events per hour. In other words, observers report an average of about 15 trails for every 20 exposures [6.21]. Jahn [6.26] describes how to determine the brightness of a recorded meteor. A *fireball* (or bolide) often leaves along its path a luminous trail which quickly disintegrates. They are of quite different brightnesses and sometimes can be recorded with a 40 s exposure at an aperture ratio of $f/2.8$, ISO $100/21°$. Faint trails need up to 4 minutes at aperture $f/1.8$, ISO $1000/33°$. The camera orientation is immediately found using a ball-and-socket which is attached to a tracking mounting or on a separate tripod; the use of the latter of which will produce star trails.

6.4.4 Eclipses

Solar and lunar eclipses offer many opportunities for photography with stationary cameras. Wide-angle photographs on high-speed film show, for instance, the deep red-colored Moon in front of a star-studded background, or, from a series of short exposures, the progress of the eclipse by the Earth or Moon. The disks of the Sun and Moon appear at a size of about 1/110 of the focal length used, that is, for a 300-mm telephoto lens (300/110 = 2.7 mm). To plan multiple exposures at certain intervals, the observer has to take into account the fact that Sun and Moon need somewhat over $2\frac{1}{2}$ hours to cross diagonally the field of a 35-mm frame with a 50-mm lens because of the apparent rotation of the sky. Photographs with such a lens can thus begin one hour before mid-eclipse and continue for about the same time interval afterwards. If the mid-eclipse does not occur near the meridian, then the changing altitude of the Sun or Moon also has to be considered when aiming the camera [6.59]. The individual exposures will be adequately spaced when made at intervals of 5 or 6 minutes.

For a lunar eclipse at aperture ratio $f/5.6$ and ISO 400/27° film, exposure times may range from about 1/500 second for the partial phase to 2–20 seconds for totality; these values should be taken only as guidelines. 10 seconds corresponds to about the maximum exposure time for an untrailed picture with a stationary 50-mm lens. In any event, the observer should vary the exposure times since the brightness of the *umbra* varies substantially from one eclipse to another. The total duration of a lunar eclipse may reach a maximum of 3.8 hours, and it can be recorded as a trail. The Moon moves about 50° during this time, which is covered by a 28 mm lens at $f/22$ and with ISO 100/21° film.

Photographing the partial phases of a solar eclipse demands a bit more care and sophistication. Owing to its enormous brightness, the Sun, even when partially eclipsed, should be photographed only with a strong filter. With, for instance, a filter of density 5 (i.e., transmission 10^{-5}) and a film speed of ISO 64/19°–100/21°, the film should be exposed 1/125 second at aperture ratio of $f/8$. To capture the prominences as they appear from the beginning of totality, remove the filter and open the aperture to $f/5.6$. The inner corona is to be exposed about 1/15 second, the outer corona about 1/2 second. Again, these values are to be used only as a guide [6.61]. Immediately after the second and before the third contact, the solar *chromosphere* is visible for a few seconds as a thin crescent. It can then be spectrographed simply with a objective prism. This so-called *flash spectrum* shows a variety of bright emission lines, in particular the D_3-line which led to the discovery of the element helium.

Apart from the eclipse itself, there are numerous *terrestrial* phenomena which are known to accompany a solar eclipse and which can be noted and photographed. In addition to the approach and retreat of the umbra and, near mid-eclipse, the appearance of the colored horizon, there are fast-moving, low-contrast shadow bands on the ground, which can be recorded only on a bright, homogeneous background with an exposure time shorter than 1/100 second (the aperture setting depends upon the film speed). Also, every small hole casts an image of the eclipsed solar disk on the ground, an effect often seen under trees and which can be replicated using a "pinhole" camera. It is quite simple to build such a camera. The image size is 1/110 of the "focal length," which corresponds to the length of a tube with an aperture at the front end 500–800 times smaller than the tube length. At the other end, the tube is closed by a screen or the camera body of a single-lens reflex camera. The exposure times required are certainly much longer, but the results may be well worth the effort.

6.4.5 Artificial Satellites

Nearly every astrophotographer will, during her or his lifetime, take at least one photograph of the sky that has been unintentionally "contaminated" by the trail of an Earth-orbiting satellite which happened to pass through the field of view. Such trails appear as lines of constant or periodically changing brightness. When a satellite (or a space shuttle in orbit) emerges from the Earth's shadow, the trail becomes continuously brighter over a rather short pathlength, perhaps superposed with brightness variations due to the satellite's rotation. Its color, normally that of sunlight, can be seen on color film to change from deep red to white as a consequence of the reddening of light passing through the Earth's atmosphere (cf. Sect. 6.4). Such additional features on a photograph may be bright (up to a magnitude of -4), and if they are undesired, the observer should be on the alert for satellites approaching the field of the photograph, and, if one does appear, should cover the lens as the field is traversed. This applies particularly during those seasons (for an observer at temperate latitudes the summer months) when most of the satellite's orbit above the observer is outside the Earth's shadow.

Two photographs taken simultaneously from two sufficiently separated places (about 100 km apart) may serve to calculate the orbit of the satellite in a manner analogous to that for finding the heights of meteors in the atmosphere [6.28]. For tracking satellite orbits special optical systems are used such as the Baker–Nunn, Hewitt, or AFU-75 cameras. A 50-mm lens with aperture ratio $f/1.8$ and ISO $1000/33°$ film reaches satellites down to about magnitude 4 or 5. The angular speed of a satellite decreases with altitude. Furthermore, it is difficult to directly track a moving satellite with a camera (without computer-control on a normal equatorial mounting), so the path is almost always recorded as a trail on the photograph.

This is not the case, however, for *geosynchronous* satellites. These orbit at much greater distances (over 36 000 km) and are therefore much fainter (as of 1987, the brightest was about magnitude 11), and require a telephoto lens with an aperture of more than 50 mm, or, better still, a telescope. Once the satellite position in the sky is known [6.29], it is a simple matter to photograph it [6.30,31]. For a camera equipped with a cable release, mounted on a stable tripod, and using ISO $400/27°$ film, an exposure time of 5 minutes with a 57-mm-aperture lens will suffice. If the position of the geosynchronous satellite over the Earth's equator is known, its topocentric coordinates, corrected for refraction, can be calculated by the formulae given by Welch [6.29]. Since the horizontal parallax of a geosynchronous satellite is $8.7°$, these formulae are not strictly valid but will suffice to pinpoint the camera field.

6.5 Tracking Cameras

6.5.1 Focal Lengths $f \leqq 500$ mm

Without tracking, photographs taken by cameras with focal lengths over about 85 mm serve little purpose unless special effects are intended. When attaching a camera to a telescope mount, care should be taken that the telescope tube, dewcap, or counterweight does not obstruct parts of the field of view. The attachment should be stable with respect to flexure and torque, so that a camera with a heavy, long-focus lens does not tilt or rotate during the exposure—an effect which can be particularly troublesome when photographing regions located far from the meridian.

6.5.2 Lunar Halos

Nocturnal halos around the Moon are much fainter than those visible around the Sun during daylight hours. For this reason and because of most observers' lack of interest in a cirrus-covered sky, photographs of such halos are rarely sought after by astronomers [6.172]. Yet their presence in the night sky is often worth photographing. The exposure time at $f/2.8$ using ISO $100/21°$ film is about 20 seconds. Thus the star trails obtained using a 35-mm wide-angle lens without tracking are short. The diameters of the more common halos are $44°$ and $92°$, and thus require focal lengths of 35 mm down to 20 mm for a photograph of the whole phenomenon. Only the *moondogs* show color to the eye. Use of poloroid filters may show some unexpected effects: some kinds of halos increase in contrast, others are unchanged. The $22°$ halos show little or no polarization, but with the filter they appear $15'$ smaller! The $46°$ halos, on the other hand, are found to be polarized up to about 20%. *Coronae* are concentric colored rings, sometimes numbering over half a dozen depending on cloud structure. They easily fit into the field of a 50-mm objective and are often much brighter than the halos. With the same camera and settings mentioned above, a 5-second exposure will suffice to capture them on film.

6.5.3 Planetary Moons

When at mean opposition, Jupiter's moons pose no problems for current moderate-speed film. With a telephoto lens of $f = 135$ mm, the distance on the film from Jupiter of Io is 1/11 mm, of Europa 1/7 mm, of Ganymede 1/4 mm, and of Callisto almost 1/2 mm. Owing to extreme faintness, no other Jovian moons are within reach of moderate telephoto lenses. Saturn's brightest moon Titan lies up to $3'17\rlap{.}''3$ from the planet; i.e., on a photograph taken with a 35-mm telephoto lens, it is about 1/8 mm away from the planet. By contrast, the separation of Iapetus is about 1/3 mm. With $f = 300$ mm, they, along with Rhea and Dione, can be reached, and with $f = 500$ mm, so can Tethys. The five brightest moons of Uranus as well as Triton, Neptune's brightest satellite, require more instrumental power: apertures of at least 150–200 mm.

6.5.4 Comets and Minor Planets

While naked-eye comets are rare, there are usually at least one and often several of them in the sky at any given time with brightnesses that can be reached by photography. Published data cannot precisely predict the the actual surface brightness, particularly that of the tail. Comets sometimes experience outbursts which increase their brightness by up to eight magnitudes (e.g., comet P/Schwassmann-Wachmann 1), and thus become observable with moderate-sized telescopes. The comet hunter will usually expose the plate as long as the sky background permits [6.55]. This is often only a short time interval since comets frequently appear in the twilight sky. Details in the brighter parts of the comets are more easily seen with short exposures, so experimenting with a range of exposure times may prove worthwhile. While color film may show color differences within the comet, black-and-white emulsions may also permit enhancement of specific details if used in conjunction with color filters. Thus, the ion or gas tail is enhanced by a blue filter (e.g., Kodak Wratten 47A), the dust tail by a yellow filter (e.g., Wratten 9 or 21); see Fig. 6.14 and [6.54,56].

Series of photographs of the comet taken with polaroid filters permit investigation of the physical nature of the dust grains. Rapidly changing structures such as jets, envelopes, and "shadows of nuclei" in the intensely greenish coma require a focal length of at least 1000 mm. Short exposures ($\approx$ 1–10 minutes) and hard-contrast emulsions (e.g., Kodak TP2415 developed in Kodak D19) are advised.

When photographing extended objects, such as comets, the f/ratio of the optical system turns out to be an important quantity, whereas the limiting magnitude reached in photographing minor planets depends primarily on the aperture and focal length. Both comets and minor planets have in common that some of them—often the especially interesting ones—exhibit a significant motion. The short trail of a minor planet may look fascinating on a photograph, but the movement reduces the limiting magnitude and, for comets, smears out details. If the comet has a well-defined core or the asteroid is bright enough, then it can be guided upon directly. Faint objects with rapid motion, on the other hand, require guiding on a star as it drifts along the crosshairs: The guide star is moved after calculated periods of time by small increments across the crosshairs of the guiding eyepiece opposite to the direction of motion of the object to be photographed. Thus the comet will be at rest in the telescope while the stars are trailing. The step width is chosen just below the angular resolution A of the film/telescope combination. The period of time and the direction of motion of the telescope are computed as follows [6.57]. Find the motion in right ascension $\Delta\alpha$ (in seconds of time per minute) and in declination $\Delta\delta$ (in seconds of arc per minute) from the ephemeris of the comet or asteroid. Then the motion of the object in seconds of arc per minute of time is

$$M = \sqrt{(15\, \Delta\alpha \cos\delta)^2 + (\Delta\delta)^2}, \tag{6.7}$$

and the position angle of its motion P_M being

$$P_M = 90° + \arctan\left(\frac{\Delta\delta}{15\Delta\alpha \cos\delta}\right). \tag{6.8}$$

The time span by which the comet or minor planet moves by the amount of the instrumental resolution A can now be found by dividing A by P_M.

Example: If $A = 8\rlap{.}''3$ (1/50 mm with $f = 500$ mm) and $P_M = 1\rlap{.}''85$ per minute, then every $8.3/1.85 = 4.5$ minutes, the guide star must be shifted by $8\rlap{.}''3$.

In the case of comets, the stepwise shift by the full amount of A is not detrimental, and very small steps are manageable only automatically. The direction of the shift can be set by a calibrated scale externally on the drawtube or in the field of view of the guiding eyepiece, indicating the position angle, and is, as stated, opposite to the motion of the object. The correct quadrant of the direction of motion can be ascertained by tapping on the top of the eyepiece (telescope moves north → position angle $P_M = 180°$, telescope moves south → position angle $P_M = 0°$) or stopping the drive (star moves west → position angle $P_M = 270°$). A timer which repeats the set time interval or a portable Walkman with a prepared cassette tape will expedite the timing procedure [6.58].

Arbour [6.10] and Everhart [6.187] describe different methods for comet guiding. The camera body is moved by the calculated amount and position angle by a stepper-motor driven screw while the telescope is tracked at the sidereal rate on a star.

Photographs of comets and minor planets are of serious astrophysical interest. The development of tail structures, detachments, and anti-tails (or their absence) can be studied with powerful optics. The rotational light curve exhibited by a minor planet shows up on serial exposures. In the case of a star occulted by a fainter minor planet, the star trail will be briefly interrupted [6.64]. A second gap in the trail would indicate another occultation (which is exceedingly unlikely) or suggest the presence of an asteroid's satellite (provided, of course, that a flaw in the emulsion can be ruled out).

6.5.5 Eclipses

When photographing the Moon as it enters the Earth's umbra, or the Sun's prominences or its corona, the very large brightness changes need to be considered. Only precisely timed exposures will reveal the many shades of brightness and tint in the umbra or in the corona. Thus the data in Table 6.4 are to be used only as guidelines and should be bracketed by shorter and/or longer exposure times.

To photograph the inner and the outer corona simultaneously, filters with a density range decreasing from center to edge have been used with good success. Otherwise the method of unsharp masking, as described in Sect. 6.4.5, may be employed. The coronal light is polarized up to 40%, and thus a series of photographs taken with different orientations of the polaroid filter in front of the lens will prove instructive.

The Earth's shadow at the distance of the Moon has an apparent diameter of $83'$, and thus a telephoto lens with focal length up to $f = 500$ mm will show impressively the motion of the Moon through the Earth's shadow when using the sidereal rate to track the camera on a star. This effect is clearly shown on multiple or long-exposed pictures. The latter require very dark neutral-density filters [6.61]. Using a stereo viewer to combine two pictures of the eclipsed Moon taken about 10 minutes apart and guided on a star, the Moon seems to "hover" above the stellar background.

Table 6.4. Data for photographing lunar and solar eclipses at $f/11$.

Object	ISO (ASA/DIN°)	Filter density	Exposure (s)
Moon, partially eclipsed:	100/21°	–	1/30
totally eclipsed:	100/21°	–	40
Sun, partially eclipsed:	25/15°	4	1/500
prominences:	100/21°	–	1/30
inner corona:	100/21°	–	1/4
outer corona:	100/21°	–	2

6.5.6 Deep Sky

Observations of all objects outside the solar system are subsumed under the heading "Deep Sky" (DS). Included are constellations, star clusters and associations, HII regions, planetary nebulae and supernova remnants, double and variable stars, dark clouds, and galaxies and clusters thereof. As pointed out in Sect. 6.5.4, photography of point-like sources and extended nebulae must be distinguished regarding the f/ratio and the aperture of the optics. Every lens has a particular application in astrophotography: superwide-angle or "fisheye" lenses for wide-field photographs of the Milky Way, 300–500-mm telephoto lenses for planetary and gaseous nebulae (diameter $\geq 5''$) and the nearer galaxies or clusters of galaxies.

DS objects are often extremely faint and offer little contrast to the sky background. To improve the limiting magnitude, one first chooses a substantially darker observing site and the most transparent nights. Solely increasing the exposure time will not necessarily result in a "deeper" photograph (although a longer focal length will help). Additionally, a high-contrast film, such as sensitized (spectroscopic) emulsions of the Kodak 103a series for selected color ranges, or Kodak TP2415, can be used. Finally, the most successful measure is to use filters which block the interference of nightglow and artificial light pollution ($\lambda = 557.7, 630.0, 646.3, 589.0$, and 589.6 nm), but transmit the light of the DS object with negligible loss. Of course, a filter works well only if the spectra of the celestial object and the light sources are quite different. The low-pressure, sodium- and mercury-vapor lamps commonly used for illumination emit at discrete wavelengths of $\lambda = 366, 404.6, 435.8, 546.1, 589.0$, and 589.6 nm, where stars, reflection nebulae, and galaxies are also relatively bright. Hence, for these objects the filter indiscriminately blocks the light from both the object and the background so that there is very little net gain.

The situation is quite different for *emission* nebulae like the Orion and Helix Nebulae, whose primary emissions lie in the blue–green ($\lambda = 495.9, 500.7$ nm) and red ($\lambda = 656.3$ nm) spectral ranges, sufficiently apart from the brightest parts of the sky background spectrum. A suitable choice of the emulsion/filter combination can thus dramatically enhance the contrast [6.62,63]. High-pressure sodium lamps, recognized from their less intensely yellow and slightly greenish-tinged light, render this kind of filtering much less effective since broad emission bands appear on both sides of the sodium absorption; these bands cannot be removed without a simultaneous loss of the sought nebular lines. Special [OIII] nebular line filters will separate this particularly intense emission of many supernova remnants, planetary and gaseous nebulae well from the sky background. Unfortunately, most black and white films are rather insensitive in the green spectral range. *Nebular filters* combined with color films give so strong a tint that they should be considered only as a last resort. The color-separation technique, which uses three black-and-white films for recording even faint nebulae without color shift, will be mentioned in Sect. 6.8.4. A much simpler method, which uses standard color films and often giving an equally acceptable impression, is to use color-compensating (Kodak CC) filters, which are available in a wide variety of different colors and densities. The filter best-suited for a particular task depends on the spectral senitivity of the emulsion, exposure time, and observing site, and must be decided by trial-and-error. Magenta filters, for example, transmit red and blue but absorb green.

The sky background is brighter in blue than in red. As there are no blue spectral lines comparable in intensity to $H\alpha$ or [OIII], filtering is much less effective at shorter wavelengths. Even under optimum conditions, the longest useful exposure time with a blue filter is always shorter than with the use of a red sharp cut-on band-pass filter which eliminates interfering light directly on the short wavelength side of $H\alpha$ [6.173].

6.5.7 Spectrography

Spectrography is a rather complex subfield of celestial photography. On the other hand, spectrographic equipment is not excessively difficult to build using prisms, gratings, or combinations of the two [6.45]. As these devices work best with parallel rays of light, they are usually placed in front of the lens. Since the price increases rapidly with size, these optical parts will be relatively small, thus necessitating a reduction in the aperture and hence the light-gathering power. The dispersing optical elements inevitably waste some light owing to absorption, reflections, and, in the case of gratings, higher-order spectra. Dispersing the light into colors additionally reduces the intensity of incoming light on the film; the contrast relative to the background is diminished since the latter is not similarly dimmed. The preceding comments refer to the technique of *slitless spectroscopy*, which is especially suited to studying the light of a point-like or a line source, including meteors, satellites, and stars trailing in the field of an unguided camera. Extended sources which can be treated as point-like sources include emission-line objects (planetary and gaseous nebulae, comets) with diameters below the angular resolution of the optics/film combination used, or for which the images in each emitted wavelength (i.e., monochromatic images) are not too strongly superposed; a clearly defined spectrum of images is then revealed (Fig. 6.8). Finally, point-like sources include stars or minor planets photographed with a guided camera, or geosynchronous satellites taken with a stationary camera.

The prism or grating is oriented in front of the lens in such a way that it produces a spectrum perpendicular to the direction of motion of the object, a procedure which is of course not always possible for meteors. The line sources as defined above furnish

Fig. 6.7. The Scorpius-Ophiuchus nebular complex is noteworthy owing to the rich colors of the numerous reflection and few emission nebulae. West of the bright red supergiant Antares is the globular cluster M4. On the right is the hottest star in the region, σ Scorpii (spectral class B1), which excites part of the surrounding nebulosity to emit the typical red Hα light (656.28 nm). The nebulae to the north reflect the light of the blue star ρ Ophiuchi. A yellow–red tint is contributed by the reflection nebula north of Antares; part of the red-emitting HII region around τ Scorpii is at the bottom. Photograph by B. Koch in Namibia/SWA with a $5\frac{1}{2}$-inch Schmidt camera at $f/1.65$ (140/140/225 mm), exposed 45 minutes on color slide film Fuji Fujichrome RD100 film with a magenta color compensating filter (Kodak CC10M). The original slide was contrast-enhanced by copying with a slide-copier on the color slide film Kodak Ektachrome 64 using a tungsten lamp (color temperature $\approx$3400 K).

the widening of their spectra by their own intrinsic motions. If the object is at rest relative to the guided camera, slight periodic motions in right ascension achieved by switching off the drive or changing the drive rate should result in a slight widening (to about $w = 0.3$ mm) while avoiding all deviations in declination which would otherwise smear the spectrum. The angle β corresponding to a widening w at the focal length f is obtained from

$$\beta = 206\,265\,\frac{w}{f}, \tag{6.9}$$

with β in arcseconds, w and the focal length f in millimeters.

Fainter stars are recorded by repeated superposition of their spectra to sum up the light [6.39,40,41,43]. With high-speed films 5th-magnitude stars are within reach after 30- to 40-minute exposure times with a lens diameter of from 50 to 60 mm. The construction and application of a grating spectrograph are described by Bourge, Dragesco, and Dargery [6.42]. With a 135-mm telephoto lens and a grating of 630

Fig. 6.8. A slitless spectrum of the Ring Nebula M57. The continuum parts of the spectra of the field stars are imaged as unstructured lines, while the discrete emissions of the planetary nebula generate a sequence of monochromatic images. The left, long-wavelength end of the spectrum shows the Hα image (656 nm), while the center is dominated by the [OIII] emission (500 nm). In the short-wavelength range, the lines Hβ (486 nm) and Hγ (434 nm) follow. The different size and structure of the Ring Nebula in Hα and in [OIII] is clearly visible. Photograph by B. Koch: the convergent light beam of a 14-inch Schmidt–Cassegrain telescope was transferred by an achromatic lens ($f = 150$ mm) parallel through a three-part Amici prism, and then imaged with a 50-mm objective in the film plane of a 35-mm camera.

lines/mm, spectra over the wavelength range 380–680 nm are obtained which are about 25 mm long, but are not in focus over this entire range, except for apochromatic or mirror systems. Prisms and gratings deflect light by angles of many degrees and thus a spectrograph has to be attached to the telescope tilted at the deflection angle, or else must be equipped with a separate finder in order to simplify the centering of the objects. Slitless spectrographs are seeing-limited at focal lengths of about $f = 1000$ mm. To avoid this drawback, or to obtain spectra of extended objects (e.g., comets, the Orion Nebula, the night sky) a slit applied to the focal plane of the telescope is followed by a collimator and prism or grating. Finally, an optical element (e.g., a 50-mm normal lens) images the spectrum onto the film. At its position in the beam, the prism (or grating) can be much smaller than when placed in front of the telescope. This more expensive construction utilizes the full aperture of the telescope and permits spectroscopy of fainter objects [6.44,45,46]. "Blazed" gratings concentrate the light into one first-order spectrum and thus improve the limiting magnitude.

6.5.8 Zodiacal Light and Gegenschein

Although photographs of the *zodiacal light* are possible with a stationary camera, tracking is still recommended in order not to further diffuse the edges of its light cone. Only powerful wide-angle lenses (f-ratio ≤ 4) with focal length shorter than about $f = 28$ mm give sufficient image angles in 35-mm format so that the outer boundary of the zodiacal light stands out against the sky background. For ISO 400/27° film, about a 15-minute exposure time will suffice [6.47]. To calculate the angle between ecliptic and horizon during of exposure, refer to Meeus [6.168].

The much fainter *gegenschein* is more difficult to photograph. Wide-angle lenses with $f < 21$ mm should be pointed so that the gegenschein is not exactly in the center, superposed by vignetting, but slightly shifted to the edge. A sequence of photographs taken 10 days apart reveals that the gegenschein advances slowly along the ecliptic. It is favorably placed for viewing from mid-January to mid-March, and from late September to late November [6.48,49], high enough in the sky for northern-hemisphere observers but away from the band of the Milky Way.

6.6 Long-Focus Astrophotography

6.6.1 Instrumental Requirements

This section covers the use of telescopes with focal lengths over about $f = 1000$ mm, for above this value, the influence of the seeing on Kodak TP2415 is always noticeable. The photographic resolution is then limited to an average of between $4''$ to about $1''.5$. Photography of bright objects (e.g., the Moon, Sun, planets) is to be distinguished from deep-sky photography. In the case of bright objects, the light intensity is usually sufficient, so that technically feasible exposure times up to about 10 seconds will be sufficient to capture fine detail with a long focal length and the resulting (unfavorable) aperture ratios of $f/20$ to $f/100$. Visual observation is therefore superior to photography, as the eye is capable of taking advantage of the usually short phases of steadiness in the atmosphere. Deep-sky objects which are fainter by orders of magnitude require photographically better-suited aperture ratios up to $f/20$, and yet the exposure times are in the range 1–100 minutes. Here the light-integrating ability of the emulsion is of advantage over the time-limited sensitivity of the eye. Poor seeing conditions during the entire exposure time result in a photographic resolution which is always below the theoretical limit (refer to Sect. 6.9).

Aperture ratios in excess of about $f/20$ are obtained by means of *positive (eyepiece) projection*. An eyepiece inserted directly behind the focal plane of the objective projects a magnified focal image onto the emulsion plane. By varying the chosen focal length of the eyepiece and its distance from the focal plane the effective focal length f_{eff} obtained changes over wide ranges. It is calculated from

$$f_{\text{eff}} = (p - f_{\text{eyp}}) \frac{f_{\text{obj}}}{f_{\text{eyp}}}, \qquad (6.10)$$

Table 6.5. Exposure times for features near the lunar terminator at $f/50$ for two different film speeds.

Lunar phase	ISO 64/19°	ISO 100/21°
2 Days after New Moon	6 s	4 s
6 Days after New Moon	3	2
10 Days after New Moon	1/2	1/4

where p is the projection distance between the principal plane of the eyepiece and the film plane, and f_{obj} and f_{eyp} are the focal lengths of the objective and eyepiece, respectively. Often the distance p has not been accurately determined [6.51], but the small uncertainty associated with it can be neglected. For a given combination of eyepiece focal length and projection distance, the effective focal length in mm can be determined experimentally from

$$f_{\text{eff}} = 206265 \, \frac{d}{\phi''}, \qquad (6.11)$$

with d the separation in millimeters measured on the film between two points of known angular separation ϕ'' in arcseconds $\phi < 1°$).

6.6.2 Moon, Sun, and Planets

The myriad details on the Sun, Moon, and planets visible in even a small telescope frequently inspire observers to photography. The Moon is best suited for first attempts as it is bright (although not excessively so), permitting short exposures and easy focusing. The best focus is achieved on sharply contrasted details such as the lunar limb or a crater at the terminator (see Table 6.5). As the Moon has very few color tinges, color photography would only reproduce the tint of the film itself. The search for luminescent gas clouds from possible volcanic eruptions (transient lunar phenomena, or TLPs) with selected interference filters has been suggested. Occultations of stars by the Moon are handled in the same manner as deep-sky photography, but when choosing the exposure time allowance is made for the irradiation by the bright lunar disk and the Moon's fast motion relative to the star. The lunar motion may be compensated for by adjusting the speed of the tracking, if necessary, in both right ascension and declination.

Venus, the planet with the largest apparent diameter (sometimes exceeding 60''), is imaged by $f = 10$ m with a diameter of 2.9 mm (Jupiter 47'' = 2.3 mm, Saturn 42'' = 2 mm including rings). Also at $f = 10$ m, the diameter of Mars at opposition varies between 1.2 and 0.7 mm. Except for Mercury and Venus, which exhibit their changing phases, the outermost planets, such as Uranus, Neptune, and Pluto, appear at best as small featureless disks. As most telescopes, in particular amateur instruments, usually have focal lengths of less than 10 m, positive projection is employed. Table 6.6 gives the exposure times to be expected.

Table 6.6. Exposure times for planets at $f/50$ for two different film speeds.

Planet	ISO 64/19°	ISO 100/21°
Mercury/Mars	1/2 s	1/4 s
Venus	1/15	1/30
Jupiter	2	1/2
Saturn	6	4

Experience with the use of filters to enhance color details in visual observation can also be applied to photography. In principle, all filters should be tried on all planets in order to ascertain which objects are optimally imaged in selected colors. UV filters reveal the very tenuous shades in the Venusian atmosphere. Such filters are nearly opaque in the visual spectral range, and thus focusing is best done with a filter of equal thickness but high visual transmission (e.g., a green filter). Reflectors are preferred in this wavelength range on account of their achromacy. Photographs of Mars through blue filters tend to reveal atmospheric phenomena (violet clearing, clouds) while orange and red filters enhance surface features. The Great Red Spot on Jupiter is enhanced by using blue or green filters. The use of filters seems to have little effect on observing Saturn's disk, but it does uncover color differences in the ring system. Transits of Mercury and Venus across the Sun are handled according to the methods of solar photography.

Photographing the Sun can be quite exciting. Even a 60-mm refractor will show spots and faculae, while a 100-mm lens is required to see the granulation [6.69,70]. The immense brightness of our nearest star necessitates the use of especially dense objective filters (metal-coated mylar or glass filters). The filter density should permit exposure times between 1/250 and 1/1000 second. With $f/50$, film speed ISO 50/18°, and with an objective filter of density 3 (transmission 10^{-3}) the exposure time will be about 1/250 second. Lacking a suitable objective filter, the Sun's image can be photographed directly off the projection screen (note: do not use solar projection with Schmidt–Cassegrain telescopes because of intense heating of the secondary mirror!). Distortion of the solar image is a drawback of this method. For solar photographs in various spectral regions, the $H\alpha$ spectral line or the region around the H and K lines of CaII (singly ionized calcium) may be considered. At $f/10$ on Kodak TP2415 with ISO 100/21° the prominences can be photographed in $H\alpha$ with interference filters using exposure times of between 1/2 to 1 second; for surface features, the exposure time should be about 1/250 second. When using coronagraphs, the exposure times must be found by trial-and-error owing to the very different total transmission of the optical system. A combination of UG1 and BG38 Schott filters shows the large-scale CaII structure of the upper chromosphere [6.2]; an exposure of about 1/125 second at $f/11$ on Kodak TP2415 should be adequate.

6.6.3 Moons of Planets

More than 65 satellites of planets are known in the solar system, not counting unconfirmed objects and satellites of minor planets, e.g., the double asteroids including the probable Apollo (Earth-crossing) asteroid 1989PB [6.178]. This number, moreover, provides ample opportunities for present-day astrophotographers! The relatively low brightness of the satellites and their proximity to their parent planets mandate the use of telescopes which are large with respect to both objective diameter *and* focal length. An aperture diameter of 150–200 mm is a good lower limit. Long-focus reflectors with small secondary mirrors and refractors are certainly preferable owing to their superior contrast when photographing satellites which are located close to their parent planets. A very narrow strip of neutral-density gelatin filter pasted onto the field stop of an eyepiece helps to reduce the irradiation by the planet. As this technique dims only the planetary disk, the satellites can easily be photographed with undiminished brightness using the projection method. A substantial fraction of irradiation originates by multiple reflection of light within the emulsion, so a small hole punched into the film will serve the same purpose though it is much more difficult to make. The requisite precision of pointing is afforded by a carefully and accurately adjusted guidescope.

Not all satellites require such particular measures. For example, Jupiter's moons VI and VII with $m_V = 14.8$ and 16.7 are at greatest elongation $62\rlap{.}'8$ and $64\rlap{.}'2$, respectively, from the planet's center at times of mean opposition, whereas Saturn's satellite Phoebe with $m_V = 16.5$ is $34\rlap{.}'9$ distant; hence they are not excessively difficult to photograph. Corresponding values for the moons of Mars are $24\rlap{.}''7$ and $64\rlap{.}''4$, and for the four brightest satellites of Uranus, they range from $14\rlap{.}''5$ to $44\rlap{.}''2$ [6.66]. Needless to say, photography of the inner moons is best performed at the times of greatest elongation. *The Astronomical Almanac* provides the necessary data.

6.6.4 Comets

In addition to what was written in Sect. 6.5.4, it should be noted that longer focal lengths ($f \geq 1000$ mm) reveal events and structures in the coma surrounding the core of the comet. These include the rapidly varying jets and envelopes as well as the fragmentation of the nucleus, all of which have occasionally been observed. Sometimes a long "shadow" of the nucleus can be noticed on the side of the coma opposite to the Sun. All this detail is of very low contrast and therefore its reproduction requires high-contrast films and contrast-enhancing processing. The core itself is usually very faint; photographs taken with long-focus instruments may serve to determine its position. The comet's rotation is evident from positional measurements of the jets from series of photographs.

6.6.5 Deep Sky

The bright inner parts of nebulae, spiral arms and dark lanes in galaxies, medium- and small-sized planetary and bipolar nebulae, and variable stars in open and globular clusters present a challenge to what can be technically achieved in astrophotography.

The cooperation of perfect telescope optics and mechanics, extensive experience with its handling, and elaborate darkroom techniques can often result in impressive results [6.67,68,179]. The basics have been dealt with earlier, but the observer would do well to heed the following advice regarding the use of filters. Owing to properties of photographic emulsions (Sect. 6.7.3), a certain threshold of intensity must be exceeded in order to form a latent image. The slower the telescope and the darker the sky background, the later will this threshold be reached. The use of filters, which are most often employed to reduce the sky background, is especially advised for astrophotography under city lights. Suppose, for example, that an observer wants to capture a faint nebula of, say, magnitude 13. At a moderate f-ratio of $f/10$ and under the darkest skies, the negative may show no trace of the nebula after a one-hour exposure, because of the low combined intensity of the nebula and the sky background below the threshhold intensity. But someone else working with an identical instrument, film, and exposure time, but using a red filter, may capture the nebula in what seems to be a much less favorable site with a brighter sky background (e.g., near a large urban area).

The light from some objects, in particular bipolar nebulae (e.g., the Egg Nebulae), is so strongly polarized that the effect of different orientations of the polarizing filter can be seen visually [6.165]. This may also be of interest to astrophotographers.

6.7 Films for Astrophotography

6.7.1 Film Formats

The black-and-white films which are used for do-it-yourself processing are available in three major formats: magazines of 35-mm film, roll film, and sheet film. The variety of films is so overwhelmingly large that this report will be restricted to those commonly used for astrophotography. The popular 35-mm film is perforated on both sides and wound around a core, and permits 12, 24, or 36 exposures, but can also be purchased in bulk quantities with a resulting reduction in cost per exposure. The roll film, 60 mm wide, is wound together with paper around a core and has a capacity of 8 to 16 exposures, depending on picture format (roll film 120). Double-length roll film is tagged "220." For single pictures, there are sheet films or glass plates mainly in 4×5-inch size.

6.7.2 Physical Composition of the Film and Formation of the Latent Image

6.7.2.1 Components of the Film. The light-sensitive layer of the film is called the *emulsion*, and is composed of crystals of different silver halides (silver chloride Ag Cl, silver bromide Ag Br, silver iodide Ag I) embedded in gelatin. Clusters of crystals are called *grains*; their composition, size, and distribution determine the photographic properties of the emulsion, which is also influenced by the kind of binding and other additives used. The thickness of the emulsion is less than 0.01 mm. One square meter

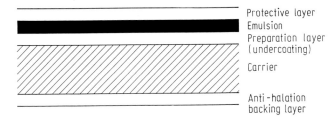

Fig. 6.9. Physical composition of the emulsion of a black-and-white film (not to scale) [6.32].

of film contains between 1 and 10 grams of silver. The emulsion lies atop a *carrier* of cellulose-acetate or polyester of thickness between 0.07 and 0.20 mm, depending on the film (see Fig. 6.9).

The emulsion does not in general adhere well to a cellulose film, and therefore the carrier is coated with a substratum adhesive layer prior to the application of the emulsion. In addition, a protective coating of a transparent anti-abrasive substance is applied over the emulsion to prevent scratches and other mechanical damage. Finally, the back of the film is dyed with an *anti-halation backing* to reduce reflected or scattered light; otherwise, fringes (halos) around bright stars would diminish the contrast and the sharpness of the image. This dye is eventually dissolved by the developer [6.32]. The *spectroscopic films* of the Kodak 103a series had an additional anti-halation backing of carbon which had to be removed before developing or after fixing the film [6.75]; see Sect. 6.7.10.

6.7.2.2 Latent Image. Incident light converts silver halides into dark metallic silver. After a short exposure, the grains embedded in the emulsion carry a very small number of silver atoms forming an invisible or *latent* image which is enhanced a million-fold by the developer. After fixing, all remaining, undeveloped silver halides are soluble so that they are removed in the subsequent rinsing. Only the blackened silver remains. The dark parts of the film now represent the highlights of the original.

When the emulsion is exposed to a luminous flux, the latent image forms and is rendered visible by the developing process. How is this sequence of events physically interpreted? Usually, the physical band model [6.71] is cited, although there is a chemical explanation of the photochemical reactions too.

1. A photon is absorbed by an electron in the valence band of the silver halide crystal. The electron absorbs sufficient energy to be lifted into the conduction band.
2. A positive hole remains in place of the electron in the valence band.
3. Electron and hole move within the crystal lattice. That is, when meeting, they can recombine by emitting a photon. Extremely pure silver halide crystals would not be able to store a permanent latent image as electron and hole would recombine within 1 μs. To prevent the recombination which destroys the image formation, electron and hole must be separated by reactions. Electrons can be captured chemically or physically by means of impurities or crystal defects. The positive hole may migrate to the surface of the grain and react with the gelatin.

4. The captured electron neutralizes mobile Ag^+ ions and leaves Ag atoms in the crystal structure.
5. The effect of the electron trap is now enhanced by the presence of the silver atom which can capture more electrons in the conduction band. Thus, additional silver atoms can form and become enriched a hundred-fold in the silver halide grains.
6. The developer converts the remaining Ag^+ ions in all grains containing Ag atoms to pure silver. The Ag atoms present in each silver halide crystal serve as catalysts, so that grains containing more silver atoms than others are reduced faster, while grains with little atomic silver react with a slower rate of development.
7. Three to six silver atoms per grain are sufficient to reduce it; low-speed films need more. The development which converts the latent image into a visible image is performed until all silver-containing grains are reduced, but without influencing other grains containing no Ag atoms. Random reductions of unexposed grains happen by other, partly statistical processes and are responsible for the chemical *fog* on the emulsion.

The modern "silver-free" *chromogenic* black-and-white films follow entirely different rules [6.72]. The typical chromogenic black-and-white film, such as Ilford XP1 400 (ISO 400/27°), is composed of emulsion layers of different sensitivities. Each layer contains, between the silver halide crystals, uncolored dye-couplers which, when exposed, become attached to a latently exposed crystal. The color developer generates black, neutral silver from the exposed crystal parts. The oxidation products of the development then dye the couplers. They appear reddish or brownish depending on make. The bleach fixer removes the entire black silver and the unexposed silver halides. Only the dye molecules remain unchanged and form the negative. Most "silver-free" black-and-white films are developed with the same chemicals and procedures as color negative films. The designation "silver-free" characterizes not the unexposed film but rather the exposed and processed negative.

6.7.3 The Characteristic Curve

The response of film to incident light is shown by its *characteristic curve*. A negative should be well traced between the highlights and the shadows, and this depends upon the combination of film and developer. A gray scale and stepped exposure times are advised to obtain quantitative estimates. The density of the negative (or positive) is plotted against the logarithm of the exposure (Fig. 6.10). The characteristic curve is thus a measureable criterion for understanding and comparing properties of films.

Light incident on the film (negative) generates a density which is graded according to exposure. Ideally, cause (light) and effect (density) would be proportional; i.e., twice the exposure results in twice the density. Actually, such proportionality exists only approximately in the linear part of the characteristic curve. Above the base density (fog), an approximately linear part of the characteristic curve follows and merges into saturation at maximum density. How can this be explained? Specifically, with the base fog at very low density, why does the density not increase in proportion to the intensity at very low light levels? The answer is that at low light levels, photons hit a silver halide crystal only occasionally, so that the information is already cancelled

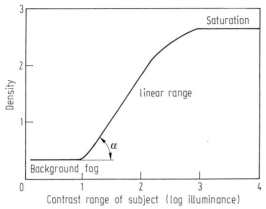

Fig. 6.10. The characteristic curve of a black-and-white negative film. The abscissa graphs the logarithm of the exposure (= intensity × exposure time, unit *lux*). Each division represents an exposure increase by a factor of 10. A change by one f-stop (exposure doubled or halved) corresponds to 0.3 logarithmic units in exposure. The ordinate shows the density (= $\log(1/T)$, where T = transmission). In general, the curve can be divided into three ranges: base fog, linear part, and saturation. The slope of the linear part γ ($\gamma = \tan \alpha$) expresses the contrast between highlights and shadows on the film.

before another photon hits the same crystal. This effect is in general quite beneficial as the film is not affected by very faint sources, ionizing radiation, etc. In other words, a light source which does not expose the film in, say, a few hours, will never do so!

On the other hand, why does the density not increase proportionally at high exposures? The reason is that the cross section of the grain is simply too small to register all incident photons.

The linear portion of the characteristic curve explains the characteristics of a film. At a low slope (low-contrast film), a larger latitude of the emulsion is recorded by a small change in density. This is desirable in lunar and planetary photography. If the linear part is steeply sloped (high-contrast film), the same latitude of the emulsion will result in a large change in density, a fact which is of importance for deep-sky pictures.

The more contrasted the negative, the steeper is the linear portion of the characteristic curve. F. Hurtler and V.C. Driffield published the results of their research on characteristic properties of photographic plates in 1890, and ever since, the so-called *gamma value* (γ), which they defined, has been the universal measure of the contrast. γ represents the increase in density of the linear portion of the characteristic curve ($\gamma = \tan \alpha$; see Fig. 4.10), and some of its uses are described in Table 6.7 below.

γ strongly depends on the kind of developer and the the time and the temperature of development. The recently introduced *contrast index* is significant for films which exhibit only a very short linear portion. It is defined as the slope of the characteristic curve covering 1.5 log exposure units from density 0.1 above base fog. This portion is important in most astronomical photography.

Table 6.7. γ for different uses.

γ	Use
0.7	Normal photography
1.0–1.5	Planetary photography
4	Deep-sky objects

6.7.4 Film Sensitivity

6.7.4.1 Film Speed. The International Standards Organization (ISO) gives the film speed in ASA/DIN°, combining the ASA and DIN data; for example: Kodak T-MAX 400 is rated as ISO 400/27°. ASA and DIN are related by the following formulae [6.73]:

$$\text{ASA} = 25 \times 2^{(\text{DIN}-15)/3}, \tag{6.12}$$

$$\text{DIN} = 15 + \frac{3 \, \log(\text{ASA}/25)}{\log(2)} \approx 1 + 10 \, \log(\text{ASA}). \tag{6.13}$$

The stated film speeds serve only as guidelines in astrophotography as they also depend on various factors. First, developers containing phenidone (e.g., Ilford Microphen) increase the effective film speed, while fine-grain developers (Kodak Technidol, Tetenal Ultrafin, etc.) diminish it. Second, low-intensity reciprocity failure (LIRF) plays a major role, particularly for long exposures times. An ISO 100/21° film may, after a certain exposure time, become more sensitive than an ISO 400/27° film at an equal exposure of the same target (cf. Sect. 6.7.6).

6.7.4.2 Spectral Sensitivity. The technical progress in photographic emulsions with respect to both higher speed and broader spectral sensitivity was strongly related to the discovery and production of chemical additives. The original, "unsensitized" photographic emulsion which consists only of silver halides is purely blue-sensitive (below 500 nm). Added dyes sensitize the emulsion to about 600 nm. This blue–green sensitive film is termed *orthochromatic*, while a film sensitized into the red range ($\approx$ 650 nm) is *panchromatic*. *Infrared films* are sensitive up to 900–1000 nm in the near infrared. As examples, spectral sensitivity curves of three spectroscopic films (Fig. 6.11) and of Kodak TP2415 (Fig. 6.12) are graphed [6.62,75,137,157].

The orthochromatic 103a-G film is sensitive to almost 600 nm while the panchromatic 103a-E/F goes beyond Hα. Common to both of them is the so-called "green gap" which is avoided only by using 103a-D film (not shown in the graph). The ordinate scale in Fig. 6.11 is linear.

6.7.5 Astrophotography with Filters

Filter photography may be used to obtain specific information on the target object and to diminish the influence of light pollution. Special techniques such as the *tri-color*

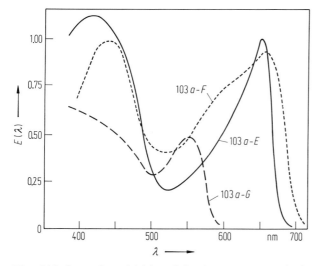

Fig. 6.11. Spectral sensitivities of the three spectroscopic films Kodak 103a-E, 103a-G, and 103a-F. The relative sensitivity $E(\lambda)$ is plotted linearly [6.62].

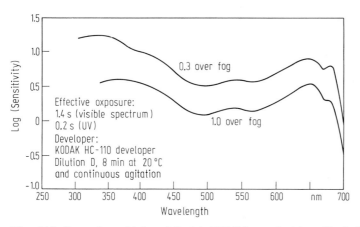

Fig. 6.12. Spectral sensitivity of Kodak TP2415, graphed logarithmically. Note the relative maximum in the red at the Hα line (656.28 nm) [6.157]. The abscissa in the figure is wavelength (nm) and the ordinate is log (sensitivity). The upper curve is 0.3 above fog, the lower is 1.0 above fog. The effective exposures are 1.4 seconds (visible spectrum) and 0.2 seconds (UV). The developer is Kodak HC-110, Dilution D, 8 minutes at 20°C and continuous agitation.

process (Sect. 6.8.4.6) operate with filters that select three different spectral ranges. The sky photographer depends even more on another aspect of filtering methods whose importance in reducing light pollution has recently increased. According to Lord Rayleigh, the background intensity due to scattering off oxygen molecules in the atmosphere is proportional to λ^{-4}, where λ is the wavelength. Thus, violet light at 350 nm is scattered 16 times more than red light at 700 nm. Ideally, a dark night sky

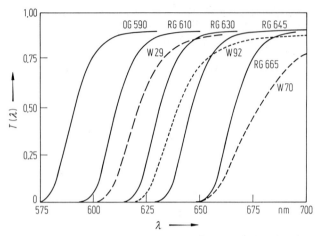

Fig. 6.13. Transmission curves and cutoff wavelengths of various Schott glass filters and Kodak Wratten gelatin filters. $T(\lambda)$ is the transmission relative to that without a filter ($T(\lambda) = 1$). OG = orange glass, RG = red glass, W = Wratten filter. The name OG 590, for instance, means a Schott glass filter whose transmission drops to $T = 0.5$ at $\lambda = 590$ nm [6.62].

would have little background luminescence due to nightglow; in reality, the artificial light radiated into the atmosphere by urban regions creates severe light pollution. This problem affects not only amateur observers but also large, professional observatories such as Mounts Wilson and Palomar in southern California. Modern observatories are therefore usually located on remote mountain tops with dark skies and a dry, dust-free atmosphere.

Contrast enhancement through filters is of benefit primarily for black-and-white photography. Photographs on color slide or color negative films suffer color shifts from filters and can thus be satisfactorily obtained only under dark sky conditions. Figure 6.13 shows the transmission curves and the cutoff wavelengths of some selected Schott and Kodak-Wratten filters transmitting the red Hα line at 656.28 nm [6.62,76,77], which dominates in HII regions such as the Orion Nebula and the North American Nebula.

Suppression of the sky brightness caused by city lights [6.74,173] may be achieved with 103a-E or TP2415 film in combination with light red (W29, OG590) or dark red filters (RG630,RG645,W92). Experience has shown that RG645 combined with powerful lenses (to $f/2.8$, 90 minute exposure) gives very good results [6.62]. The optimal filter thickness yielding the highest contrast seems to be about 6 mm (Fig 6.15). When used with common f-ratios of $f/5.6$, this dark filter requires unacceptably long exposure times and therefore lighter red filters will be preferred. For the green spectral range, 103a-G is used in combination with the W44 filter, and for the blue range, 103a-O, gas-hypersensitized Ektagraphic HC (without filter) or 103a-F with a W47B blue filter.

Figure 6.14 is a photograph of Halley's comet taken on Kodak TP2415 with blue filter W47B. The gas tail is prominent while the dust component is suppressed.

Fig. 6.14. Comet Halley on 1986 April 14 at 23^h13^m UT. Photograph by N. Sommer in Namibia/SWA, 60-minute exposure with $5\frac{1}{2}$-inch Schmidt camera, $f/1.65$ (140/140/225 mm) on gas-hypersensitized Kodak TP2415 with blue filter W47B.

The film/filter sensitivity is graphed as in Fig. 6.15 by combining the spectral sensitivity curve of the film with the selected filter curve.

In Fig. 6.15, the narrower the selected curve is around the Hα line, the more contrasted will be the negative. A "deepest" photograph with the filter RG645 requires a much longer exposure time than, for instance, OG590, which transmits a good deal more background light and reaches the same density in much shorter times. The latter, however, is achieved at the expense of contrast. In Table 6.8 are the Contrasts of Kodak 103a-E film with selected filters by Schott and Kodak given. Column 2 gives the cutoff wavelengths where the filter transmission is 50%. Column 3 lists the contrast of the 103a-E/filter combination relative to OG590. Column 4 gives maximum exposure times for various 103a-E/filter combinations relative to 103a-E/OG590, assumed to

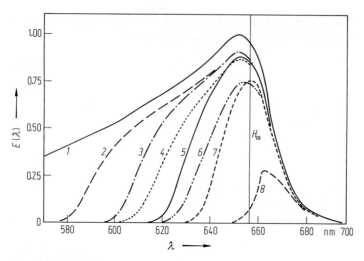

Fig. 6.15. Combining spectral sensitivity curve of the emulsion with a filter curve gives the spectral sensitivity of the chosen film/filter combination [6.62]. The curves shown represent the following: (*1*) 103a-E without filter, normalized to $E(\lambda) = 1$ at maximum, (*2*) with OG590, (*3*) with RG610, (*4*) with W29, (*5*) with RG630, (*6*) with W92, (*7*) with RG645, and (*8*) with RG665. The thickness of all Schott-glass filters is 3 mm.

Table 6.8. Contrast [6.62] of Kodak 103a-E film with selected filters by Schott [6.77] and Kodak [6.76] from Fig. 6.15.

Filter	λ_{cutoff} (nm)	Contrast	t_{max} (min at $f/2.4$)
OG590	590	1.00	≈ 30
RG610	610	1.21	37
W29	620	1.41	44
RG630	630	1.69	52
W92	638	1.93	68
RG645	645	2.40	83
RG665	665	0.87	>330

be 30 minutes at $f/2.4$. The values t_{max} are to be taken merely as guidelines and may be increased by up to about 30% under very dark skies.

6.7.6 Low-Intensity Reciprocity Failure (LIRF)

The photochemical *reciprocity law* of Roscoe and Bunsen states that a given illuminance, which is the product of intensity and exposure time, will always generate the same density on the film. In other words, halving the light intensity while doubling the exposure time should give the same result. In equation form,

$$I_1 t_1 = I_2 t_2. \tag{6.14}$$

All photographic emulsions deviate from this rule, however, when the range of exposure times of 1/1000–1 second in normal photography is exceeded either way. LIRF results in a drop in the film speed at exposure times of over 1 second. Equation (6.14) may therefore be generalized by incorporating the *Schwarzschild exponent p* to

$$I_1 t_1^p = I_2 t_2^p. \tag{6.15}$$

Some other phenomena which cause the film to depart from its ideal behavior are the *Intermitting effect, Clayden effect, Villard effect, Herschel effect, Sabattier effect*, and *solarization*. Others originate in the development (*edge, Eberhard*, and *Kostinsky effects*) [6.75].

The exponent p falls in the range 0.6–0.7 for many films, and may, for special emulsions and procedures (cold-camera and hypersensitization; see Sect. 6.7.11), reach $p = 0.90$–0.99. Table 6.9 lists values of p for some films common in astrophotography.

As an illustration of Eq. (6.15), an example to follow will demonstrate that, for exposure times longer than a certain critical value, a lower-speed film may actually be more sensitive than a high-speed film if the former has a higher Schwarzschild exponent p. In general, for any two films 1 and 2 with Schwarzschild exponents p_1 and p_2,

$$\text{Film 1:}\ E_1(t) = E_1(0)\,\frac{t^{p_1}}{t},$$

$$\text{Film 2:}\ E_2(t) = E_2(0)\,\frac{t^{p_2}}{t},$$

where $E(0)$ and $E(t)$ denote the film speeds in ASA at the beginning of the exposure and at time t, respectively.

Example: At what time t will Film 1 (103a-E, $E_1(0) = 125$ ASA, $p_1 = 0.9$) become more sensitive than Film 2 (Tri-X, $E_2(0) = 400$ ASA, $p_2 = 0.72$)?
Since

$$E_1(t) = E_2(t)$$

it follows that

$$t \geq \left(\frac{E_2(0)}{E_1(0)}\right)^{1/(p_1-p_2)}$$

In this example, $t \geq 640$ s ≈ 11 min. That is to say, for exposure times up to 10 minutes, the Tri-X is faster, but after that time, the apparently "slower" film 103a-E has caught up. Evidently, the film sensitivity by itself does not indicate suitability for long time exposures.

The situation is more complex for color films. A color slide film consists of three color-sensitive layers, each with different LIRFs. Therefore almost all films show a color shift with increasing exposure time, but this effect need not be detrimental. For instance, a green-sensitive layer which is slower than the red layer will damp the light pollution and simultaneously enhance the contrast on the Hα-emitting nebulae. Undesired color tinges may be corrected with color compensating filters (Kodak CC filter) [6.76].

6.7.6.1 A Simple Method to Measure the Schwarzschild Exponent. Equation (6.15) links the stellar light intensity I with exposure time t, while the relation between intensity and stellar magnitude is given by

$$\frac{I_1}{I_2} = 10^{-0.4(m_1-m_2)}, \tag{6.16}$$

with m_1, m_2 referring to stellar magnitudes. Assuming that two exposures have been made with the same equipment but with different exposure times t_1 and t_2, then Eqs. (6.15) and (6.16) are solved for Δm to give:

$$\Delta m = m_2 - m_1 = 2.5 p \, \log(t_2/t_1). \tag{6.17}$$

If the magnitudes m_1, m_2 are now interpreted as the *limiting* magnitudes $m_{1,\text{lim}}$, $m_{2,\text{lim}}$ of the two exposures, then solving for p yields

$$p = 0.4 \frac{m_{2,\text{lim}} - m_{1,\text{lim}}}{\log(t_2/t_1)}. \tag{6.18}$$

Example: Exposure 1: $t_1 = 1$ min, limiting magnitude $m_{1,\text{lim}} = 9\overset{m}{.}0$. Exposure 2: $t_2 = 10$ min, limiting magnitude $m_{2,\text{lim}} = 11\overset{m}{.}3$. From Eq. (6.18), it follows that $p = 0.92$.

In practice, the limiting magnitudes of several exposures are graphed as functions of the exposure time, and the slope of the function $m_{\text{lim}}(\log t)$ gives the exponent p via a *least-squares fit*.

6.7.7 Resolving Power

Size, density, shape, and arrangement of the silver halide crystals are the parameters that determine the resolving power as well as the speed of the film; the larger the grains, the faster the emulsion. The size of a developed silver grain is not identical with that of the original crystal. During development, the crystals change shape, and several may even form a conglomerate. Depending on density, the silver grains may be embedded in several different layers and scatter light in all directions. The irregular arrangement of silver grains can be seen under high magnification, and is known as *graininess*. It limits the suitability of the negative for enlargements. Graininess also depends on exposure; intense illumination leads to dense negatives whose image reaches throughout the emulsion layer. High-energy developers (such as Kodak D19 and Tetenal Dokumol) generate grainier but more deeply developed and thus "deeper" negatives than do fine-grain developers (Kodak D76, Tetenal Ultrafin).

The resolving power of the film is measured by the number of lines per millimeter which can be distinguished. The necessary test cards with various black-and-white patterns can be obtained from a local camera shop or direct from Kodak. The resolving power also depends on the contrast. Usually the film suppliers state the resolving power for contrast ratios of 6:1 (sometimes 1.6:1) and 1000:1. Table 6.9 lists resolving power and Schwarzschild exponents for several selected films.

Table 6.9. Data on selected films.

Film	ISO Speed (ASA/DIN°)	Resolution** (lines/mm)	Schwarzschild Exponent p	Reference
Black-and-White Film				
AGFA				
AGFA Ortho 25	25/15°	350	0.85	[6.32]
AGFA Pan 100	100/21°	145	0.64	[6.32]
AGFA Pan 400	400/27°	110	0.70	[6.32]
ILFORD				
Pan F	50/18°		0.73	[6.32]
XP1 400	400/27°		0.73	[6.32]
KODAK				
T-Max 100	100/21°	63/200		[6.189]
T-Max 400	400/27°	40/125	0.90*	[6.189]
T-Max P3200	3200/36°	40/125		[6.162,181]
TP2415	100/21°	125/320	0.70	[6.86]
HSIR 2481	25/15°	32/80	0.79	
103a-Series	≈ 125/22°	80	0.98	[6.86]
Color Slide Film				
FUJI				
Fujichrome RD50D	50/18°			[6.180]
Fujichrome RD100	100/21°	50/125		
Fujichrome RH400	400/27°	40/125	0.72	
Fujichrome P1600	800/30° – 3200/36°			
KODAK				
Ektachrome 100	100/21°			
Ektachrome 200	200/24°			
Ektachrome 400	400/27°	40/80	0.71	[6.75]
Ektachrome P800/1600	800/30° – 1600/30°	40–60	0.73	[6.150]
SCOTCH				
3M 1000	1000/31°		0.82	[6.78,148]
AGFA				
Agfachrome 1000RS	1000/31°		0.70	[6.56,163]
Color Negative Film				
KODAK				
Ektar 25	25/15°			
Ektar 125	125/22°			
Ektar 100	100/21°			
KONICA				
SRV 400	400/27°			[6.174]
SR 6400	1600/30°			[6.174]
SR-V 3200	3200/36°			[6.78,174,182]

* Measurement by the authors.
** Remarks: The resolution is sometimes expressed for the contrast ratios of 1.6:1 or 1000:1, respectively.

6.7.8 Recommended Films

Among the numerous available photographic products, the ones listed below are especially recommended [6.176]. Black-and-white: Kodak T-Max 100 and T-Max 400 [6.162], TP2415 (high red sensitivity!). Color slide films: Ektachrome 100, Ektachrome 200 (high red sensitivity!), Fujichrome 400 [6.151], Ektachrome 400 [6.149,151], Agfachrome 1000RS, 3M 1000 [6.148,151], and Fujichrome P1600. Color negative films: Kodak Ektar, Konica (see Table 6.9).

The ISO 400/27° films are generally recommended because of their high initial speed. The neophyte in astrophotography benefits from simple standard development. The TP2415 is considered the best all-around film for moderate speed and superior resolution. It was developed from the earlier SO115 and SO410 films and now replaces the high-contrast copy (HCC) by Kodak for reproductions. Additionally to TP2415 35-mm film there is the 120/220 roll film TP6415, and the $4'' \times 5''$ TP4415 with the same characteristics; even at an enlargement of $25\times$, no graininess appears. The film can be developed in Technidol^{TM} and POTA as low-contrast film for planetary photographs, but also in D19, HC110, Microphen, Dektol, and Dokumol as high-contrast film for deep-sky photographs [6.81]. Depending on the kind of developer, its speed varies from ISO 25/15° and ISO 200/24°. A gas-hypersensitized TP2415 will maintain its initial speed without a substantial increase in fog for over one year at a deep-freeze storage of about $-20°C$. The spectroscopic emulsions of the Kodak 103a series are gradually being superceded by fine-grain emulsions. In particular, the 103a-E (now out of production) did rival the gas-hypersensitized TP2415, which is somewhat slower but resolves details better and distinctly improves the limiting magnitude.

6.7.9 Storage of Films

The warranty date stamped on every film package provides an indication of the age of the film. Although a film can be used past the indicated date, aging may alter speed and gradation and may increase the film fog. Color films may suffer from color shift. Various factors during storage can affect how fast a film ages. An exposed, latent image which is not promptly developed may deteriorate in that the film sensitivity drops, the fog increases, and thus the gradation levels off.

High temperatures accelerate the aging of a film, and therefore films should be stored in the original packaging at a temperature under $+20°C$. Films taken from refrigerator storage at below $+10°C$ need one or two hours, and from the deep freeze (below $0°C$) up to six hours, to adapt to the ambient temperature. As dew will condense on cold film if it is placed immediately in the camera, the film should always be left to adjust its temperature in the plastic canister in which it was purchased.

High humidity can be even more damaging than high temperature, but the original plastic canister usually affords sufficient protection. Other films may be stored in closed polyethylene bags with perhaps a dessicating chemical (silicate gel) added. After removal from the original package, films can be exposed to a relative humidity of 50–60% for a limited time. Gases from wood varnish, disinfectants, cleaners, and formaldehyde from pressed wood all may damage the unprotected film even when in the camera. Exposed films should therefore not be stored too long before development [6.79]. Slides will fade when exposed to intense light (particularly UV-radiation), high temperature, or high humidity. To avoid this, the following precautions should be taken in storage:

- Store exposed film materials at a temperature under $+20°C$ and at a relative humidity less than 60%
- Store slides in darkness. Even then, the dyes may change over the course of time.

6.7.10 Film Development

6.7.10.1 Development of Black-and-White Film. Many recipes for film development are recommended but they should be considered only guidelines when trying to test a film/filter combination for a particular astrophotographic application. Note that the following comments on film/developer combinations are not aimed at completeness.

When processing films of the spectroscopic 103a series, and irrespective of which developer is chosen, the anti-halation backing should be carefully removed. This can be done before development by placing the film into a bleach solution (pH=9). The advantage is that carbon particles in the anti-halation backing cannot stick to the emulsion side during the development process. The other option is to remove it very cautiously after rinsing using the fingers to rub the carrier side (and perhaps also the emulsion side in case particles are deposited there). Table 6.10 lists the standard developers for those black-and-white films which are most frequently used in astrophotography.

6.7.10.2 Development of Color Films. Most of the color slide films used in astrophotography are intended for development by the E6 process (Kodak, Tetenal), which can be performed without much expense. The advantage of E6 over other processes is that the reversal of the image after the primary and color development is not attained by an intermediate exposure but rather chemically in a bleaching bath. The development can be carried out by hand in a tank at constant temperature provided by a water bath, or by inserting the tank into a processor. The *Color Kit* by Tetenal, for instance, contains the primary developer, color developer, bleach fixer, and stabilizer. Only during the primary development must the solution be kept carefully at $+38°C \pm 0.5°C$ by a water bath. The steps in the process are less critical and allow tolerances in the temperature from $\pm 1°C$ to $\pm 3°C$ [6.32].

6.7.11 Sensitization of Films

Increasing the speed when preparing the film material is called *sensitization* and can be accomplished using chemical and optical techniques. Chemical sensitization takes place by a ripening process which increases the grain size and forms "germs" of reduction. Optical or color sensitization renders the photographic emulsion sensitive to light of longer wavelengths. Hence, orthochromatic yellow/green-sensitive and panchromatic red-sensitive emulsions are obtained.

Table 6.10. Black-and-white films and developers.

Film	Developer	Concentration* (parts)	Time (min at 20°C)	Agitation (s)	Reference
Deep-Sky					
103a-Series	ILFORD Microphen	undiluted	12	30	[6.74]
103a-E	ILFORD Microphen	undiluted	15 (14°C)	30	[6.62]
TP2415	TETENAL Dokumol**	1 + 9	4-8	30	
T-MAX-Emulsions	KODAK T-MAX dev.	1 + 4	6-10	30	[6.189]
Sun					
Agfa Ortho 25	KODAK HC110	1 + 25	10	10	[6.80]
Agfa Ortho 25	ILFORD Neophin-Blue	1 + 10	12	10	[6.80]
Agfa Ortho 25	AGFA Rodinal	1 + 50	12	10	[6.80]
TP2415	KODAK HC110	1 + 25	8	10	[6.80]
TP2415	ILFORD Neophin-Blue	1 + 10	8	10	[6.80]
TP2415	AGFA Rodinal	1 + 50	10	10	[6.80]
Moon, Planets					
Agfa Pan 25	KODAK HC110	1 + 32 (42)	8	10	[6.80]
Agfa Pan 25	ILFORD Neophin-Blue	1 + 10	8 (7)	10	[6.80]
ILFORD Pan F	KODAK HC110	1 + 32 (42)	8	10	[6.80]
ILFORD Pan F	ILFORD Neophin-Blue	1 + 10	10 (8)	10	[6.80]
TP2415	KODAK HC110	1 + 42 (50)	6	10	[6.80]
TP2415	ILFORD Neophin-Blue	1 + 10	7 (6)	10	[6.80]
TP2415 (Mars)	AGFA Rodinal	1 + 100	14	–	[6.183]
TP2415 (Jupiter)	AGFA Rodinal	1 + 25	14	–	[6.183]

* Note: Concentration 1 + 9 means 1 part developer to 9 parts water.
** Used by the authors.

Table 6.11. Resolving power of Kodak TP2415 [6.81].

Contrast Ratio	KODAK HC110 Dil. D, 8 min at 20°C	KODAK Technidol LC 15 min at 20°C
	Resolution	Resolution
1000:1	320 lines/mm	400 lines/mm
1.6:1	125	125

In contrast to these sensitizing techniques which alter the emulsion are the processing techniques which aid further in increasing the speed. When employed before the exposure, these methods are called *hypersensitization*; they increase the basic sensitivity of the film material, and slow down the decrease in speed with time. The goal is therefore to retain the original speed by reducing the LIRF [6.130,137].

A method developed by F. Dersch and H.H. Dürr in 1937 increases the sensitivity after the exposure, and is called *latensification* (from latent image intensification) but

Table 6.12. Contrast values of Kodak TP2415 [6.81].

Contrast	Gradation	KODAK Developer	Developing Time (min at 20°C)	Film Speed ISO
High	2.50	Dektol	3	200/24°
	2.25 – 2.50	D19	2 – 8	100/21° – 200/24°
	1.20 – 2.10	HC110 (B)	4 – 12	100/21° – 250/25°
	1.00 – 2.10	D76	6 – 12	50/18° – 125/22°
	0.80 – 0.95	HC110 (F)	6 – 12	32/16° – 64/19°
Low	0.40 – 0.80	Technidol LC or POTA	7 – 18	25/15° – 32/16°

it is no longer of interest. Another hypersensitizing technique which increases speed during the exposure is called *cooled-emulsion photography*.

The following list summarizes the various hypersensitizing methods:

1. Bathing in Ag NO$_3$ before or after exposure [6.123,127,128].
2. Dry methods (hypersensitizing by gas or Hg vapor before the exposure.)
3. Pre-flashing by diffuse illumination (still practiced).
4. Cooling during exposure.

6.7.11.1 Cooled-Emulsion Photography. The cooling of an emulsion effects changes in the intermolecular structure of the sensitive layer. The very low light levels which are typical of exposures of faint astronomical objects produce mostly short-lived "subgerms" which generally cannot be preserved in development. Lower temperatures reduce the ion conductivity of the silver bromide, and the subgerms become more stable. Depending on the emulsion, the optimal temperatures are between −40°C and −70°C. Decreasing the temperature from 293 K to 194 K (−78.5°C, the temperature of dry ice) achieves an increase in durability of the subgerms by about a factor of 1000, favors formation of conglomerates from two silver atoms, and leads to a higher effective speed at low intensity levels [6.82,83].

6.7.11.2 The Cold-Camera Technique. Nowadays this technique is practiced only for color photographs as the sensitivity of black-and-white films is more efficiently increased by the simpler gas hypersensitization.

E.S. King noted as early as the turn of the century that sky photographs taken on cold winter nights reached fainter limiting magnitudes than ones obtained on warm nights; the effect may amount to $0.^{\mathrm{m}}5$. The cold camera for color photographs originated from research by A.A. Hoag at the U.S. Naval Observatory in Flagstaff, Arizona at the beginning of the 1960s; his results have since been supplemented by numerous studies, including contributions by amateurs. Only observers already versed in the basics of astrophotography should advance to this technically demanding subject. The low temperature reduces the low-intensity reciprocity failure and thus improves the color balance. Normal 35-mm cameras cannot be used at low temperatures since film cooled from the rear would immediately become completely dew-covered. Moreover,

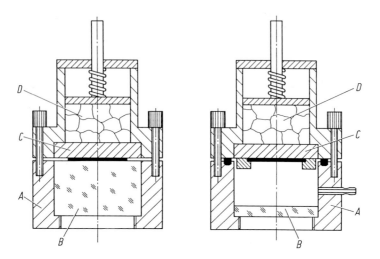

Fig. 6.16. Two possible constructions of a cold camera designed by Stättmayer [6.82]: with glass cylinder (*left*) or vacuum chamber (*right*). (*A*) = plastic case, (*B*) = glass cylinder or window, (*C*) = cooling plate, (*D*) = cooling agent (dry ice).

the cooled film is very fragile and cannot be readily advanced through the camera. The cold camera thus consists of a dry-ice chamber and an evacuated or dry-gas chamber with an optical window for the purpose of preventing condensation of water on the film [6.169].

In principle, there are two different cold-camera constructions. One is the camera with an evacuated or gas-filled chamber, and the other is a device which presses the emulsion side of the film against a thick glass cylinder, thus protecting the emulsion from moisture. The requisite temperature of below $-40°C$ is obtained electrically (the *Peltier effect*), or with coolants, such as dry ice (frozen CO_2), which is easy to obtain and handle. Cameras which substitute dry gas in place of the vacuum have been less successful. When using a glass cylinder, even one which is well-insulated, a weak heating is needed to prevent the formation of dew on its front end. Dust which gets between glass and film may scratch the emulsion. On the other hand, the film is held perfectly flat and need not be cut into small pieces [6.84]. When building or buying a cold camera, all the advantages and drawbacks of the two constructions have to be considered. Figures 6.16 and 6.17 show the design of a cold camera according to Stättmayer [6.82].

6.7.11.3 Using the Cold Camera. For color photography, the cold camera is superior to hypersensitization owing to the improved color balance, but at the expense of greater efforts before, during, and after the exposure since the cooling must be done precisely at these times. Compared with hypersensitized emulsions, the exposure time can be reduced by half. Overall, the cold-camera method is one to be tackled only by skilled observers, and some of the precious clear night hours may be lost to technical preparations necessitated by this procedure. This is shown by the following list of

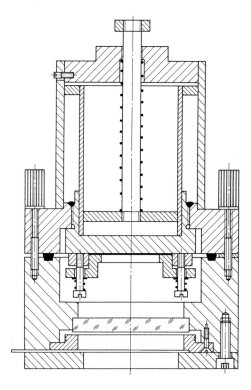

Fig. 6.17. Cross section of the vacuum cold camera after Stättmayer [6.82]. The dimensions are diameter 90 × 135 cm, and the material costs about $60.

operations for taking photographs using the less-complicated camera camera with a glass cylinder:

1. Set on the object.
2. Connect the camera.
3. Focus.
4. Remove camera.
5. In the dark, insert film and close camera.
6. Connect dry ice chamber.
7. Fill dry ice.
8. Let cool for 2 minutes.
9. Add telescope cap.
10. Connect camera with telescope.
11. Check tracking.
12. Open camera cap and telescope cap.
13. Expose.
14. Replace telescope cap and camera cap.
15. Remove camera and take into darkroom.
16. Remove dry ice.
17. Take film out.
18. Clean and dry camera thoroughly.

6.7.11.4 Gas Hypersensitization. Treatment of films with either pure gases or commercially supplied gas mixtures is nowadays among the standard procedures of advanced sky photography. Some optical companies[2], are licensed to perform the gas hypersensitization of photographic films, but observers may wish to explore the possibility of treating the film themselves. It is recommended that the reader consult [6.86,101,102,103] in order to become thoroughly acquainted with this subject. Advanced amateurs will find important suggestions in [6.86,108,109,110].

A pressure- and vacuum-stabilized tank can be built simply, or, for lack of adequate facilities, ready-to-use equipment can be purchased. The gas mixtures can be obtained at reasonable prices from various suppliers. The Lumicon Company[3], for example, offers several different hypersensitizing kits which include disposable gas cylinders. Gases which are used are pure hydrogen (H_2), or the less-dangerous *forming gas* consisting of 92% N_2 and 8% H_2. Since even 4% H_2 in air is inflammable, open flames and sparks should definitely be avoided in the vicinity of the gas nozzle. The treatment with pure hydrogen is performed at a temperature of around 30°C, which, however, must be kept constant [6.190]. Treatment with forming gas should be at 55°C to 60°C lasting from 4 to 24 hours depending on film type and gas pressure, the latter rarely exceeding 3 bars. The residual air pressure is, for practical purposes, not very critical [6.86]: films treated at 24 mb residual air pressure are only a few percent less sensitive than those treated at about 0.03 mb.

The procedure is as follows. The film is wound (in the darkroom) onto an open reel and placed into the hypersensitizing tank. By means of a hand pump or water-flow pump the chamber is evacuated down to a few millibars of pressure, and is then flooded with the gas (or perhaps flushed with gas beforehand). The tank is then heated (by automatically controlled internal heating or in an oven) including a warm-up period of 30–60 minutes. It is then allowed to cool back to room temperature, and the film is rewound onto the core of the film cartridge. Unless the film is to be exposed within a few hours, it should be stored in its container in the refrigerator or freezer. Figure 6.18 demonstrates the difference between untreated and treated TP2415. The left-hand image shows only stars but the right-hand image distinctly depicts the North American Nebula.

Each film responds differently to the parameters hydrogen content of the gas, the temperature, and the gas pressure. Therefore, the data presented in Table 6.13 are to be used only as guidelines.

TP2415 (TP4415 sheet film, TP6416 roll film) is currently *the* film to use for amateur astrophotography [6.103,122,185]. Without treatment and at a basic speed of ISO 25/15° to ISO 200/24°, depending on development (see Table 6.12) and owing to its red sensitivity ([6.157], Fig. 6.12), it is eminently suited for solar photography in $H\alpha$ light and for lunar and planetary photography in white light [6.183]. After hypersensitizing, the increased Schwarzschild exponent up to $p \approx 0.99$ makes it the ideal film for deep-sky work; even solar and planetary photographs profit from shortened exposure times [6.183].

---

2 e.g., University Optics, P.O. Box 1205, Ann Arbor, Michigan 48106 USA.
3 Lumicon, 2111 Research Dr. No. 5S, Livermore, California 94550, U.S.A.

Fig. 6.18a, b. Comparison of photographs of the Milky Way west of Deneb, untreated (**a**), and with gas-hypersensitized TP2415 (**b**), exposed under identical conditions 10 minutes each with a telephoto lens at $f/2.8$, $f = 180$ mm. While in **a** only stars are visible, **b** reveals the typical shape of the North America Nebula. Photos by P. Stättmayer.

Table 6.13. Data on gas hypersensitization.

Film	Duration of Treatment	Pressure (bars)	Temperature (°C)	Relative Sensitivity	Schwarzschild Exponent p
TP2415[7]	3^d	0.7	30	1.0	—
TP2415[1]	$7 - 10^d$	1.0	30	—	—
TP2415[2]	24^h	1.2	60	$0.8^{(2)}$	$0.99^{(4)}$
TP2415[2]	11^h	2.5	60	$1.3^{(2)}$	$0.99^{(4)}$
TP2415	8^h	≈ 1.5	66	$2.9^{(15)}$	—
TP2415	3^h	≈ 1.5	22	$2.9^{(16)}$	—
103a-O	$4 - 5^d$	1.0	20	$2.0^{(6)}$	—
T-MAX 100 Prof.	$\approx 5^h$	≈ 1.5	66	$1.6^{(9)}$	—
T-MAX 100 Prof.	$2^h 20^m$	≈ 1.5	22	$1.7^{(10)}$	—
T-MAX 400 Prof.	$3^h 30^m$	≈ 1.5	66	$3.1^{(11)}$	—
T-MAX 400 Prof.	$1^h 40^m$	≈ 1.5	22	$1.6^{(12)}$	—
T-MAX 3200 Prof.	$3^h 30^m$	≈ 1.5	66	$7.3^{(13)}$	—
T-MAX 3200 Prof.	$1^h 30^m$	≈ 1.5	22	$7.7^{(14)}$	—
Ektachrome 200	6^h	1.2	60	$0.7^{(3)}$	—
Ektachrome 200[8]	15^h	1.1	50	—	—
Ektachrome 400	$2 - 3^d$	1.0	30	$0.5^{(6)}$	—
Ektachrome 400[8]	15^h	1.1	50	—	—
Fujichrome 50D	2^h	2.0	60	—	—
Fujichrome RD100	2^h	$1.3^{(5)}$	60	$0.7^{(6)}$	—
Fujichrome RD100[8]	10^h	1.1	50	—	—
Fujichrome RD400[8]	10^h	1.1	50	—	—

Notes: (1): 8% H_2 [6.103], (2): 10% H_2 [6.108], (3): 8% H_2 [6.108], (4): [6.86], (5): 10% H_2 [6.117], (6): [6.136], (7): 100% H_2 [6.124], (8): 8% H_2 [6.184], (9): 8% H_2, $4 \times 5''$ film, untreated film speed 0.73 compared to untreated 103a-O = 1 [6.195], (10): 100% H_2, data as before [6.195], (11): 8% H_2, $4 \times 5''$ film, untreated film speed 1.58 compared to untreated 103a-O = 1 [6.195], (12) 100% H_2, data as before, (13): 8% H_2, 35-mm film, untreated film speed 3.44 compared to untreated 103a-O = 1 [6.195], (14): 100% H_2, 35-mm film, data as before [6.195], (15): 8% H_2, $4 \times 5''$ film, untreated film speed 0.32 compared with untreated 103a-O = 1 [6.195], (16): 100% H_2, data as before [6.195].

Gas-hypersensitized TP2415 should be handled with great care. The thin carrier may cause the film in the camera body to warp by nearly 0.2 mm at the optical center, a behavior which is also observed with color film [6.108]. Storing the film for a while may reduce the warping. Sharp bending may cause blackening of parts of the emulsion, visible on the developed negative. Additionally, the dried film may be exposed by electrical discharges. Gas-hypersensitized TP2415 loses about 10% of its sensitivity when stored for 3 months in the refrigerator at 8°C, and in addition the fog increases. In the deep freeze at −20°C, practically no loss of speed occurs even after one year of storage. Color films are more sensitive to storage. Sealed Kodak Ektachrome 200 keeps its initial speed without color tinges for one month at 8°C.

6.7.11.5 Construction of a Gas-Hypersensitizing Unit. Designing a hypersensitizing tank requires some precautions regarding its rigidity, since the lid and walls may be subjected to forces well over 1000 N. For reasons of sturdiness and also to achieve

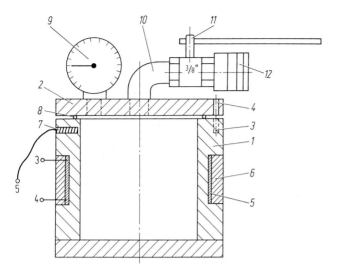

Fig. 6.19. Construction schematics for gas hypersensitization with an electronic temperature control, after Stahlhut [6.164]. To safeguard against pressures over 3 bars, a safety valve is recommended. (*1*) Film chamber of steel or aluminum, outer dimensions 140 mm×100 mm, inner dimensions 100 mm×70 mm. The tank accommodates two reels for 35 mm or one for medium-format film 120. (*2*) Dimensions: 140 mm×15 mm; two connections 3/8″ for gas valve and manometer. (*3*) Threads 3 × 1/4″ at 120°, 15-mm deep, on radius 120 mm. (*4*) Hole through the lid, diameter 6.5 mm at 120°. (*5*) Heater: nichrome wire 0.975 Ω/m, length 3 m, power 50 W (at 12 V). (*6*) Insulation. (*7*) Temperature sensor. (*8*) O-ring, diameter 110 mm×2 mm. (*9*) Manometer: 3/8″-thread, range 0–3 bar. (*10*) Curved pipe 3/8″. (*11*) Valve 3/8″. (*12*) Fast clutch 3/8″.

a high heat capacity, the walls should be from 5 to 10 mm thick. Another important point is the handling of pure hydrogen or gas mixtures containing hydrogen. Hydrogen is a highly inflammable and explosive gas. Working with pure hydrogen requires the highest degree of caution as the range of ignitability with air ranges from 4%–70% hydrogen. The density of hydrogen is much lower than that of air, so it readily escapes into the atmosphere and will not accumulate near the ground. Figure 6.19 shows the construction of a gas-hypersensitizing tank after Stahlhut [6.164], and Fig. 6.20 sketches the outer mechanical connections.

The following parts are needed: a vacuum pump (water-flow, hand pump), manometer (0–3 bars), 2 valves, hoses and connectors, regulated thermometer, and insulating material; a 35-liter gas cylinder from Lumicon, with valve. Figure 6.21 explains the electronic temperature regulation.

The distinct advantage over tanks without heating and temperature regulation is that hypersensitizing can also be performed, for instance, with a car battery (12 V DC). Of course, the tank can also be put into an oven. With a tank wall about 10 mm thick, the heat capacity is sufficiently high that temperature changes in the oven of ±10°C are reduced to a mere ±1°C in the tank, which is more than adequate.

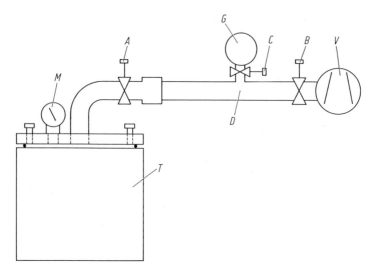

Fig. 6.20. Chamber for gas hypersentization with hose connections. (*T*) Film chamber, (*D*) triple-hose connection, (*G*) gas cylinder, (*V*) vacuum pump (water-flow or hand pump), (*A*) and (*B*) valves (hose clamp), (*C*) fine-adjustment valve for the gas cylinder or pressure-release valve for small steel bottles, (*M*) manometer.

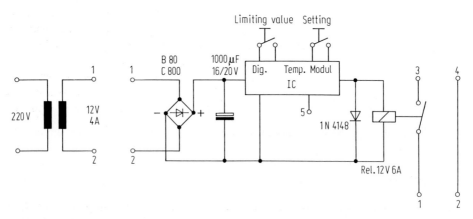

Fig. 6.21. Circuit diagram for electronic temperature regulation with 110 V(220 V)/12 V transformer, rectifier, digital readout for temperature, and relay switch after Stahlhut [6.164]. (5) is the temperature sensor, (3) and (4) are connected with the nichrome wire (see Fig. 6.19).

6.8 Advanced Darkroom Techniques

6.8.1 Darkroom Equipment

The space required for a darkroom is a well-darkened room with electrical and water connections. If there is no running water from faucets, then the water in the vessels must be changed regularly. Even a bathtub can serve well [6.100]. The trays containing the photochemical solutions should be placed well apart to avoid transmission of vibrations and splashes to the enlarger. The enlarger is usually the centerpiece of the lab. Its carrier should be stable enough that vibrations will affect neither the lens nor the lamp while exposing. In principle, the enlarger lens should have at least as good optical quality as the camera lens in order not to create additional optical aberrations. Oversized enlargements may be projected onto the floor or, with a tilted enlarger, onto the wall. Autofocus is of minor importance. It is important, however, that the film be kept perfectly flat by pressing it against a smooth glass plate, preferably one of a sort of glass that does not produce *Newton's rings* [6.32].

Sharpness over the entire field is obtained only when the paper lies entirely flat. There are printing easels with adjustable brackets which hold the paper only at the edges and thus produce a white frame. If frameless enlargements are desired, a printing easel with a glass plate is advised. This requires extreme cleanliness; otherwise, every speck of dust between the glass and paper will show up on the print. Large-sized polyethylene (PE) paper over 30×40 cm can be used without a printing easel because it usually lies very flat. Nevertheless, the last resort is a printing easel fitted with a a a device to evacuate air between the frame and the paper. The enlarger is coupled with a timer to regulate the exposure; analyzers are of lesser importance.

At least three trays and one rinsing basin are needed, and they should preferably be differently colored to avoid confusion during processing. Each chemical is always used in the same vessel, which should be fitted with a spout for pouring out the solution.

6.8.2 Black-and-White Paper

Traditional black-and-white photopapers use felt papers for the emulsion carrier, covered with a layer of barium sulfate (barite) and gelatin. This prevents the emulsion from penentrating into the felt, and it forms a highly reflective surface for glossy prints. It is upon this barite layer that the emulsion is set.

Modern PE-papers are quite similar in their physical composition. A thin felt paper is covered on both sides with waterproof polyethylene layers. The rear is often frosted so that it can be labeled, while the front is specially prepared to bind with the emulsion. Depending on the paper type, silver chloride, silver bromide, or mixtures thereof are embedded in gelatin. For the purpose of protecting the image from mechanical damage, the outer layer of the paper consists of a thin coating with hard gelatin and additives. One square meter contains about 4 grams of silver bromide, 10–20% of which is used to build the image. The remaining silver is removed by the fixer.

Table 6.14. Black-and-white paper grades.

Name	Abbrev.	Gradation
Extra Soft	ES	0
Soft	S	1
Special	SP	2
Normal	N	3
Hard	H	4
Extra Hard	EH	5
Ultra Hard	UH	6

6.8.2.1 Paper Grades. The grade of a photopaper is a measure of the difference in *density* for a given difference in intensity of two exposures. A paper is called "hard" when a small exposure difference creates a large density difference on the paper, whereas "soft" paper responds with only a small density difference. Thus, the grade of a paper is a measure of the gradual reproduction of a given contrast range of the original negative. The best print is obtained when the negative density range equals the density range of the paper. Thus, for a high-contrast negative, a soft paper is employed, and, conversely, for a low-contrast negative, a hard paper should be used. In astrophotography, lunar and solar photographs usually require soft papers, while for deep-sky photography, where it is desirable to reveal the faintest details, hard papers are preferred.

In contrast to film emulsions, paper sensitivity is not normalized and is of minor importance in practice. What *is* significant for photopapers is their extremely unfavorable low-intensity reciprocity failure, a drawback which becomes all the more evident when enlarging very dense negatives. The paper tint ranging from warm-black to neutral-black to bluish-black is primarily conditioned by the particular make. One need not purchase quantities of every paper grade since there are *multigrade papers* whose gradation can be adjusted by exposing the paper through yellow and magenta filters of different densities.

6.8.2.2 Paper Codes. The paper code is a three-digit number, with the last digit characterizing the surface.

```
0  = self-generated glossy
1  = glossy (when dried for high gloss)
2  = half-dull
2a = velvet
3  = dull
4  = pearl
5  = rough dull
6  = linen
7  = silken engraved screen
8  = fine-grain dull (fine grain)
9  = fine-grain glossy (crystal)
```

The second to the last digit indicates the surface tint:

1 = white
2 = chamois
3 = ivory

The third to last digit gives the carrier and its thickness:

- = paper thickness
1 = cardboard
2 = extra-strong cardboard
3 = medium, polyethylene-covered cardboard

Example: PE 310 EH = Polyethylene paper with white, high-gloss surface, extra hard.

The above example has been chosen purposely because it is one which is frequently used in deep-sky astrophotography. This paper affords the deepest black and highest contrast.

6.8.3 Color Processing

The processing of negative or positive color films, plates, or papers demands generally more time and materials than does black-and-white processing. This is because, in addition to the correct evaluation of the brightness, numerous steps must be taken to effect the reproduction of the distinct colors. Color analyzers especially are indespensible here. Except for the production of color prints, the development of color films with the commercially available kits for the *E6 Process* (Kodak, Tetenal) is safe and simple.

As an example, the major steps in the development of color slide film with the Tetenal color-reversal process UK6 are described. The color image originates as follows: The *first developer* used in color-reversal processes has the same function as the black-and-white developer. It generates a black-and-white negative of metallic silver in all three layers of the film, and it reduces only the silver crystals without reacting with the dye couplers. The *color developer* has two functions: It reduces the silver crystals while oxidizing itself. The oxidization product forms dyes with the color couplers embedded in the three emulsion layers, one layer with yellow dye, one with magenta, and one with green. In the *bleach fixing bath*, metallic silver produced by the two previous developers is removed because it would cover the color image; only the picture of the dyes remains. The silver is transformed into silver halide and is subsequently removed by the fixer. A subsequent *stabilizing bath* completes the procedure.

6.8.4 Special Techniques

6.8.4.1 Dodging and Burning, Correcting for Vignetting. The trickiest problem in printing is to transfer a certain density range of the negative onto the photopaper, which always has a lesser range. Sect. 6.8.4.5 will show that only *unsharp masking* solves this problem adequately.

It is relatively simple to correct for *vignetting*, i.e., the diminution in brightness radially outward from the center of the image, by additional exposure of the outer parts. Corners of the negative are less dark than the central part and thus appear darker on the print. A completely exposed print can be obtained by "dodging" the lighter parts of the negative with a piece of cardboard. Stepping down the aperture of the enlarger lens not only reduces additional vignetting, but also increases the exposure time sufficiently that dodging can be done deliberately and without haste. Additional exposure (burning) is necessary for the densest parts of high-contrast negatives. In particular, the bright parts of gaseous nebulae often appear "burned-out" on prints. A suitably shaped piece of cardboard with a hole and attached to a wire is moved during the exposure in the unsharp range (i.e., sufficiently off the paper) through the beam so that sharp contours generated by the cardboard do not appear on the final print.

6.8.4.2 The Composite Printing Method. High magnification reveals the graininess of the negative, thus hiding fine detail (e.g., in planetary photography). The principle of *composite printing* is to enlarge two or more negatives of the same object onto one sheet of photopaper [6.113]. The graininess of the resulting print is less compared to that of a single negative. The individual negatives should be of about equal density. A superimposition frame with a lid is needed, as well as a white sheet of paper attached to the lid. First, a sheet of photopaper is placed into the frame and the lid closed. The enlarger is then switched on, and some stars or prominent features on the negative will be marked on the white sheet of paper on the lid. The first partial exposure follows. Then the other negatives are inserted and exposed successively onto the same photopaper after the frame has in each case been placed exactly into the position as indicated by the markings. The correct exposure time is determined as follows: First, find the exposure time for a single negative. Divide this time by the number of negatives. This partial exposure time is increased for the first negative by 25%, for the others by 10% each. Rather than utilizing the paper grade normally used for printing a single negative, the next harder paper grade is used instead.

6.8.4.3 The Sandwich Method. This procedure is favored particularly when demonstrating the positional changes of moving objects such as minor planets and comets. Two negatives of the same star field, exposed at different times, are, with a slight displacement, mounted one atop the other and then printed.

6.8.4.4 Contrast Enhancement. The subtle nuances of surface brightness which occur, for instance, in comets and gaseous nebulae, can scarcely be separated by the traditional methods of printing. Negatives or color slides can be distinctly contrast-enhanced by simply duplicating them with a slide copier onto high-contrast film. This technique is well known [6.92,186,187], but the disadvantage is that the picture grains as well as the background fog grains are enhanced simultaneously.

While working at the Anglo-Australian Observatory, photographer D.F. Malin developed a method [6.91,93,94,97,156] for revealing extremely faint detail in deep-sky objects while largely suppressing the background fog. When the object is of low intensity, practically only the upper part of the emulsion is exposed, and it is there that the latent image is formed. The background fog, on the other hand, is formed of randomly distributed grains in the entire depth of the emulsion layer. The coarser the film grain, the deeper the layers in which the emulsion is exposed. The ideal film combines high sensitivity with very fine grains as well as anti-halation backing to diminish the light scatter in the emulsion. Gas-hypersensitized TP2415 (TP6415 roll film 120 and TP4415 4 × 5″ sheet film) comes closest to meeting all of these requirements.

Contrast enhancement is achieved by making a contact copy of the original negative onto an extra-hard document film: by means of a diffusing glass plate, the original negative is pressed with its emulsion side against that of the copy film, which is lying on a black sheet of paper to avoid reflections. The copy film is then diffusely illuminated through the original negative. Only a narrow density range centered near the density of the image information at the emulsion surface is copied with only minor contamination by the background density from the deeper layers of the emulsion. It is important that the illumination be uniformly diffused, such as is obtainable using a diffusing glass plate. In practice, the procedure is as follows: The light source is the enlarger with a timer. The orthochromatic copying film is commercially available as sheet film in sizes 4″ × 5″ and larger. Kodak HDU1P, Agfa O811P, and Fuji Fujilith RO100 are well suited. The sheets of film can be processed under red darkroom illumination, so handling them is quite easy. The emulsion surface can easily be identified by its brighter reflection of the darkroom light. The film is processed as normal photopaper in three vessels. After the exposure of the contact copy, developing follows in standard developers for document films, for instance Kodak D19, 1–2 minutes at 20°C, but the development time is not critical. The usual stopbath and rapid fixer follow. After developing under red light, the film still appears milky white, it will quickly become clear and transparent with increasing time in the fixer. This completes the fixing process.

To fully exploit the gradation of the copy film, a densitometer may be employed, although with some practice the best-exposed contact positive can often be selected by merely looking. In any event, the range of exposure is narrow. The developed contact positive is now enlarged on extra-hard paper or, if it is to be used for a slide presentation, is copied once again onto the same copy film, resulting in either case in a negative. Further copying is not recommended since the contrast scarcely increases while the film graininess becomes more prominent. As an example of contrast enhancement, Fig. 6.22 shows a photograph of the very faint supernova remnant Simeis 147 in Taurus.

6.8.4.5 Unsharp Masking. The method of unsharp masking allows the printing of negatives with a contrast range of from 30:1 to 1000:1. The intensity differences in most bright nebulae exceed the contrast range of all photopapers. For instance, if the faint outer parts of the Orion Nebula are well traced, the center will be entirely burned out. Sometimes, the contrast range reaches up to 4 density steps [6.89].

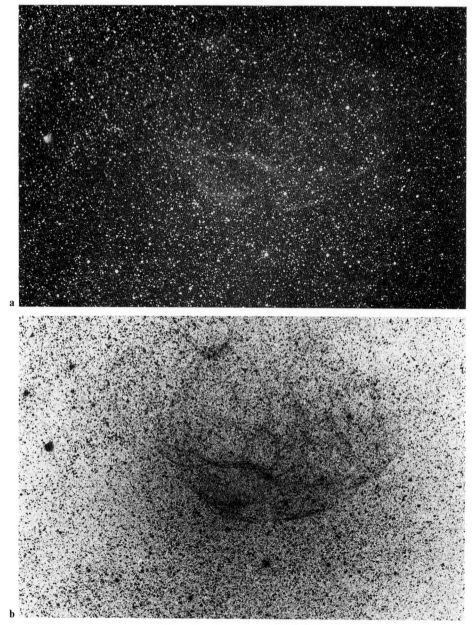

Fig. 6.22a, b. Contrast enhancement of a photograph of supernova remnant S147 in Taurus. The gain is distinctly visible, the graininess increases somewhat, and a slight vignetting of the camera is apparent. Photograph by B. Koch and N. Sommer at the INAG/CERGA Observatory (France), exposed 60 minutes on 103a-E film with red filter W92 with 5.5-inch Schmidt camera at $f/1.65$ (140/140/225 mm).

The density D is related to the transmission T (normalized from 0 to 1) through the equation $D = \log(1/T) = -\log(T)$. Therefore,

Transmission 100% : 1.00 → $D = 0$,
Transmission 10% : 0.10 → $D = 1$,
Transmission   1% : 0.01 → $D = 2$,   etc.

The procedure is as follows: first, a low-contrast positive is copied from the negative. Thereby a slight unsharpness must be achieved by separating the original negative from the copy film by a thin glass plate. Then, for printing, the original negative is sandwiched by its positive unsharp copying mask. Thus, reducing the contrast range of the negative to that of the printing paper is done by increasing the density in the brighter parts of the negative without affecting its darker regions. The unsharpness of the mask serves to avoid double contours.

Example: A negative has $D = 4.0$ in the densest parts and $D = 0.8$ in the sky background. This gives a density range of the subject of 3.2. The mask to be prepared should be entirely transparent ($D = 0$) in the densest parts of the object and should have the background density $D = 1.7$. The original-plus-mask composite thus has $D = 4.0$ in the densest parts and $D = 0.8 + 1.7 = 2.5$ in the background, so that the density range is reduced to $D = 4.0 - 2.5 = 1.5$ (contrast ratio 32:1), which can be easily printed on soft paper.

6.8.4.6 Three-Color Composites. Application of the *tri-color method* makes possible the color portrayal of very faint celestial objects which would not ordinarily be imaged on color slide or negative films because of their low sensitivity and the lack of filtering [6.114,115,116]. The principle employed is that a color image can be constructed from three black-and-white negatives filtered in the basic colors blue, green, and red. The filtered black-and-white negatives of an object are successively made on the three spectroscopic films 103a-O (blue sensitive), 103a-G (green sensitive), and 103a-E (red sensitive). The film/filter combinations are: 103a-O + W2B (UV-blocking filter), 103a-G + W8, and 103a-E + W29. The exposure time of each filtered picture should relate to that of an unfiltered one by the following factors: 1× for 103a-O, 1.2× for 103a-G, and 1.8×–2× for 103a-E. These factors should be taken only as guidelines, as they will vary somewhat depending on emulsion and storage.

With a single film, TP2415, tri-color photography is performed with the Kodak filters W47 (blue), W57A (green), and W23A (red). The three black-and-white negatives are now copied (each with its specific filter) onto a single sheet of Ektacolor film to produce a final color positive film which can easily be enlarged onto Cibachrome positive paper.

6.9 Photographic Limiting Magnitude

6.9.1 Recorded Photographic Limiting Magnitude m

The recorded magnitude limit m of point-like sources (whose angular diameter is below the resolution of the telescope/film combination, i.e., stars, small planetary nebulae, etc.) is primarily a function of the lens diameter D, the exposure time t, the

Schwarzschild exponent p, the film sensitivity E, eventually a filter factor k, and a constant c to be found experimentally:

$$m = 5 \log D + 2.5 \log t^p + 2.5 \log E - 2.5 \log k + c. \tag{6.19}$$

The constant c comprises the resolving power of the film, the imaging quality of the optics, and the transparency. Stättmayer [6.56] quotes $c = 2^m$ under perfect sky conditions, where D is in centimeters, t in minutes, and E in ASA. In what follows, stars are referred to as point-like objects, and, for the sake of simplicity, $k = 1$ (no filter) is assumed.

6.9.2 Maximum Photographic Magnitude Limit $m_{\lim}$

Apart from instrumental factors, such as the aperture D and focal length f, the size of the scattered light disk on the film will also determine the parameters of the photograph. The various factors which potentially determine the diameter b of the stellar disk on the film are:

- Diffraction of light at the aperture → b_0.
- Limited resolving power of the film (graininess) → b_1.
- Influence of seeing (at long focal lengths) → b_2.
- Precision of guiding → b_3.
- Aberrations introduced by the lens → b_4.

The total image diameter b recorded on the film is approximately

$$b = b_0 + b_1 + b_2 + b_3 + b_4. \tag{6.20}$$

Some comments follow:

1. The diameter of the *Airy diffraction disk* b_0 depends on the wavelength λ and the aperture ratio f/D: $b_0 = 2.44 \times \lambda \times (f/D)$.
2. The limited resolving power A of the film ranging from 50 to 320 lines/mm gives, for the diameter of the scattered light disk $b_1 \approx 1/A$, values in the range 3–20 μm.
3. At longer focal lengths, the atmospheric influence becomes noticeable. Presuming steady air, the angular diameter α of the seeing disk is below $1''$, but in turbulent air or near the horizon α is in the range $5''$–$10''$. The diameter of the seeing disk on the film is therefore $b_2 = k \times \alpha \times f$ (with $k = 5 \times 10^{-6}$ when α is in arcseconds).
4. Similarly, the guiding-limited disk is $b_3 = k \times \beta \times f$, where β is the guiding accuracy in arcseconds.
5. Spherical aberration and coma in reflectors and spherical and chromatic aberrations in refractors cause imperfectly recorded stars, particularly near the corners of the field.

The criterion that a star just barely be recorded is that the brightness of its scattered light disk plus sky background be greater than or nearly equal to the brightness of the sky background alone, which in turn results from night glow, light pollution, etc. This criterion will be used in the following. (If the negative is contrast-enhanced, then even a star brightness of 1% over the background will suffice.)

The photographic limiting magnitude on the film is given by Knapp [6.99]:

$$m_{\lim} = 5 \log(f/b) - 8^{m}\!.5 + m_{sky}, \qquad (6.21)$$

where b is the diameter of the scattered stellar disk on the film, f is the effective focal length, m_{sky} is the sky background brightness in magnitudes per square degree (mag/$\Box^2$), and $m_{\lim}$ is the limiting magnitude.

The brightness m_{sky} under dark sky conditions is at least $4^m/\Box^2$ and rises to $2^m/\Box^2$ or more in urban areas [6.196]. Other factors relating to sky *transparency* enter into m_{sky}. This value must be experimentally determined for a given observing site. The attainable limiting magnitude also depends on the ratio f/b, that is, on the resolving power of the optics, with b nearly equal to the sum given in Eq. (6.20). It is seen that the limiting magnitude does not directly depend on the film sensitivity nor upon the lens diameter. With all influences added, the result is the following expression:

$$\frac{f}{b} = \frac{1}{b_1/f + 2.44\lambda/D + k(\alpha + \beta) + b_4/f}, \qquad (6.22)$$

where α and β are both measured in arcseconds.

Special Cases

1. *Short-focus lenses* (wide-angle or normal lens) and coarse-grain film of low resolution, ideal guiding, no distortion: $b_1 \gg b_0, b_2, b_3, b_4$.

$$m_{\lim} = 5 \log(f/b_1) - 8^{m}\!.5 + m_{sky}. \qquad (6.23)$$

   Choosing the same film (b_1 = constant), the limiting magnitude depends logarithmically only upon the focal length. *Example:* $f = 50$ mm, $f/D = 1.8$, $A = 50$ lines/mm (color slide or 103a-E), $\lambda = 500$ nm, seeing $\alpha \leq 5''$. It is found that $b_0 = 2$ μm, $b_1 = 20$ μm, $b_2 = 1$ μm. Under perfectly dark skies ($m_{sky} = 4^m$), the limiting magnitude is $m_{\lim} = 12^{m}\!.5$, which is entirely attainable.

2. *Long-focus lenses* ($f = 2000$ mm, $f/D = 10$), fine-grain film of high resolution ($A = 300$ lines/mm), no distortion ($b_4 = 0$), ideal seeing ($\alpha = 0''\!.5$), guiding tolerance $\beta \approx \pm 1''$. This results in $b_0 = 12$ μm, $b_1 = 3$ μm, $b_2 = 5$ μm, $b_3 = 5 \times 10^{-6} \times 2000$ mm$\times 2 = 20$ μm. This example demonstrates that the diameter of the scattered light disk depends essentially upon the guiding accuracy β; on the other hand, if β is sufficiently small then the recorded image is, in the case of sufficient anti-halation backing of the emulsion, diffraction limited. At a night sky brightness of 4^m, the limiting magnitude becomes $m_{\lim} \approx 19^{m}\!.0$. If the guiding is noticeably inferior (e.g., $\beta = \pm 5''$), then the limiting magnitude is reduced to $16^{m}\!.6$.

3. *Influence of seeing* ($b_2 \gg b_0, b_1, b_3, b_4$)

$$m_{\lim} - m_{sky} = 18^{m}\!.1 - 5 \log \alpha'', \qquad (6.24)$$

where α'' is the diameter of the seeing disk in arcseconds. For $m_{sky} = 4^m$, the dependence on the seeing α is shown in Table 6.15.

Table 6.15. Dependence of limiting magnitude on the seeing.

α	m_{lim}
1″	$18.^{m}1$
5	14.6
10	13.1
20	11.6

6.9.3 Maximum Exposure Time t_{max}

The question often arises, "With which maximum exposure time can the limiting photographic magnitude be reached?" Equating (6.19) and (6.21) yields

$$t_{\text{max}} = \frac{\text{const} \times (f/D)^{2/p}}{E}, \tag{6.25}$$

where t_{max} is in minutes, E in ASA, and p is the Schwarzschild exponent. The constant "const" is determined by the sky background and can reach the value 500 under ideal sky conditions [6.56].

6.9.4 The Standard Sequence for Determining the Limiting Magnitude

To find the magnitude limit, star field No. 51 in the *Atlas of Selected Areas* [6.119] can be used (Schaefer [6.188] has published a chart of M67 reaching down to magnitude 21). A chart from the *SAO Star Atlas* [6.118] is shown in Fig. 6.23 to find the field with stars down to magnitude 9. Figure 6.24 shows the field SA 51 with a size of 60′ × 60′; the photographic limiting magnitude is $m_{\text{lim}} = 12.^{m}5$ (in blue).

The dashed part of Fig. 6.24 is shown enlarged in Fig. 6.25, with a photovisual limiting magnitude of $20.^{m}5$ [6.120]. The stars common to both Figs. 6.24 and 6.25 show the difference between the *blue* (or *photographic*) *magnitude* and the *photovisual magnitude*. Which of them applies best to one's own photograph depends upon the spectral range in which the picture was made. Panchromatic films, which are sensitive into the red range, correspond better to the photovisual magnitude, whereas photography on orthochromatic film (Kodak 103a-O) represents the blue (or photographic) magnitude.

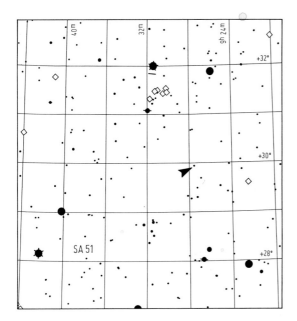

Fig. 6.23. The field SA 51 excerpted from the *SAO Star Atlas* [6.118]. The field size is about $5° \times 5°$. The star shown by an arrow is SAO 79445, $m_V = 9\overset{m}{.}1$, with the coordinates $7^h 27\overset{m}{.}5$, $+29°56'$ (1950.0), or $7^h 30\overset{m}{.}6$, $+29°50'$ (2000.0).

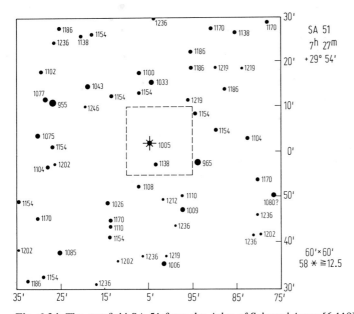

Fig. 6.24. The star field SA 51 from the *Atlas of Selected Areas* [6.119]. The field of $60' \times 60'$ contains 58 stars with blue magnitudes brighter than $m_{pg} = 12\overset{m}{.}5$. North is at the top. Stellar magnitudes are to two decimal places with the decimal point omitted (e.g., $1005 = 10\overset{m}{.}05$). The bright star in the center is SAO 79445. The dashed insert represents the field of Fig. 6.25.

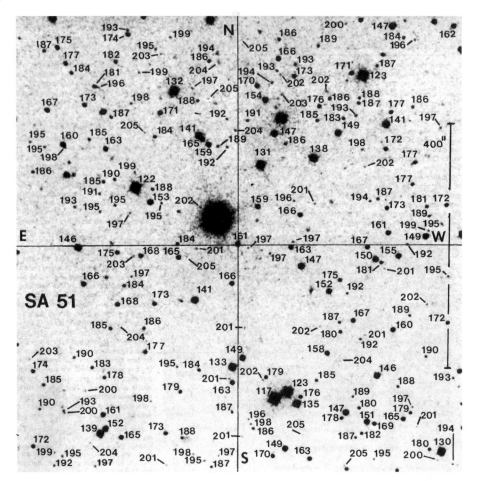

Fig. 6.25. The field SA 51 on an exposure by E. Everhart [6.120] with a 40-cm Newtonian telescope, $f/5.5$, on gas-hypersensitized TP2415. North is at the top, and the field measures $12' \times 12'$. The bright star in the center is SAO 79445. The photovisual magnitudes were measured with an iris photometer and go down to about $20\overset{m}{.}5$. They are given to one decimal place with the decimal point omitted (e.g., $201 = 20\overset{m}{.}1$).

6.10 Further Reading

Besides the more important references on astrophotography, from English- and German-language astronomical periodicals, the following are popular magazines which often contain contributions to the area of astrophotography.

Astronomy, Astro Media Corp., 625 East St. Paul Ave., Milwaukee, WI 53502, USA.

Ciel et Espace, L'Association Francaise d'Astronomie, Observatoire du Parc Montsouris, 17 rue Émile-Deutsch-de-la-Meurthe, 75014 Paris, France.

Orion, Zeitschrift der Schweizerischen Astronomischen Gesellschaft, Zentralsekretariat der SAO, Hirtenhofstr. 9, 6005 Luzern, Switzerland.

Sky & Telescope, Sky Publishing Corporation, 49 Bay State Road, P.O. Box 9102, Cambridge, MA 02238-9102, USA.

Sterne und Weltraum, Verlag Sterne und Weltraum, Dr. Vehrenberg GmbH, Portiastrasse 10, 81545 Munich, Germany.

References

6.1   Riepe, P.: Kleinbildoptiken in der Astrofotografie. *Sterne und Weltraum* **24**, 280 (1985).
6.2   Martinez, P.: *Astrophotography II*, Willmann-Bell, Richmond 1985.
6.3   Dragesco, J.: Images stellaires obtenues en photographie du ciel profond. *Ciel et Espace* **215**, 62 (1987).
6.4   Schur, C.: Experiments with All-Sky Photography. *Sky and Telescope* **63**, 621 (1982).
6.5   Schur, C.: The Zodiacal Light in Color. *Sky & Telescope* **64**, 199 (1982).
6.6   Sloan, J.: More on All-Sky Photography. *Sky & Telescope* **66**, 70 (1983).
6.7   Vielmetter, H.: (1627) Ivar am Laufenden Band. *Sterne und Weltraum* **24**, 536 (1985).
6.8   Vielmetter, H.: Halley am Stadtlicht-Himmel. *Sterne und Weltraum* **25**, 608 (1986).
6.9   Liller, B., Mayer, B.:*The Cambridge Astronomy Guide*, Cambridge University Press, Cambridge New York 1985.
6.10 Arbour, R: A Camera that Tracks Comets. *Sky & Telescope* **74**, 428 (1987).
6.11 di Cicco, D.: Photography Through a Telescope. *Sky & Telescope* **72**, 569 (1986).
6.12 Sinnott, R.W.: Editor's Note on E.S. King's Method of Photographic Polar Alignment. *Sky & Telescope* **78**, 542 (1989).
6.13 Riepe, P.: Astrophotographie mit transportablen Geräten. *Sterne und Weltraum* **22**, 89 (1983).
6.14 Del Vo, P.: Conviene guidare su una stelle equatoriale? *L'astronomia* **35**, 65 (1984).
6.15 Sidgwick, J.B.: *Amateur Astronomer's Handbook*, Enslow Publishers, Hillside 1980.
6.16 Gleanings for ATMs: Neutralizing Atmospheric Dispersion. *Sky & Telescope* **43**, 388 (1967).
6.17 Wedel, B.: Das atmosphärische Spektrum und seine Beseitigung. *Sterne und Weltraum* **10**, 339 (1971).
6.18 Paul, H.E.: *Outer Space Photography for the Amateur*, American Photographic Book Publishing Co., Inc, Garden City 1976, p. 118.
6.19 Rummel, W.: Der Polsucher, ein neues Justierinstrument. *Sterne und Weltraum* **7**, 186 (1968).
6.20 Ahnert, P.: *Kalender für Sternfreunde 1970*, J.A. Barth, Leipzig 1969.
6.21 Filimon E.: Meteorstrombeobachtung: Amateurprogramm. *Astro-Magazin*, 140 (1986).
6.22 Jahn, J.: Vergleich der Meteoranzahl bei verscheidenen Astrokameras. *Kometen, Planetoiden, Meteore* **2**, No. 4, 36 (1987).
6.23 L'Équipe "Perséides 80": Le Météorographe. *Ciel et Espace* **182**, 46 (1981).
6.24 Dannemann, A.: Fotografische Meteorbeobachtung und deren Auswertung. *Kometen, Planetoiden, Meteore* **1**, No. 1, 15 (1986).
6.25 Reimann, I.: Anmerkungen zum Artikel von A. Dannemann in KPM 1. *Kometen, Planetoiden, Meteore* **1**, No. 2, 19, (1986).
6.26 Jahn, J.: Bestimmung der wahren Helligkeit eines Meteors auf einer Aufnahme. *Kometen, Planetoiden, Meteore* **2**, No. 4, 29 (1987).
6.27 Levy, D.H., Edberg, S.J.: *Observe Meteors. The Association of Lunar and Planetary Observers Meteor Observer's Guide*, published by the Astronomical League, Washington 1986.

6.28 Bohrmann, A.: *Bahnen Künstlicher Satelliten*, Bibliographisches Institut, Mannheim 1966.
6.29 Welch, D.L.: Observing Geosynchronous Satellites. *Sky & Telescope* **71**, 606 (1986).
6.30 Maley, P.D.: Photographing Earth Satellites. *Sky & Telescope* **71**, 563 (1986).
6.31 King-Dele, D: *Observing Earth Satellites*, Van Nostrand Reinhold Company, New York 1983.
6.32 Tetenal: *Richtig entuickeln.* (34th edn.) 1984.
6.33 Lovi, G.: A Look at Seeing. *Sky & Telescope* **70**, 577 (1985).
6.34 Walker, M.F.: How Good Is Your Observing Site?. *Sky & Telescope* **71**, 139 (1986).
6.35 Cox, R.E.: Some New Illuminated Finders. *Sky & Telescope* **49**, 183 (1975).
6.36 Röhr, T.: "Richtig" fokussierte Astro-Fotos: Eine neue Lösung für ein altes Problem. *Mittlungen der Volkssternwarte Köln* **30**, No. 3, 15 (1986).
6.37 Greenler, R.: *Rainbows, Halos, & Glories*, Cambridge University Press, Cambridge 1980.
6.38 Meinel, A. & M.: *Sunsets, Twilights, and Evening Skies*, Cambridge University Press, Cambridge New York 1983.
6.39 Pollmann, E.: Spektroskopische Veränderlichenbeobachtung. *Sterne und Weltraum* **24**, 340 (1985).
6.40 Wagner, B.: Astrospektrographie mit einfachen Mitteln. *Sterne und Weltraum* **25**, 218 (1986).
6.41 Waber, R., McPherson, R.: Photographing Star Spectra. *Sky & Telescope* **33**, 322 (1967).
6.42 Bourge, P., Dragesco, J., Dargery, Y.: *La Photographie Astronomique D'Amateur*, Publications Photo-Cinéma P. Montel, Paris 1979.
6.43 Patterson, J., Michaud, P.: Photographing Stellar Spectra. *Astronomy* **8**, 39 (1980).
6.44 Gebhardt, W., Helms, B.: Ein Selbstbau-Prismenspektograph zum Gebrauch am Celestron-8. *Sterne und Weltraum* **15**, 58 (1976).
6.45 Sorensen, B.: A Simple Slit Spectrograph. *Sky & Telescope* **73**, 98 (1987).
6.46 Albrecht, C.: Astrospektrographie mit Spiegelteleskopen. *Sterne und Weltraum* **11**, 195 and 243 (1972).
6.47 Solberg, Jr., H.G.: Photographing the Zodiacal Light. *Sky & Telescope* **29**, 323 (1965).
6.48 Solberg, Jr., H.G., Minton, R.B.: Photographing the Gegenschein. *Sky & Telescope* **31**, 380 (1966).
6.49 Viewing the Zodiacal Light and Gegenschein. *Sky & Telescope* **23**, Observer's Page (1962).
6.50 Boney, W.H.: Depths of Space. *Sky & Telescope* **70**, 503 (1985).
6.51 Hückel, P.: Nützliche Tips für die Okularprojektion. *Sterne und Weltraum* **24**, 42 (1985).
6.52 Valleli, P.A.: The Focal Reducer as a Telescope Accessory. *Sky & Telescope* **46**, 405 (1973).
6.53 Gee, A.E.: How to Design Telecompressors. *Sky & Telescope* **67**, 367 (1984).
6.54 Edberg, S.J.: *IHW Amateur Observer's Manual for Scientific Comet Studies*, pp. 6–4, Enslow Publishers, Hillside, NJ/ Sky Publishing Corporation, Cambridge MA 1983.
6.55 di Cicco, D.: Shooting Halley. *Sky & Telescope* **71**, 23 (1986).
6.56 Stättmayer, P.: Hinweise zur Kometenphotographie. *Sterne und Weltraum* **24**, 476 (1985).
6.57 Stättmayer, P.: Eine einfache Methode zur indirekten Kometennachführung. *Sterne und Weltraum* **13**, 132 (1974).
6.58 Lüthen, H. and Schröder, K.-P.: Nachführung auf den Kometen mit dem GA-2 Nachführansatz. *Kometen, Planetoiden, Meteore* **1**, No. 2, 29 (1986).
6.59 An Observer's Kit for This Month's Lunar Eclipse. *Sky & Telescope* **49**, 280 (1975).
6.60 Janesick, J., Blouke, M.: Sky on a Chip: The Fabulous CCD. *Sky & Telecope* **74**, 238 (1988).
6.61 A Midsummer's Night Eclipse. *Sky & Telescope* **62**, 391 (1981).
6.62 Celnik, W.E., Riepe, P.: Photographie extrem schwacher HII-Regionen Teil 1. *Sterne und Weltraum* **23**, 458 (1984).
6.63 Celnik, W.E., Riepe, P.: Photographie extrem schwacher HII-Regionen Teil 2. *Sterne und Weltraum* **24**, 100 (1985).
6.64 Dunham, D.W.: May's Pallas Occultation A Success. *Sky & Telescope* **66**, 270 (1983).

6.65 The Polarized Corona in Color. *Sky & Telescope* **63**, 210 (1982).
6.66 Beatty, J.K.: Planetary Satellites: An Update. *Sky & Telescope* **66**, 405 (1983).
6.67 Riepe, P. et al.: Galaxienphotographie Teil 1. *Sterne und Weltraum* **26**, 34 (1987).
6.68 Riepe, P. et al.: Galaxienphotographie Teil 2. *Sterne und Weltraum* **26**, 155 (1987).
6.69 Remmert, E.: Die Sonnenphotographie und ihre Probleme Teil 1. *Sterne und Weltraum* **24**, 158 (1985).
6.70 Remmert, E.: Die Sonnenphotographie und ihre Probleme Teil 2. *Sterne und Weltraum* **24**, 606 (1985).
6.71 Kitchen, C.R.: *Astrophysical Techniques*, Adam Hilger, Bristol 1984.
6.72 Langford, M.: *Die große Fotoenzyklopädie*, Christian-Verlag, München 1983.
6.73 Beck, R., Hänel, A.: Grundlagen der Astrofotografie. *Festschrift Sternwarte Bonn*, p. 45.
6.74 Dreyhsig, J., Leder, N.: Astrofotografie in der Großstadt. *Sterne und Weltraum* **22**, 438 (1983).
6.75 Kodak Brochure P-315, *Plates and Films for Scientific Photography*.
6.76 Kodak Brochure B3, *Kodak Filters for Scientific and Technical Uses*.
6.77 Schott Brochure, *Optische Glasfilter*.
6.78 di Cicco, D.: Skyshooting with the Fastest Color Film. *Sky & Telescope* **74**, 558 (1987).
6.79 AGFA-Gevaert (Phototechnical information): Die Haltbarkeit fotografischer Filme.
6.80 Hornung, H., Hückel, P.: Sonnen-, Mond- und Planetenphotographie mit Amateurteleskopen. *Sterne und Weltraum* **23**, 393 (1984).
6.81 Martinez, P.: Aide-toi et le film t'aidra. *Ciel et Espace* **199**, 39 (1984).
6.82 Stättmayer, P.: Tiefkühlfotografie. *Sterne und Weltraum* **22**, 144 (1983).
6.83 Leue, H.J., Tomoscheit, D.: Zur Praxis der Tiefkühlphotographie. *Sterne und Weltraum* **12**, 473 (1973).
6.84 Newton, J.: A Cold Camera for Astrophotography. *Astronomy* **9**, 39 (1981).
6.85 Iburg, N.: The Shoot Out: Cold Camera vs. Gas Hypering. *Astronomy* **9**, 61 (1981).
6.86 Höbel, P.: Theorie und Praxis der Hypersensibilisierung von Filmemulsionen. *Sterne und Weltraum* **22**, 430 (1983).
6.87 Lightfood, D.: Making the Most of Black-and-White Astronegatives. *Astronomy* **10**, 51 (1982).
6.88 Garner, W., Meaburn, J.: The Combination of Unsharp Masking and High-Contrast Copying. *AAS Photo-Bulletin* **20**, 3 (1979).
6.89 Malin, D.F., Zealy, W.J.: Astrophotography with Unsharp Masking. *Sky & Telescope* **57**, 355 (1979).
6.90 Kriete, A.: Kontrastverstärkung durch lichtoptische Filterung und photochemische Kantendifferenzierung an Astroaufnahmen. *Sterne und Weltraum* **17**, 296 (1978).
6.91 Malin, D.F.: Photographic Amplification of Faint Astronomical Images. *Nature* **276**, 591 (1978).
6.92 Baumgardt, J.: Enhancing Astronomical Photographs with a Slide Copier. *Sky & Telescope* **66**, 574 (1983).
6.93 Koch, B.: Photographische Hochkontrastverstärkung astronomischer Negative. *Sterne und Weltraum* **24**, 156 (1985).
6.94 Koch, B., Sommer, N.: Capturing Faint Nebulae with Contrast Enhancement. *Sky & Telescope* **69**, 83 (1985).
6.95 Coco, M.J.: Enhancing Color Photographs With Filters. *Sky & Telescope* **70**, 215 (1986).
6.96 Mette, V.: Eine alternative Methode der Kontrastverstärkung. *Sterne und Weltraum* **25**, 598 (1986).
6.97 Malin, D.F.: Photographic Enhancement of Direct Astronomical Images. *AAS Photo-Bulletin* **21**, 4 (1981).
6.98 Gorski, A.B.: Enhancing Astronomical Photographs. *Sky & Telescope* **58**, 184 (1979).
6.99 Knapp, H.: Über die Reichweite von Objektiven bei Astroaufnahmen mit kleinen Montierungen. *Sterne und Weltraum* **3**, 262 (1964).
6.100 Tetenal: *Die Schwarzweiß-Positiv-Technik*.
6.101 Vehrenberg, H.: Hypersensibilisierung. *Sterne und Weltraum* **20**, 193 (1981).

6.102 Vehrenberg, H.: Hypersensibilisierung von s/w- und Farbfilmen. *Sterne und Weltraum* **20**, 246 (1981).
6.103 Becker, P., Bojarra, U.: Erste Erfahrungen mit hypersensibilisierten Filmen. *Sterne und Weltraum* **21**, 34 (1982).
6.104 Sliva, R.: Hypersensitizing, Part 1. *Astronomy* **9**, 39 (1981).
6.105 Sliva, R.: Hypersensitizing, Part 2. *Astronomy* **9**, 48 (1981).
6.106 Healy, D.: Experiments with Gashypered Film. *Sky & Telescope* **61**, 174 (1981).
6.107 di Cicco, D.: Notes on Gas Hypersensitizing. *Sky & Telescope* **61**, 176 (1981).
6.108 Stättmayer, P.: Gas Hypersensibilisierung von Filmmaterial. *Sterne und Weltraum* **21**, 532 (1982).
6.109 Höbel, P.: Untersuchungen von Filmemulsionen. *Sterne und Weltraum* **13**, 205 (1974).
6.110 Smith, Å.G., Hoag, A.A.: Advances in Astronomical Photography at Low Light Levels. *Ann. Rev. Astron. Astrophys* **17**, 43 (1979).
6.111 Babcock, T.A.: A Review of Methods and Mechanism of Hypersensitization. *AAS Photo-Bulletin* **24**, 3 (1976).
6.112 Babcock, T.A., Ferguson: A Novel Form of Chemical Sensitization Using Hydrogen Gas. *Photographic Science and Engineering* **19** (1975).
6.113 Jones, S.E.: Methods, Advantages, and Limitations of Compositing Photographic Images. *AAS Photo-Bulletin* **11**, 15 (1976).
6.114 Vehrenberg, H.: Photographs of Deep-Sky Objects. *Sky & Telescope* **55**, 295 (1978).
6.115 Alt, E., Rusche, J.: Indirekte Astrofarbenfotografie nach dem modifizierten Dreifarbenverfahren. *Orion* **33**, No. 148, 67 (1975).
6.116 Marling, J.B.: Advances in Astrophotography. *Sky & Telescope* **67**, 582 (1984).
6.117 Laepple, L.: Astrofotografie in der Schule. *Sterne und Weltraum* **25**, 586 (1986).
6.118 *Smithsonian Astrophysical Observatory Star Atlas*, Smithsonian Institution, Washington DC 1966, 1971.
6.119 Brun, A., Vehrenberg, H.: *Atlas of Selected Areas* (3rd edn.), Treugesell-Verlag, Düsseldorf 1980.
6.120 Everhart, E.: Finding Your Telescope's Magnitude Limit. *Sky & Telescope* **68**, 28 (1984).
6.121 Everhart, E.: Adventures in Fine-Grain Astrophotography. *Sky & Telescope* **61**, 100 (1981).
6.122 Everhardt, E.: Hypersensitization and Astronomical Use of Kodak Technical Pan Film 2415. *AAS Photo-Bulletin* **24**, 3 (1980).
6.123 Walker, P.E.: Hypersensitization of Kodak Technical Pan Film 2415 by Bathing in Silver Nitrate Solution. *AAS Photo-Bulletin* **24**, 7 (1980).
6.124 Marling, J.B.: Gas Hypersensitization of Kodak Technical Pan Film 2415. *AAS Photo-Bulletin* **24**, 9 (1980).
6.125 Vidal, N.V.: Hypersensitization of Kodak IIIa-J and 103a-D Emulsions with Forming Gas. *AAS Photo-Bulletin* **21**, 3 (1979).
6.126 Schumann, J.D.: A Hypersensitization Process Faster than Nitrogen Soaking. *AAS Photo-Bulletin* **20**, 13 (1979).
6.127 Jenkins, R.L., Franell, G.C.: The Hypersensitization of Infrared Emulsions by Bathing Treatments. *AAS Photo-Bulletin* **17**, 3 (1978).
6.128 Schoening, W.E.: Hypersensitizing Infrared Plates with Silver Nitrate Solution. *AAS Photo-Bulletin* **17**, 12 (1978).
6.129 Lapointe, M.: Which Films are Worth Hydrogenating? *Sky & Telescope* **55**, 401 (1978).
6.130 Mutter, E.: Die Technik der Negativ- und Positivverfahren. *Die wissenschaftliche und angewandte Fotografie*, Band V, Wien 1955.
6.131 Barrai, C.: Hypersensibilisation des films à l'hydrazote. *Ciel et Espace* **209**, 29 (1986).
6.132 Dragesco, J.: Progrès spectaculaires en astrophotographie d'amateur. *Ciel et Espace* **185**, 44 (1982).
6.133 Heudier, J.L., Sim: *Astronomical Photography 1981*, CNRS INAG.
6.134 Heudier, J.L.: Le renouveau de la photographie astronomique I. *L'Astronomie* **91**, 313 (1977).

6.135 Heudier, J.L.: Le renouveau de la photographie astronomique II. *L'Astronomie* **91**, 341 (1977).
6.136 Marling, J.B.: Gas Hypersensitization of 35-mm B& W and Color Film - III. *Deep Sky Monthly* **3** (1981).
6.137 Heudier, J.L.: Astronomical Photography: Its Present Status. *AAS Photo-Bulletin* **26**, 3 (1981).
6.138 Brandt, L.: Zur Geschichte der Himmelsfotografie. *Sterne und Weltraum* **19**, 122 (1980).
6.139 Ilford: *Labortechnik*.
6.140 VDS-Arbeitsgruppe Astrofotografie: *Astrofotografie-Eine Einführung in die Himmelsfotografie*.
6.141 Conrad, C.M. et al.: Evaluation of Nine Developers for Hypersensitizied Kodak Technical Pan Film 2415. *AAS Photo-Bulletin* **38**, 3 (1985).
6.142 Smith, A.G.: Reciprocity Failure of Hypersensitized and Unhypersensitized Kodak Technical Pan Film 2415. *AAS Photo-Bulletin* **31**, 9 (1982).
6.143 Everhardt, E.: Color Negative Films for Astrophotography. *AAS Photo-Bulletin* **30**, 9 (1982).
6.144 Smith, A.G.: Comparison of the Absolute Sensitivity of Kodak Technical Pan Film 2415 with Standard Astronomical Emulsions. *AAS Photo-Bulletin* **30**, 6 (1982).
6.145 Becker, K.: Der Farbdiafilm Fujichrome RH400. *Sterne und Weltraum* **25**, 106 (1986).
6.146 Jacobs, G.: Der Farbnegativfilm Fujicolor HR1600. *Sterne und Weltraum* **25**, 107 (1986).
6.147 di Cicco, D.: Film Notes for Astrophotographers. *Sky & Telescope* **68**, 371 (1984).
6.148 Celnik, W.E. et al.: Der 3M 1000 in der Astrophotographie. *Sterne und Weltraum* **24**, 218 (1985).
6.149 Riepe, P. and Celnik, W.E.: Astrophotographie mit dem Kodak Ektachrome 400. *Sterne und Weltraum* **21**, 317 (1982).
6.150 Riepe P., Ransburg, W., and Celnik, W.E.: Im Astrotest: Der Kodak Ektachrome P800/1600. *Sterne und Weltraum* **24**, 542 (1985).
6.151 di Cicco, D.: Another Superspeed Color Film. *Sky & Telescope* **66**, 506 (1983).
6.152 di Cicco, D.: ASA 1000 and Color Too. *Sky & Telescope* **65**, 215 (1983).
6.153 Gordon, B.: *Astrophotography*, Willmann-Bell, Richmond 1985.
6.154 Griesser, M.: *Himmelsfotografie*, Hallwag-Verlag, Stuttgart Bern 1982.
6.155 Covington, M.: *Astrophotography for the Amateur*, Cambridge University Press, Cambridge 1985.
6.156 Malin, D.F., Murdin, P.: *The Colours of the Stars*, Cambridge University Press, Cambridge London New York 1986.
6.157 KODAK Data Sheet P-A4.
6.158 KODAK Date Sheet P-B8.
6.159 KODAK Data Sheet P-B6.
6.160 Brodkorb, H. et al.: Das Dreifarbenverfahren. *Orion* **31**, 55 (1973).
6.161 Healy, D.: Films for Deep Sky Photography. *Deep Sky* **2**, No. 3, 16 (1984).
6.162 KODAK Data Sheet T-MAX-Films.
6.163 Celnik, W.E. et al.: Der Farbdiafilm Agfachrome 1000 RS in der Astrophotographie. *Sterne und Weltraum* **26**, 287 (1987).
6.164 Stahlhut, J.: Private Communication.
6.165 Ney, E.P.: The Mysterious "Egg Nebula" in Cygnus. *Sky & Telescope* **49**, 21 (1975).
6.166 Dorst, F.: Die Leuchtenden Nachtwolken am 7./8. Juni 1976. *Sterne und Weltraum* **15**, 328 (1976).
6.167 Janesick, J., Blouke, M.: Sky on a Chip: The Fabulous CCD. *Sky & Telescope* **74**, 238 (1987).
6.168 Meeus, J.: *Astronomical Formulae for Calculators* (3rd edn.), Willmann-Bell, Richmond 1985, p. 48.
6.169 Newton, J., Teece, P.: *The Guide to Amateur Astronomy*, Cambridge University Press, Cambridge New York 1988.
6.170 Wallis, B.D., Provin, R.W.: *A Manual of Advanced Celestial Photography*, Cambridge University Press, Cambridge New York 1988.

6.171 di Cicco, D.: Sky Photography Near the Arctic Circle. *Sky & Telescope* **73**, 343 (1987).
6.172 Schaaf, F.: A Field Guide to Atmospheric Optics. *Sky & Telescope* **77**, 254 (1989).
6.173 Entrop, H.A.: Photographing the Deep Sky Amid City Lights. *Sky & Telescope* **77**, 108 (1989).
6.174 Riffle, J.: Comparing Color Films for Astrophotography. *Sky & Telescope* **76**, 207 (1988).
6.175 Hallas, T., Mount, D.: Enhanced-Color Astrophotography. *Sky & Telescope* **78**, 216 (1989).
6.176 Olson, D.W.: Who First Saw the Zodiacal Light? *Sky & Telescope* **77**, 146 (1989).
6.177 Phillips, J.C.: Results from the Ashen-Light Campaign. *Sky & Telescope* **79**, 108 (1990).
6.178 Ostro, S.J.: Radar Reveals a Double Asteroid. *Astronomy* **18**, 38 (1990).
6.179 di Cicco, D.: Astrophotography Now and Then. *Sky & Telescope* **76**, 463 (1988).
6.180 Lodriguss, J.: "Slow" Films for Astrophotography. *Sky & Telescope* **76**, 207 (1988).
6.181 Keene, G.T.: Sky Mapping With Kodak T-MAX P3200 Film. *Sky & Telescope* **76**, 436 (1988).
6.182 Peronto, J.: Telephoto Lenses for Astrophotography. *Sky & Telescope* **76**, 438 (1988).
6.183 Reynolds, M., Parker, D.: Hyperer Film for Planetary Photography. *Sky & Telescope* **75**, 668 (1988).
6.184 Iburg, B.: Adventures in Gas Hypering. *Sky & Telescope* **73**, 110 (1987).
6.185 Maury, A.: A Hypersensitization Primer. *Sky & Telescope* **75**, 586 (1988).
6.186 Hunter, T.B.: Improved Astrophotos by Copying. *Sky & Telescope* **74**, 326 (1987).
6.187 Everhart, E.: Tracking Comets with a Stepping Motor. *Sky & Telescope* **73**, 208 (1987).
6.188 Schaefer, B.: How Faint Can You See? *Sky & Telescope* **77**, 332 (1989).
6.189 Kodak Data Sheet P-A5.
6.190 Cole, D.C.: Temperature Control of a Hypering Tank. *Sky & Telescope* **78**, 658 (1989).
6.191 di Cicco, D.: Astrophotography Now and Then. *Sky & Telescope* **76**, 463 (1988).
6.192 Saulietis, A., Maley, P.: A Compact Image Intensifier. *Sky & Telescope* **76**, 632 (1988).
6.193 McLean, I.S.: Infrared Astronomy's New Image. *Sky & Telescope* **75**, 254 (1988).
6.194 Sinnot, R.W.: CCDs and the Amateur. *Sky & Telescope* **76**, 546 (1988).
6.195 Smith, A.G.: Gas Hypersensitization of 14 Kodak Films and Use of the Films in Astronomy. *Kodak Tech Bits*, No. 3, 3 (1989), Publication No. P3-89-3.
6.196 Skiff, B.: Keeping up with Science (Scanning the Literature). *Deep Sky* **8**, 43 (Winter 1989–90).

7 Fundamentals of Spectral Analysis

R. Häfner

7.1 Introduction

Our knowledge of celestial bodies results largely from the analysis of the radiation which they emit, absorb, or reflect. Apart from radiative particles, there are electromagnetic waves that obey the well-known relation $c = \nu\lambda$, where $c = 2.9979250 \times 10^8$ m s^{-1} is the speed of light in vacuum, λ the wavelength, and ν the frequency. The electromagnetic spectrum spans the range from the ultra-short-wavelength γ-rays through ultraviolet (UV), visible, and infrared ranges to the meter-long and longer radio waves. The bulk of this spectrum is not observable from the surface of the Earth, because the enveloping atmosphere is transparent only within certain wavelength "windows." This chapter is concerned with the radiation reaching the observer through the "classical" optical window between $\lambda = 300$ nm to around $\lambda = 1000$ nm. (The much wider "radio window" is discussed in Chap. 9.) With the aid of suitable instruments, this radiation can be decomposed into its spectral constituents, and the spectrum analyzed. The following sections provide a survey of the theory of spectra and the objects, instruments, and several methods of analysis which are accessible to amateurs and astronomers at small colleges.

7.2 The Theory of Spectra

7.2.1 The Laws of Radiation

The three fundamental laws of spectroscopy ascribed to G. Kirchhoff (1859) are:

1. An incandescent solid or liquid body or a gas of sufficient absorption (under high pressure) emits a *continuous spectrum* containing all wavelengths.
2. An incandescent gas under low pressure emits a discrete or *emission-line spectrum*, that is, light at a finite number of specific wavelengths which are characteristic for that gas.
3. A cool gas through which is passed white light (continuous spectrum) absorbs those wavelengths from the continuous light which it would itself emit if it were sufficiently hot. These wavelengths are thus missing or weakened, thereby resulting in the appearance of dark *absorption lines* in an otherwise continuous spectrum.

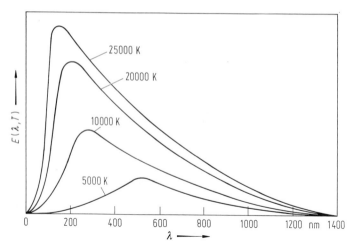

Fig. 7.1. Planck energy distribution curves.

The Sun and most stars display a continuous spectrum. They consist of gases, since their temperatures are much too high for ordinary matter to exist in the liquid or solid state. The spectra also usually show dark absorption lines, indicating that the stars are surrounded by a cooler gaseous "atmosphere." The latter absorbs from the radiation certain parts characteristic for the composition of the gas. Emission-line spectra are observed in certain objects, and lead to the conclusion that these objects consist of—or at least are surrounded by—rarefied gases.

That stars appear to have distinctly different colors illustrates that the maximum intensity of emission occurs at different wavelengths, a fact which finds its quantitative and qualitative interpretation in *Planck's law of radiation* and in relations derived from it. Strictly speaking, they are valid only for the so-called *blackbody*, which has certain idealized properties:

- It is expected to be in thermal equilibrium;
- The entire isotropic radiation field is generated by thermal excitation of atoms (see below);
- Radiation does not enter from nor does it escape to the environment.

The gases composing a star have a high *opacity*, that is to say, they very effectively absorb the radiation which is generated deep in the interior through thermonuclear reactions and which slowly flows outward. Although the existing temperature gradient within the star and the emission of radiation from its surface—obviously without which the star would not be visible—conflicts with the postulated condition of thermal equilibrium, a star can nevertheless, in first approximation, be considered a blackbody.

The radiation field of the blackbody was successfully interpreted theoretically by Max Planck (1900) through the assumption that radiant energy is *quantized*, that is, exists in discrete parcels or *photons* of energy $hc/\lambda = h\nu$ (Planck's constant $h = 6.626\,196 \times 10^{-34}$ [J·s]). The radiated energy emitted by 1 cm^2 of the surface of the blackbody per second and per unit wavelength at temperature T into unit solid angle is

$$E(\lambda, T) = \frac{2hc^2}{\lambda^5(e^{hc/\lambda kT} - 1)} = \frac{2h\nu^3}{c^2(e^{h\nu/kT} - 1)}, \quad (7.1)$$

where $k = 1.380\,622 \times 10^{-23}$ [J·K^{-1}] is the Boltzmann constant. Figure 7.1 shows the energy distribution for several astrophysically relevant temperatures. It is seen that the amounts of energy emitted at all wavelengths differ widely depending on temperature. The higher the temperature, the higher the total energy emitted and the greater the shift of the maximum intensity toward shorter wavelengths (hence, hot stars appear blue while cool stars appear red). The latter property is formulated in *Wien's displacement law*, which gives the amount of wavelength shift as a function of temperature:

$$\lambda_{max} = CT^{-1}, \quad (7.2)$$

where λ_{max} is the wavelength of the maximum intensity. The constant C equals 0.2898 when λ is given in cm and T in K. Adding the contributions from all spectral regions gives the total energy emitted per second per square centimeter over all wavelengths at a specific temperature T:

$$E(T) = \sigma T^4 \quad \text{(Stefan–Boltzmann law)}, \quad (7.3)$$

where $\sigma = 5.66961 \times 10^{-8}$ [J m^{-2}s^{-1}K^{-4}] is the Stefan–Boltzmann constant. Both laws can be exactly deduced from Planck's law, although they had been found earlier by empirical studies. Two noteworthy approximations of Planck's law, which were also previously known, are:

- for short wavelengths ($h\nu/kT \gg 1$), *Wien's law*,

$$E(\nu, T) = \frac{2h\nu^3}{c^2} e^{-h\nu/kT}, \quad (7.4)$$

holds with sufficient precision.
- for the long-wave spectral range ($h\nu/kT \ll 1$), the *Rayleigh–Jeans law*,

$$E(\nu, T) = \frac{2\nu^2 kT}{c^2}, \quad (7.5)$$

applies.

7.2.2 The Line Spectrum

J. Fraunhofer (1814) was the first to find the dark lines named after him in the spectrum of the Sun and also to investigate them in the spectra of Venus and several bright stars. He determined their positions in the spectrum and assigned letters to the strongest lines (e.g., the sodium "D-line"), a notation which is still used today along with more exact spectroscopic codes. The interpretation of the lines was necessarily postponed until the 20th century when atomic physics supplied the theoretical principles. The atomic model developed by N. Bohr (1913) will here suffice to explain the basic facts. An atom consists of a central *nucleus* surrounded by a system of *electrons*. The entire mass is practically concentrated in the nucleus, which consists of *protons*, each carrying the positive elementary electrical charge, and usually also of uncharged *neutrons*. It is the number of protons which determines the chemical element. The simplest atom, that of hydrogen, for instance, has 1 proton. The oxygen atom has 8, and that of uranium 92 protons. A nonionized (see below) atom has zero total charge. The nucleus containing Z protons must therefore have Z electrons, each carrying the one negative elementary charge, somehow distributed around the nucleus. The mass of one electron is about 1/1836 of the mass of either a proton or a neutron. The radius of the nucleus is about 10^{-15} m, five orders of magnitude smaller than that of the atom. The features of the simple hydrogen atom (1 proton, 1 electron) will be more closely described here as they may help to understand more complicated atoms. The hydrogen lines appear in certain series whose patterns had already been represented by the *Rydberg formula* (1890):

$$\frac{1}{\lambda} = R \left(\frac{1}{n^2} - \frac{1}{m^2} \right), \tag{7.6}$$

where $R = 1.097\,373\,12 \times 10^{-2}$ [nm^{-1}] is the Rydberg constant and n and m are integers such that $m > n$. For each fixed n, a series is generated with m equal to any higher integer. The best-known series, which by no coincidence lies in the visible part of the spectrum, derives from $n = 2$. It is named after the scientist Balmer, who, in 1885, first gave the special formula for it. Following Bohr, the pattern is explained by assuming that an electron can revolve around the nucleus in certain permitted orbits at various distances without emitting energy. Higher energy levels of the electron correspond to orbits farther from the nucleus. Absorption or emission of electromagnetic radiation can initiate the transition of the electron between these permitted orbits or levels. As the emitted radiation consists of *photons* whose energy depends on their wavelength, only those wavelengths whose energy equals the energy difference between permitted levels can be absorbed or emitted. If the permitted energy levels are numbered from 1 to n beginning nearest the nucleus, all transitions beginning or ending at the first or *ground level* ($n = 1$) produce the *Lyman series* in absorption or in emission; transitions starting or ending at the second level ($n = 2$), the *Balmer series*; and so on (see Fig. 7.2). Even in this simplest of cases, that of atomic hydrogen, quantum mechanics demonstrates that the real features are not quite so simple. In fact, when helium, the next simplest atom with 2 protons, normally 2 neutrons, and 2 valence electrons is considered, Bohr's theory is already inadequate to describe the observations. Nevertheless, the Bohr theory is still an excellent approximation to the full theory so long as only one-electron sytems are considered (e.g., alkali metals, or ions with one valence electron).

Heavier atoms show more complex features. With increasing nuclear charge number and a correspondingly increasing number of electrons, the various energy levels become gradually filled with electrons. The maximum number z of electrons "fitting" into a certain level depends on its level number,

$$z = 2n^2, \tag{7.7}$$

that is to say, the lowest level ($n = 1$) is maximally occupied by 2 electrons, the next ($n = 2$) by 8 electrons, and so on. This so-called *exclusion principle* was empirically derived by Wolfgang Pauli in 1925. In heavier atoms, such as iron, the various levels are not filled successively; rather, an outer level may start to build before the lower

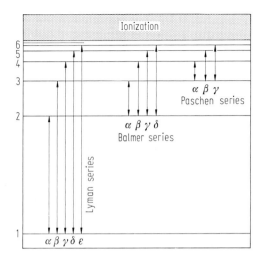

Fig. 7.2. Schematic energy-level diagram for hydrogen.

level is filled. The outer electrons may then perform complicated transitions between levels that are not filled or only partly filled, sometimes within the same level, and thus generate a variety of spectral lines characteristic of the atom considered. Often, special groups of related lines are created. For instance, sodium always shows two closely neighboring lines (called "doublets"), the best known being the yellow D-lines, while magnesium has singlet and triplet lines.

Many astronomical objects show so-called *forbidden lines* whose origin involves long-lived energy levels. Normally, an electron reaching a higher level will leave it again spontaneously after only $\sim 10^{-8}$ s to return to a lower level. However, there also exist energy states where the electron may dwell for seconds, minutes, or years. Under "normal" conditions, the electron in such a state will, via interactions with other particles or with photons, lose its excitation energy (see below) long before the atom has a chance to radiate the forbidden lines. Only within the extremely rarefied environments of certain gaseous nebulae in space does this collisional interaction become so minimal that the forbidden line radiation can occur. To distinguish forbidden from permitted lines, the elemental codes (see Appendix Table B.29 in Vol. 3) are displayed in square brackets (e.g., [OIII], [FeIV]).

The most complicated spectra of all are those emitted by *molecules*, which are aggregates of two or more atoms. Here, in addition to the jumps in the joint electron shell, vibrations between nuclei may occur as well as rotations around certain axes. Quantum theory postulates that, as with the electron transitional energies, the vibrational and rotational energies are quantized. The corresponding energy differences are very low, meaning that there are numerous closely adjacent levels, and transitions between them generate lines in the infrared. The superposition of electron jumps, rotations, and atomic vibrations generates the observed molecular spectrum in the visible and UV range. Instead of a single line corresponding to the electron jump, there appears a "band" of many closely adjacent lines which are often not entirely resolved. Usually, the lines comprising the band, depending on the type of molecule and the electron transition, crowd together at the short-wave or long-wave end to form the so-called "head" of the band.

7.2.3 Excitation and Ionization

At low temperatures, almost all atoms of a gas are in the ground state; their valence electrons reside in the lowest possible energy level. Absorption of suitably energetic photons or collisions with other particles may "bump" the electrons to higher levels. The atoms are then said to be *excited*. To generate, for instance, the Balmer series, some fraction of the existing hydrogen atoms must have their electrons in the $n = 2$ level. How many atoms are in this specific level depends in general on the temperature of the gas.

The formal representation of this relation is based on the results of investigations by L. Boltzmann and is therefore known as the *Boltzmann equation*:

$$N_s \sim N_0 \, e^{-\chi_s/kT}, \tag{7.8}$$

where N_0 is the number of atoms in the ground state, N_s that at the excited level, and χ_s the corresponding excitation energy. T is the excitation temperature, which, in thermal equilibrium, coincides with the gas kinetic temperature (proportional to the kinetic energy of the particles) and with the radiative temperature. (In general, however, thermal equilibrium does not hold and these temperatures are not quite equal.) The conclusions are, of course, of a statistical nature. Each electron will, after residing a certain time in the excited state, tend to return to the ground level, although in some other atoms the electrons will assume these excited levels. Evidently, even at higher temperatures, N_s cannot exceed N_0. Only at very high temperatures do the two values become approximately equal. In the solar atmosphere with a temperature of around 6000 K, for instance, only about one hydrogen atom out of 10^8 is excited to the second level and thus able to produce a Balmer line; hence, at any given moment the overwhelming majority of hydrogen atoms do not contribute to the generation these lines.

As the temperature increases, an atom may gradually reach higher and higher excitation states until finally the electron becomes detached from the parent atom. The atom is then said to be *ionized*. The minimum energy needed to ionize an atom from the ground level is called the *ionization energy*. If the energy provided lies above the ionization minimum, it will, after the ionization process, be transmuted into kinetic energy of the free electron. The number N_i of ionized atoms (ions) of a kind which exist at a given temperature results from the Boltzmann statistics in connection with quantum theory, and is described by an equation derived in 1920 by M.N. Saha:

$$N_i \sim T^{3/2}(N_0/N_e)e^{-\chi_i/kT} \quad \text{(Saha equation)}, \tag{7.9}$$

where N_0 is a number of neutral atoms, χ_i the ionization energy, N_e the number of free electrons in the gas, and T the ionization temperature, which again coincides with the other temperatures only in the case of thermal equilibrium. The factor $T^{3/2}$

causes N_i to grow without limit as temperature increases, so that at a sufficiently high temperature, most of the atoms will be ionized. Besides the temperature, the electron density is the most important quantity in the ionization process. At a given temperature, the ionization is stronger when the number of existing free electrons in the gas is low. The ionization energies differ for various kinds of atoms; for instance, sodium and potassium are easier to ionize than magnesium or silicon. The expectation is therefore that, at given T and N_e, one element can be multiply ionized (has lost several electrons) while another element is still essentially neutral. The energy levels of an ion are very different from those of the neutral atom. Specifically, in each ionization state, specific levels exist and generate spectral lines characteristic of that kind of ion when the remaining electrons perform the corresponding transitions.

7.3 The Objects of Spectral Studies

There are fundamental differences between objects which contain internal energy sources and thereby generate electromagnetic radiation (stars), those which basically reflect the radiation from other sources (planets, moons lacking atmospheres), and those which emit or absorb energy through interaction with radiation (emission nebulae, planets and moons with atmospheres, comets) or with the Earth's atmosphere (meteors). The following gives a brief survey of basic spectroscopic features of astronomical objects and especially those which may be explored by astronomers at small colleges or by amateurs. More detailed descriptions appear in Vols. 2 and 3.

7.3.1 Stars

It is found that stellar spectral features, the absorption lines in particular, display substantial differences with respect to number, intensity, and type which are in principle directly correlated with the position of maximum intensity of radiation (and hence with the temperature through Wien's displacement law). As the shift of maximum intensity is not easy to determine, the classification of stars according to certain features of their spectral lines is a valuable tool.

In view of the concepts mentioned in the previous section, it may be surprising that dark lines appear in stellar spectra at all, as every excited atom reradiates its absorbed energy again after a normally very brief time interval. This is especially true in *resonance lines* (transitions between ground and first excited level), where the electron cannot but return to the ground level by emitting the same spectral line previously absorbed. (The atom may, during its brief residence in the excited state, again be excited by absorption of a photon or by collision with another particle, but this mechanism cannot account for all absorption features.) The explanation of this paradox is that the spontaneous re-emission for the various atoms occurs *isotropically* (i.e., uniformly in all directions), and therefore only a certain small fraction of the energy is directed toward the observer. Except for resonance lines, the electron may also take a path over various intermediate levels and emit different spectral lines. In each case, the net effect is a reduction of the continuum at the position of the

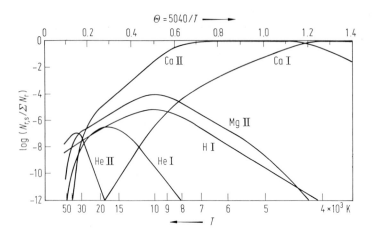

Fig. 7.3. The dependence of ionization and excitation on temperature (after Unsöld).

absorption wavelength, i.e., a "dark line." Temperature, chemical composition, and, to a much lesser extent, the gas pressure of the stellar atmosphere—extended some 10^{-3} to 10^{-2} of the stellar radius—are the factors determining the features of the stellar spectrum. Assuming that the chemical composition of atmospheres of all stars is by and large the same, the foremost decisive quantity is the surface temperature.

A basic grouping of stars according to intensities of lines of various elements has long been known under the name "Harvard Classification." With the aid of Saha's ionization theory, it became possible to interpret intensities of the various lines *quantitatively*. Figure 7.3, which displays the results of applying Eqs. (7.8) and (7.9) for several elements, shows the number of atoms of the element in a certain excited state relative to the total number of those atoms (in all states) as a function of temperature at an average electron density. As is the custom in astrophysics, the spectrum of a neutral atom is characterized by a Roman numeral "I" after the element code, and higher ionization stages by correspondingly higher numbers (e.g., neutral calcium = Ca I, singly ionized calcium = Ca II, etc.; see Appendix Table B.30). The occurrence of a higher ionization state is a function, as mentioned, of the ionization energies which differ for various elements (the ionization energy for calcium, for instance, is relatively low, while for helium it is quite high). Retaining the traditional letter codes of spectral classes, but reordered according to decreasing temperatures (about 30 000 down to about 3000 K), the following sequence was obtained:

O–B–A–F–G–K–M .

Each class is further subdivided decimally.

The still-used terms "early-type" for the hot stars of classes O, B, and A, and "late-type" for the cool spectral types K and M are rooted in astronomy history; they are not related to the true age of the stars.

The basic spectral features and the corresponding surface temperatures are compiled in Table 7.1.

Table 7.1. Characteristics of the major spectral classes.

Spectral Type	Color	Atmospheric Temperature (K)	Distinguishing Criteria	Typical Example	
O	blue	> 25 000	Lines of highly ionized atoms, primarily HeII, SiIV, NIII; H-Balmer series very weak.	ζ Pup 10 Lac	O5 O9
B	blue	11 000 – 25 000	HeII vanishes between B0 and B5; HeI stengthens but disappears by B9; also lines of SiII, SiIII, OIII, MgII; Balmer series intensifies from classes O to A.	α Vir (Spica) β Ori (Rigel)	B1 B8
A	blue	7500 – 11 000	HeI absent; Balmer series reaches maximum intensity at A2; lines of singly ionized elements such as MgII, SiII, FeII, TiII, CaII, etc. present (maximum at A5); lines of neutral metals very weak.	α Lyr (Vega) α CMa (Sirius)	A0 A1
F	blue–white	6000 – 7500	Balmer series decreases in intensity; lines of ionized and neutral metals about equally intense; CaII strong.	α Car (Canopus) α CMi (Procyon)	F0 F5
G	yellow–white	5000 – 6000	Balmer series continues to wane; CaII very strong; lines of neutral metals very strong; first appearance of CH-bands.	α Aur (Capella) Sun	G0 G2
K	orange–red	3500 – 5000	Lines of neutral metals with low excitation energies very strong; CN- and CH-bands present, TiO-bands appear at K5; Balmer lines virtually absent.	α Boo (Arcturus) α Tau (Aldebaran)	K2 K5
M	red	< 3500	TiO-bands very strong; lines of neutral elements with low excitation energies (e.g., CaI) very strong.	α Sco (Antares) α Ori (Betelgeuse)	M1 M2

A graph of the absolute magnitudes of stars versus their spectral types reveals that the vast majority of stars fall into a narrow band, the so-called *main sequence*, running diagonally from the intrinsically very bright O-stars to the intrinsically faint M-stars. This diagram is named the *Hertzsprung–Russell diagram* (or simply H–R diagram) after the astronomers who first investigated it independently in 1907 and 1913. Figure 7.4 presents such a graph, the vertical scatter resulting from an uncertainty in the luminosity determinations. The main sequence and, mostly above it, other groups of data points are present. It is directly apparent that stars of the same spectral class can have different absolute magnitudes and therefore, by inference, must radiate different amounts of energy even at identical surface temperatures. This is possible only if

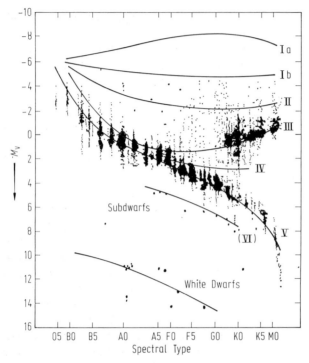

Fig. 7.4. The Hertzsprung–Russell diagram.

the brighter star has a larger surface area and hence a larger radius. This fact lays the foundation for the following terms used by Morgan and Keenan (MK) in a finer subdivision of star types into *luminosity classes* characterized by roman numerals according to the different sizes:

Ia	Brightest supergiants	V	Main-sequence stars (dwarfs)
Iab,Ib	Less bright supergiants		
II	Bright giants	VI	Subdwarfs
III	Giants	VII	White dwarfs, also designated
IV	Subgiants		by "D"

Notable examples of this MK classification are β Ori (Rigel) B8 Ia, the Sun G2 V, α Boo (Arcturus) K2 III, α Ori (Betelgeuse) M2 Iab. Compare the objective prism spectra of main-sequence stars with the examples of other luminosity classes given in Fig. B.2 in Vol. 3.

Spectra of main-sequence stars, giants, and supergiants of the same spectral class differ from each other in small, but noticeable, details. Luminosity criteria include primarily the width of Balmer lines (narrower in high-luminosity stars) and intensity ratios of lines of certain ions to those of neutral atoms (ionic lines are stronger and neutral lines weaker in high-luminosity stars). As the gas densities in the atmosphere diminish for high-luminosity stars, the pressure broadening of lines (H-lines) also diminishes, while the degree of ionization increases (metallic lines; compare with Saha equation).

Over 99% of the known stars fit into this two-dimensional scheme. The remainder have either anomalous abundances of certain elements in their atmospheres, or different modes of line formation. For example, they may have strong magnetic fields or high rotational velocities associated with mass loss. At the hot end of the spectral sequence there are also the *Wolf–Rayet stars* (Type W), the Be and shell stars, and stars with anomalous He or N abundances. Among the A-type stars, there are those with strong magnetic fields and those with metal anomalies. Among the late classes after type K, the S-types and the C-types (carbon stars, formerly designated R- and N-stars) with abundance anomalies branch off as side sequences.

The large group of physically variable stars not fitting the normal sequence will not be discussed here. These objects are often too faint to be observed spectroscopically with small instruments and with the necessary time resolution. Only the *novae* will be discussed here, as interesting spectroscopic observations may on occasion be possible for amateurs. Novae are white dwarfs in close binary systems whose brightness increases by as much as 16 magnitudes in a short time (hours to months) and declines thereafter over years to the original brightness. Though the time scales are different, the spectra of novae in the various stages of eruption are similar enough to be combined in a separate class Q subdivided by the sequence 0 to 9.

The spectrum in the pre-nova state usually resembles that of an early-type star. The pre-maximum halt, about 2 magnitudes below maximum, is followed by a slower ascent showing an A-type spectrum (Q0) with blue-shifted absorption lines (see below) originating in a rapidly expanding shell (up to 1000 km s^{-1}). At the maximum and shortly thereafter, the spectra (Q1, Q2) match F-supergiants. The first descent to about $3.^m5$ below the maximum (Q3) shows strong emissions of H and metal ions with absorption edges at the short-wavelength end. Later on, broad, diffuse absorptions of H, OII, NII, and NIII (Q4, Q5) from a second shell ejected at higher speed (sometimes several shells and up to several thousand km s^{-1}) appear. The Q6 spectrum is characterized by emissions of He and [NII]. In the following transition, novae behave individually. They may display strong brightness fluctuations or a minimum. All novae return then to a brightness of about 6 magnitudes below maximum. In this transition, the "nebular spectrum" Q7, a pure emission spectrum of HeI, FeII, FeIII, CII, OII, etc., develops. The well-known forbidden green oxygen lines [OIII] appear and reach a maximum in the phase of slow descent (Q8, Q9). The post-nova shows a continuous O-type spectrum, superposed by narrow emissions of H, HeI, and [OIII].

7.3.2 The Sun

The proximity of the Sun permits detailed spectroscopic studies of many individual features, a procedure not possible for other stars. The "Quiet Sun" has a normal G2 V spectrum, very rich in lines. The Rowland Tables list about 24 000 lines over the range $\lambda\lambda$ 293.5–877.0 nm [7.1]. About 20 000 of them are identified, and their primary data, such as intensity and excitation energy, are compiled. In the red spectral range, absorptions by the Earth's atmosphere interfere increasingly. These *telluric lines* make up about 60% of the lines observed in the solar spectrum between $\lambda\lambda$ 600 and 900 nm. The tables also list the lines from sunspots, which, as lower-temperature regions (about 1600 K cooler than outside these regions), show a spectrum similar to a K0 star; that is, the lines of neutral elements, particularly Ca, V, Cr, and Ti, are enhanced compared with the normal solar spectrum, and the ionic and Balmer lines weakened. Molecular bands primarily of MgH, TiO, and CaH and lines of Li, Ru, and In also appear. Occasionally, the Balmer lines appear in emission, especially when light bridges cross the spots. Spectra taken of the solar limb also show characteristics typical of lower-temperature stars.

During the course of a total solar eclipse, the *flash spectrum* of the chromosphere is observed, which consists only of emission lines since the tangential viewing of the solar limb removes the absorptions and leaves only the re-emissions. About 3000 lines are known, but they are not simply a reversal of the normal solar spectrum. Lines of ions and of neutral atoms with high excitation energy appear enhanced, while neutral lines of low energy are weakened. Also observed are strong He I lines, in particular the so-called D3-line (587.6 nm). All of this is strongly indicative of a high chromospheric temperature.

Studies show the temperature to rise from the surface discontinuously to about 10^6 K at 10^4 km above the surface, where the transition into the corona begins. The latter appears at total eclipses as an extended, whitish ring several 10^6 km wide and forms the outermost part of the solar atmosphere. In the optical spectrum, the radiation of the K-, L-, and F-coronae are superposed. While the F-spectrum is a normal solar spectrum reflected by interplanetary dust some distance away from the Sun (at the transition to zodiacal light), the K-spectrum consists of a pure continuum generated by scatter of sunlight by free electrons directly above the chromosphere. The thermal velocities of electrons are so high, owing to the high temperatures caused (most likely) by shock waves from the chromosphere, that the lines are completely smeared out by the Doppler effect (to be discussed later). Finally, the spectrum of the L-corona represents the "true" light of the corona. It consists of forbidden emissions of highly (9 to 15 times) ionized elements such as Fe, Ca, V, Cr, Mn, K, and Co. Over 30 lines, in particular in the short-wavelength range, have been identified. The most notable are the forbidden green line at 530.3 nm from [Fe XIV], the yellow line at 596.4 nm from [Ca XV], and the red corona line at 637.4 nm of [Fe X]. Collisions with fast electrons cause the high ionizations, and the near-vacuum density then allows forbidden lines to appear in the same manner as in emission nebulae.

Other spectral features of solar phenomena will not be dealt with here, as their observation requires large and expensive instrumentation.

7.3.3 Planets and Moons

All planets excepting Mercury and perhaps Pluto have gaseous atmospheres. In most cases, low-lying thick clouds reflect or scatter the sunlight, which subsequently penetrates part of the outer atmosphere where it suffers absorptions. A planetary spectrum consequently consists of a solar spectrum plus additional absorption bands from the gases in the planet's atmosphere. As the radiation must furthermore pass through

Earth's atmosphere to reach the detector, the spectrum is often contaminated by telluric lines due to the familiar constituents found there. The observed planetary bands lie mostly in the red and infrared spectral range and are ascribed to the following molecules:

Venus: CO_2, H_2O, HCl, HF;
Mars: CO_2, H_2O;
Jupiter and Saturn: H_2, CH_4, NH_3;
Uranus and Neptune: H_2, CH_4.

Not all gases have been spectroscopically identified, as many lines of potential constituents do not occur in the accessible range at the temperatures of planetary atmospheres.

Saturn's large satellite Titan is the only case of Earth-based identification of a permanent atmosphere containing CH_4 (methane). Other satellites, including Earth's Moon and also Saturn's rings, show, after eliminating telluric lines, a purely reflected solar spectrum.

7.3.4 Comets

At large distances from the Sun, a comet consists only of a tiny nucleus which has a spectrum of reflected sunlight. As it approaches the Sun, parts of the nucleus evaporate, are further broken up by photochemical processes, and form a diffuse gaseous shell called the *coma*, which the sunlight excites to fluorescence or resonance light. The spectrum shows, apart from sunlight, emission bands of molecules, radicals, and radical ions: OH, NH, NH_2, CH, CN, C_2, C_3, OH^+, CH^+, and, at close approach, atomic lines of Na and perhaps of Ni, Fe, and O.

In the space roughly inside the orbit of Mars, the comet develops its characteristic *tail*, generated by the radiation pressure of the Sun and by the solar wind (ions and electrons traveling outward from the Sun at speeds of around 400 km s^{-1} at the orbit of Earth). Two basic types of tails are distinguished:

Type I: a long, narrow tail with emission bands primarily of CO^+, and additionally of CO_2^+, CH^+, CN^+, C^+, OH^+, N_2^+, and H_2O^+ generated by short-wavelength solar radiation (additional ionization).

Type II: a broad, diffuse tail consisting of colloidal particles which merely reflect the sunlight.

7.3.5 Meteors

Small, solid particles (with masses mostly between 10^{-3} to 10^{-5} kg) penetrating the Earth's atmosphere cause the luminous phenomenon known as *meteors*. Collisions ionize atmospheric molecules and evaporate the surface material of the intruding body (or even the entire body), with a resulting emission of light. The characteristic spectrum generated depends somewhat on the orbit and the entrance velocity. Basically, it consists of emission lines, sometimes superposed over a weak continuum. Low-excitation lines of FeI, NaI, CaI, MgI, and CaII-H and CaII-K appear, but lines of AlI, SiI, CrI, MnI, NiI, and even Hα and Hβ have been observed. Ionic lines have been found in fast meteors with relative velocities of over 30 km s^{-1}, and also in slower meteors toward the end of the path. The N$_2$-bands occasionally seen are of telluric origin. The forbidden green [OI]-line at 557.7 nm sometimes seen shortly after immersion into the atmosphere for 1 or 2 seconds has not yet been explained. Meteor spectra can be divided according to dominant lines and numbers of lines into four types and four classes:

> Type X: dominated by NaI-D or by MgI (518.4 and 383.3 nm).
> Type Y: dominated by CaII-H and -K.
> Type Z: dominated by FeI or CrI.
> Type W: fits none of the above cases.
> Class a: > 49 lines
> Class b: 20 – 49 lines
> Class c: 10 – 19 lines
> Class d: 1 – 9 lines.

Most spectra obtained are of Type Z and Class d.

7.4 Spectroscopic Instruments

The following will emphasize general aspects of spectroscopic equipment, while special instruments will be discussed only when they are accessible to amateurs or professional astronomers at small colleges.

7.4.1 The Methods of Spectral Dispersion

The heart of any spectral device is a dispersing element from which the incident light emerges, separated according to its various wavelengths, in different directions. To achieve dispersion, prisms and various types of gratings are usually employed. An interferometer is another possibility.

7.4.1.1 The Prism. Light obliquely incident on a glass surface suffers a change of direction owing to the different velocities of propagation in air and in the glass medium; it is thus said to be *refracted*. Refractive behavior is governed by the *law of refraction*, more commonly called *Snell's law*, which states that

$$\frac{\sin i}{\sin i'} = \frac{c_1}{c_2} = \frac{n_2}{n_1} = n, \tag{7.10}$$

where i and i' are the angles of incidence and of refraction with respect to the normal (i.e., perpendicular) to the surface, c_1 and c_2 the propagation speeds in air and in medium, and n_1 and n_2 the corresponding refractive indices. Usually, merely the relative index n of the medium against air is needed. Passage through a prism causes two changes in direction. The amount of the deflection angle φ changes with the angle of incidence i and angle of the refracting edge (apex angle) γ of the prism. The

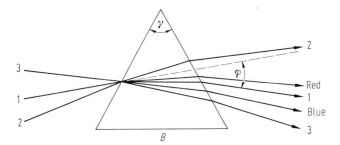

Fig. 7.5. Transmission and dispersion by a prism (schematic).

deflection is a minimum when the ray passes the prism symmetrically, parallel to the base B, and thus incident and exiting rays form the same angles with respect to the normals to the prism surfaces. For small incident angles and small γ, the angle of deflection is in general given by

$$\varphi \approx (n-1)\gamma. \qquad (7.11)$$

Symmetric transmission gives strictly, for any γ,

$$\varphi = 2\ \arcsin\left[n\sin(\gamma/2)\right] - \gamma. \qquad (7.12)$$

The refractive index depends on wavelength ($n_{red} < n_{blue}$), and blue light is more deflected than red light. The change of φ with wavelength λ is called the *angular dispersion* of the prism. Like the refraction index, it is a material constant which in addition depends on γ. Differentiating Eq. (7.12) with respect to wavelength yields the angular dispersion for symmetric passage,

$$\frac{d\varphi}{d\lambda} = \frac{2\sin(\gamma/2)}{\left[1 - n^2 \sin^2(\gamma/2)\right]^{1/2}} \frac{dn}{d\lambda}. \qquad (7.13)$$

For a 60° prism of standard glass with $1.4 < n < 1.6$, this relation simplifies to:

$$\frac{d\varphi}{d\lambda} \approx n \frac{dn}{d\lambda}. \qquad (7.14)$$

The change of refractive index with wavelength, $dn/d\lambda$, can be computed with the data from Table 4.7. Of course, the angular dispersion of a given prism increases for an asymmetric passage, but with the drawbacks that the imaging quality as well as the resolving power A diminish. The latter specifies how much of a wavelength difference $\Delta\lambda$ at a given λ can be separated, and should be as large as possible. In a fully illuminated prism and symmetric transmission, the resolving power is given by

$$A = \frac{\lambda}{\Delta\lambda} = B\frac{dn}{d\lambda}. \qquad (7.15)$$

Note that the resolving power does not depend on the angle γ but only on $dn/d\lambda$ and the base length B of the prism, and is effectively diminished in nonsymmetric transmissions. For a standard prism with a base length of 10 cm, A is of the order of 10^4.

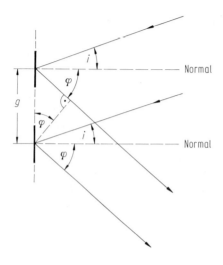

Fig. 7.6. Geometry of reflection grating.

To obtain a high dispersion at low deflection, a *direct-vision prism*, which is usually composed of three or five prisms, all with opposing orientations and dissimilar refractive indices, is employed. The disparate refractive and dispersive properties of the materials (e.g., crown and flint glasses) make such designs possible.

7.4.1.2 Diffraction Gratings. The term *diffraction grating* characterizes a large number of fine grooves etched in a parallel and equidistant fashion on a plane or concave surface (e.g., glass). The distance between two grooves or between two "unetched" surface strips is called the *grating constant g*. There are both transmission and reflection gratings, but this distinction is, in theory, immaterial. In practice, reflection gratings are more widely used. When parallel rays of light enter the grating, the unetched surface strips between grooves act as transparent or reflecting diffraction slits. The combined action of many such slits enhances the light of a certain wavelength only in specific directions; this phenomenon is called *constructive interference*. The position of the intensity maximum is given by the condition

$$g(\sin i + \sin \varphi) = m\lambda, \tag{7.16}$$

where i is the angle of incidence, φ that of diffraction, and λ the wavelength. This formula is known as the *grating equation*. The order m can have the values 0, ± 1, ± 2, etc., that is to say, apart from the case $m = 0$ (that is, reflected or transmitted without diffraction for $i = \varphi$), the path difference of neighboring rays is an integer multiple of the wavelength when the light is to propagate into direction φ. A negative value for m occurs when incident and diffracted rays are not on the same side of the normal and $|\varphi| > i$.

Since the maximum condition Eq. (7.16) holds for any λ, white (i.e., multicolored) light is separated into wavelengths. In contrast to the prism, the deflection is here proportional to wavelength (the so-called "normal spectrum"), being stronger for red than for blue light. The angular dispersion is nearly independent of λ:

$$\frac{d\varphi}{d\lambda} = \frac{m}{g\cos\varphi}. \tag{7.17}$$

The dispersion is evidently higher the higher the order m used for observation and the smaller the grating constant. In each order, an entire spectrum of practically constant dispersion (except for the change in $\cos\varphi$) is generated, but at higher orders more overlap occurs, with the violet end of the following spectrum falling onto the red end of the preceding one. For instance, 800 nm in the first order coincides with 400 nm in the second order. In order to achieve high resolving power with the grating, the spectral range of interest may be pre-selected by a prism or a filter. Of course, m cannot be arbitrarily large; the maximum number is given by $m_{max} = 2g/\lambda$. In practice, usually the first or second order is used.

The resolving power of a grating is given by

$$A = \frac{\lambda}{\Delta\lambda} = mZ, \tag{7.18}$$

where Z is the total number of grooves. Commercially available grating dimensions reach resolving powers of about 10^4 to 10^5. The quality of the grating is assumed to be adequate; for instance, grooves which are irregular in shape or spacing do not yield good interference patterns, and periodic division errors produce annoying spurious maxima called "ghosts."

Theoretically, the intensity of spectra should diminish with increasing order. This is not the case for all gratings, but depends on the shape in which the lines are etched. Suitable shapes can even concentrate almost the entire intensity (as in a prism) into one direction, i.e., into *one* spectrum of a particular order, and additionally to achieve a high resolving power. Such a device is called a *blaze grating*. There is also the *echelle grating*, which uses the narrow sides of rectangular grooves at a high grating constant and achieves high resolving power (up to 10^6) at the expense of a very narrow, overlap-free spectral range.

7.4.2 The Design of a Spectral Instrument

This section will describe the basic arrangements of spectral devices, while only briefly mentioning the most important special constructions.

7.4.2.1 Slitless Spectral Instruments. The simplest spectral instrument, the *objective prism*, is obtained by mounting a prism of small refracting angle in front of the telescope objective at the end of the tube so as to cover the entire objective. The great advantage of this design is in obtaining simultaneously the spectra of all objects in the field of view. The angular relation for symmetric transmission gives the angle $\beta = \varphi/2$ between the base of the prism and the optical axis of the telescope. With f as the objective focal length, the wavelength-dependent spectral dispersion d in the

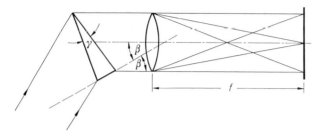

Fig. 7.7. Principle of the objective prism.

focal plane becomes

$$d = f \frac{d\varphi}{d\lambda} \tag{7.19}$$

$$= f \frac{2\sin(\gamma/2)}{\left[1 - n^2 \sin^2(\gamma/2)\right]^{1/2}} \frac{dn}{d\lambda}. \tag{7.20}$$

It is the reciprocal value of d which is usually given and—not entirely correctly—called "dispersion" (normally in nm/mm or Å/mm). Combining the formulae for resolving power and for angular dispersion gives the resolution Δl of the entire device in the focal plane at wavelength λ:

$$\Delta l = \frac{2 f \lambda \sin(\gamma/2)}{B \left[1 - n^2 \sin^2(\gamma/2)\right]^{1/2}}. \tag{7.21}$$

Use of an objective grating is, because of the low intensity of diffraction images, normally restricted to a combination with an objective prism (see below).

7.4.2.2 Spectral Equipment with a Slit

General Aspects. More demanding but advantageous in many ways is the slit design, whose basic composition appears in Fig. 7.8. At the focus of the objective (aperture D, focal length f) is the slit S onto which the seeing or diffraction disk of the object to be observed is imaged. The *collimator* (D_1, f_1) induces the light to fall parallel onto the dispersing element E, which may be a grating or one or several prisms. The slit is at right angles to the direction of dispersion. The camera lens (D_2, f_2) finally images the spectrally separated light into the focal plane of the camera. The spectral instrument is fitted to the telescope when

$$f/D = f_1/D_1 \tag{7.22}$$

and hence when the dispersing element is optimally illuminated. Imaging the slit into the focal plane of the camera also makes apparent the imaging properties of the dispersing element in the direction of the dispersion; that is to say, a monochromatic beam changes its diameter Φ_1 after passage through the element into Φ_2 in the

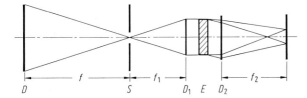

Fig. 7.8. Principle of slit design.

direction of dispersion. The ratio of these diameters is the factor V, given by

$$V = \frac{\Phi_1}{\Phi_2} = \frac{\cos i}{\cos \varphi}, \tag{7.23}$$

which is exactly 1 for prisms with symmetric transmission and usually of the order of 1 for gratings. This gives, for the imaging by the slit,

$$b' = bV(f_2/f_1), \tag{7.24}$$

where b is the slit width and b' the width of the imaged slit in the focal plane of the camera.

The same expressions, (7.20) and (7.21), are obtained for dispersion d and resolution Δl in the focal plane by a prism as for the objective prism design, but with f replaced by f_2. For a 60° prism and $1.4 < n < 1.6$, the relations are simplified thus:

$$d \approx n f_2 \frac{dn}{d\lambda}, \tag{7.25}$$

$$\Delta l \approx n f_2 \frac{\lambda}{B}. \tag{7.26}$$

Corresponding formulae for the grating are

$$d = f_2 (m/g) \cos \varphi, \tag{7.27}$$

$$\Delta l = f_2 (\lambda/g) Z \cos \varphi. \tag{7.28}$$

At perpendicular incidence, the cosine of the deflecting angle is given by the simple relation

$$\cos \varphi = (1 - \sin^2 \varphi)^{1/2} = \left[1 - \left(\frac{m\lambda}{g}\right)^2\right]^{1/2}. \tag{7.29}$$

Special Mountings. Various kinds of mountings are feasible for prisms and gratings. The *Littrow mounting* offers a compact design. It normally uses either half of a 60° prism with a mirrored rear side or else a plane reflection grating in such a way that the incidence angle equals the deflection angle, and therefore the collimator can also serve as a camera lens (auto-collimation). The slit in this arrangement is on the side, and the light to be analyzed is fed into the slit by a 90° prism.

The *Ebert mounting* is also an auto-collimator design, distinguished by low imaging errors, that is, by subdued coma and virtually absent astigmatism. The light from the slit reaches a concave mirror which illuminates a reflection grating. The spectrally separated light reflected by the grating is then focused by the same mirror in its focal plane. Slit and focus then are usually arranged one on top of the other.

Most mountings with concave gratings use the well-known arrangement of the *Rowland circle* in its many variations. The imaging function by the grating itself makes collimator and camera objective redundant, which has the advantage that a large wavelength range is imaged simultaneously. The strong astigmatism can be partly removed by a cylindrial lens. Slit, grating, and the curved spectrum lie on a circle whose diameter equals the radius of curvature of the grating. The most compact design of this kind, the *Eagle mounting*, is obtained in auto-collimation as a modified Littrow mounting. A secant of the Rowland circle then forms the axis of the spectroscope with grating and spectrum at opposite ends. The slit is either slightly above or below the position where the spectrum appears, or the light is entered sidewise by a 90° prism. A detailed discussion of special mounting designs can be found, for instance, in Thorne [7.2].

7.4.3 Radiation Detectors

The various spectral instruments have been given different names, depending on the detector used to record the spectrum. Visual inspection is said to be made with a *spectroscope*, and scale readings by a measuring mark (crosswire) with a *spectrometer*. The simplest device is the *eyepiece spectroscope*, consisting of a direct-vision prism and a cylindrical lens which replaces the eyepiece at the telescope. The lens serves to widen the "line-thin" spectrum obtained from the focal image of the object. The serious observer will not be content to merely look at a spectrum but will ultimately obtain a *spectrograph*, which uses photographic film or, even better, plates as detector. Photographic emulsions have substantial advantages over direct viewing: (a) in contrast to the eye, information is accumulated, so that longer exposure times reach fainter objects; (b) a larger wavelength range can be obtained at the same time; (c) the exposed emulsion is a long-lived document suitable for later analysis. The most serious drawback is the nonlinear response of photographic emulsions to light intensities, which thus necessitates, for certain analyses, a calibration. Chapter 6 studies the specific properties of photographic materials more closely and explains terms like "sensitivity," "density," "gradation," etc. There is also the "scanning" spectrometer, or *monochromator*, which sequentially scans the spectrum in the focal plane of the camera photoelectrically, and records the information on tape or diskette. This method, like other developments in modern electronics (e.g., image-tube, diode arrays), was for a long time too demanding to be widely used, particularly with respect to data reduction. With the advances made in the area of personal computers and the availability of large computers capable of image processing, such technologies are also increasingly coming within the reach of the small college and amateur astronomers. For instance, Buil [7.3] reports on his own experience with the use of a charge-coupled device (CCD) as a spectroscopic detector. CCDs have already been used for some time in professional astronomy since their high quantum efficiency, linearity, and dynamics are advantages over photographic emulsions. The following discussion, however, will be based on the use of photoplates and films as detectors (see also Chap. 6.7).

7.4.4 Designs, Hints for Operation, and Accessories

7.4.4.1 Objective Prisms. The use of an objective prism offers the simplest and least expensive way to generate spectra. Reflectors, refractors, or simply telephoto objectives can be used. An upper limit in aperture is reached when the prism, which should completely cover the objective, becomes too large and too heavy for the mounting and generates problems of stability and flexure. Equatorial mountings and automatic tracking are advantageous and even indespensible for long exposure times. The seeing disk acts as a quasi-entrance slit, which has the advantage of high light efficiency, but also the drawback that seeing fluctuations combined with defective tracking substantially diminish the quality of the spectra. The greatest benefit of this method, as mentioned, is in simultaneously obtaining spectra of all objects in the field, although they may be superposed on each other. Since the objects being "spectrographed" are of different brightnesses, well-exposed spectra as well as underexposed ones will be obtained. To reach all objects to the limiting magnitude of the instrument may thus require several exposures. With specific data on the design and on the photo-material, the limiting magnitude will depend primarily on the width of the spectrum:

$$m_{\lim} \approx 18.5 + 5 \log f - 2.5 \log \left[h'(\Delta\lambda/d^{-1})(\Delta P)^2 \right], \tag{7.30}$$

where f is the objective focal length in meters, h' the width of the spectrum, $\Delta\lambda$ the wavelength range photographed (for blue plates normally between $\lambda\lambda 400$ and 500 nm), d^{-1} the reciprocal dispersion, and ΔP the resolution of the photographic emulsion (on the average 0.02 mm). This relation holds as long as ΔP and not the seeing is the limiting factor for the resolving power of the instrument. Under many atmospheric conditions, a seeing disk of 2×10^{-5} rad, corresponding to a disk of $2 \times 10^{-5} f$ in the focal plane, can be reckoned with. The focal length therefore should not be too large, and should also, if possible, satisfy the condition

$$\Delta P \geqq 2 \times 10^{-5} f. \tag{7.31}$$

In most modest-sized telescopes, the diameter of the diffraction disk is smaller by about one order of magnitude and need not be considered in this context.

Without taking any other measures, the spectra produced are quite skinny (width $\approx f\, 2 \times 10^{-5}$) and must therefore be "widened" for detailed investigations. This is easily accomplished when, in an equatorial mounting, the refracting edge of the prism is oriented in right ascension (thus the spectra are dispersed in declination), and then the tracking speed is varied by slow motion so that the guide star moves back and forth between two wires in the guiding scope. To use the object itself for guiding requires offsetting the guidescope by the angle $\phi = 2\beta$ (since the telescope forms the angle β with the base of the prism), a fact which is to be born in mind when attaching the prism.

The choice of prism material depends somewhat on the given focal length of the telescope. For short focal lengths, it is desirable (but not always possible) to have an increased refracting angle and also a higher refractive index as well as a higher value of $dn/d\lambda$. With the data of the entire unit, the actual wavelength resolution is

$$\Delta\lambda = d^{-1}\Delta P, \tag{7.32}$$

and the length L of the spectrum is

$$L \approx f \Delta n \frac{2 \sin(\gamma/2)}{[1 - \bar{n}^2 \sin^2(\gamma/2)]^{1/2}}, \qquad (7.33)$$

where Δn is the difference of refractive indices for the limiting wavelengths and $\bar{n}$ is their mean value.

The previously mentioned limiting magnitude is not always reached since the fog caused by sky background limits the effective exposure time. Hence, an important quantity for the spectrograph is its *speed I*, formally defined as the ratio of energy incident on the area $h' \Delta x$ of the photographic layer ($\Delta x = 0.1$ nm) to the energy entering in the same wavelength interval through the telescope objective:

$$I = \frac{D^2}{h'd}, \qquad (7.34)$$

where again D is the objective aperture, d the dispersion, and h' the width of the spectrum, and assuming that $f < \Delta P/2 \times 10^{-5}$. The exposure time is inversely proportional to I and is best determined by trial and error. It also depends on light losses due to absorption, reflection, and refraction at the optical elements, and on obstructions in the light path (e.g., the plate holder in a Schmidt camera) and of the speed of the photographic emulsion, all of which would be virtually impossible to comprise in a single equation. As a first crude orientation, a relation by Gramatzki [7.4] may be used to obtain the exposure time t in seconds:

$$\log t = 4.2 + 0.4\, m - \log(O/S), \qquad (7.35)$$

with m the magnitude of the object, O the effective aperture area of the telescope, and S the area of the spectrum.

The experiences of amateur observers in stellar spectroscopy using various prism/telescope or prism/telephoto objective combinations are reported in some detail by Waber and McPherson [7.5], Albrecht [7.6], Pollmann [7.7], Sorensen [7.8], and Wagner [7.9].

To venture into the observationally challenging realm of meteor spectroscopy requires extremely powerful (large objective diameter, small f-ratio) equipment with a wide field of view and very sensitive (i.e., fast) emulsions. Though the meteor phenomenon has an average duration of about 1 s, the extremely rapid motion reduces the effective exposure time to less than 10^{-2} s. If, by fortuitous circumstance, the path of the meteor is perpendicular to the dispersion (i.e., parallel to the refracting edge of the prism), the spectrum is widened according to the segment of the trail recorded. Some observers leave the camera shutter open while awaiting an unpredictable meteor event, but this may cause background fog. This can be avoided with a *photoelectric meteor detector*, which opens the camera shutter at the very instant a meteor appears in the field. (Contact address for information on meteor spectra: NASA/Langley Research Center, Hampton, Virginia, USA.)

Comet spectroscopy also requires powerful equipment because of the typically low surface brightness of a comet. The refracting edge of the prism is oriented approximately parallel to the comet tail, which forestalls a possible superposition of the coma and tail spectra. Exposure times are considerably increased above the values given in Eq. (7.35).

Spectra of the solar chromosphere are photographed only during total solar eclipses. Shortly before the second and also just after the third contact, the chromosphere flashes on for about one second as a narrow crescent. With the refracting edge of the prism in a powerful instrument parallel to the crescent, the *flash spectrum* with its characteristically curved emission lines is photographed; the crescent acts as an infinitely distant entrance slit of which monochromatic images in the corresponding wavelengths are formed. Because of irregularities of the lunar limb (e.g., lunar mountains), some photospheric light may leak through and manifest itself in the the spectrum as continuous bands containing absorption lines. By using a plateholder which is movable perpendicular to the direction of dispersion and a slit which lets only the central part of the crescent-shaped lines pass, one can attempt, by continuous or stepped motion displacement of the plateholder during exposure, to catch the transition from photospheric to chromospheric spectrum, i.e., a reversal from absorption to emission lines. To obtain the coronal spectrum during totality requires a powerful slit spectrograph.

7.4.4.2 Slit Spectrographs. Slit spectrographs are, in some respects, less economical than objective prisms because, for instance, they produce the spectrum of only that object which illuminates the slit and they also require long exposure times to fully utilize the spectral resolution. Since the slit eliminates most of the influence of the sky background, such lengthy exposures then become practicable. The slit width should be fairly small for the purpose of good resolution, but still large enough to cover the seeing disk in order to avoid unduly long exposures; thus $b \lesssim 2 \times 10^{-5} f$. To guide the object during exposure on the slit requires a slit viewer by which the light reflected by the somewhat slanted slit jaws can be observed. The slit jaws can be composed of polished metal or, even better, glass which has been vapor-coated with aluminum, leaving a correspondingly narrow strip free. The widening can be accomplished by using the slow motion in right ascension or declination, or with a motor-driven plane-parallel glass plate in the light path which changes tilt and displaces the beam in a parallel fashion. For reasons of stability, it is less feasible to move the plateholder on a slide perpendicular to the direction of dispersion. The needed minimum slit length h can be calculated from the relation $h = h' f_1/f_2$, where h' is the desired spectrum width and f_1/f_2 is the ratio of focal lengths of telescope and camera.

The following considerations assume that a powerful telescope ($D \approx f/2$) is available, and that a dispersing element has been obtained from one of the numerous suppliers (e.g., Spindler and Hoyer, Phywe, Ealing, Edmund Scientific Co.) according to the conditions mentioned, the desired observing application, and of course the financial constraints. Gratings are normally used only for solar spectroscopy, as the requisite high light intensity is then available. Owing to the distribution of intensity into various spectral orders, they require long exposure times that are not feasible in many applications. On the other hand, good blaze gratings concentrate most of the dispersed light into a single order, but may be too expensive to acquire. For some work in grating spectroscopy, inexpensive replicas of coarser gratings may suffice (e.g., 100 lines/mm corresponding to $g = 10^4$ nm, a size of 2.5 cm corresponding to $Z = 2.5 \times 10^3$), but even then some of the theoretical resolving power will be lost.

Grating foils are less recommended as they often have irregular distances between the grating grooves. The better light yield by having all the intensity in one spectrum and of course the lower price makes prisms better suited for small college/amateur work in stellar spectroscopy. To reach higher dispersions and better resolving power, several prisms can be arranged in sequence, each one in the minimum of deflection, but again this is achieved at the expense of longer exposure times.

With a given telescope and dispersing element, the other quantities of the spectrograph may be calculated. The collimator diameter D_1 is

$$D_1 = Zg \cos i = Zg \qquad \text{(Perpendicular incidence) Grating,} \qquad (7.36)$$

$$D_1 = \frac{B \cos[(\gamma/2) + \beta]}{2 \sin(\gamma/2)} \qquad \text{(Minimum of deflection) Prism.} \qquad (7.37)$$

The condition of optimum illumination gives for the focal length f_1 of the collimator

$$f_1 = \frac{f}{D} D_1. \qquad (7.38)$$

Making the slit image b' about equal to the resolution ΔP on the plate fixes the focal length f_2 of the camera objective (normally then $\Delta l < \Delta P$; see Eqs. (7.21) and (7.28)):

$$f_2 = \Delta P \, (f_1/f) \, 2 \times 10^{-5} V_{\lim}, \qquad (7.39)$$

where $V_{\lim}$ holds for the maximum wavelength (the red limit) of the recorded spectrum. A prism with a symmetric light path has $V_{\lim} = 1$.

The camera lens should be close to the dispersing element as the lens diameter $D_2 \gtrsim D_1$ increases with the distance. For reflection gratings, it should be outside the incident light path in order to avoid vignetting. The lens diameter can be determined graphically by using the grating equation (7.16) for the limiting wavelength of the range to be recorded. The actual resolution of wavelength again follows Eq. (7.32). The length of the spectrum produced by a prism is given by Eq. (7.33) upon replacing f by f_2. For a 60° prism with $1.4 < n < 1.6$, the relation simplifies to become

$$L \approx f_2 \bar{n} \Delta n, \qquad (7.40)$$

again with Δn as the difference of refractive indices for the limiting wavelengths, and $\bar{n}$ their mean value. For gratings, the relation is

$$L \approx m f_2 (\Delta \lambda / g) \cos \varphi_{\bar{\lambda}}, \qquad (7.41)$$

where $\Delta \lambda$ is the difference of limiting wavelengths and $\varphi_{\bar{\lambda}}$ is the deflection angle at the mean wavelength $\bar{\lambda}$.

Provided the seeing disk $f\alpha$ (where α is the angular size of the disk in radians) is contained entirely in the slit, then, as for the objective prism, Eq. (7.34) gives the speed for this configuration. However, the light losses will be higher owing to the normally larger number of optical parts, and particularly when using a grating. When the slit width is less than $f\alpha$, then the speed is

$$I = \frac{D^2 b}{d h' f \alpha}. \qquad (7.42)$$

Depending on the combination of dispersing element and optical imaging system, the resulting focal surfaces may be tilted (refractor) or curved (Schmidt camera, spherical mirror, concave grating). In the latter cases, photographic plates are pressed against an appropriately shaped surface. Fewer problems will arise when using film. The adjustment is best made by taking several trial exposures.

A simple slit spectrograph designed for a Celestron-8 is described by Gebhardt and Helms [7.10]; somewhat more expensive grating devices are described by Schroeder [7.11] and Sorensen [7.12]. Because of the extreme intensity of solar radiation, the combination slit–collimator–grating–camera (or –eyepiece) will suffice for solar spectra. Two very simple and easily copied constructions (transmission gratings and reflection gratings in the Littrow design) are described by Christlein [7.13]. A simple arrangement with a direct-vision prism is given by Schmiedeck [7.14], and one with a grating by Delvo [7.15]. With the facility to vary the slit width, a spectrograph permits observations of prominences, for example, in the light of the Hα-line and with a wide-open slit. Design suggestions and experiences have been given, for instance, by Newton [7.16].

7.4.4.3 Accessories. Following the stage of pure observation and data recording, there are a few crucial steps which should be taken to effect even a comparatively modest processing of the spectra. Such processing will mandate the use of some indispensible accessories, to be described below.

The determination of radial velocities (see below) requires that a spectrum of *comparison lines* of laboratory origin be located in, or better still, adjacent to, the stellar spectrum. For objective prism work, the simplest technique is to place a transparent vessel containing a watery solution of neodymium chloride in front of the prism. Thus, in every spectrum, an absorption line at $\lambda 477.28$ nm is generated independent of the temperature of the solution. The mixing ratio should be 1:6; at higher concentrations, the line becomes asymmetric. Slit spectrographs are best used with a mask or diaphragm placed directly behind or in front of the slit; the idea is to cover that part of the slit which generates the object spectrum, but leave adequate space for the comparison spectrum above and below. The exposure is best made with the diffuse light of a spectral lamp containing a sufficient number of lines (e.g., He or Hg), and taken twice, before and after the object exposure. The sharpness of comparison lines then serves as a preliminary indicator of potential errors originating during the object exposure.

To measure the lines, a microscope with a micrometer slide is used which—if not at hand—can be readily constructed with a commercially available chart recorder (Albrecht [7.17]).

For many applications, a density curve must be derived by way of calibrating equipment. Density marks of known relative intensities are exposed on the same plate, or at least on a plate from the same batch, and these calibration plates are developed together with the spectral plates. The exposure times should be by order of magnitude the same as that for the spectra. It is simplest to use a wedge which, for instance, may be a glass plate coated lengthwise with platinum in continuously- or stepwise-increasing density. The plate is exposed through the homogeneously illuminated wedge, perhaps with a broad-band filter to fix the desired spectral range.

The required exposure times can be adjusted by varying the voltage at the light source or by inserting a neutral density filter. The relation between transparencies (i.e., ratios of transmitted to incident intensities; cf. photometers in Chap. 8) from the density marks thus obtained and the known ratios of incident intensities generates a pseudo-density curve sufficient to deduce the relative intensities of the object spectrum.

Building a *sensitometer* is more laborious but less expensive. It consists of a metal plate with circular apertures of different, known diameters. In homogeneous and diffuse illumination (provided by a ground-glass screen) the light is diminished in the ratio of the relative sizes of the apertures. From the separate apertures, tubes of equal width and blackened on the inside lead to a second metal plate with equally large round apertures and which is pressed against the photoplate. Keep in mind that the tubes are parallel to each other and that their length is large compared with the size of the apertures. After exposure, circular markings of equal size but different density are obtained.

A combination of objective prism and objective grating can be used to make a calibration using the starlight itself. In this method, a coarse grating (e.g., of wires) is placed in front of the prism so as to have the dispersion directions of grating and prism at right angles. Each object in the field generates—though at the expense of exposure time—a series of adjacent spectra with regularly diminishing intensities to the sides of the central spectrum. Expressed in magnitudes and referred to a wavelength λ, this gives

$$m_n - m_0 = 2.5 \log \left[\left(\frac{bn\pi}{b+a} \right)^2 / \sin^2 \left(\frac{bn\pi}{b+a} \right) \right] \tag{7.43}$$

$$n = 1, 2, 3, \ldots,$$

with the indices 0 and n representing the zero- and n-order spectra; a and b are the widths of gratings and spaces respectively. Usually one makes $a = b$ so that interference cancels the spectra of even order and Eq. (7.43) is correspondingly simplified. Mostly only the first-order spectra are pronounced. Their separation from the zero-order spectrum is given by $f\lambda/(a+b)$. Normally, the intensity ratio or the magnitude difference is determined empirically since the theoretical value is rarely reached owing to imperfections in the grating. For this purpose, spectra of stars of known magnitudes and preferably equal spectral types (e.g., the Pleiades) are needed. The transparencies of zero-order and first-order spectra at a given wavelength are determined and graphed against the magnitude of the star, thus yielding two curves displaced from each other exactly by the amount Δm to be determined. This Δm is also called the *grating constant* K of the objective grating. When the data points are not numerous enough, several exposures with various exposure times may be needed. When K is known, the density curve for an object is determined in the following manner:

The transparency differences $\Delta_i = T_i^1 - T_i^0$ between zero-order and first-order spectra are read at various wavelengths and are graphed against the transparencies T_i^0. The curve obtained is graphically smoothed, as in Fig. 7.9a. The measured points should not range too far in wavelength, as the density curve depends on λ. If necessary, readings from several closely adjacent spectra on the plates are made to obtain a

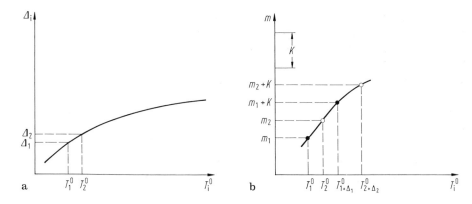

Fig. 7.9. a Curve of differences. b Pseudo-density curve.

sufficient number of data points. Then, a conveniently chosen transparency T_1^0 is assigned an arbitrary magnitude m_1 as a zero-point. In the Δ_i curve, the transparency T_1^0 then belongs to a difference Δ_1 which again corresponds to the grating constant K and leads to a second pair $(T_{1+\Delta_1}^0, m_1 + K)$. These pairs (T_i^0, m_i) are graphed in a diagram and the two data points connected. Any intermediate point (T_2^0, m_2) can be chosen to which a second point $(T_{2+\Delta_2}^0, m_2 + K)$ is determined. The procedure is continued until the required pseudo-density curve $T^0(m)$ is represented by a smooth set of points, as in Fig. 7.9 b.

A *photometer* is employed to measure transparencies. The principle is that the light of the photometer lamp (to be kept constant) shines through a condensor and illuminating objective through the plate area to be photometrically measured. The projecting objective images the area onto a measuring slit which is adjustable in height and width and behind which lies a field lens and, at its focus, a photosensitive element. It generates a current proportional to the incident light intensity and measureable with a sensitive microammeter. The current intensity thus measures the transparency. Instructions for building a somewhat more convenient recording photo-densitometer, where the plate table is moved by a synchronous motor and the transparencies are continuously written by a chart recorder, have been given by Pollmann [7.18].

With the pseudo-density curve known, recorded line profiles can then be translated into relative intensities. The areas segmented out of the continuum by the lines are proportional to the line intensity. Measuring the total intensity is best done with a *planimeter*, but a serviceable (albeit crude) alternative is to graph the line profiles on graph paper and count squares within the profile.

7.5 The Analysis

This section will provide suggestions for an initial analysis of the recorded spectra. Those methods not requiring intensity calibrations will be dealt with first.

7.5.1 Spectral Classification

The rather low dispersion as usually obtained in objective prism exposures (20–30 nm/mm at Hγ) is sufficient to classify stellar spectra. The first step is to become acquainted with the appearance of stellar spectra, a task most simply done by recording an A-type spectrum with its characteristic Balmer lines in the blue range. These strong lines are used to determine the dispersion curve, which relates the wavelength of a line to its position in the spectrum, the latter quantity being strongly wavelength dependent in the prism. The easiest method is to use a measuring microscope, but photographic enlargements or projections onto a wall (making sure that the optical axis of the projector is kept at right angles to the wall) may also be measured. When the dispersion curve (which may depend on the position on the plate) is fixed, fainter lines may be identified with the aid of tables by Moore [7.19] or Zaidel et al. [7.20]. Comparing the line intensities with those of standard stars from Seitter [7.21] obtained with the same equipment, the stars to be investigated can then be classified according to type and perhaps also to luminosity class. With sufficient experience, and with the aid of the criteria and literature presented in Sect. 7.3, the spectra of comets and meteors may be classified and chromospheric lines and coronal spectra identified.

7.5.2 Line Changes

Drastic changes in spectral lines, such as the transition from absorption to emission lines during a nova eruption, can be easily followed visually, or better, with a recording photometer using photographic spectra. The observation of periodic line splittings in spectroscopic binary stars usually demands a high dispersion. The quantitative study of these events, however, requires an intensity and a radial-velocity calibration.

7.5.3 Radial Velocities

The radial velocity of a star is the velocity component of its spatial motion along the line of sight from the Earth to the star. It is expressed in km s^{-1} and counted positive when the star recedes from the observer, negative when approaching. It is found by measuring the displacement, suffered through the Doppler effect, of the stellar lines in the spectrum relative to the comparison lines. For velocities small compared with the speed of light, the following approximation can be used:

$$\Delta\lambda = v\lambda/c, \tag{7.44}$$

where v is the radial velocity, c the speed of light, λ the laboratory wavelength, and $\Delta\lambda$ the measured shift. Since the motion of the star as well as of the observer enters into $\Delta\lambda$, the measured radial velocities must be corrected for the influence of the Earth's motion. The corrected velocity is thus referred to the Sun as a point at rest. Superposed upon the radial velocity owing to spatial motion may be velocity components caused by relative motions *within* the observed system, such as atmospheric motions in pulsating variables or orbital motions in binaries. Periodic events, repeatedly observed at suitable intervals, provide the necessary data for the construction of the *radial velocity curve*, which shows the relative changes with respect to the systemic velocity. In this context, the complex motions which occur in the shells of WR-, Be-, and P Cygni stars must also be mentioned.

For those stars with high-velocity shell expansions and binaries with sufficiently large radial-velocity amplitudes (see Appendix Table B.31), objective-prism photographs also permit the determination of *relative* radial velocities, that is, differences of radial velocities of absorption and emission lines, or double lines. With the neodymium chloride line as reference point, the ascertainment of *absolute* radial velocities may be attempted. The dispersion should not be too low, and the dispersion curve needs to be determined more precisely than in the case of mere line identifications. It can be derived from standard stars of known radial velocities. Elaborate, complicated methods (e.g., the reversion method, Fehrenbach prisms) reach accuracies of 5–30 km s^{-1}, depending on the spectral type. Amateur observers using the neodymium chloride method will have to be satisfied with less than this precision. In general, the accuracy improves with higher dispersion, greater widening of the spectra, and more frequently repeated measurements, perhaps on separate exposures (at the same phase for variable stars). Compromises will often have to be made because of the required exposure times and—in this case—also because of the time variation of the object. For instance, the exposure time should not exceed 1/10 of the period for variable stars. Best suited for such measurements, however, are slit exposures with a comparison spectrum. Using a measuring microscope, prism spectra can be processed with the dispersion relation

$$a - a_0 = \frac{C}{(\lambda - \lambda_0)^k}, \tag{7.45}$$

where a is the reading at the recorder (or microscope screw). From the constants a_0, λ_0, C, and k, and making use of the known wavelengths in the comparison spectrum, the (displaced) wavelength of the stellar spectrum can be found. For grating exposures, a linear or quadratic interpolation will suffice. Each reading should be made repeatedly by moving the wire toward the line center, always from the same side, and then remeasuring the spectrum but in a position reversed 180° for possible periodic errors of the screw. The room temperature should remain constant during the entire set of readings. The differences (measured wavelength) − (laboratory wavelength) then give from the Doppler formula the radial velocities, which are then to be reduced to the Sun. The component of the Earth's orbital velocity in the direction of the star, to be added to the measured radial velocity, is given by

$$v_E = - \left[(\cos \alpha \cos \delta) \Delta x + (\sin \alpha \cos \delta) \Delta y + (\sin \delta) \Delta z\right] A, \tag{7.46}$$

where α and δ are the right ascension and declination of the star, and Δx, Δy, and Δz the daily changes of the rectangular, equatorial coordinates of the Sun tabulated in astronomical almanacs and which can be considered velocity components of the Earth expressed in astronomical units (AU) per day. The factor $A = 1.731 \times 10^3$ converts this velocity into km s^{-1}.

The influence of the Earth's rotation is maximally 0.5 km s^{-1}, and it, as well as the very small correction due to the orbital motion with the Moon, need not be considered here.

The reduction methods in the ensuing sections require previous intensity calibration.

7.5.4 Color Temperatures

Stellar spectra recorded with the right-angle objective grating/prism combination offer the facility of determining the *color temperature*, defined as that temperature which a blackbody of the same relative intensity distribution in the spectral range considered would have. To correct the energy distribution for errors due to extinction, optics, and plate properties, a comparison star of known color temperature should be photographed on the same plate. Describing the energy distribution by the Wien approximation (accurate to 1% when $\lambda T \leq 0.3$ [cm·K]), the color temperature T_* is

$$T_* = \left(\frac{1}{T_C} + \frac{\delta \Phi}{c_2}\right)^{-1}, \tag{7.47}$$

where T_C is the color temperature of the comparison star, $c_2 = hc/k = 1.4388$ [cm·K], and $\Delta\Phi$ is the differential quotient $0.921\, d(m_* - m_C)/d(1/\lambda)$. The transparencies of the continuum in the star and comparison spectra are determined at various wavelengths and are translated into magnitudes through the pseudo-density curve. Graphing the corresponding magnitude differences versus $1/\lambda$ (λ in cm) should result in a straight line whose slope, multiplied by the factor 0.921, represents the quantity $\Delta\Phi$ from which the color temperature of the star under study is found. Fluctuations of extinction by the Earth's atmosphere during exposure are negligible at zenith distances of $< 30°$. Of course, differential interstellar extinction must always be considered. In case Wien's approximation does not apply, the relation

$$T_* \left(1 - e^{-c_2/\bar{\lambda}T_*}\right) = \left(\frac{1}{T_C(1 - e^{-c_2/\bar{\lambda}T_C})} + \frac{\Delta\Phi}{c_2}\right)^{-1} \tag{7.48}$$

follows from the Planck function, and is solved by iteration. Here, $\bar{\lambda}$ is the mean wavelength of the range used to determine $\Delta\Phi$. α Lyrae (Vega) can serve as a comparison star. Its color temperatures for some wavelength ranges are listed in Table 7.2.

7.5.5 Equivalent Widths and Line Profiles

The term *equivalent width* is defined as the width of a rectangular absorption feature with central intensity zero and whose area equals that bounded by the spectral line under study. Its formal expression is

$$W = S d^{-1} I_0^{-1}, \tag{7.49}$$

Table 7.2. Color temperatures for α Lyrae for four different wavelength ranges.

$\lambda\lambda$ (nm)	T (K)	$\lambda\lambda$ (nm)	T (K)
420 – 465	17 700	420 – 670	17 900
420 – 550	17 500	465 – 550	17 600

where S is the measured area, I_0 the difference (background fog) − (undisturbed continuum at the position of the line), and d^{-1} again the reciprocal dispersion. The equivalent width, usually given in the wavelength scale, characterizes the strength or intensity of the spectral line, which in turn is essentially a function of the number of absorbing atoms and of excitation conditions in the stellar atmosphere. To achieve optimal accuracy, the determination of equivalent widths of spectral lines should be effected with as high a dispersion as possible. In objective prism spectra, only the strongest lines (often only the Balmer lines) can be used.

The amateur can, for instance, monitor changes in the equivalent width, particularly of Balmer and strong metal lines, occurring in many variables such as δ Cephei-type stars. Measuring equivalent widths is simpler and more reliable than determining the true line profiles, since they are independent of the so-called "instrumental profile." This term encapsulates all influences which modify the true, original profile of the line. The culprits are primarily the spectrograph and photometer slits, whose images superpose on the true profiles. It would be necessary to determine (best empirically) how the profile of an infinitely narrow line finally appears in the recording. In practice, a very faint line of the comparison spectrum is examined with a photometer, and the measured stellar line can then be mathematically rectified. Such rectified profiles give much more information on the structure of the star's atmosphere than does the equivalent width. But this exceeds the frame of the small college/amateur facilities, and, in particular, requires a high degree of mathematical sophistication as well as the use of large computers.

Several strong effects can be studied qualitatively without rectification procedures. There are, for example, the sometimes very high rotational velocities (up to 500 km s^{-1}) found in particular in many B-type stars, and which lead through the Doppler effect to very broad, dish-shaped line profiles. The width of the lines is a function of the rotational velocity of the star, or rather of the projection of this speed in the direction of the observer, who in general is not in the equatorial plane of the star. With the possibility of working on spectra of high dispersion (better than 3 nm/mm), the analysis can also be made quantitative. From determined equivalent widths of weak lines and of as many elements as possible, the so-called *curves of growth* can be constructed. These observationally obtained curves are compared with corresponding theoretical curves in order to derive, in a relatively simple way, various stellar

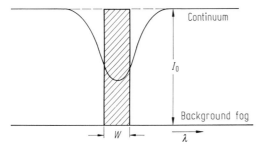

Fig. 7.10. The definition of equivalent width.

quantities such as abundances of the corresponding elements, excitation and ionization temperatures, microturbulence velocities, and damping parameters. This comparison requires extensive information from atomic physics, for instance, excitation energy (Moore [7.19]) and transition probabilities (Kurucz and Peytremann [7.22]). Even a summary discussion of this application would exceed the frame of the present introduction. Readers interested in gaining insights into this fascinating subject may wish to consult Voigt [7.23] or Aller [7.24], which, among other sources, also survey this subject in an approachable fashion.

References

7.1 Moore, C.E., Minnaert, M.G.J., and Houtgast, J.:*The Solar Spectrum* 2935 " to 8770 ", National Bureau of Standards Monograph 61, Washington 1966.
7.2 Thorne, A.P.: *Spectrophysics*, Chapman and Hall Ltd., London 1974.
7.3 Buil, C.: A Charge-Coupled Device for Astronomy. *Sky and Telescope* **69**, 71 (1985).
7.4 Gramatzki, H.J.: *Hilfsbuch der astronomischen Photographie*, Ferd.Dümmlers Verlag, Berlin Bonn 1930.
7.5 Waber, R., McPherson, R.: Photographing Star Spectra. *Sky and Telescope* **33**, 322 (1967).
7.6 Albrecht, C.: Astrospektrographie mit Spiegelteleskopen. *Sterne und Weltraum* **11**, 195 (1972).
7.7 Pollmann, E.: Sternspektroskopie. *Sterne und Weltraum* **16**, 296 (1977).
7.8 Sorensen, B.: An Objective-Prism Spectrograph. *Sky and Telescope* **65**, 460 (1983).
7.9 Wagner, B.: Astrospektroskopie mit einfachen Mitteln. *Sterne und Weltraum* **25**, 218 (1986).
7.10 Gebhardt, W., Helms, B.: Ein Selbstbau-Prismenspektrograph zum Gebrauch am Celestron 8. *Sterne und Weltraum* **15**, 58 (1976).
7.11 Schroeder, D.J.: A Grating Spectrograph for a College Observatory. *Sky and Telescope* **47**, 96 (1974).
7.12 Sorensen, B.: A Simple Slit Spectrograph. *Sky and Telescope* **73**, 98 (1987).
7.13 Christlein, G.: *Physikalische Experimente im Bild*, Bayer. Schulbuchverlage, München 1968.
7.14 Schmiedeck, W.: A Simple Technique for Recording the Sun's Spectrum. *Sky and Telescope* **57**, 395 (1979).
7.15 Delvo, P.: A Spectroscope with a Holographic Grating. *Sky and Telescope* **54**, 65 (1977).
7.16 Newton, J.B.: A Spectroscope Attachment for Viewing Solar Prominences. *Sky and Telescope* **39**, 120 (1970).
7.17 Albrecht, C.: Astrospektrographie mit Spiegelteleskopen II. *Sterne und Weltraum* **11**, 243 (1972).
7.18 Pollmann, E: Spektroskopische Veränderlichenbeobachtung. *Sterne und Weltraum* **24**, 340 (1985).
7.19 Moore, C.E.: *A Multiplet Table of Astrophysical Interest*, NBS Technical Note 36, U.S. Department of Commerce, Washington 1959.
7.20 Zaidel', A.N., Prokof'ev, V.K., Raiskii, S.M., Slavnyi, V.A., Shreider, E.Y.: *Tables of Spectral Lines*, IFI-Plenum, New York London 1970.
7.21 Seitter, W.C.: *Atlas für Objektivprismenspektren*, Dümmler, Bonn 1970.
7.22 Kurucz, R.L., Peytremann, E.: *A Table of Semiempirical gf-Values*, Special Report 362, Smithsonian Astrophysical Observatory, Cambridge MA 1976.
7.23 Voigt, H.H.: *Abriss der Astronomie*, B.I.-Wissenschaftsverlag, Mannheim Wien Zürich 1980.
7.24 Aller, L.H.: *Astrophysics, The Atmospheres of the Sun and Stars* (2nd edn.), The Ronald Press Co., New York 1963.

8 Principles of Photometry

H.W. Duerbeck and M. Hoffmann

8.1 Introduction

8.1.1 General Description and Historical Overview

Radiation from space is practically the only source of information on stars and other distant celestial objects for the observational astronomer. The incident radiation is described completely by its flux, its direction, its color (wavelength), its polarization, the time of arrival, and—in certain cases—also by the location of the receiver. Photometry involves the measurement of radiation summed over a more or less extended wavelength region. Depending on the receiver of radiation and the method of observation, one can discriminate between visual, photographic, and photoelectric photometry, point and surface photometry, narrow-band, intermediate-band and broadband photometry, photometry with high time resolution, etc.

The device used to measure the radiation is called a *photometer*. The light-sensitive part of such a device, which transforms the infalling light to be measured into another quantity, such as an electric current, is the *detector*. The signal from the detector undergoes a series of transformations before the result is stored permanently. Various kinds of detectors can be used for astronomical observations: the eye, photographic emulsions, photomultiplier tubes (PMTs), semiconductor photodiodes, television-type detectors such as vidicons, solid state image detectors like the charge-coupled device (CCD), image intensifiers, or electronographic detectors. On the assumption that the use of the aforementioned equipment is not too complicated for the observer, such equipment will be discussed in Sect. 8.3.1.

Photometry has a long tradition in astronomy: in the second century B.C., the Greek astronomer Hipparchus introduced the concept of *stellar magnitudes* and used them for expressing the brightnesses of the stars. This nearly logarithmic scale of stellar magnitudes is still in use today—defined as a purely logarithmic scale—in optical as well as in near-infrared astronomy.

The rise of astrophysics and the observation of variable stars in the 19th century created a type of photometry which surpassed the simple arrangement of stars into 6 magnitudes. The "step-estimate" method was developed and visual photometers were constructed and employed. In the following decades, photography, photoresistors, and photocells came into astronomical use.

An ever increasing number of professional and amateur astronomers started to engage in visual photometry of variable stars during the last century, especially after an

"Appeal to Friends of Astronomy" was made by F.W.A. Argelander in 1844. Today, amateur astronomers are organized into many associations all over the world; more than one hundred thousand visual brightness estimates of variable stars are made each year. In the course of the 1980s, photoelectric photometry has won more and more advocates, since photometers can be constructed rugged and cheaply using state-of-the-art electronics. Photometers are commercially available and easy to use. Moreover, constructing a do-it-yourself-type photometer presents no insurmountable problems. Photoelectric observations of high accuracy of a large series of relatively bright stars are of significant value, and since professional astronomers today are finding increasingly less time to make regular photoelectric observations for weeks and months, the amateur astronomer with a small telescope makes valuable contributions to this area of astronomy. This fact has led to a flood of publications in recent years. The International Amateur–Professional Photoelectric Photometry Association (IAPPP) is especially active, and publishes books and a journal in this field. Here we will attempt to give a general overview of the topic, explaining the possibilities of photometry of different astronomical objects, and will explain the basic facts of photometer construction. Details, such as plans for the construction and wiring of photometers, can be found in the advanced literature.

8.1.2 Units of Brightness

This section will show the relations between astronomical and physical units. Its contents will be useful for the following sections, but are not absolutely necessary, and can be skipped during a first reading.

A light source (e.g., a star) radiates a certain amount of energy into space; its radiated power L (in astronomy, this is called *luminosity*) indicates the amount of energy per second streaming through an imaginary surface which completely surrounds the star. It should be noted that the radiated power remains the same no matter how large the surrounding surface: an increase of surface (as it surrounds the star at a larger distance) is accompanied by a corresponding decrease in the illumination power; the former quantity increases with the square of the distance between the star and the observer, while the latter *decreases* by the same amount.

The amount of light per unit time radiated, not into complete space, but into a unit solid angle (which can be envisaged as an infinitely long cone with an aperture angle of 65.5 degrees) is called the *luminous flux*; it is measured in lumens (lm). 1 lm is radiated by a light source of radiated power 1 candela (1 cd) into the unit solid angle. 1 cd is the radiated power which is emitted vertically from an area 1/600,000 m^2 of the surface of a blackbody at the melting point of platinum (2042 K).

Since a light source is, in general, extended, the radiation emerging from its surface has a brightness or *luminance* B. It is the ratio of the luminous flux I of a radiating body and its surface area A,

$$B = \frac{I}{A}, \tag{8.1}$$

and is measured in cd m^{-2}. In earlier times, the unit *stilb* (sb) was also used, where 1 sb = 1 cd cm^{-2}; the luminance of clear sky is in the range 0.2–0.6 sb.

The *illuminance E* of an object is also important in astronomy. It is directly connected with the concept of the stellar magnitude and is defined as the *luminous flux* Φ which arrives at an area A, i.e.,

$$E = \frac{\Phi}{A}. \tag{8.2}$$

If 1 lm arrives vertically at 1 m², the illuminance is measured as 1 lux (lx).

The magnitudes used in astronomy must be connected with the system of physical units. A direct luminous flux calibration of the stars α Lyr and 109 Vir by means of a platinum furnace was carried out by Tüg and collaborators [8.1]. A star of apparent visual magnitude $0\overset{m}{.}0$ has an illuminance of 2.54×10^{-6} lx (if the effect of light absorption of the Earth's atmosphere is neglected). A star of *absolute* visual magnitude $0\overset{m}{.}0$ generates a luminous flux of 2.45×10^{29} cd. The luminance of a "model" sky with one star of 0^m apparent visual magnitude per square degree is 0.84×10^{-10} cd m^{-2}.

Since stars or other objects (e.g., radio or X-ray sources) emit radiation to Earth, which is only vaguely understood as "illuminance," astronomers also use the concept of *monochromatic radiation flux* (flux), which is the illuminance per unit wavelength or unit frequency. Tüg et al. [8.1] found that the flux radiated at wavelength 555.6 nm (green light[1]) from α Lyr amounts to

$$\Phi_\lambda = 3.47 \times 10^{-11} \text{ J m}^{-2} \text{ s}^{-1} \text{ nm}^{-1} \tag{8.3}$$

at the Earth. This corresponds to a *monochromatic photon flux* of

$$N_\lambda = 9.7 \times 10^7 \text{ photons m}^{-2} \text{ s}^{-1} \text{ nm}^{-1}. \tag{8.4}$$

The above relations are important if physical parameters are derived from astronomical observations; in general, astronomical photometry is a method of comparing "illuminances" E or "radiation fluxes" Φ_λ of different stars.

When two stars observed with the same photometer yield the radiation fluxes Φ_1 and Φ_2, their magnitude difference is

$$\Delta m = m_1 - m_2 = -2.5 \log_{10} \left(\frac{\Phi_1}{\Phi_2} \right), \tag{8.5}$$

where a magnitude difference of 1^m corresponds to a ratio of the radiation fluxes of 2.512. The minus sign in the formula takes into account that smaller radiation fluxes correspond to larger stellar magnitudes.

8.1.3 The Receivers

This section will discuss the properties of the human eye, of the photographic emulsion, and of different photoelectric devices as receivers of radiation. More details are found in the cited literature.

---

[1] 1 nm (nanometer) = 10^{-9} m; it should be added that in astronomy the unit 1 Å = 0.1 nm = 10^{-10} m is still in use.

8.1.3.1 The Eye. In amateur astronomy, a large number of photometric observations —called estimates—are carried out by eye, and hence a knowledge of the properties of the human eye is desirable. A good description of the processes occurring in the eye is found in Schnapf and Baylor [8.2].

The eye is approximately of spherical shape. Light penetrates the lens, which is limited by the iris, and enters the interior, the vitreous humor. The image is formed on the backlying choroid coat, which is composed of two different types of vision cells, about 125 million *rods* and 6 million *cones*. The less light-sensitive cones are employed for color vision during the daytime or with strong illumination (daytime vision, color vision, direct or foveal vision). These receivers are concentrated in the central region of the choroid coat, the so-called "yellow spot" or fovea. The angular resolution is about 1/2 arcminutes. The rods are about 10 000 times more sensitive than the cones; they can be employed only for black-and-white vision (indirect vision, extrafoveal vision). Faint astronomical objects are most easily recognized by indirect vision, that is, by looking past the object and thus employing the rods. It is not advisable to examine two stars at the same time to obtain an estimate, because different areas of the choroid coat have different sensitivities.

The power of sensation S for light of different intensity depends, within certain limits, on the logarithm of the strength of stimulus, E (the Weber–Fechner law):

$$\text{constant} \times (S - S_0) = \log_{10}\left(\frac{E}{E_0}\right). \tag{8.6}$$

The time of adaption from bright to dim is about 15 to 45 minutes. A dim red light, which just allows the keeping of a logbook or the study of star maps, is recommended when observing in darkness.

The reaction of the human eye to red light is different from its response to other colors. When magnitude differences between white and red stars at different intensities (e.g., with telescopes of different sizes) are determined, major deviations may occur (the Purkinje phenomenon). Visual magnitude estimates of red stars should be made by choosing other red stars for comparison.

To allow a comparison with other receivers, when a 6th magnitude star is observed, 200 photons enter a fully adapted eye every 1/20 of a second. The quantum efficiency[2] of the human eye, which is indeed able to detect single photons, may be as high as 15% (Hecht, Shlaer, and Pierenne [8.3]) and thus has a better efficiency than the photographic emulsion. It does not, however, possess the latter's storage capacity.

The limits of visual observations are also investigated by Lukas [8.4], where additional references are given.

The contrast sensitivity of the human eye is indicated by the *contrast step* in magnitudes; the smaller the step, the greater the contrast sensitivity. The human eye has, at a surface brightness of 2^m per square arcsecond ($2^m/\beta''$), a contrast sensitivity which corresponds to a contrast step of $0\overset{m}{.}05$. At a surface brightness of $16^m/\beta''$, this step size decreases to $0\overset{m}{.}2$. When looking at closely neighboring bright and dark sequences, the resolving power enters the contrast sensitivity in the form of the modulation transfer

---

[2] The quantum efficiency is defined as the percentage of photons incident on a receiver which produce detectable events.

function (MTF). The diameter of the pupil varies with surface brightness, and thus the MTF is brightness-dependent. When observing double stars, the resolving power (i.e., the smallest resolvable distance in arcseconds) for best contrast conditions (given by suitable magnitudes of the primary component), ideal seeing conditions, and using the finest terrestrial telescopes, depends on the magnitude difference of the components according to an empirical formula[3]:

$$\log_{10} d'' = -0.914 + 0.1913 \Delta m. \tag{8.7}$$

Here, we assume that the distinguishability of the two neighboring light points depends only on the properties of the eye, and not on the diffraction of the telescope.

The effective pupil diameter, which corresponds to the resolving power of the pupil, is 1.8 mm for pupil diameters larger than 2.5 mm, corresponding to the resolving power of the naked eye of 70″. Below diameters of 1.5 mm, the effective pupil diameter is equal to the true one.

The maximum pupil diameter for dark-adapted eyes is a function of the age of the observer. A rule of thumb is: For a person between 20 and 70 years old, the maximum diameter decreases from a value of 8 mm by 1 mm every 10 years.

8.1.3.2 The Photographic Emulsion. The radiation receivers in a photographic emulsion are silver halide crystals with sizes of about $1\,\mu\text{m}$, which are generally embedded in a thin gelatine layer. The emulsion is attached to a base such as a sheet of film paper or a glass plate. The exposure to light causes the formation of "germs" of metallic silver; during development the exposed crystals are converted completely into metallic silver, which in its finely distributed form appears black (photographic negative). The formation of "germs" occurs only when the radiation is energetic enough (i.e., of short wavelength or blue light); the addition of certain organic substances increases the sensitivity of the emulsion to light of longer wavelengths. Detailed reviews will be found in Chap. 6 and in [8.5]. James [8.6] gives a detailed description of the general processes.

For astronomical applications, emulsions for light or infrared radiation up to a wavelength of 1100 nm (e.g., Kodak I-Z) are used; emulsions which are sensitive up to 1500 nm are being tested. At the short wavelength end, the limitation arises because the Earth's atmosphere completely absorbs UV quanta below 320 nm; also the gelatine and the optical elements are not transparent to UV radiation (UV-sensitive photographic emulsions are not embedded in gelatin). Generally, the quantum efficiency is below 1%.

The brightness information of a photographic emulsion emerges in a scale of density. A density interval corresponds to a small brightness interval when the gradation of the emulsion is high (= high-contrast emulsion). This is important for the attempted photometric accuracy, which depends on the scatter of density and the size of the measuring diaphragm. The scatter of density depends on the type of emulsion and the value of the density itself. When the light is scattered through a photograph in all possible directions (a cone with an aperture angle of 180°), the measured density is

---

3 See Brosche, *Astr. Nachr.* **288**, 33 (1964).

Table 8.1. Scatter of diffuse density for different blue-sensitive emulsions used for astronomical purposes. The nominal resolving power is 80 line pairs per millimeter for the 103a-O emulsion, 87 for the IIa-O, and 200 for the IIIa-J (from Furenlid [8.7]).

Diffuse density	Scatter (rms) for an area of 1000 μm^2		
	103a-O	IIa-O	IIIa-J
0.2	–	0.017	0.007
0.4	0.037	0.023	0.011
0.6	0.039	0.028	0.014
0.8	0.041	0.033	0.017
1.0	0.046	0.038	0.019
1.2	0.051	0.042	0.022
1.5	0.060	0.048	0.025
2.0	0.079	0.058	0.031
2.5	0.101	0.068	0.036
3.0	0.129	0.075	0.038

called *diffuse density*. On the other hand, the density measured for a bundle of rays transmitted perpendicular to the plate in a cone of aperture angle 7° is called *specular density*. As is seen in Table 8.1, fine-grain emulsions possess not only a steeper gradation, but also a smaller scatter of diffuse density. The photometric accuracy achieved with such emulsions is much better than that obtained with "faster," but coarser grain emulsions. If possible, one should choose fine-grain emulsions. This often necessitates pre-exposure sensitization of the emulsion (e.g., by baking in dry atmospheres or bathing in special solutions).

8.1.3.3 The Photomultiplier Tube (PMT). When light hits a cathode (generally composed of alkaline oxides) inside a vacuum tube, electrons are released. This is called the external photoelectric effect. The electrons are directed, by a potential difference, onto a dynode, where they are able to release more electrons, which are then directed to a second dynode, and so on (Fig. 8.1). Amplifications by a factor 10^6 are reached when about 10 dynodes are employed. The quantum efficiency can be as much as 20%. The PMTs are preferentially blue-sensitive, but special cathodes can extend the sensitivity up to 1.1 μm (S-1 photocathode). Figure 8.2 shows the sensitivity function of typical cathodes. PMTs have been standard equipment in professional astronomy for more than 40 years, and they have also come into use in amateur astronomy in recent years.

PMTs need a stabilized high-voltage power supply (over 1000 V), and are sensitive to overexposure of light and mechanical stress. The sensitivity in the blue and visual region is very good, the time resolution in photon-counting mode excellent. A thorough discussion of the properties of PMTs is given in [8.8,9,10].

PMTs are available from various manufacturers (RCA, EMI, ITT, Hamamatsu) at various prices depending on the quality. The "classical" PMT, the 1P21 of RCA, costs about $100; a tube of somewhat poorer performance, the 931A, about $20. The

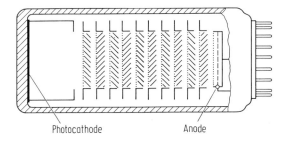

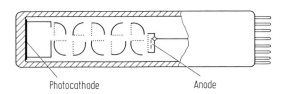

Fig. 8.1. Structure of a photomultiplier tube: Light incident on a photocathode releases an electron that is accelerated by the potential difference to the first dynode, releasing there more electrons. The "electron avalanche" is measured as a pulse at the anode. Different types of photomultiplier tubes are in use, depending on the arrangement of dynodes, e.g., the "Venetian blind" type (*above*), or the box and grid type (*below*). From Young [8.11].

"miniature PMTs" of the Hamamatsu Company are also of interest: the R869 tube has a spectral response similar to that of a 1P21; the cathode diameter is about 5 mm, which is sufficient for use in a photometer, and reduces the dark current, as compared to other PMTs. A socket with built-in voltage divider is often offered as an accessory.

PMTs can be used either at the ambient temperature or cooled; in the latter case the dark current is reduced by several orders of magnitude. The cooling is achieved with dry ice or by the use of a *Peltier element*. On the negative side, cooling invites—aside from higher operating costs—problems caused by the formation of dew on the optical parts in the photometer.

8.1.3.4 The Photodiode. A semiconductor such as silicon or germanium which has been "doped" with special other sorts of atoms can become conducting because of the released negatively charged electrons (n-type) or the positive "holes" they leave behind (p-type). A *diode* is made up of a combination of a p- and an n-layer, and lets a current pass only in one direction.

Light that is absorbed in the p–n boundary layer releases electrons and changes the resistance. Diodes specially designed for the measurement of light have a layer (I) between the p- and the n-region, where light is absorbed and where the formation of electron–hole pairs by thermal effects is reduced. Such diodes are called *PIN-diodes*.

The use of PIN-photodiodes is relatively simple. The power is usually supplied by a battery, the costs are low, and the detector is insensitive to overexposure. The sensitivity in the red and infrared regions is good. Disadvantages are that the relatively weak signals must be highly amplified, that the blue sensitivity is low, and that the

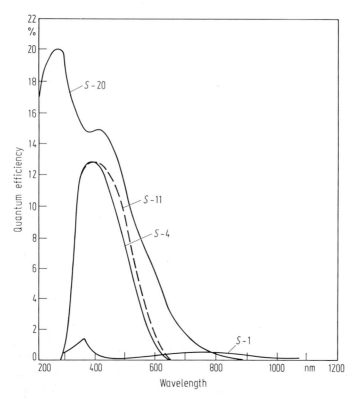

Fig. 8.2. Spectral sensitivity of different photocathodes. The quantum efficiency indicates what percentage of the incident photons at a given wavelength release an electron. The characteristics of the S-20 cathode shown here are due to the UV-transparency of the entrance window, which makes it sensitive to the near-ultraviolet. The S-1 cathode has a low quantum efficiency, but it is sensitive to near-infrared radiation.

time resolution is low. The use of photodiodes among amateur astronomers is fairly widespread. A detailed description of the use of semiconducting detectors is given in [8.12]. The sensitivity of a blue-sensitized PIN-photodiode is shown in Fig. 8.3.

8.1.3.5 Two-Dimensional Radiation Receivers. Charge-coupled devices (CCDs) for one- or two-dimensional image recording have been used for photometric purposes in astronomy in recent years. In the two-dimensional case, a silicon chip of area 100 mm^2 carries a grid of about 250 000 light-sensitive elements. The exposure produces electron-hole pairs in the silicon layer. The formed electrons are stored and can, after exposure, be measured via an exit diode [8.13,14]. The voltage produced there is digitized and stored on a video tape or in a computer. Like the photodiodes, the CCDs are red-sensitive and can be used for short-wavelength radiation only with special coatings. The quantum efficiency is at maximum 80%. The use of CCDs in amateur astronomy is described in [8.15,16,17].

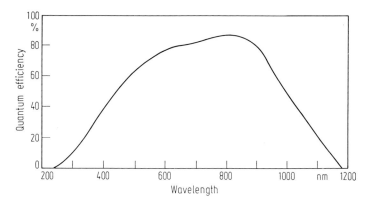

Fig. 8.3. The spectral sensitivity of a photodiode. It is sensitized for the blue spectral region and is generally sensitive to the near-infrared.

8.2 Limits and Accuracies of Photometric Measurements

The radiation of celestial bodies is influenced in many ways while en route to the observer. The radiation of fixed stars is partly scattered and absorbed when it penetrates the interstellar matter. This interstellar absorption has different strengths at different wavelengths (i.e., it is selective), which leads to a "reddening" of the light. A similar effect occurs in the Earth's atmosphere, and is there called (selective) extinction. Part of the incident radiation is scattered over the whole sky (the blue color of the daylight sky is due to the enhanced scattering of blue sunlight; see also Fig. 8.4). The scattered light of the Moon also increases the sky brightness by several orders of magnitude, and makes impossible the photometry of faint stars (Fig. 8.5). As another example, the brightness of the dark night sky increases by as much as 10% when Venus is above the horizon.

Turbulent elements in the air randomly deflect incoming radiation from its original path, thereby causing a net widening of the bundle of light rays. Since the elements are displaced rapidly, a wavelength-dependent scintillation of intensity and direction is observed. The normal bending of light in the atmosphere—the refraction—also tends to widen the incident light bundle. At large zenith distances, each pointlike stellar image is widened to a noticeable spectrum owing to the wavelength dependence of refraction.

The light collected by a telescope is also changed by the optical elements. Wavelength-dependent reflection losses are produced at mirror surfaces; radiation is absorbed when passing through glass, and another part is lost by reflection.

After an additional weakening and partial blockage by color filters, the remaining light reaches the detector, which registers, depending on its efficiency, only a fraction of the incident radiation. The final, measured result is obtained only after further processing of the signal, where additional information losses occur owing to, for example, amplifier noise. Sect. 8.5 describes reduction methods where these effects are taken into account. Most effects are internally compensated because the light from

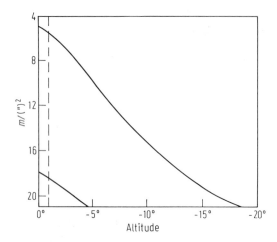

Fig. 8.4. Visual brightness of the sky near the zenith, given in magnitudes per square arcsecond, for different (negative) altitudes of the Sun and the Moon, reckoned from the horizon. The *dashed line* indicates the horizon, corrected for refraction and diameter of the celestial body.

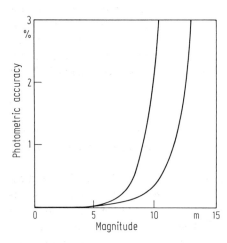

Fig. 8.5. Photometric accuracy (in %) for stars of different magnitudes observed with a telescope of 200 mm aperture for an integration time of 15 seconds, with dark sky (*right line*) and with full moon (*left line*).

the target star is compared with that from one (or more) comparison stars; both rays travel the same path into the measuring device.

The natural and instrumental sources of error can be summarized in the case of the photoelectric photometer:

Natural:
- fluctuations of sky brightness and transparency;
- inclusion of faint background stars in the measuring diaphragm;
- "overflow" of the stellar image when bad seeing occurs and a small diaphragm size is used.

Instrumental:
- vignetting in the radiation beam of the telescope and the photometer;
- nonlinearities in the amplifier electronics;
- dead times in the detector and the amplifier;
- missing magnetic or electric shieldings;
- varying temperature of the detector;
- polarization effects in the optics;
- measuring errors caused by different centering of two stars in the diaphragm (if no Fabry lens is used).

8.3 Astronomical Color Systems

The diligent reader will have noticed that "visual" magnitudes were discussed in Sect. 8.1.2, and that radiation fluxes can be measured in different wavelength regions. Thus, stellar magnitudes can be defined in different wavelengths regions. The use of photography led to the introduction of "photographic magnitudes," sometimes called blue magnitudes, because the first photographic emulsions were sensitive only to radiation of $\lambda \leq 490$ nm. The difference of two magnitudes measured in different wavelength regions is called a *color index*. The most frequently used color index is a measure of how much more light a given star radiates in the blue than in the green, compared to a normal star (defined in general as an unreddened star of spectral type A0). Such a color index provides a simple measure of the energy distribution of a star between the two wavelengths. The color index of a star is thus defined quantitatively as:

$$m(\lambda_2) - m(\lambda_1) = -2.5 \log \left(\frac{a \, \Phi_2}{b \, \Phi_1} \right), \tag{8.8}$$

where λ_1 and λ_2 are the effective wavelengths at which the radiation fluxes Φ_1 and Φ_2 are measured, and a and b are constants which are characteristic for a given measuring device.

The development of accurate photometers was accompanied by the design of astronomical color systems: Before the light falls onto a photocell, it generally passes through a color filter, which may be one of a sequence (color sequence), or it is dispersed[4] so that color measurements in different bands can be made simultaneously. More precisely, the combination of telescope, filter, and sensitivity function of the receiver defines a *bandpass*. A set of color filters, in combination with a given telescope type and a characteristic curve of a certain cathode type, defines a particular color system.

Broadband systems (bandpass widths typically 100 nm), intermediate band systems (30 nm), and narrow band systems (10 nm) are in use. Each type has specific uses in astronomy. The most widely known, which will also be discussed here, are the *UBV* system, developed by H.L. Johnson and later expanded to longer wavelengths (R, I, J, K, L, M) (Fig. 8.6, Table 8.2), and the *uvby* system, developed by B. Strömgren.

---

[4] In a measuring sequence or simultaneously by a prism, a grating, or dichroitic filters.

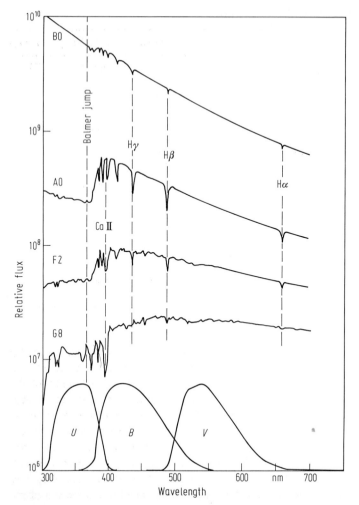

Fig. 8.6. Transmission curves of the U, B, and V filters of the Johnson UBV system (*bottom*) and radiation fluxes of main-sequence stars of different spectral types (*top*). It is seen that the slope of the stellar continuum, which depends on the surface temperature, can be determined by the $(B-V)$ index. The Balmer jump, which depends for a given temperature on the luminosity of the star, can be determined by the $(U-B)$ index. From Henden and Kaitchuck [8.8].

In the UBV system, the brightness in the visual "V" region is roughly equal to the visual magnitude; this magnitude is designated simply as V. The B magnitude is approximately equal to the photographic magnitude, and U is the ultraviolet magnitude, which should be measured with a reflecting telescope. The color indices of the UBV system (see Eq. (8.8)) are $U-B$ and $B-V$.

The standard UBV filters are designed for use in a reflecting telescope and a PMT with an S-4 cathode sensitivity. The U filter may have an additional transmission in the red, the so-called "red leak," which becomes noticeable when red stars are observed.

Table 8.2. Properties of the *UBVRI* system [8.18].

Filter	Colored Glass	Cathode	λ_{eff} (nm)	$\Delta\lambda_{\text{eff}}$ (nm)
U	Corning No. 9863	S4	366.0	70
	UG2 (2 mm)	S4		
	UG12 (2 mm) + BG18 (2 mm)	S20ext		
	UG2 (1 mm) + CuSO$_4$–Solution	S1		
B	Corning No. 5030 + GG13 (2 mm)	S4	440.0	97
	BG12 (1 mm) + GG13 (2 mm)	S4		
	BG12 (4 mm) + BG18 (2 mm) + GG4 (1 mm)	S20ext		
	BG12 (1 mm) + BG18 (1 mm) + GG385 (2 mm)	S1		
V	Corning No. 3384	S4	553.0	85
	GG11 (2 mm)	S4		
	GG14 (3 mm) + BG18 (2 mm)	S20ext		
	GG495 (2 mm) + BG18 (1 mm)	S1		
R	OG550 (2 mm) + RG6 (1 mm)	S20ext	700.0	21
	OG570 (2 mm) + KG3 (2 mm)	S1		
I	BG3 (1 mm) + RG610 (2 mm)	S20ext	880.0	22
	RGN9 (3 mm)	S1		

The transmission of the original *V* filter is limited toward longer wavelengths by the declining efficiency of the S-4 cathode. If one wishes also to observe red and infrared magnitudes with a single PMT, then the transparency at long wavelengths of the standard *UBV* filters must be reduced by additional color filters, as indicated in Table 8.2. For the *U* region a liquid filter, consisting of a copper sulfate solution, can be used. All filters can or could be supplied by the Schott Co., unless otherwise indicated. However, some are no longer available (e.g., UG2 must be replaced by UG1). Some have been given new designations (e.g., GG13 = GG385, GG11 = CG495). If in doubt, the Schott filter catalogue should be consulted, and equivalent wavelengths should be determined for the chosen filter combinations and the employed receiver sensitivity characteristics. These wavelengths should be close to the effective wavelengths given above. The equivalent wavelength is calculated by

$$\lambda_{\text{eq}} = \frac{\sum \lambda\, \theta(\lambda)}{\sum \theta(\lambda)}, \tag{8.9}$$

where $\theta(\lambda)$ is the transmission curve of the color glass combination.

Magnitudes and colors of selected stars in the *UBVRI* system are given in Appendix Table B.28 (in Vol. 3). Supplementary catalogues are:

1. A.A. Henden, R.H. Kaitchuck: *Astronomical Photometry*, Van Nostrand Reinhold Co., New York 1982. Table C1/C2, p. 292 contains *UBV* magnitudes of more than 150 stars, some bright, some faint.
2. B. Iriate, H.L. Johnson, R.I. Mitchell, W.K. Wisniewski: Five Color Photometry of Bright Stars: The Arizona-Tonantzintla Catalogue. Magnitudes and Colors of 1325 Bright Stars. *Sky and Telescope* **30**, 21 (1965). This catalogue contains *UBVRI* magnitudes of bright stars over the whole sky.

3. A.A. Hoag, H.L. Johnson, B. Iriate, R.I. Mitchell, K.L. Hallam, S., Sharpless: Photometry of Stars in Galactic Cluster Fields. *Publications of the US Naval Observatory* (Washington), Second Series, 17, Part 7 (1961). This catalogue contains UBV photometry of stars in 70 open clusters, and additional finding charts.
4. V.M. Blanco, S. Demers, G. G. Douglass, P.M. Fitzgerald: Photoelectric Catalogue, Magnitudes and Colours of Stars in the UBV and U_cBV Systems, *Publications of the US Naval Observatory (Washington)*, Second Series, 21 (1968). This catalogue contains UBV data of 20,705 stars.
5. B. Nicolet: Catalogue of Homogeneous Data in the UBV Photoelectric System. *Astronomy and Astrophysics Supplement Series* **34**, 1 (1978). This catalogue contains UBV data of 53 000 stars.

In most of the broadband systems in use today, the magnitude of Vega (α Lyr) is set to 0^m at all wavelengths. The color indices of Vega in these color systems are therefore also equal to 0. It should be noted that equal color indices of a star do not necessarily mean a constant (flat) energy distribution; it means only that the energy distribution is close to that of an unreddened A0 V star.

From the color indices $(U - B)$ and $(B - V)$ of a normal star, its temperature and spectral type can be determined, and—with restrictions—also its luminosity class. The interstellar reddening E_{B-V} changes the color indices of stars. If E_{B-V} is known, the *intrinisic* or *unreddened colors* $(U - B)_0$ and $(B - V)_0$ can be calculated:

$$(B - V)_0 = (B - V) - E_{B-V}, \tag{8.10}$$

$$(U - B)_0 = (U - B) - 0.72 \, E_{B-V}, \tag{8.11}$$

The interstellar absorption in the V region is

$$A_V = 3.2 \, E_{B-V}. \tag{8.12}$$

From the measured color indices a reddening-independent quantity Q can be calculated:

$$Q = (U - B) - 0.72 \, (B - V). \tag{8.13}$$

Table 8.3 gives the intrinsic colors $(U - B)_0$ and $(B - V)_0$ and the Q-values for stars of different spectral types and luminosity classes. Possible spectral types and luminosity classes can be derived from the Q-value of a star. Although not uniquely defined, the following plausibility criteria are of help: bright (and hence nearby) main-sequence stars (luminosity class V) are hardly reddened, while faint (and hence distant) supergiants (luminosity class I) are usually heavily reddened.

8.4 The Technique and Planning of Observations

In the planning of an observation or series of observations, one should first consider whether the accuracy (spectral resolution, time resolution, photometric accuracy) which can be achieved with a given photometer is sufficient for the tackling of the problem at hand. Second, the use of a special color system (i.e., the use of certain filters) should be investigated. One should not forget to observe suitable standard

Table 8.3. Intrinsic colors and Q-values of stars of different spectral types and luminosity classes. After Straizys [8.19].

Spectral type	Luminosity class	$(U-B)_0$	$(B-V)_0$	Q
O5	I	−1.15	−0.33	−0.91
	III	−1.15	−0.33	−0.91
	V	−1.15	−0.33	−0.91
B0	I	−1.09	−0.23	−0.92
	III	−1.09	−0.30	−0.87
	V	−1.08	−0.30	−0.86
B5	I	−0.77	−0.09	−0.70
	III	−0.59	−0.16	−0.48
	V	−0.57	−0.17	−0.45
A0	I	−0.25	0.01	−0.35
	III	−0.06	−0.02	−0.05
	V	−0.02	−0.02	−0.01
A5	I	−0.08	0.09	−0.14
	III	0.13	0.15	0.02
	V	0.09	0.15	−0.02
F0	I	0.16	0.20	0.02
	III	0.10	0.38	−0.11
	V	0.02	0.30	−0.21
F5	I	0.32	0.39	0.03
	III	0.09	0.43	−0.25
	V	−0.01	0.44	−0.36
G0	I	0.48	0.75	−0.18
	III	0.15	0.65	−0.39
	V	0.06	0.58	−0.42
G5	I	0.88	1.08	−0.16
	III	0.63	0.86	−0.14
	V	0.20	0.68	−0.39
K0	I	1.34	1.35	−0.07
	III	0.86	1.01	−0.10
	V	0.45	0.82	−0.29
K5	I	1.86	1.60	0.04
	III	1.80	1.51	0.17
	V	1.12	1.15	0.00
M0	I	1.93	1.69	−0.03
	III	1.88	1.56	0.16
	V	1.23	1.44	−0.20
M5	I	1.60	1.80	
	III	1.45	1.56	−0.14
	V	1.24	1.58	−0.49

stars or standard fields with known magnitudes and colors together with the "program stars." The recording of the precise time is eminently important when stellar occultations or minima and maxima of variable stars are to be observed. In any case, a time measurement is associated with each photometric measurement, and will later be used to derive corrections for the influence of the Earth's atmosphere.

Additional effects, such as differential extinction or the light-travel time effect (see Sect. 8.6), have to be considered as the accuracy of measurements is increased. The quality of atmospheric conditions plays an eminent role; it is for this reason that the large national and international observatories are located in some of the remotest corners of the Earth. Nevertheless, the observer who owns a small and/or portable telescope may be consoled to know that he possesses a major advantage in carrying out long-term programs or in following special events.

8.4.1 Point Photometry and Surface Photometry

Two different types of photometric observations can be distinguished, *point photometry* and *surface photometry*. When carrying out point photometry (a typical application being photoelectric stellar photometry), a diaphragm is used to isolate a "point" in the sky. The diaphragm can sometimes cover a noticeable sky area without, however, resolving it. When carrying out surface photometry, on the other hand, a region of sky is "scanned" in either one or two dimensions with a certain resolution. The resolution is dependent on the size of the measuring diaphragm, the size of the picture elements of the detector, or on the resolution of the telescope. The generally employed methods of point photometry include absolute photometry, differential photometry, high-speed photometry, photometry of occultations, and interferometry. In the following sections, those methods which are of particular importance in amateur astronomy will be discussed.

8.4.2 Visual Photometry: Differential Observations

The best known method for the visual determination of stellar magnitudes is Argelander's *step-estimate method*. The magnitude of a star (usually a variable) is determined by comparing it with other stars in the field whose magnitudes are known. In the following, the variable is designated with the letter a, the comparison star with b. The variable whose brightness is to be estimated should appear in the instrument as bright as a star of magnitude $0\overset{m}{.}5$ to $3\overset{m}{.}5$ (the Fechner region) with the naked eye. The required apertures of the instrument are given in Table 8.4.

The brightness difference of the two stars is given in "steps," which are defined according to Argelander (1799–1875) as follows:

0: "If both stars appear to be of equal brightness, or if I estimate sometimes the first, sometimes the second a little bit brighter, I call them of equal brightness and I note a 0 b."
1: "If, at first glance, both stars appear of equal brightness, but if careful observing (repeatedly shifting from a to b and back) shows a always, or with rare exceptions, to be just noticeably brighter, I call a one step brighter than b and note a 1 b; however, if b is the brighter one, I note b 1 a, so that always the brighter is in front and the fainter behind the number."

Principles of Photometry

Table 8.4. Necessary apertures for brightness estimates. The limiting magnitude for good estimates is about 1 magnitude brighter than the (intrinsic) limiting magnitude.

Magnitude range of the variable	Aperture of the instrument (in mm)
$0.^m5$ to $4.^m5$	7 mm (eye)
2.5 to 6.5	20 mm (opera glasses)
4.5 to 8.5	50 mm (2-inch or binoculars)
6.5 to 10.5	100 mm (4-inch)
8.5 to 12.5	250 mm (10-inch)

2: "If one star always and without doubt appears brighter than the other, this difference is taken as two steps and designated $a\ 2\ b$ if a is the brighter one, and $b\ 2\ a$, if b is the brighter one."

3,4: "A difference which is noted at first glance is taken as three steps and is noted as $a\ 3\ b$ or $b\ 3\ a$. Finally, $a\ 4\ b$ denotes an even more pronounced difference."

It is not advisable to estimate in more than 4 or 5 steps. One step corresponds roughly to one-tenth of a magnitude. At least two comparison stars should always be used, one of which (b) should be brighter, and the second (c) fainter than the variable (a). One then records the date[5], time[6], program object and comparison stars, and the relative step-brightnesses as, for example, "$b\ s_1\ a\ s_2\ c$," where s_1 and s_2 are step values. When observing variable stars, special finding charts with the identification of the variable and data on visual magnitudes of comparison stars are quite useful. Such a chart is shown in Fig. 8.7.

If the variable is "bracketed" between two comparison stars b, c of known brightness, and if m_1 is the magnitude of the brighter star b, m_2 that of the fainter comparison c, then the magnitude of the variable m_V can be calculated according to the following formula.

$$m_V = m_1 + \left(\frac{m_2 - m_1}{s_1 + s_2}\right) s_1, \qquad (8.14)$$

or, after a simple transformation,

$$m_V = m_2 + \left(\frac{m_2 - m_1}{s_1 + s_2}\right) s_2, \qquad (8.15)$$

The step value, whose reciprocal value indicates the slope in the equations, is $(s_1 + s_2)/(m_2 - m_1)$.

A more elaborate, numerical reduction for the use of four comparison stars is given by Jahn [8.20], and a graphical reduction for five comparison stars is given by

---

5 In order to eliminate later uncertainties, one may note the date in the following way: 1987 June 5/6 Sa/Su.

6 One should note the used time (e.g., mountain standard time (MST) eastern daylight saving time (EDT), or universal time (UT)). The use of universal time is recommended, because it is not subject to hourly shifts in the course of the year. A count of hours beyond 24 is often useful (1^h00 in the morning of the following day = 25^h00 of the previous day).

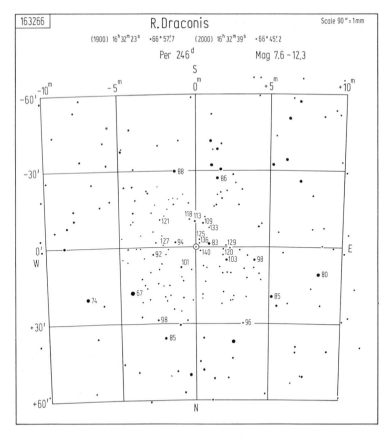

Fig. 8.7. AAVSO finding chart of the Mira variable R Dra. The field has a size of about $2° \times 2°$, and south is up. The variable is in the center of the chart. The magnitudes of the comparison stars are given to a tenth of a magnitude; decimal points are omitted to avoid confusion with stars. From [8.21], with kind permission of J.A. Mattei, director of the AAVSO.

Hoffmeister [8.23]. The light curve of the Mira star R Dra, derived from visual estimates of members of the American Association of Variable Star Observers (AAVSO), is shown in Fig. 8.8.

Collections of charts for the visual observation of variable stars are:

- Charts of the American Association of Variable Star Observers (AAVSO), 25 Birch Street, Cambridge, MA 02138, USA (charts for about 2000 variables with visual magnitude scales for step estimates).
- Charts of the Bundesdeutsche Arbeitsgemeinschaft für veränderliche Sterne e.V. (BAV), Munsterdamm 90, 12169 Berlin, Germany (charts for about 700 eclipsing binaries; additional charts for short- and long-period pulsating variables).
- Charts for Southern Variables, published by Astronomical Research Ltd., P.O. Box 3093, Greerton, Tauranga, New Zealand.

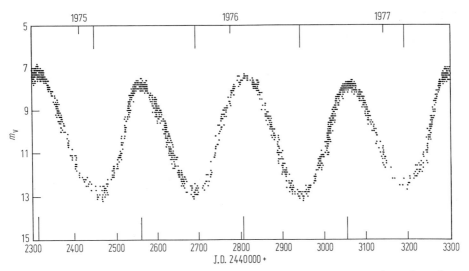

Fig. 8.8. Light curve of the Mira variable R Dra, derived from visual observations of members of the AAVSO. From [8.24], with kind permission of J.A. Mattei, director of the AAVSO.

A useful star atlas for these purposes is: E. Scovil: AAVSO Variable Star Atlas (Sky Publishing Corporation, Cambridge, Massachusetts, USA 1981). It contains all variable stars which are brighter than $9\overset{m}{.}5$ (visual) at maximum and whose amplitudes are larger than $0\overset{m}{.}5$.

A detailed description of techniques of visual observations can be found in Hübscher et al. [8.22].

8.4.3 Photoelectric Photometry: Differential Observations

The *method of differential observations* is the easiest way to obtain useful observations and results with a photoelectric photometer, especially when observing variable stars of relatively short period and small amplitude. The method is applicable to all types of photometers: A variable star and two stars in its vicinity are chosen. The two stars should satisfy the following criteria:

1. Their distance from the variable should be small enough that they can be seen together with the variable in the finder of the telescope. This makes it possible to point easily to the various objects. (If a telescope with computer-controlled positioning is available, this condition is somewhat relaxed, but the angular distances should not be so large as to minimize differential extinction effects by the Earth's atmosphere.)
2. They should have colors and magnitudes similar to those of the variable. A useful atlas for selecting comparison stars is one that also gives the spectra (or at least colors) of all stars, such as the *Atlas Borealis/Eclipticalis/Australis* by A. Becvar, issued by Sky Publishing Corporation, Cambridge, MA.

3. They should not be variable. They should not, for instance, be listed in the *Catalogue of Variable Stars*. However, even a star not listed there might be variable, often with quite a small amplitude. The constancy of any of the comparison stars can be checked using two stars whose brightness difference should be monitored. This yields information regarding not only their constancy, but also of the quality of the night, performance of the instrument, etc.

If the variable is designated with V, the two comparison stars with C1 and C2, and the sky alone with (S), the following measuring sequence can be carried out (using a *UBV* filter set for each step, if desired):

$$(S) - C1 - V - (S) - V - C1 - C2 - (S) - C2 - C1 - V - (S) - V - C1 - \cdots$$

Because of the symmetry in the sequence of observations, the following brightness differences can be calculated from the ratios of neighboring measurements:

$$\Delta m \ (V - C1) = -2.5 \log_{10} \left(\frac{V + V - 2S}{C1 + C1 - 2S} \right) \tag{8.16a}$$

and

$$\Delta m \ (C2 - C1) = -2.5 \log_{10} \left(\frac{C2 + C2 - 2S}{C1 + C1 - 2S} \right). \tag{8.16b}$$

The symmetrical arrangement of measurements has the advantage that *slow* atmospheric changes are of little consequence, since they influence the variable and the comparison star by the same amount. Two measurements of two stars yield one magnitude difference. If one calculates from each variable star measurement a magnitude difference (e.g., by linearly interpolating the count rate or deflection of the comparison star), the two magnitude differences would not be independent of each other. It is often better and more convenient to work with fewer measurements of higher precision than to deal with many measurements with larger scatter. There are, however, exceptions, such as the observation of rapidly varying objects (e.g., cataclysmic variables).

The results obtained by this method must be corrected for differential extinction of the first and second order (see Sect. 8.5.1) and for light-travel time (see Sect. 8.5.4).

If a two-channel photometer is used, the variable object will be observed in the first channel, and the comparison star in the second one. From time to time, the telescope is offset slightly to measure the sky brightness. Since the measurements are done simultaneously, a measuring sequence is not necessary, but the influence of differential extinction must still be taken into account.

A BASIC program for the reduction of differential photoelectric observations is given by Guinan et al. [8.18].

8.4.4 Absolute Photoelectric Photometry

The field of *absolute photoelectric photometry* is much less attractive in amateur circles because, in general, no useful results can be obtained. It is important—and difficult—to transform the magnitudes from the local (instrumental) color system to a standard color system. Stringent demands for the stability of sky transparency and the stability of the equipment are required. Since many stars and star clusters have already been observed down to quite faint magnitudes in conventional color systems, it is left to the amateur to check, for instance, the color-magnitude diagram of an open cluster [8.25].

8.4.5 Photometry of Occultations

When the changes of brightness occur with time scales of seconds to milliseconds and they are not periodic, the observer is forced to omit the comparison star measurements (unless he has a two-channel photometer at his disposal). Such rapid brightness changes occur when a body of the solar system occults a star or another member of the solar system. Instead of a comparison star, one has two reference magnitudes: the combined brightness of occulting and occulted object outside occultation, and the residual brightness of the occulting object during occultation. Possible occulting objects include the Moon, the planets, and planetoids; occulted objects are planets, planetoids, and fixed stars. The reader is also referred to Chap. 17 on stellar occultations by the Moon. The speed of the apparent motion of occulting objects and the diameters of the occulted objects determine the brevity of the observed phenomena. Typical data are:

Motion of the Moon:	$0\rlap{.}''5 \; s^{-1}$
Motion of planetoids:	$0.01 \times$ Moon motion
Motion of Neptune:	$0.001 \times$ Moon motion
Diameter of a planet:	$10''$
Diameter of a bright planetoid:	$0\rlap{.}''1$
Diameter of a fixed star:	$< 0\rlap{.}''05$ (Antares)

When a comet or a planet enshrouded by an atmosphere occults a fixed star, the structure of the light curve can be most intriguing, because it shows how different atmospheric layers are transparent to the light of the fixed star. Times of occultation are only roughly defined.

Structured light curves can also be observed when a star disappears or reappears behind an object without atmosphere, such as the Moon or a planetoid. These effects are caused by diffraction effects. For the Moon, the effects begin to occur when the object is 0.01 arcseconds from contact. 20 milliseconds elapse between the first and the second diffraction minimum if the occulation is a central one. The amplitude of brightness fluctuations is as large as $0\rlap{.}^m5$ (Fig. 8.9). The smaller apparent motion of a typical planetoid is compensated by its greater distance, so that the time scales of the diffraction effects are of the same order of magnitude.

Ten years ago, the observation of lunar occultations was still a domain of professional astronomy, because the use of photometers with high time resolution (less than 10 milliseconds) was not yet widespread. The situation has changed considerably in recent times. Chen [8.27] describes a portable photometer for the use of occulation observations.

If a circular planet with an atmosphere centrally occults a star, the so-called *central flash* of the occulted star can be measured at the instant the star is exactly behind the

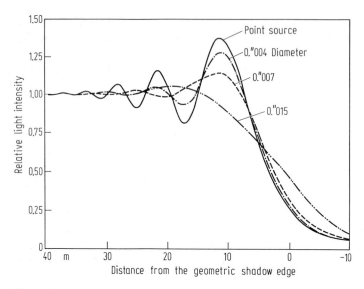

Fig. 8.9a. Theoretical diffraction patterns for lunar occultations of stars of different angular diameters. Higher amplitudes of the oscillations correspond to smaller angular diameters. From Evans [8.26].

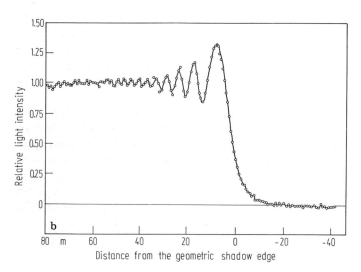

Fig. 8.9b. Photoelectric observation of the lunar occultation of β Sco. An analysis of the diffraction pattern yields that the star is a binary with a separation of $0.''0006$.

center of the planet. The flash is caused by light refraction at the planet's limb. An area photometer which can resolve the limb would register a bright ring along the planet's limb during the time of the flash.

In some cases, occultations by planetary objects are accompanied by unforeseen phenomena, such as occultations by unknown planetary satellites or rings. It is therefore advisable to start the observations well ahead of the pre-calculated "normal" occulation, and to continue them for quite some time after the event.

Some estimates of the obtainable accuracy should be given: Using a 20-cm telescope with a photometer having an integration time of 10 milliseconds, a statistical error of 5% per integration can be expected for stars of $6\overset{m}{.}5$ and occulting planetoids; for objects of $10\overset{m}{.}5$ and an integration time of 1 second, an error of 3% can be anticipated.

Occultations in daylight are easiest to observe visually, since the low contrast makes other observations quite insensitive. If one uses a photometer with a diaphragm diameter of $30''$, and observes with a sky brightness of 2^m per square arcsecond, the observed brightness jump of an occulted star of 0^m is only $0\overset{m}{.}01$, which is at the limit of detectability.

Within certain limits, one can also obtain useful results with visual observations. The aim is mainly the determination of the time of the start or end of the occultation. A tape recorder is often used for the recording of the time signals of a radio station and the acoustical signal which is produced by the observer with the aid of a horn or bell at the moment of the event. Special time signal receivers are commercially available. Often a radio with shortwave band is sufficient to receive the time signal stations at 5, 10, 15, or 20 MHz.[7] The results should be corrected for the human reaction time. It is dependent on the time of the day, the contrast circumstances, and whether one observes the start or the end of the occultation. Some prior experience is usually needed to determine this correction.

More precise measurements can be carried out by photoelectric photometry. For example, the photometer signal can be recorded by a strip-chart recorder. The time signals can be mixed into it through a second input, or recorded by a second pen. Data can also be digitally recorded by means of a microcomputer; arrangements with mixed analog and digital signals are also possible. For example, the photometer signal is converted into an analog signal by a voltage–frequency converter and recorded together with the analog signal of a time station using a stereo cassette recorder. The recording can later be analyzed with the aid of an analog–digital converter.

8.4.6 Visual, Photographic, and Electronic Surface Photometry

Often a project demands a specific angular resolution and a certain angular coverage. Whether or not a given photometer meets the requirements is a question of the properties of the telescope as well as of the detector. In many cases, a large sky area can be covered by a mosaic of single observations. However, the temporal and spatial variation of the photometric conditions can pose problems. The resolving power can

---

7 See also Chap. 4.7.9.

also depend on wavelength. The possibilities and problems are complex, and so only two examples will be discussed here:

1. The tail of a bright comet should be measured in its totality. Comet West, for example, had a maximum area of 500 to 1000 square degrees. A possible solution, if one has a simple photoelectric photometer at one's disposal, is the following: the size of the diaphragm is increased, if possible, by observing a demagnified image of the sky using a secondary image produced by a collimator. The area brightness is homogenized by defocusing in order to avoid selection effects due to background stars. Single points or strips are measured at equal distances in the tail region. Extinction and sky background are measured in a comet-free area and by a later scan of the observed area when the comet has disappeared, with the same sky and focus settings as before.
2. Magnitudes of the components of double stars should be measured by "scanning." The resolution is improved by measuring at shorter wavelengths (if possible), and by magnifying secondary imaging. One can also use a slit-shaped diaphragm or a blade which can be moved directly inside the photometer, or by moving the telescope with the slow motion control across the object [8.28,29,30].

8.5 Reduction Techniques

8.5.1 Reduction of Photometric Measurements—Generalities

None of the values which are directly read from a photoelectric photometer, from an iris diaphragm photometer, or estimated by visual means are true "stellar magnitudes." Each recorded measurement of brightness (and its corresponding time; see Sect. 8.5.4) must be free from falsifying local influences before a true stellar magnitude can be obtained.

One must especially take into account the influence of the Earth's atmosphere, as was discussed in Sect. 8.2., and the behavior of the measuring device. Starlight is scattered and absorbed in the atmosphere to a degree which increases when the path in the atmosphere is longer (the astronomer uses the concept of *air mass*), and which also increases toward shorter wavelengths.

The observer thus measures different magnitudes of the same star, depending on how high it is in the sky. One can determine from a series of observations, taken at different elevations (or zenith distances), how bright the star *would* appear without the falsifying influence of the Earth's atmosphere. The following formulation is used:

$$m_0 = m - (k' + k''c)X, \qquad (8.17)$$

where m is the measured magnitude of the star, m_0 the magnitude of the star outside the atmosphere, k' the extinction coefficient of first order (Fig. 8.10), k'' the extinction coefficient of second order, c the color index of the star, and X the air mass. At the zenith, the air mass is $X = 1.0$, while outside of the Earth's atmosphere it is $X = 0.0$.

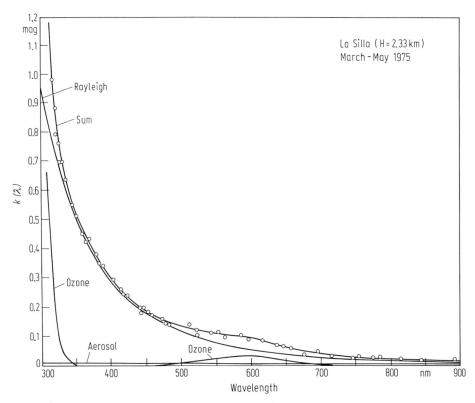

Fig. 8.10. Coefficient k of the atmospheric extinction as a function of wavelength (in nm) for the site La Silla, Chile, 2330 m above sea level. Note that the extinction is composed of different components: small, mutually absorbing dust particles (aerosols), broad molecular absorption bands (predominantly of ozone), and air molecules which cause Rayleigh scatter (in proportion to λ^{-4}). The extinction increases towards small wavelengths; the use of broadband filters yields mean values of k. From Tüg et al. [8.1].

A good approximation is:

$$X = \sec z, \tag{8.18}$$

where z is the zenith distance of the star and the trigonometric function sec (secant) is the reciprocal of the cosine. From the equation for the zenith distance follows the relation

$$X = \frac{1}{\sin \varphi \sin \delta + \cos \varphi \cos \delta \cos h}, \tag{8.19}$$

where φ is the geographic latitude of the observer's station, δ is the declination of the star, and h is the hour angle of the star (h = sidereal time minus right ascension of the star). It is apparent that the air mass of the star is smallest when $h = 0$, that is, when the sidereal time is equal to the right ascension. The star is then on the meridian. Observations of "extinction stars" far away from the meridian are also necessary

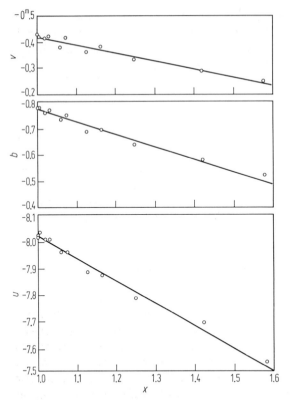

Fig. 8.11. Extinction (in magnitudes) for the spectral bands U, B, and V as a function of air mass X. From Henden and Kaitchuck [8.8].

in order to properly determine the extinction coefficients, while program stars (e.g., variables) should be observed close to meridian transit, where measurements of highest photometric accuracy are attainable.

To determine true star colors and magnitudes, the observer must measure a multitude of extinction stars of various colors under good atmospheric conditions (good, stable transparency of the atmosphere—no haze, no cirrus, no clouds!). Appendix Table B.25 gives a selection of several bright stars of different colors distributed evenly over the whole sky. One obtains a system of equations of the type (8.17) for the single stars, and obtains graphically (Fig. 8.11) or by the method of least squares (see Chap. 3) the extraterrestrial magnitudes m_0 in the instrumental system. If a series of stars of known color and magnitude is measured in a color system, then transformation equations of the type

$$m_{\rm st} = m_0 + \beta c + \gamma \tag{8.20}$$

can be established which permit the transformation of the instrumental magnitude m_0 into into a standard magnitude $m_{\rm st}$. Here, c is the color index of the star in the standard system (since this is not known in the beginning, it has to be assumed and determined by iteration), and β and γ are, respectively, the color coefficient and the zero-point constant of the instrumental system. Transformation programs for *UBV* photometry in BASIC are given in the books of Henden and Kaitchuck [8.8, p. 339], and Ghedini [8.31, p. 20]. The correction for differential extinction between the variable V and comparison star C can be calculated for each hour angle of the variable by determining for both stars the air masses from Eq. (8.19), the difference ΔX, and, according to Eq. (8.17), the correction

$$\Delta m = (k' + k'' \Delta c) \Delta X. \tag{8.21}$$

The second term can often be neglected. In some cases, however, this term is not only important, but also a function of time, namely when a variable changes its color noticeably (e.g., during the deep minimum of an Algol system). This correction must be applied to all magnitude differences, which are described in Sect. 8.4.3.

8.5.2 Reduction of Photographic Observations—Generalities

If the observer is interested in the determination of stellar magnitudes, then he or she must, as in the case of photoelectric photometry, observe several *standard stars* according to a particular "schedule." Similarly, in the case of photographic photometry, one must observe a *magnitude sequence*, i.e., a group of stars of known magnitudes. Ideally, the photograph of the field of interest includes such a sequence, e.g., a well-observed open cluster or a Selected Area [8.32]. In general this will not be the case, and one must then, for the transposition of a magnitude scale, take two exposures which fulfill the following criteria:

- Use the same photographic emulsion (cut from the same plate or the same film, or from the same plate batch).
- Use the same instrument.
- Use the same exposure time, expose both the field of interest and the field with the standard sequence in close succession.
- Develop both photographs under similar conditions.
- Take both fields with similar air masses (the most convenient time of observation for the field and the standard field can easily be calculated when preparations are being made for the observations).
- Sky background should be the same (no moonlight, no clouds).

Methods of reduction for photographic photometry are sorted below according to increasing accuracy:

1. Visual estimates of the stellar images on the photographic negative, performed with the Argelander method of visual photometry;
2. Determination of diameters of stellar images on a high-contrast, highly magnified copy (or of equidensitometry rings on a copy made with contour film); when preparing such a copy, one must take care to ensure for a homogeneous illumination of the field.

3. Measurement of the plate on an iris diaphragm photometer (design, e.g., after W. Becker [8.33]) or on a microdensitometer. Such measuring machines are, however, available only in some astronomical institutes. Some amateurs have attempted to build a *film scanner*, which consists of a step-motor-driven, rotating drum holding the film, and a photocell, driven in the same way along an axis which measures the brightness of a photograph in various positions [8.34]. Another device for the scanning of one-dimensional images (e.g., spectra) in combination with a C64 computer is described in [8.35].

Some studies do not require an exact magnitude scale. Only a relative scale in steps is needed for the derivation of variable star light curves and periods. This scale is defined by several stars in the vicinity of the variable, and is estimated by means of the Argelander method (Sect. 8.4.2) on a series of plates. A photograph of a field with known stellar magnitudes and taken with the same instrument is usually sufficient for a coarse determination of steps.

8.5.3 Reduction of Digital Images—Generalities[8]

Astronomical images can be reduced effectively only if they are available in digital form. The method of digitization is dependent on the type of detector employed. When using a CCD, the video signal must be read out by means of a digitizer. If an image on photographic paper or film is available, the image formation must be digitized by means of a *scanner*. Such a device for amateur astronomers is described by Peelen [8.34].

A digitized image consists of an arrangement of discrete points, called picture elements or—abbreviated—*pixels*. Each pixel corresponds to a numerical value, and thus a digitized image can be regarded as a matrix of X or Y numbers. Such an image can be processed on a home or personal computer. The storage capacity of the computer limits the processable image size.

Each of the goals of image processing can be reached by using reliable computer programs. While it is impossible within the framework of this chapter to list the programs in detail, the following descriptions should serve as useful illustrations:

1. *Background subtraction.* A constant value is subtracted from all values, so that the sky background is, on the average, equal to zero. If the image was obtained with a CCD, then the logarithm of the sum of the numbers for a star, reduced to zero sky background, is directly proportional to stellar magnitude. When a photograph is scanned, the "density sums" of stars of known magnitude must be used to define a calibration curve.
2. *Star positions.* "Marginal sums" of a star in X and Y are computed; that is, the data are summed in columns and rows in a small square of numbers which contains the brightness data of the star. These X- and Y-histograms of summed data are approximated by Gaussian fits. The central coordinates of the Gaussian fits yield the X- and Y-coordinates of the star considered, which then have to be converted into sky coordinates by means of transformation equations (see Chap. 3.4).

---

[8] V. Gericke and M. Nolte collaborated with the authors on this section.

3. *Filtering*. Each image contains *noise*, i.e., statistically distributed disturbances which are recorded along with the signal of the astronomical object of interest. The causes of such disturbances are manifold. Fluctuations of the seeing during observation and of the power supply of the readout electronics are two of many possible sources. These disturbances can be filtered out by mathematical methods. Thus one does not process the numerical values of the original $X \times Y$ matrix, but instead first assigns to each pixel a new numerical value. These filtered values are obtained by assigning to each pixel a mean value, calculated from the original value and the values of the surrounding pixels. As an illustration, consider a one-dimensional image with 1×10 pixels.

Pixel No.: 1 2 3 4 5 6 7 8 9 10
Value: 2 3 5 6 9 6 9 4 2 1

This image is now filtered by assigning to each pixel the arithmetical mean of the pixel value and the values of both neighboring pixels. We at once encounter a difficulty, which occurs at each filtering: image points at the edges cannot be filtered. For our example, the filtered image looks like this (rounded to integer numbers):

Pixel No.: 1 2 3 4 5 6 7 8 9 10
Value: 3 5 7 7 8 6 5 2

One clearly notices that the image is smoothed by this process. The values rise monotonously to pixel no. 6 and then decline in the same fashion. Of course, filtering is not limited to this single and simple technique, but can be done in very different ways. More neighboring pixels can be "weighted" by assigning higher weight to closer pixels, and lower weight to more distant ones. Many different functions can be chosen as the weighting function, for example, a step function or a Gaussian curve. The choice of filtering depends on the problem of image processing. There is even the danger of information loss by too strong a filtering.

4. *Isodensity plots*. All receivers and display systems have a specified dynamic range. If the dynamic range of the screen is insufficient for a clear display of the image, an isophote display can be used to advantage, as described below:

If the pixel values are 8-bit numbers, there are, in total, 256 possible ($= 2^8$) gray steps. If, however, the screen can display only 16 different gray steps, much detail is lost. It is possible, for the first 16 gray steps of the original image, to assign directly those of the screen, and to assign, however, for the subsequent step of the image the value 0 of the screen, and to continue in this manner. Such an image with a "sawtooth" contour resembles a photographic equidensity picture (cf. Fig. 8.12).

A description of the possibilities of image processing with home computers is given by Peelen [8.34].

8.5.4 The Reduction of Time: the Heliocentric Correction

Upon consulting a clock, the time of observation is written down, or, if the data are recorded with the aid of a computer, they are stored concurrently with each datum

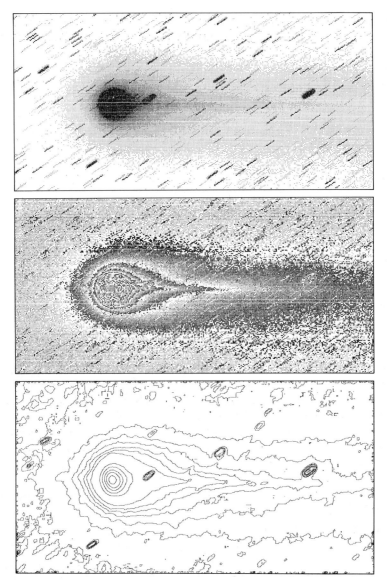

Fig. 8.12. Computer-processed images of Comet Halley. A photographic plate taken with the 0.40-m GPO astrograph of the European Southern Observatory was digitized with the PDS microdensitometer of the Astronomical Institute Münster. From top to bottom: display with 8 density steps, with "sawtooth" contours, and with lines of equal intensity.

of measurement. Any deviation of the time of the local clock is determined before and after the observation, and the correction is applied after the observation during the reduction stage. If, for example, the observer intends to determine the time of minimum light of an eclipsing binary, then he must take care to register precisely the times of each brightness measurement. The resulting time of minimum light (or of any other time) is, in the final data set, not given in universal time: because of the motion of the Earth around the Sun, events in deep space are registered on Earth sometimes earlier, sometimes later, depending on which side of the Sun the Earth is found. Indeed, in 1676 O. Roemer used observations of eclipses of Jupiter's moons at different positions of the Earth in its orbit to determine the velocity of light. In order to compensate for effects of the finite velocity of light, the time of observation should be transformed to the time of observation of a fictitious observer placed at the center of the Sun. This is called *heliocentric time*.

When observing a variable star for an extended period of time, the time is indicated, instead of by year, month, day and time of the day in the Gregorian calender, by the so-called *Julian date* (JD). This period started as day number 1 at 12 o'clock universal time on January 1 in the year -4712 (i.e., 4713 B.C.). The time of the day is given in fractions of a day. Table 8.7 is useful when one wishes to transform the date into the Julian day number.

Thus, the reduction of times into *heliocentric Julian date* (JD hel.) may be performed using one of the following methods:

1. Using tables (e.g., R. Prager: Tafeln der Lichtgleichung. *Kleine Veröffentlichungen der Univ.-Sternw. Berlin-Babelsberg* **12** (1932); A.U. Landolt and K.L. Blondeau: Tables of the Heliocentric Correction. *Publications of the Astronomical Society of the Pacific* **84**, 784 (1972)).
2. Using the scheme listed below, which is easily programmable (L.E. Doggett, G.H. Kaplan, P.K. Seidelmann: *Almanac for Computers for the Year 1978*, Nautical Almanac Office, Washington, D.C. (1978); A.A. Henden and R.M. Kaitchuck [8.8]).

The Julian date of the day of observation can be found in Appendix Table B.9 in Vol. 3.

(a) First calculate the universal time (UT). For observers in Great Britain, through which passes the Prime Meridian, the formula is quite trivially

$$UT = WET \text{ (Western European Time)},$$

or, during the summer months when WEST (Western European Summer Time) is in effect,

$$UT = WEST - 1^h$$

In North America, several times zones exist (see Sect. 2.5.4). The corresponding standard times include Atlantic (AST), Eastern (EST), Central (CST), Mountain

(MST), Pacific (PST) and Yukon (YKT). Thus,

$$\begin{aligned} \text{UT} &= \text{AST} + 4^h \\ &= \text{EST} + 5^h \\ &= \text{CST} + 6^h \\ &= \text{MST} + 7^h \\ &= \text{PST} + 8^h \\ &= \text{YST} + 9^h \end{aligned}$$

In Alaska and Hawaii, UT is obtained by adding 10^h to the local standard time. During the months when daylight-saving time is in effect, 1^h should be *subtracted* from the values given above (e.g., for the eastern U.S., UT=EDT+4^h).

(b) Calculate the time of the day as a fraction of the day:

$$\text{fraction of day} = \text{UT (in hours and fractions of hours)} \div 24$$

(c) Calculate the Julian date JD. Note that the Julian day starts at noon. Therefore, subtract 0.5 from the fraction of the day obtained above.

To find the heliocentric Julian date, first calculate the relative Julian century from

$$T = (\text{JD} - 2415020)/36525. \tag{8.22}$$

The mean longitude of the Sun is calculated from

$$L = 279.696678 + 36000.76892T + 0.000303T^2 - p, \tag{8.23}$$

where p is given by

$$p = [1.396041 + 0.000308(T + 0.5)](T - 0.499998). \tag{8.24}$$

The value of p is the precession from 1950 to the day of the observation; it is subtracted from the actual length in order to get the length for 1950.0.

The mean anomaly of the Sun is

$$G = 358.475833 + 35999.04975T - 0.00015T^2. \tag{8.25}$$

The values X and Y (for the year 1950) are calculated by the series

$$\begin{aligned} X = {} & 0.99986 \cos L - 0.025127 \cos(G - L) + 0.008374 \cos(G + L) \\ & + 0.000105 \cos(2G + L) + 0.000063T \cos(G - L) \\ & + 0.000035 \cos(2G - L), \end{aligned} \tag{8.26}$$

$$\begin{aligned} Y = {} & 0.917308 \sin L + 0.023053 \sin(G - L) + 0.007683 \sin(G + L) \\ & - 0.000057T \sin(G - L) - 0.000032 \sin(2G - L). \end{aligned} \tag{8.27}$$

Then, with α and δ denoting, respectively, the right ascension and declination of the object,

$$\Delta t = -0.0057755 \left[(\cos \delta \cos \alpha) X + (\tan \varepsilon \sin \delta + \cos \delta \sin \alpha) Y \right], \tag{8.28}$$

where Δt is in days and ε is the obliquity of the ecliptic, its time dependency being given by

$$\varepsilon = 23°\!.4523 - 0°\!.0130125T. \tag{8.29}$$

Therefore, the heliocentric Julian date is

$$\text{JD(hel)} = \text{JD} + \Delta t. \tag{8.30}$$

8.5.5 Determination of Minimum Light Times and of Periods

The establishment of the times of minimum (or maximum) light of variable stars serves for the exact determination of their periods, and for the determination of possible period changes, which are important for our understanding of the evolution of variables (eclipsing binaries as well as pulsating variables). The same formalism can be used for the determination of times of minima and maxima. Organizations of amateur astronomers, such as the BAV (German Study Group of Variables), the BBSAG (Observers of Eclipsing Binaries of the Swiss Astronomical Society), and the AAVSO, are very active in this field.

The visual, photographic, or photoelectric observations yield for the minimum or maximum of the light curve N pairs of values (t_i, m_i), $i = 1, 2, \ldots, N$, representing the heliocentric times and magnitudes (or magnitude differences with respect to a comparison star). All discussed methods are also applicable in the case of step estimates of magnitudes. The interval between two adjacent observations can lie in the range of minutes for short-period eclipsing binaries or pulsating RR Lyrae stars, but it lies in the range of several days for long-period Mira stars. The series of observations can be analyzed by various methods, which are discussed below and illustrated in Fig. 8.13.

8.5.5.1 The Curve-Intersecting Line (Pogson's Method). Using graph paper, the magnitudes (Y-axis) are plotted versus time (X-axis). Subsequent points are connected by straight lines. The descending and ascending branches of the light curve are connected by several straight lines (their number should not exceed $N/2$). The central point of each of these lines is determined. If all X-values of the central points scatter around a mean value, this mean value is assumed to be the time of minimum light. Its standard deviation provides a reliable gauge of the accuracy of the minimum time determination. If the points show a systematic trend to one side, the light curve is asymmetric. Here, the intersecting point of a line, which is drawn "by eye" through the central points or determined by the method of least squares (see Chap. 3), with the light curve is assumed to be the time of minimum light.

A BASIC program of this method is given by Ghedini [8.31, p. 64].

8.5.5.2 The Kwee–van–Woerden–Method. This method, which was often employed in earlier times, has fallen out of favor because it can be applied only to points which are equally spaced in time (which often have to be obtained by means of interpolation of the actual data), and because it is not easily programmable. The reader interested in this method should consult the original paper [8.36].

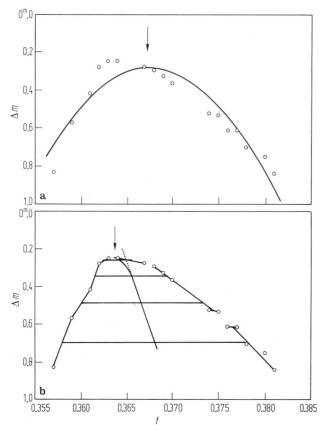

Fig. 8.13. Determining the time of maximum light by (**a**) curve bisection and (**b**) approximation by a polynomial. The abscissa t is in fractions of a day.

8.5.5.3 The Polynomial Fit. A method relatively free of mathematical problems is the "best fit" method, where the measuring points are approximated by a polynomial of nth degree, i.e., a function of the form $y = \sum a_i x^i$, where $i = 0, 1, \ldots, n$. The x-value of the minimum of the approximated curve is assumed to be the minimum time. If the segments of the curve are symmetrical with respect of the minimum time, it is sufficient to approximate it by a parabola ($n = 2$). Asymmetries are taken into account by uneven terms, i.e., the cubes and higher powers.

A BASIC program, which automatically determines the "best" power of the polynomial, and carries out the approximation is given by Ghedini [8.31, p. 53].

8.5.5.4 The Polygon Fit. This method, which is advantageous if the data points on the two branches of the light curve are unevely distributed, is discussed by Ghedini [8.31, p. 59].

8.5.5.5 Period Determination—Generalities. If a minimum time was determined, a comparison with other minima of the same object is warranted. The minima of an eclipsing binary should occur in a very regular way, assuming no physical processes occur which influence the length of a period (mass flow from one component of the system to the other, mass loss from the system, gravitational influence of a third body, etc.) Periodically occurring events are also found in other measurements, where, however, they may vanish after some time (from a few minutes to several years).

As examples, consider the light changes of asteroids caused by their rotation or the "flickering" phenomenon in cataclysmic variables. If the period is constant and the time of the extreme values of the brightness fluctuation is measured with an error which is small compared with the length of period, the period is given as the greatest common measure of the interval lengths, which results from the combinations of all available minimum (or maximum) times. The period can be determined by trial and error or by the method of least squares. For a single interval, the period is the quotient of the interval length and the number of periods lying in this interval (= difference of epochs). An algorithm, for instance, that of the greatest common measure, is not suitable because of the inherent errors of the minimum times, or because of a possible slight period variation. One should beware of apparent periodicities, especially when the true period and the length of a day, or the true period and the repetition period of the measurement, show a simple ratio. Differences also occur if there are "beat periods," caused by the superposition of several periods (e.g. in pulsating variables). It is sometimes helpful in the period-finding process if one utilizes in the analysis not only the extrema but also some other characteristic parts of the light curve. The accuracy of the period determination is given by the function

$$\text{Total duration} = f \text{ (number of periods } E\text{)}, \tag{8.31}$$

whose slope is equal to the period. It is also dependent on the number of data points (observed minima), on the total length of the interval, and on the accuracy of the individual data points.

If there exist large time intervals during which no observations were made, the connection of different series of observations will pose problems, because the number of periods elapsed in the gap is unknown. Systematic (e.g., computer-aided) tests are suggested. If the deviations, the so-called (O−C) values (see below), of the individual data points for a given period have a periodic structure, the period determination is likely to be erroneous. A critical study of several "best" solutions seems advisable. More details and references are found in Fullerton [8.37].

Ghedini [8.31, p. 95] gives two programs for period analysis. The first one is based on the Lafler–Kinman method [8.38], and chooses the best arrangement of observations; i.e., the correct period is the one for which the polygon through the data points—arranged according to phase—is of minimum length. The second one is based on Fourier analysis.

8.5.5.6 The Period Behavior. When the elements of the light changes of a variable star, the period P and the zero epoch E_0, are known, the times (in JD(hel)) of the occurring minima (for eclipsing binaries) or maxima (for pulsating stars) can be calculated for all epochs of the past and the future:

$$E_n = E_0 + nP, \qquad (8.32)$$

where E_0 and P can be taken from the *General Catalogue of Variable Stars* [8.39].

Is it rewarding to observe and determine minimum or maximum times also in such cases? Most certainly: because all "celestial clocks" show irregularities, their minima and maxima can occur earlier or later than anticipated. When planning such observations, one should, however, always consider whether the accuracy obtained is sufficient for a study of possible period changes.

To investigate the accuracy of the "clock rate" of a variable, the (O–C) diagram is used: Compare the *o*bserved minimum time with the *c*omputed one, according to Eq. (8.32). If several minimum times have been observed (or collected from the literature), then an (O–C) diagram can be constructed to display the time differences as a function of time (JD should be used) or as a function of epoch E. The course of the differences reveals something of the nature of the elements of the variable:

1. If the sequence of points runs through the origin and shows a rising (or declining) tendency, the zero epoch is correct. The period length is, however, too short (or too long).
2. If the sequence of points does not run through the origin, showing, however, a horizontal course, then the zero epoch must be corrected, the period being correct.
3. If the sequence of points shows a course that cannot be approximated by a straight line, the elements of the light variation of the star are themselves variable, as follows:
   - If the points lie approximately on a parabola, the period changes permanently by the same amount, this change can be taken into account by the addition of a quadratic term aP^2 in Eq. (8.32).
   - If the points can be approximated by a polygon, the period shows sudden changes; often, lengthenings and shortenings of the period follow in an irregular sequence.
   - If the sequence of points shows a step-like appearance, the star has a constant period, but exhibits phase jumps.

As an example, Fig. 8.14 shows an (O–C) diagram of the eclipsing binary RZ Cas. This system shows sudden period changes, as is clearly seen in the (O–C) values derived from photoelectrically determined minimum times. The minimum times which were derived from visual observations do not reveal this behavior because of their low accuracy.

A detailed discussion of the (O–C) diagram is given by Willson [8.41].

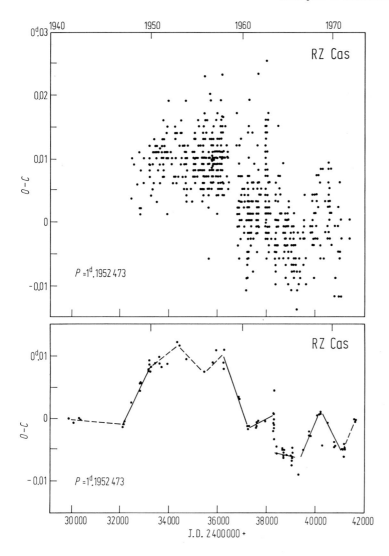

Fig. 8.14. (O−C) diagram of the Algol system RZ Cas for the years 1940–1973. The upper diagram shows the results of visual minimum times, the lower one those based on photoelectric observations. The period is unstable and shows "jumps," leading to lengthenings and shortenings of the period. From Herczeg and Frieboes-Condé [8.40].

8.6 Photometry of Different Astronomical Objects—Generalities

8.6.1 Photometry of Solar System Objects

The wide variety of objects in the solar system demands the use of a correspondingly extensive collection of methods by which to obtain photometric data. Except for the fluorescence of cometary matter and the radiation of meteorities entering the Earth's atmosphere, it is always reflected sunlight which is measured. This radiation carries information regarding the reflection properties of the surfaces.

In this context, the following geometrical quantities are quite important: the orientation of the pole of the object, as given by the coordinates of the planetary north pole, and the planetocentric coordinates of the subsolar and the subterrestrial points. The coordinates which are defined by these three points and the angles between them determine, together with the relevant distances (Sun–object and object–Earth), the geometry of the process of reflection and scattering. The geometry is known for most of the large bodies of the solar system, and it is described by the position angle of the rotational axis of the planet, the phase of rotation, the phase of illumination, the incident angle of radiation, and the aspect.

For each observation, we can write down an equation which contains known parameters (the geometrical parameters and the measured value) and unknown ones (the physical properties of reflection such as the shape of the object (in the case of a planetoid), the albedo, and the scattering property (the roughness of the surface at all scales)). If, as in the case of planetoids and small, distant planetary satellites, the geometry of reflection is not completely known, additional unknowns, such as the phase of rotation and the position of the polar axis (possibly influenced by precession), enter the equations. These parameters must be determined by a manifold of independent observations. One should try to determine beforehand during the planning stage of observations—and not in the reduction stage—in what sense the planned observations are to cover "new aspects" and which observing times are most informative.

If seasonal or long-term changes of the reflectional properties occur (especially on planets and moons with atmospheres, by sand storms on Mars, or by vulcanism on Io, etc.), a morphological study becomes difficult or even impossible. A series of photometric observations can then be used for meteorological investigations. The most important causes of brightness variations of solar system objects are listed below:

1. Effects of varying distance.
2. Rotational modulation because of bright–dark surface features and/or irregular shape of the object.
3. Phase effects.
4. "Seasons" and meteorology.

It is most important to take into account the influence of the varying distance on the measured brightness. In order to be compared with one another, the magnitudes are converted into those that would be observed at a *mean heliocentric opposition distance*

r_0. This distance is, for Jupiter and its satellites, 5.208 AU, for Saturn 10.529 AU, for Uranus 19.191 AU, and for Neptune 30.071 AU. For the calculation of this correction, the geocentric distance d and the heliocentric distance r for the time of observation are needed. They are tabulated in the annual *Astronomical Almanac*. The value of d is found in Section E under "True Geocentric Distance," $r(t)$ under "Radius Vector." The distance correction of magnitude Δm_d is calculated from

$$\Delta m_d = 5(\log r_0 + \log(r_0 - 1.0) - \log d - \log r), \tag{8.33}$$

where all distances are given in AU.

Phase effects can be noted in the case of the Earth's moon when comparing the total magnitude at full and new moon. However, subtle effects can be detected when comparing the different surface magnitudes of an illuminated region at full moon and at first or last quarter.

Magnitude corrections for the phase angle of the four large moons of Jupiter, of Saturn's moons Titan and Rhea, and of the planets Uranus and Neptune are given by Lockwood [8.42].

If the Earth's moon did not rotate in a bound, synchronous state, we would note a distinct rotational light variation between the "near" side, which is to a large extent constituted by dark mare regions, and the "far" side, which is predominantly bright. If, for example, the Mare Imbrium were the only dark region on the lunar surface, then the total light of the Moon's side that contains this region in a central position would appear fainter by $0\overset{m}{.}02$ than the side without it. Because of similar causes, the amplitudes of rotational light variations of Mercury, Mars, and the large, regularly shaped minor planets (see Appendix Table B.19) are of the same order of magnitude. The effects are often much more pronounced for Jupiter's moons Europa, Ganymede and Callisto and for Saturn's moon Iapetus because of the large differences of the surface brightness of ice and rock (Fig. 8.15).

8.6.1.1 Photometry of Minor Planets. The largest planetoid, Ceres (1), has a diameter of 940 km and its distance from the Earth at opposition is 1.77 AU. It then appears at an angular diameter of just $0\overset{''}{.}7$. In general, planetoids are observed as pointlike objects, and thus photometric measurements can be performed in a manner similar to that for of fixed stars—aside from the fact that the former change their position in the sky and that they have a different scintillation behavior. The simple empirical fact that "planets do not twinkle" reflects that the statistical fluctuations of scintillation are averaged out across the extended planetary disks. Effects of this sort can be noted even in the case of the major planetoids. The amplitude of scintillation of Ceres (Fig. 8.16) amounts to one-third that of a pointlike object.

Because of the continuously changing distance of a planetoid from the Earth and Sun, similar corrections, as discussed in the previous section, must be applied to the photometric data. The following parameters are needed, the first of which can be taken from *The Astronomical Almanac*, the others from the ephemeris for minor planets (*Efemeridy Malykh Planyet*):

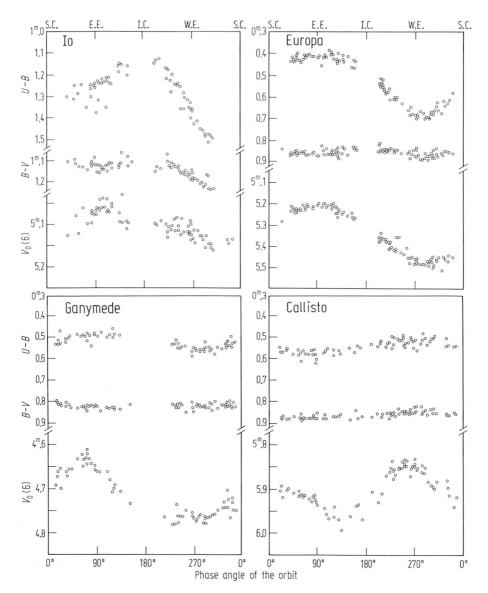

Fig. 8.15. V-magnitudes and $(U-B)$ and $(B-V)$ color indices of Jupiter's four bright satellites, corrected to mean opposition distance and phase angle $\alpha_p = 6°$, given as a function of the phase angle of their orbit. The repeatability of the light curve indicates a bound rotation of the satellites. The measurements were made with different telescopes of the Lowell Observatory (apertures of 0.53–1.07 m). Note the large scatter and the red color of the satellite Io. From Millis and Thompson [8.43].

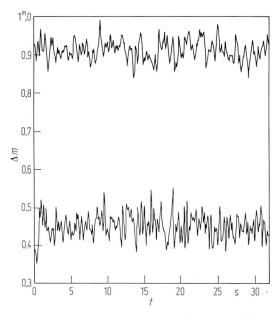

Fig. 8.16. Amplitude of scintillation for the asteroid Ceres and for a fixed star of equal brightness. The amplitude of the asteroid is only one-third that of the pointlike fixed star.

R = distance between Earth and Sun (in AU),
d = distance between Earth and planetoid (in AU),
r = distance between Sun and planetoid (in AU),
r_0 = semimajor axis of the planetoid's orbit (in AU).

The light-travel time must be taken into account. Light travels the Sun–Earth distance of 1 AU in 8.317 minutes and the planetoid–Earth distance in $8.317d/R$ minutes. In order to get the time at which the planetoid had the observed brightness, the value $8.317d/R$ minutes must be subtracted from the observed time.

The magnitude must also be "normalized" to one at unit distance. The observed magnitude of a planetoid is dependent on d and r. In order to obtain comparable magnitudes of a planetoid at different times, or magnitudes among planetoids, the observed magnitude m_{obs} is transformed into one for which the planetoid has distances $d = 1$ and $r = 1$, both in AU, from the Earth and Sun. This *reduced magnitude* m_r is given by:

$$m_r = m_{\text{obs}} - 5(\log r + \log d). \tag{8.34}$$

Sometimes the observed magnitudes m_{obs} are reduced to the mean opposition distance $(r_0 - 1)$. The magnitude m_0 is calculated from

$$m_0 = m_{\text{obs}} - 5(\log r + \log d) - \log[r_0(r_0 - 1)]. \tag{8.35}$$

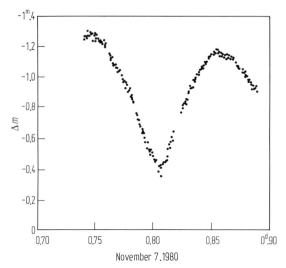

Fig. 8.17. B light curve of the asteroid (216) Kleopatra, observed with a two-beam photometer of the 1-m telescope of Hoher List Observatory. The period is 5^h4 and the amplitude is $0.^m87$. From [8.45].

The *phase angle* must also be considered. The phase angle α_p is the angle at the position of the planetoid in the triangle Sun–asteroid–Earth. It is calculated from the relation

$$\cos \alpha_p = \frac{d^2 + r^2 - R^2}{2rd}, \qquad (8.36)$$

or, with higher accuracy, from the formula for $\tan(\alpha_p/2)$ (see Chap. 3.5).

If one plots the magnitudes, corrected for light-travel time and transformed to a constant distance, as a function of α_p, a brightness increase is generally noted when α_p becomes smaller. The extrapolation of the brightness curve yields the brightness for the phase angle $\alpha_p = 0°$. The magnitude can be given by linear regression as a function of α_p. The formula is

$$m_0 = 5 \log(rd) + H - 2.5 \log[(1 - G)\phi_1 + G\phi_2], \qquad (8.37)$$

where

$$\phi_1 = \exp\left\{-3.33 \left[\tan\left(\frac{\alpha_p}{2}\right)\right]^{0.63}\right\}$$

and

$$\phi_2 = \exp\left\{-1.87 \left[\tan\left(\frac{\alpha_p}{2}\right)\right]^{1.22}\right\}.$$

This formula is valid for $0° \leq \alpha_p \leq 120°$. The well-observed magnitude variation of some planetoids shows that the brightness depends not only upon the phase angle, but also, because of the irregular surface, on rotation. The periods are of the order of several hours, and the amplitudes can be as high as several tenths of a magnitude

Table 8.5. Surface brightnesses and luminances of objects in the solar system.

Object	m_V/β''	B (sb)
Sun	$-10\overset{m}{.}5$	
Mercury	1.5 to 4.5	0.6
Venus	0.5 to 1.5	2.0
Moon	3 to 7	0.6 (full moon)
		0.1 (first and last quarter)
Earth light	14 (on new moon)	
Earth light	22 (on Venus at inferior conjunction)	
Mars	4 to 5	0.2
Jupiter	5 to 5.5	0.070
Saturn	6.5 to 7	0.028
Uranus	8	0.0037

(Fig. 8.17). Long-term as well as short-term programs of planetoid photometry are obviously of interest. More extensive information may be found in Binzel [8.44].

8.6.1.2 Photometry of Major Planets. As a rule of thumb, we give, in Table 8.5, typical surface brightnesses (visual magnitudes per square arcsecond) and luminances (sb).

The human eye is capable of discriminating between very small differences in the luminances of bright areas and is therefore well suited for the photometry and structure recognition of small areas on planetary surfaces. Figure 8.18 shows how strong the brightness contrast must be in order to be perceived by the eye. The luminances noted by the eye can be magnified or decreased by means of the telescope magnification: the larger the magnification, the lower the luminance which is produced by an illuminated area on the retina. It is therefore advisable to transpose, by a suitable choice of magnification, the luminance into the range of largest contrast sensitivity.

8.6.1.3 Photometry of the Moon. Quantitative brightness measurements of the lunar surface by means of photoelectric photometry are still of some interest, despite the fact that very accurate measurements of a few lunar regions had been made during the Apollo landings. Of particular interest is the comparison of intensity and color of known regions with those not yet explored. Lunar photometry entails—if one does not have at one's disposal a two-dimensional detector—the integral measurement of different regions. The diaphragm size should be of the order of 5″.

Each region has a characteristic *brightness function* which depends on the phase angle ψ. The radiance of a region (i.e., the radiation of a projected unit surface into unit angle toward the direction to Earth) is thus determined by

- the *albedo*, which provides information on the geological and surface characteristics (albedo = total light reflected into all directions ÷ total infalling light);
- the phase angle (the angle between the Sun and Earth as measured on the point observed);

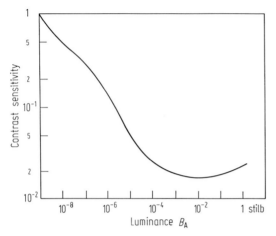

Fig. 8.18. Sensitivity of contrast of the human eye as a function of luminance B_A. Luminances of the order of magnitude 10^{-3}–10^{-1} allow the recognition of magnitude contrasts of $0\overset{m}{.}018$.

- the longitude (the angle between Earth and the direction normal to the lunar surface, measured at the observed point in the plane determined by the Sun and Earth);

Some of the different projects that can be carried out are:

- general albedo mapping (in different colors);
- determination of brightness functions of unusual regions;
- searches for transient lunar phenomena (TLPs).

The brightnesses of the various regions, which are measured with a diaphragm of known diameter (in square arcseconds), can be converted into magnitudes per square arcsecond through observation of a suitable fixed star (of spectral type similar to that of the Sun: G2); thus one obtains *absolute* lunar photometry. Another possibility is to observe a "standard region" in the vicinity (*relative* lunar photometry).

More detailed contributions have been written by Hedervari [8.46] and Westfall [8.47].

8.6.1.4 Photometry of the Sun. Solar photometry can be at once both easy and complicated. In general, no comparison stars are available, so that different areas on the Sun can be compared only in a differential way. Since a copious amount of light is available, on the other hand, a small telescope and a relatively insensitive detector (e.g., a PIN photodiode) will amply suffice for the observations.

In a review article by Chapman [8.48] two areas of research suitable for amateur astronomers are suggested: the study of solar activity, especially the observation of sunspots, and the study of the limb darkening of the Sun.

The photometry of the intensity variations of sunspots in different color regions can be performed with a telescope with a large image scale and by means of a photometer with a small diaphragm size; the telescope remains stationary while the photometer carries out a scan through the region of interest. An even better way of obtaining data

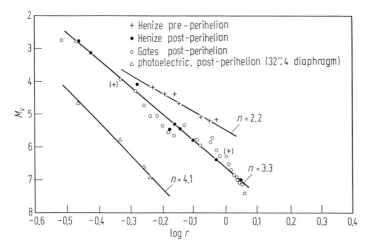

Fig. 8.19. Absolute magnitude M_V of comet Kohoutek (1973f), reduced to a distance of 1 AU from Earth, as a function of distance r from the Sun. The change of brightness before perihelion (*crosses*) is different from that afterwards (*circles* and *dots*). *Triangles* indicate photoelectric observations of the cometary nucleus after perihelion. From Angione et al. [8.49].

is to employ a one- or two-dimensional receiver like a CCD or a reticon. The umbra has a surface brightness of from 10 to 20% of the undisturbed solar surface. The scatter of light in the solar and the terrestrial atmosphere as well as in the telescope must be taken into account.

Similarly, the limb darkening of the Sun can be determined in different colors using a stationary telescope and photometer with a narrow diaphragm. The data collection of both limb darkening and sunspots can be carried out by means of a strip-chart recorder or by digital data acquistion.

8.6.1.5 Photometry of Comets and Meteors. In contrast to the relatively sharp disks of planets, planetary satellites, and especially planetoids, the photometry of other objects in the solar system is made difficult because of their extended, often diffuse appearance. This applies to comets, meteors, and the zodiacal light.

Two magnitudes can be defined for comets: the total magnitude and the magnitude of the nucleus. Visual estimates of the total magnitudes are often published in the literature in spite of their shortcomings. Even when determined by experienced observers, differences of up two magnitudes for the same instant are not uncommon. The estimates of surface brightness depend strongly on the diameters of the entrance and exit pupils of the instrument used, and the tail magnitudes often do not show a regular brightness decline as a function of the distance from the nucleus. Therefore faint parts of the tail are often overlooked, despite the fact that they are non-negligible owing to their vast extent. The total brightness of tailless comets can be estimated quite readily by defocusing a comparison star. When publishing photoelectric magnitudes, the size of the diaphragm used (in ″) must be given, as well as the relevant information on the telescope used to obtain the data.

Figure 8.19 shows visual and photoelectric magnitudes of comet Kohoutek to illustrate the different magnitudes determined.

It is practically unfeasible to determine the magnitude of the nucleus for comets which are near the Earth or the Sun ($r \leq 1$ AU), since the nucleus is always embedded in the bright coma. A thorough introduction into the photometry of comets is given by A'Hearn [8.50], who also gives a list of filters that are especially suited for comet photometry.

The brightness of a meteor is more easily determined, but its interpretation can suffer due to many ambiguities (material, porosity, entering velocity, angle of incidence, fragmentation). Visual estimates yield relatively good results.

Photographs often register only the brighter meteors; a rotating sector allows the determination of its angular velocity. Somewhat fainter objects can be found when the camera is rotated with a typical angular speed—comparable to that of the surveyed meteors—away from the radiant, and light from the remaining solid angle is shielded from the camera.

The zodiacal light extends over nearly the whole sky with measurable brightness. The main problem is to determine the terrestrial and stellar contributions to the sky brightness, and to correct the measurements for them. One possible way to secure a zero-point determination of the sky background is to observe the dark side of the moon (taking care to correct for scattered light of the illuminated side and Earthlight). The contribution of starlight can be determined when the same region is measured at another season. It is not feasible to employ this method at a site located near a major city, since the scattering of urban light, which depends on the local weather and the season, also strongly depends on azimuth. The scattered light from a city of 100 000 inhabitants can be detected at distances of up to 100 km!

The mutual eclipses of planetary objects with their global effects (i.e., total light losses) are also noteworthy. The total magnitude of the eclipsed object decreases by a certain amount, as indicated in the following examples: On Earth, the magnitude of solar radiation is decreased by $0^{m}_{.}001$ by a transit of Venus. A transit of of the Sun by Jupiter as viewed from Saturn (which will not occur for several decades) would decrease the brightness of Saturn by $0^{m}_{.}03$. The shadow of Ganymede decreases the brightness of Jupiter by less than $0^{m}_{.}002$. In the rare case that Io, Ganymede, and Callisto all cast shadows on Jupiter at the same time, Jupiter's brightness would decline by $0^{m}_{.}004$. Such events will occur in 1997, 2004, 2032, 2038.

Eclipses of asteroids by planets are possible, though rare. During a passage through the shadow cone of a planet at a distance of 0.01 AU between planetoid and planet, the following brightness declines occur: eclipse by Mercury, $0^{m}_{.}02$, Venus $0^{m}_{.}57$, Earth $2^{m}_{.}11$, Mars $0^{m}_{.}89$, Ceres $0^{m}_{.}01$. At such a distance, the planetoid would reach nearly the tip of the umbral cone of the Earth. At the much greater distance of 0.1 AU, planetoids are fainter by $0^{m}_{.}01$ when in the shadow (penumbra) of Venus, Earth, and Mars.

The observation of occultations of solar system objects is also discussed in Sect. 8.6. Attention should also be drawn to the contributions by Millis [8.51] on eclipses of planets and planetary moons, by Harris [8.52] on eclipses of asteroids, by Blow [8.53], and also by Heintz on lunar occultations in Chap. 17.

8.6.2 Stellar Photometry

As was already mentioned, absolute photometry in well-defined color systems is a field less suited for amateurs. A project to establish (visual) scales for the visual observation of variable stars is mentioned by Henden and Kaitchuck [8.8, p. 24]. The main field of photometric research on stars by amateurs will likely be the investigation of variable stars. Different projects are briefly described.

8.6.2.1 Flare Star Patrol. The monitoring of flare stars is an activity undertaken by many amateur astronomers. Here, mechanically and electrically stable equipment should be used, in connection with good sky conditions. For most of the time, the brightness of the flare star is monitored with good time resolution. The U and B color regions are most suitable, and the recording of data can be accomplished with a strip-chart recorder or by pulse counting with high storage capacity. The monitoring is interrupted from time to time by short measurements of the sky brightness and the brightness of a comparison star in the same color filter. Flares occur fairly rarely, and one must persist for several nights on a single object to obtain a statistically meaningful result. Unfortunately, most flare stars are quite faint.

A photographic patrol of flare stars can be carried out by taking multiexposure photographs of special fields (e.g., young star clusters). Between exposures (lasting about 10 minutes each), the telescope is shifted by a certain amount, thereby generating "strings" of stellar images of the same object, which can be examined for brightness variations, and especially for stellar flares.

8.6.2.2 Short-Period Variables: δ Scuti Stars, Dwarf Cepheids, RR Lyrae Stars, and Cepheids. For the first three types of stars listed in this section heading, a few hours of observing time are sufficient to obtain a complete light curve (Fig. 8.20). The objects often show dramatic changes in both the shape and amplitude of the light curve, caused by a beat phenomenon. In such cases, longer observing campaigns are useful.

8.6.2.3 Long-Period Variables: RV Tau stars, Mira Stars. The main body of data on long-period varible stars is comprised of visual estimates made by a large number of amateurs who, distributed across the world, have been able to obtain fairly complete light curves without the usual gaps caused by poor weather. An impressive collection of such composite visual light curves of Mira stars is shown in the AAVSO Report No. 38 [8.24].

One possible method of observing such variables with a photometer in the red and infrared regions, where they emit most of their light, is described by Wing [8.56].

8.6.2.4 Eclipsing Binaries. Three groups of eclipsing binaries can be distinguished: the short period group of W UMa stars (see Fig. 8.21), and the medium- to long-period systems of β Lyr and Algol types. A catalog of 3546 eclipsing binaries is found in Wood et al. [8.58].

Often it is possible to obtain an "instantaneous" light curve of a W UMa system during a single night. Other systems require weeks, months, or years, before all phases

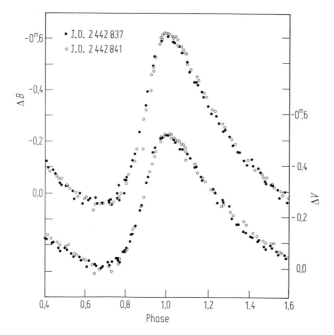

Fig. 8.20. *B* and *V* light curve of the dwarf cepheid SZ Lyn, observed with the 0.35-m telescope and photoelectric photometer of Hoher List Observatory. The amplitude in the blue spectral region is larger than in the visual. From Duerbeck [8.55].

of a light curve are covered by observations. Careful observations show that light curves undergo small but important changes. A comparison of instantaneous light curves of a W UMa system can be quite fascinating; a composite light curve of an Algol system, where the different observing runs do not fit together properly, can be annoying.

A subgroup of the Algol systems is the group of RS CVn systems, which show periods from below one day up to several days. Their light curves display a broad depression of about $0^m\!\!.1$, which slowly shifts in phase. It can be explained by the presence of "star spot" (analogous to sunspot) activity on one of the components which is not rotating in perfect synchronization with its orbital motion.

Even if one does not have the time to observe a complete light curve, one can still concentrate on the exciting observation of a complete eclipse: from such data, the time of minimum light can be derived. This is described in detail in Sect. 8.5.5. Predictions of eclipsing binary minima for a given year (and, in addition, maxima of RR Lyr stars) are given in the annual *Rocznik Astronomiczny Obserwatorium Krakowskiego* (founded by T. Banachiewicz, and edited by K. Rudnicki and his collaborators at the Obserwatorium Astronomiczne Uniwersytety Jagiellonskiego, ul. Orla 171, 30-244 Krakow, Poland).

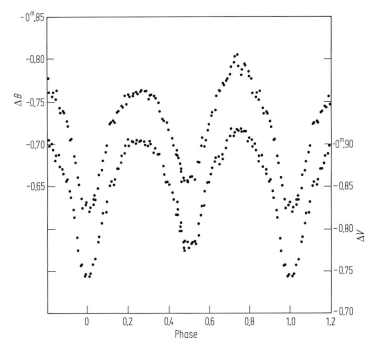

Fig. 8.21. B and V light curves of the W UMa system *i* Boo, observed with the 0.35-m telescope of Hoher List Observatory. The light curves in different spectral regions yield noticeable differences. From Duerbeck [8.57].

If a complete light curve of an eclipsing binary has been obtained, then it can be subjected to careful analysis in order to derive the characteristic parameters of the eclipsing binary. These parameters are:

- the radii of the two components A and B, r_a and r_b, in units of the radius of the orbit;
- the relative luminosities of both components, L_a and L_b in the given color region, so that $L_a + L_b = 1$;
- the inclination of the orbit (in degrees);
- the limb darkening coefficients u_a and u_b on the surfaces of the two stars.

There are classic methods, such as that of Russell and Merrill, and modern ones, such as the *synthetic light curve technique*. The following books are worthy of the reader's attention:

W.D. Heintz: *Double Stars*, D. Reidel Publ. Co., Dordrecht 1978; L. Binnendijk: *Properties of Double Stars*, University of Pennsylvania Press, Philadelphia 1960; S. Ghedini [8.31, p. 146].

8.6.2.5 Eruptive Variables: Novae, Dwarf Novae, and Supernovae. These objects are especially interesting, since they show "unique" outbursts: the supernovae, because they undergo only one outburst in their lifetime, the novae (Fig. 8.22) on historical time scales, and the dwarf novae, because an outburst never completely matches a

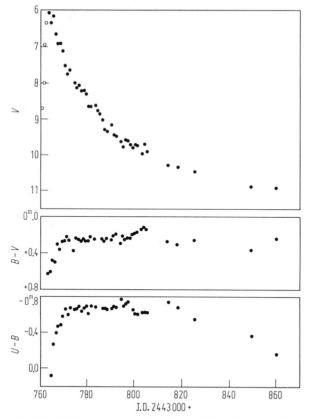

Fig. 8.22. V light curve and $(B-V)$ and $(U-B)$ color indices of Nova V1668 Cyg (1978), observed with the 0.35-m telescope of Hoher List Observatory. The brightness maximum coincides with large color indices (i.e., low temperature). From Duerbeck et al. [8.59].

previous one. Very important contributions to astronomy can be made in all of these cases. Any visual, photographic, or photoelectric observations which can be obtained will often serve as a valuable complement when spectroscopic data or data obtained in other wavelength regions (infrared, ultraviolet, X-ray) are being interpreted by professional astronomers.

8.6.3 Surface Photometry

Surface photometry of extended objects like planets, double stars, the Milky Way, and galaxies can be made with a single-channel photometer, which is kept fixed or scans a specific area in the sky in a pre-programmed way. It can be carried out with a special surface photometer (area scanner) [8.18], or with a one- or two-dimensional receiver (photographic plate, CCD).

8.7 Construction or Purchase of Receivers and Equipment for Reductions

A multitude of observational opportunities are available to the amateur astronomer who is interested in astronomical photometry which can enrich the experience in his or her hobby. S/he can, with the aid of a small or large telescope, either make estimates with the eye or carry out photoelectric observations. Nowadays, commercially available photometers are readily available, as are kits for building them at home; in addition, there exist numerous sources in the literature containing instructions for constructing one's own photometer. Furthermore, there are groups of interested amateurs and professionals who publish information on the construction, operation, and astronomical applications of photometers and the associated telescopes.

Perhaps the best source of information is the International Amateur–Professional Photoelectric Photometry Association (IAPPP), which publishes books, conference reports, and circulars on all aspects of astronomical photometry. Membership is currently $15 per year, and the publishing organization is I.A.P.P.P. Communications (subscriptions: Rolling Ridge Observatory, P.O. Box 8125, Piscataway, NJ 08854, USA; Editor: Dr. D.S. Hall, Dyer Observatory, Vanderbilt University, Nashville, Tennessee 37235, USA).

8.7.1 Advice on the Purchase of Photometers

Some of the commercially available photometers are listed below with comments.

1. Optec Solid-State Photometer SSP-3 (Fig. 8.23). Literature: [8.60,61,62]
   The SSP-3 photometer uses a PIN photodiode as receiver. The spectral sensitivity is between 300 nm and 1100 nm. The blue sensitivity is low; sensitivity maximum is in the infrared, at 850 nm. The exit voltage is produced via a shunt resistance of 50 GΩ. It is converted via a voltage-to-frequency converter into a frequency. A counter is built into the photometer housing. By means of a slide, filters can be brought into the beam of light; Johnson B and V filters are standard. The power is supplied by means of a battery. Stars of blue to infrared magnitude of 10^m (ultraviolet to 8^m) can be observed in combination with a 30-cm telescope. The cost of the photometer is below $1000. The manufacturer is Optec, Inc., 199 Smith Street, Lowell, MI 49331, USA.
2. Optec Photoelectric Photometer SSP-5. Literature: [8.63]
   The SSP-5 photometer uses a Hamamatsu R1414 miniature photomultiplier with S-5 (i.e., UV-extended S-4) spectral response. The PMT signal is fed to a voltage-to-frequency converter and a counter. Manual 2- and 6-position filter sliders can be used: a motorized 6-position filter slider as well as an IBM-PC interface for automatic data acquisition are available. UBV-photometry of stars down to 13^m with a 30 cm telescope should be possible. The cost of the standard configuration (including Johnson UBV filters) is about $2000. Strömgren filters are also available. The manufacturer is Optec, Inc., 199 Smith Street, Lowell, MI 49331, USA.

Fig. 8.23. The SSP-3 solid-state photometer of Optec, Inc. From above, the following can be seen: eyepiece, sliding mirror, filter slide, box of the PIN photodiode, box with switches for amplification and integration time selector, digital display and data output.

3. EMI GEMCON, Inc. Starlight-1 photon-counting photometer. Literature: [8.64]. This photometer is a complete photon-counting photometer with an EMI 9924A PMT, power supply, pulse amplification and discrimination, a 10 Mcps counter with analog and digital output, a diaphragm wheel, and a set of *UBV* filters.

4. HPO-photometer (Fig 8.24). Literature: [8.64].
   The Hopkins–Phoenix Observatory (J.L. Hopkins, Hopkins–Phoenix Observatory, 7812 West Clayton Drive, Phoenix, Arizona 85033, USA) supplies parts for a compact photon-counting photometer for smaller telescopes. Mr. Hopkins also offers advice on the construction of do-it-yourself photometers.

8.7.2 Advice on the Construction of Photometers

For experienced do-it-yourself astronomers, the books by Hall and Genet [8.66] and by Henden and Kaitchuck [8.8], as well as the contributions in most other books mentioned in the Supplemental Reading List give enough information on how to build for oneself a photometer head, the electronics (pulse amplifier and converter), and data acquisition. Just a few main points will be discussed in this section.

With the aid of the Schnitzer [8.67] photometer (Fig. 8.25), it is easy to illustrate the general design of such a device. The photometer is placed near the focal plane

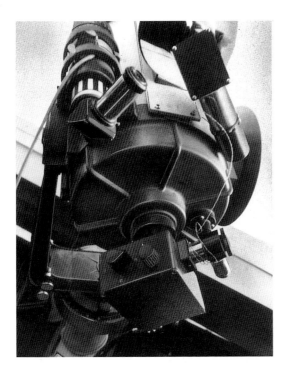

Fig. 8.24. The Hopkins–Phoenix Observatory. The PEPH-101 photometer, a homebuilt pulse-counting photometer, is connected to a Celestron C-8 telescope. With kind permission of J.L. Hopkins, Hopkins–Phoenix Observatory.

of the telescope. It consists of an optical device for the observation of the star field, including a crosswire for the centering of the object of interest, often in conjunction with a movable plane mirror. When the mirror is moved aside, the light of the star to be measured falls through a diaphragm, a filter, and a Fabry lens onto the light-sensitive detector (e.g., a PMT). The use of more dark slides which block the light path, for instance, in front of the photometer and in front of the light sensitive cell, is advisable to protect the photometer from dust and strong light.

If a number of diaphragms are available, a slide which can be placed at certain well-defined positions should carry several bore holes with different diameters. Their sizes will be determined by the focal length of the telescope, the quality of the telescope tracking, and the average seeing conditions of the observatory. Diameters of 20, 30, and 40 arcseconds are good choices. An approximate formula yields

$$d \text{ (in mm)} = \frac{d'' \text{ (in arcsec)}}{206\,365} \times f \text{ (in mm)}, \tag{8.38}$$

where d and d'' are the diaphragm diameters measured in mm and arcseconds, respectively, and f is the focal length in mm. The Fabry lens is used to produce an image of the telescope objective on the surface of the receiver, i.e., the cathode of the PMT. The image of the star thus forms an evenly illuminated spot on the cathode.

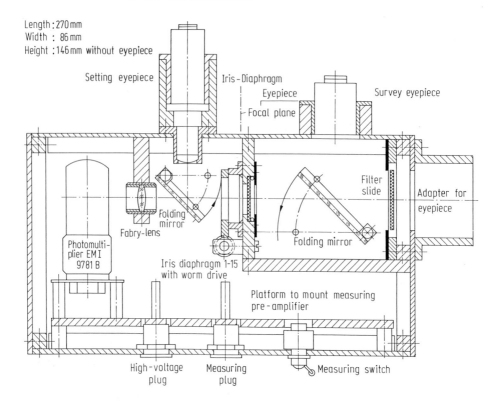

Fig. 8.25. Sectional drawing of the Schnitzer photometer (see text for details).

The *entrance pupil* of the telescope is the primary mirror or the telescope objective. In order to reduce the brightness of the sky background, the field of view of the receiver is limited by the diaphragm, which is placed in the focal plane of the telescope. If a lens is placed behind this diaphragm, the *exit pupil* of the optical system is the image of the objective as formed by this diaphragm. Each ray that passes the entrance pupil passes the corresponding point in the exit pupil; a detector positioned in the exit pupil "sees" a spot of light which is not influenced by image motion (e.g., seeing).

The geometrical relations are shown in Fig. 8.26 and are given by the following formulae:

$$\frac{1}{F} = \frac{1}{s_1} + \frac{1}{s_2} \tag{8.39}$$

and

$$\frac{d}{D} = \frac{s_2}{s_1}. \tag{8.40}$$

D is the diameter of the objective, F the focal length of the Fabry lens, d the diameter of the spot on the receiver (= exit pupil), s_1 the distance between the objective and the Fabry-lens, and s_2 the distance between the Fabry lens and the receiver. In general, the Fabry lens is placed near the diaphragm, and $d/D \ll 1$. Then $s_1 = f$ and $s_2 = F$.

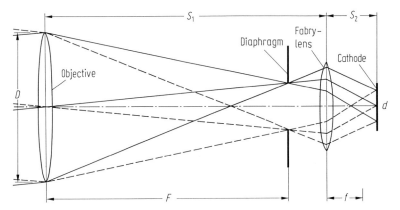

Fig. 8.26. Configuration of the Fabry lens (see text for details).

The focal length of the Fabry lens is approximately

$$F = d\frac{f}{D}. \tag{8.41}$$

When acquiring such a lens, one should take care that it has a good UV-transparency (e.g., quartz). To test the function of such a lens, one should move a bright star across a large diaphragm and check the recorded signal: it should show a well-pronounced, flat maximum region [8.66, p. 5-24; 68].

The output signal of the PMT can be processed in various ways:

1. Using a current amplifier; data recording by a strip-chart recorder; the amplifier (electrometer) should be of good quality. Hints for constructing such a device are given by Henden and Kaitchuck [8.8, p. 184] and by Oliver [8.69]. The strip-chart recorder is fairly expensive, and this type of data storage is not very popular any more, especially since reduction work, the measuring of paper strips and possible entering of data by hand into a computer, is quite tiresome.
2. Using a current amplifier and a voltage-to-frequency converter; data are displayed on a counter. The counter may have a digital output for data acquisition. Details are given by Henden and Kaitchuck [8.8, p. 195] and by Oliver [8.70].
3. Using a pulse amplifier and sharpener; data are displayed on a counter with digital output. This method is widespread in professional astronomy and also possible among amateurs. Again, details of construction are given by Henden and Kaitchuck [8.8, p. 167], by Hall and Genet [8.66], by Hopkins [8.65], and by Simons [8.71].

Depending on the quality of the electronics used, pulse-counting photometers exhibit count losses at high count rates (bright stars). For such a device the "dead time" must be determined, i.e., the time during which a second incoming pulse cannot be separated from the first one and is thus counted as a single event. One measures the count rate n (events/second), while the true count rate is N (events/second). The relation

$$n = Ne^{-t_d N}, \tag{8.42}$$

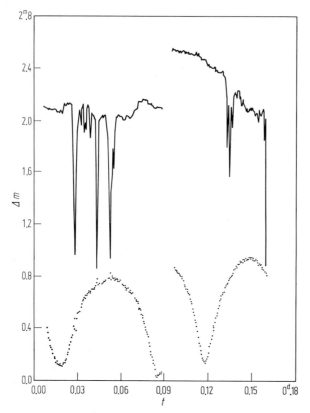

Fig. 8.27. Results of the two-beam photometer attached to the 1-m telescope of the Hoher List Observatory. The *upper* light curves show the brightness variation of the comparision star, measured in the first channel, which are caused by passing clouds and haze. The variable, measured in the second channel, is influenced by the same variations in transparency. The differential magnitudes yield a light curve which is practically not influenced by the adverse observing conditions (*lower*). The magnitude scale markings indicate $0\overset{m}{.}2$.

holds, where t_d is the dead time.

The dead time is often determined by the use of a gray filter, which can be placed into the light beam, and which weakens the radiation by a factor $1/b$. If different objects of different brightnesses are measured with and without the filter, one obtains the count rates n_F and n_0. A comparison of the observed count-rate n_0 with bn_F yields a measure for the dead time. A graph of $\ln(bn_F/n_0)$ as a function of bn_F yields a straight line of slope t. With the above formula, the determination of N can proceed in an iterative fashion. Additional information is to be found in [8.8] and [8.72].

There are virtually no limits to the activity in the realm of astronomical photometry. An observer working under the poor British skies may perhaps be interested in constructing a two-channel photometer, which measures simultaneously a variable and a comparison star, and which is able to compensate for fluctuations in the transparency of the atmosphere (Fig. 8.27). Or s/he may design an automated telescope with a

Fig. 8.28. The future of photoelectric photometry: amateur astronomers L.J. Boyd and R.M. Genet have developed completely automated photoelectric telescopes (APTs). Three telescopes of this type carry out observations of solar-type stars, RS CVn stars, and other variables. They are situated on Mount Hopkins in southern Arizona. With kind permission of R.M. Genet, Fairborn Observatory.

photometer, which is able to carry out an observing program without the physical presence of an astronomer (Fig. 8.28, [8.73]). A quick perusal of the recent issues of any popular magazine on astronomy will suffice to show the current trends in amateur astronomy. But, while it is a good practice to continually improve and update the photometric instrument and the accompanying electronics, one should not forget to observe: the value of the observational results will endure for many years.

8.8 General Literature on Photometry

The following can be recommended for general reading on astronomical photometry (see also the Supplemental Reading List for Vol. 1).

Binnendijk, L.: *Properties of Double Stars*, University of Pennsylvania Press, Philadelphia 1960. (A basic introduction to all aspects of double star research, with special emphasis on the reduction and analysis of photoelectric observations of eclipsing binaries. Slightly outdated, but still a useful introductory text.)

Carleton, N. (ed.): *Astrophysics Part A: Optical and Infrared*, Vol. 12 of *Methods of Experimental Physics*, Academic Press, New York London 1974. (Contains three contributions by A.T. Young on photomultipliers, photometers, and the reduction of photoelectric observations; a high-level text.)

Genet, R.M. (ed.): *Solar System Photometry Handbook*, Willmann-Bell, Richmond VA 1983. (A collection of articles on all aspects of photometry of the Sun, Moon, planets and planetoids, comets, and stellar occultations by the Moon, planets, and planetoids.)

Genet, K.M., Genet, R.M., Genet, D.R.: *The Photoelectric Photometry Handbook*, The Fairborn Press, Mesa AZ 1987.

Ghedini, S.: *Software for Photoelectric Photometry*, Willmann-Bell, Richmond VA 1982. (A collection of BASIC computer programs on nearly all aspects of photometry, especially that of variables.)

Hall, D.S., Genet, R.M.: *Photoelectric Photometry of Variable Stars* (2nd edn.), Willmann-Bell, Inc., Richmond VA 1989. (A good introduction to photometer construction and the use of photometers for variable star observations.)

Hall, D.S., Genet, R.M., Thurston, B.L. (eds.): *Automatic Photoelectric Telescopes*, The Fairborn Press, Mesa AZ 1986. (A collection of articles about the construction and the use of automatic photoelectric telescopes, some built by amateur astronomers.)

Hardie, R.H.: Photoelectric Reductions. p. 178 in W.A. Hiltner (ed.), *Stars and Stellar Systems Vol. II: Astronomical Techniques*, University of Chicago Press, Chicago 1962.

Henden, A.A., Khaitchuk, R.H.: *Astronomical Photometry*, Van Nostrand Reinhold Co., New York 1982. 2nd edn.: Willmann-Bell, Inc., Richmond VA 1989. (A solid introduction to photometer construction and use.)

Hübscher, J., Braune, W., Fernandes, M., Broemme, A.: *Einführung in die visuelle Beobachtung Veränderlicher Sterne*, BAV-Berliner Arbeitsgemeinschaft für Veränderliche Sterne e.V., Berlin 1983. (A good introduction to visual photometry with many worked-out examples, which is also very useful for self-study.)

Johnson, H.L.: Photoelectric Photometers and Amplifiers. p. 157 in: W.A. Hiltner (ed.), *Stars and Stellar Systems Vol. II: Astronomical Techniques*, University of Chicago Press, Chicago 1962.

Percy, J.R. (ed.): *The Study of Variable Stars Using Small Telescopes*, Cambridge University Press, Cambridge 1986. (A good collection of conference reports by amateurs and professionals.)

Sterken, C., Manfroid, J.: *Astronomical Photometry – A Guide,* Kluwer Academic Publishers, Dordrecht, Boston, London 1992. (A modern introduction to the physical background, reduction methods, and astronomical applications.)

Walker, C.: *Astronomical Observations–An Optical Perspective*, Cambridge University Press, Cambridge 1987. (Principles and methods of measurement of photometric, spectroscopic, and other observations.)

Warner, B.: *High Speed Astronomical Photometry,* Cambridge University Press, Cambridge 1988. (Technique and astronomical application of high speed photometry.)

Wolpert, R.C., Genet, R.M.: *Advances in Photoelectric Photometry, Vol. 1* (etc.), published by the Fairborn Observatory, Fairborn OH 1983. (A collection of contributions from the amateur scene, with many tips for do-it-yourself construction and data processing by personal computer.)

Wood, F.B.: *Photoelectric Photometry for Amateurs*, Macmillan, New York 1963. (A very outdated textbook, but which still contains some useful contributions.)

References

8.1 Tüg, H., White, N.M., Lockwood, G.W.: Absolute Energy Distributions of α Lyrae and 109 Virginis from 3295 AA to 9040 Å. *Astronomy and Astrophysics* **61**, 679 (1977).

8.2 Schnapf, J.L., Baylor, D.A.: How Photoreceptor Cells Respond to Light. *Scientific American* **256**, Nr. 4, 32 (April 1987).

8.3 Hecht, S., Shlaer, S., Pirenne, M.H.: Energy, Quanta, and Vision. *Journal of General Physiology* **25**, 819 (1942).

8.4 Lukas, R.: Was kann man visuell noch beobachten? *Sterne und Weltraum* **22**, 424, (1983).

8.5 Seitter, W.C., Budell, R.: Kurze Einführung in die astronomische Photographie. *Mitteilungen der Astronomischen Gesellschaft* **62**, 162 (1984).
8.6 James T.H.: *The Theory of the Photographic Process*, 4th edn., Macmillan Publ. Co., Inc., New York; Collier Macmillan Publ., London 1977.
8.7 Furenlid, I.: Signal-to-Noise of Photographic Emulsions, p. 153 in *Modern Techniques in Astronomical Photography*, R.M. West and J.L. Heudier (eds.), ESO Proceedings, Geneva 1978.
8.8 Henden, A.A., Kaitchuck, R.H.: *Astronomical Photometry*, Van Nostrand Reinhold Co., New York 1982.
8.9 Lallemand, A.: Photomultipliers, p. 126 in *Stars and Stellar Systems Vol. II: Astronomical Techniques*, W.A. Hiltner (ed.), The University of Chicago Press, Chicago 1962.
8.10 Johnson, H.L.: Photoelectric Photometers and Amplifiers, p. 157 in *Stars and Stellar Systems Vol. II: Astronomical Techniques*, W.A. Hiltner (ed.), The University of Chicago Press, Chicago 1962.
8.11 Young, A.T.: Photomultipliers: Their Cause and Cure, p. 1 in *Methods of Experimental Physics: Astrophysics, Vol. 12A*, N. Carleton (ed.), Academic Press, New York 1974.
8.12 Wolpert, R.C., Hall, D.S., Reisenweber, R.C.: IAPPP Communication No. 14, Special I.R. and Solid-State Photometry Issue (December 1983).
8.13 Kristian, J., Blouke, M.: Charge-Coupled Devices in Astronomy. *Scientific American* **247**, No. 4, 48 (October 1982).
8.14 Mackay, C.D.: Charge-Coupled Devices in Astronomy, p. 255 in *Annual Review of Astronomy and Astrophysics 24*, Annual Reviews Inc., Palo Alto 1986.
8.15 Harris, C.: Silicon Eye: A CCD Imaging System. *Sky and Telescope* **71**, 407 (April 1986).
8.16 Buil, C.: A Charge-Coupled Device for Amateurs. *Sky and Telescope* **69**, 71 (January 1985).
8.17 Bickel, W.: Ein CCD-Versuch. *Sterne und Weltraum* **25**, 40 (1986).
8.18 Guinan, E.F., McCook, G.P., McMullin, J.P.:Acquisition, Reduction and Standardisation of Photoelectric Observations, p. 79 in *The Study of Variable Stars Using Small Telescopes*, J.R. Percy (ed.), Cambridge University Press, Cambridge 1986.
8.19 Straizys, V.: *Multicolor Stellar Photometry* (Russian with English summary). Mokslas Publishers, Vilnius 1977.
8.20 Jahn, W.: Die Photometrie von Fixsternen und Planeten, p. 374 in *Handbuch für Sternfreunde*, 2. Auflage, G.D. Roth (ed.), Springer, Berlin 1967.
8.21 Mayall, M.W.: *Manual for Observing Variable Stars*, rev. edn., American Association of Variable Star Observers, Cambridge MA 1970.
8.22 Hübscher, J., Braune, W., Fernandes, M., and Broemme, A.: *Einführung in die visuelle Beobachtung veränderlicher Sterne*, 2nd edn., Berliner Arbeitsgemeinschaft für Veränderliche Sterne, Berlin 1983.
8.23 Hoffmeister, C.:*Variable Stars*, 2nd edn., Springer, Berlin 1985.
8.24 Mattei, J.A. (Hsgr.): *Observations of Long-Period Variables*, American Association of Variable Star Observers, Cambridge MA 1983.
8.25 Hörber, E., van Stephoudt, H., Veldscholten, W.: Photometrische Messungen in der Schule. *Sterne und Weltraum* **22**, 24 (1983).
8.26 Evans, D.S.: Photoelectric Observing of Occultations–II. *Sky and Telescope* **54**, 289 (October 1977).
8.27 Chen, P.C.: Portable High Speed Photometer Project, p. 10-1 in *Solar System Photometry Handbook*, R.M. Genet (ed.), Willmann-Bell, Inc., Richmond VA (1983).
8.28 Rakos, K.D.: Photoelectric Area Scanner. *Applied Optics* **4**, No. 11, 1453 (1965).
8.29 Fritze, K.: Untersuchungen zur lichtelektrischen Photometrie enger Doppelsterne. *Veröffentlichungen der Sternwarte Berlin-Babelsberg* **14**, No. 5 (1963).
8.30 Smak, J.: CE Cas a, CE Cas b, and CF Cas in NGC 7790, *Acta Astronomica* **16**, 11 (1966).
8.31 Ghedini, S.: *Software for Photoelectric Photometry*, Willmann-Bell, Richmond VA 1982.
8.32 Vehrenberg, H., Brun, A.: *Atlas of the Harvard-Groningen Selected Areas*, Treugesell-Verlag, Düsseldorf 1965.

8.33 Becker, W.: Astronomische Übungsgeräte. *Sterne und Weltraum* **12**, 145 (1973).
8.34 Peelen, R.: Image Processing with an Apple Computer, *Sky and Telescope* **67**, 177 (February 1984).
8.35 Dohnke, K.O.: Der Commodore C-64 als Registrierphotometer, *Sterne und Weltraum* **25**, 478 (1986).
8.36 Kwee, K.K., van Woerden, H.: A Method for Computing Accurately the Epoch of Minimum of an Eclipsing Variable. *Bulletin of the Astronomical Institutes of the Netherlands* **12**, 327 (1956).
8.37 Fullerton, A.W.: Searching for Periodicity in Astronomical Data, S. 201 in *The Study of Variable Stars Using Small Telescopes*, J.R. Percy (ed.), Cambridge University Press, Cambridge 1986.
8.38 Lafler, J., Kinman, T.D.: An RR Lyrae Star Survey with the Lick 20-inch Astrograph-II. The Calculation of RR Lyrae Periods by Electronic Computer. *Astrophysical Journal Supplement* **11**, 216 (1964).
8.39 Kholopov, P.N. (ed.): *General Catalogue of Variable Stars*, 4th edn., Nauka, Moskow 1985 ff.
8.40 Herczeg, T., Frieboes-Condé, H.: The Period of RZ Cassiopeiae. *Astronomy and Astrophysics* **30**, 259 (1974).
8.41 Willson, L.A.: The (O–C) Diagram–A Useful Tool, p.219 in *The Study of Variable Stars Using Small Telescopes*, J.R. Percy (ed.), Cambridge University Press, Cambridge 1986.
8.42 Lockwood, G.W.: Photometry of Planets and Satellites, p.2-1 in *Solar System Photometry Handbook*, R.M. Genet (ed.), Willmann-Bell, Richmond VA 1983.
8.43 Millis, R.L., Thompson, D.T.: UBV Photometry of the Galilean Satellites. *Icarus* **26**, 408 (1975).
8.44 Binzel, R.P.: Photometry of Asteroids, p. 1-1 in *Solar System Photometry Handbook*, R.M. Genet (ed.), Willmann-Bell, Richmond VA 1983.
8.45 Grossmann, M., Hoffmann, M., and Duerbeck, H.W.: Photometric Measurements of 216 Kleopatra. *The Minor Planet Bulletin* **8**, No. 2, 14 (1981).
8.46 Hedervari, P.: Lunar Photometry, p. 4-1 in *Solar System Photometry Handbook*, R.M. Genet (ed.), Willmann-Bell, Richmond VA 1983.
8.47 Westfall, J.E.: An Invitation to Lunar Photometry, *I.A.P.P.P. Communication No. 14*, 64 (1983).
8.48 Chapman, G.A.: Solar Photometry, p. 5-1 in *Solar System Photometry Handbook*, R.M. Genet (ed.), Willmann-Bell, Richmond VA 1983.
8.49 Angione, R.J., Gates, B., Henize, K.G., Roosen, R.G.: The Light Curve of Comet Kohoutek. *Icarus* **24**, 111 (1975).
8.50 A'Hearn, M.F.: Photometry of Comets, p. 3-1 in *Solar System Photometry Handbook*, R.M. Genet (ed.), Willmann-Bell, Richmond VA 1983.
8.51 Millis, R.L.: Occultations by Planets and Satellites, p. 7-1 in *Solar System Photometry Handbook*, R.M. Genet (ed.), Willmann-Bell, Richmond VA 1983.
8.52 Harris, A.W.: Asteroid Occultations, p. 8-1 in *Solar System Photometry Handbook*, R.M. Genet (ed.), Willmann-Bell, Richmond VA 1983.
8.53 Blow, G.L.: Lunar Occultations, p. 9-1 in *Solar System Photometry Handbook*, R.M. Genet (ed.), Willmann-Bell, Richmond VA 1983.
8.54 Fernandes, M.: Lichtelektrische Photometrie veränderlicher Sterne. *Sterne und Weltraum* **22**, 408 (1983).
8.55 Duerbeck, H.W.: New B, V Light Curves of SZ Lyncis, *Information Bulletin on Variable Stars*, 1171 (1976).
8.56 Wing, R.F.: Observation of Variable Stars in the Infrared, p. 127 in *The Study of Variable Stars Using Small Telescopes*, J.R. Percy (ed.), Cambridge University Press, Cambridge 1986.
8.57 Duerbeck, H.W.: The Variable Light Curve of 44 *i* Boo – Observations and Implications. *Astronomy and Astrophysics Supplement* **32**, 361 (1978).
8.58 Wood, F.B., Oliver, J.P., Florkowski, D.R., Koch, R.H.: A Finding List for Observers of Interacting Binary Stars (5th edn.). *Publications of the Department of Astronomy, Univer-*

sity of Florida, Vol. I, and *Publications of the University of Pennsylvania*, Astronomical Series XII, University of Pennsylvania Press, Philadelphia 1980.

8.59 Duerbeck, H.W., Rindermann, K., Seitter, W.C.: A UBV Light Curve of Nova Cygni 1978. *Astronomy and Astrophysics* **81**, 157 (1980).

8.60 Persha, G., Sanders, W.: The SSP-3 Photometer, p. 130 in *Advances in Photoelectric Photometry, Vol. I*, R.C. Wolpert and R.M. Genet (eds.), Fairborn Observatory, Fairborn OH 1983.

8.61 Persha, G.: Photodiode Detectors, p. 4-26 in *Photoelectric Photometry of Variable Stars*, D.S. Hall and R.M. Genet (eds.), International Amateur-Professional Photoelectric Photometry, Fairborn OH 1982.

8.62 Optec, Inc.: *Manual for Model SSP-3 Solid-State Photometer*, Optec, Inc., Lowell MI 1982/84; *SSP-3 Card User's Manual Rev. 1.0*, Optec, Inc., Lowell MI (1987); D.F. Figer: The automated Photometry Data Reduction Template, Optec, Inc., Lowell MI.

8.63 Optec, Inc.: *Model SSP-5 Photoelectric Photometer Technical Manual*, Optec, Inc., Lowell MI (1990); The Optec SSP-5 Photometer. *IAPPP Communications*, No. 36, 28 (1989).

8.64 Wolpert, R.C.: A Photon-Counting Stellar Photometer, p. 4-54 in *Photoelectric Photometry of Variable Stars*, D.S. Hall and R.M. Genet (eds.), International Amateur-Professional Photoelectric Photometry, Fairborn OH 1982.

8.65 Hopkins, J.L.: Low-Speed Equipment, p. 6-1 in *Solar System Photometry Handbook*, R.M. Genet (ed.), Willmann-Bell, Richmond VA 1983.

8.66 Hall, D.S., Genet, R.M.: *Photoelectric Photometry of Variable Stars*, D.S. Hall and R.M. Genet (eds.), International Amateur-Professional Photoelectric Photometry, Fairborn OH 1982.

8.67 Schnitzer, A.: *Lichtelektrische Photometrie veränderlicher Sterne für Astro-Amateure. Eine Bauanleitung*. Available from: Bundesdeutsche Arbeitsgemeinschaft für Veränderliche Sterne (BAV), Munsterdamm 90, 12169 Berlin, F.R. Germany.

8.68 Kämper, B.-C.: Zur Funktionsweise der Fabry-Linse. *BAV Rundbrief 34*, No. 1, 28 (1985).

8.69 Oliver, J.P.: DC Amplifiers, p. 5-10 in *Photoelectric Photometry of Variable Stars*, D.S. Hall and R.M. Genet (eds.), International Amateur–Professional Photoelectric Photometry, Fairborn OH 1982.

8.70 Oliver, J.P.: Voltage-to-Frequency Conversion, p. 5-21 in *Photoelectric Photometry of Variable Stars*, D.S. Hall and R.M. Genet (eds.), International Amateur-Professional Photoelectric Photometry, Fairborn OH 1982.

8.71 Simons, D.: A Lightweight Pulse-Counting Photometer. *Sky and Telescope* **72**, 295 (September 1986).

8.72 Africano, J., Quigley, R.: Deadtime. *Journal of the American Association of Variable Star Observers* **6**, No. 1, 53 (1977).

8.73 Hall, D.S., Genet, R.M., Thurston, B.L. (eds.): *Automatic Photoelectric Telescopes* (IAPPP Communication No. 25), The Fairborn Press, Mesa AZ 1986.

9 Fundamentals of Radio Astronomy

W. J. Altenhoff

9.1 Introduction

It was over a century ago that James Clerk Maxwell (1831–1879) demonstrated that the spectrum of electromagnetic radiation is not limited to visible light but extends far into infrared and ultraviolet wavelengths. In 1887, Heinrich Hertz succeeded in generating and detecting electromagnetic waves in the radio range. It seemed reasonable to investigate whether or not the Sun emits such rays. In 1890, T.A. Edison in the United States and O. Lodge in England suggested that a search be made for radio emission from the Sun. In 1896, J. Scheiner and J. Wilsing in Germany tried to detect microwave emission from the Sun, and in 1901, E. Nordmann in France searched for solar short wave radiation, all with negative results. Thus, at the turn of the last century, it seemed conclusive that no radio rays from the Universe reached the Earth.

In 1932, Karl G. Jansky, who was in the employ of Bell Laboratories in Holmdel, New Jersey, used a specially constructed directional antenna to locate all potential sources of interference for wireless communication in the wavelength range around 14.6 m. The antenna is shown in Fig. 9.1. He identified three natural sources: 1. nearby thunderstorms, 2. distant tropical thunderstorms, and 3. a constant noise or "hiss" of unknown origin. Jansky first surmised that the Sun was somehow causing this unidentified disturbance, but later he concluded that the radiation originated from the whole Milky Way. Although this discovery was widely reported in the new media and although Jansky himself also presented his results in *Popular Astronomy*, the predecessor of *Sky & Telescope*, astronomers showed little interest.

Only a young radio engineer named Grote Reber from Wheaton, Illinois recognized the significance of Jansky's find. In 1937 Reber built in his backyard a radio telescope of diameter 9.6 m (see Fig. 9.2), which was to become the prototype for all modern radio telescopes. Reber started with the assumption that, like the visible radiation from stars, radio radiation from the Milky Way is of *thermal* origin (see below). Therefore he built a receiver to observe at wavelengths of 9.1 cm. Failing in this attempt, he subsequently tried the wavelengths 33 cm and 1.87 m. Only at the longest wavelength did he manage, in 1939, to confirm Jansky's discovery while simultaneously refuting his own hypothesis on the origin of the radiation.

Reber's report in the *Astrophysical Journal* was apparently as unenthusiastically received as Jansky's results had been eight years earlier. Reber improved his receiver and systematically observed the sky. Four years later, he published a radio map of

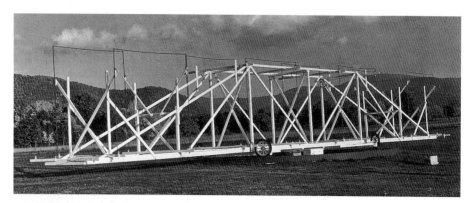

Fig. 9.1. Replica of K.G. Jansky's antenna which discovered cosmic radio radiation. It is located at the entrance to the National Radio Astronomy Observatory, Green Bank, West Virginia.

Fig. 9.2. Grote Reber's radio telescope. This historical instrument was rebuilt in the National Radio Astronomy Observatory in Green Bank, West Virginia.

the Milky Way and the first radio measurements of the Sun. This work marks the beginning of radio astronomical research; it triggered the prediction of the spectral line of hydrogen at $\lambda = 21$ cm by H. van de Hulst in 1944 and the interpretation of the then newly discovered nonthermal galactic radiation as *synchrotron radiation* generated by the electron component of cosmic rays (O. Kiepenheuer 1950). The subsequent rapid development of radio astronomy is evidenced by several prominent milestones: In 1954 W. Baade and R. Minkowski succeeded in identifying the radio source Cygnus A with the optical position of a galaxy. In 1960, A. Sandage showed the agreement of the radio position of 3C48 with a starlike object now known to be a *quasar*. M. Schmidt in 1963 determined the distance to the radio source 3C273 and J.L. Greenstein and T.A. Matthews that of 3C48 from the redshifted spectral lines, and thus they demonstrated the cosmological distances of quasars. In 1965, A. Penzias and R. Wilson found the 3° background radiation which is considered to be the best evidence in support of the *Big Bang theory*. 1967 brought the serendipitous discovery of pulsars through the research of J. Bell and A. Hewish, and in 1968, L. Snyder, P. Palmer, and B. Zuckerman detected the first organic molecule (formaldehyde, H_2CO) in space. In 1986, A. Witzel, D.S. Heeschen, C. Schalinski, and T. Krichbaum reported variability of order 40% on a timescale of less than one day for some extragalactic sources, and in 1992, A. Wolszczan and D.A. Frail detected a planetary system around the millisecond pulsar PSR1257+12 with (at least) two planets with masses comparable to that of the Earth.[1]

Technical progress was correspondingly rapid: the Sun had been at the limit of detectability in Reber's measurements, and the angular resolution obtained was a very crude 14°. Modern instruments can measure intensities 7 or 8 orders of magnitude weaker, and, by linking together telescopes over intercontinental distances, the angular resolution now reaches the incredibly minute value of $0''.0001$! As will be shown, such an achievement requires about equal expenditures in telescope/receiver and computer equipment and instrumentation. The stringency of these requirements for modern telescopes may be comprehended by the fact that the 30-m telescope in Berlin-Aldershof (limiting wavelength 30 cm) and the cylindrical telescope of size 122×183 m built into a valley in Urbana, Illinois (limiting wavelength 75 cm), to cite just two examples, have already been scrapped.

In contrast to optical astronomy, where modest instruments still contribute to serious research, the demands needed in radio astronomy for results which compete with professional research are so high that an amateur working on his or her own has little chance of making worthwhile contributions.

On the other hand, there do exist certain useful tasks which may be undertaken by an amateur radio observer; these are, in the opinion of this author:

1. Construction of a receiver system for didactic purposes.
2. Long-term monitoring of the variable radiation from the Sun and Jupiter.
3. Study of the propagation of radio waves in the atmosphere.

It is the purpose of this chapter to provide a brief introduction to the basics of radio astronomy and to establish criteria for estimating the expenditure of time and effort needed to obtain observational results. The literature should be consulted for construction designs. A comprehensive presentation of radio astronomy is given in the textbooks by J.D. Kraus [9.1] and K. Rohlfs [9.2]. A review of radio astronomical results is given in a popular book by G. Verschuur [9.3] and in greater depth by G. Verschuur and K.I. Kellermann [9.4].

---

[1] This important result still needs independent confirmation.

Fig. 9.3. Aerial view of the 100-m telescope in Effelsberg, Germany.

9.2 Radio Radiation

The long-wave range of the electromagnetic spectrum above 1 mm wavelength (frequency $\nu \leq 300$ GHz) is occupied by radio waves. Three essential properties of these rays can be estimated from their considerably longer wavelengths compared with those of light.

1. The angular resolution is determined by the ratio of instrument diameter D in meters and wavelength λ in meters. In the radio range, the half-width Θ ['] of the antenna lobe is adopted as a measure of resolution and is very well approximated by a Gaussian distribution. It is given by

$$\text{Half-width} \quad \Theta\ ['] \approx 4300\ \lambda/D. \tag{9.1}$$

   For instance, the 100-m telescope in Effelsberg (Fig 9.3) has an antennal lobe of half-width $2\rlap{.}'6$ at $\lambda = 6$ cm.
2. The energy per photon is, according to quantum theory, lower by the orders of magnitudes by which the radio waves are longer. Therefore, the unit of flux density, named the *jansky* (Jy) after the founder of radio astronomy and adjusted for typical

radio sources, is extremely small:

$$1 \text{ Jy} = 10^{-26} \text{ W m}^{-2} \text{ Hz}^{-1}. \tag{9.2}$$

As an illustration, the Effelsberg radio telescope with a collecting surface of 7854 m^2 and a receiving bandwidth of 500 MHz receives from a typical source less than 10^{-12} Watt! The power received from a nearby radar or radio transmitter is stronger by many orders of magnitude, as can be easily computed from the known transmission power and distance. Consequently, radio astronomy in wavelength ranges occupied by broadcast transmitters is not practicable.

3. The ratio of the size of a scattering particle to the wavelength permits an estimation as to the type or amount of scatter which occurs. Experience shows interstellar dust grains to be so small that all radio waves penetrate unimpeded. Raindrops and hailstones cause, in the millimeter range (where particle size is comparable with wavelength) *Mie scatter* and damping, while in the centimeter range (λ much larger than particle size) *Rayleigh scatter* and negligible damping occur. Water and ice clouds scatter the radiation originating from the Earth's surface. For longer radio waves, the water drops are so negligibly small compared with the wavelength that even Rayleigh scattering can be neglected.

9.2.1 Thermal Radiation

The radiation emitted by an ideal blackbody is described in the radio range by the *Rayleigh–Jeans approximation*, which gives for the (monochromatic) surface brightness

$$B_\nu = 2kT_b \lambda^{-2} \quad [\text{W m}^{-2} \text{ rad}^{-2} \text{ Hz}^{-1}], \tag{9.3}$$

where the Boltzmann constant is $k = 1.38 \times 10^{-23}$ J K^{-1}, and T_b is the radiation temperature. Integration of the brightness distribution over the angular area of the source gives the *flux density* S_ν. For a constant radiation temperature over the surface, a simplified expression for S_ν is

$$S_\nu = B_\nu \Omega = 2kT_b \lambda^{-2} \Omega \quad [\text{Jy}], \tag{9.4}$$

where Ω is the solid-angle size of the source in rad^2. For small angles, $\Omega = \pi R^2$, with $R =$ radius of source in radians.

Example: Calculate the surface brightness and radiation flux from Venus for a frequency of 43 GHz, a radiation temperature of 405 K, and an apparent radius of the planetary disk of 8$''$17 (solid angle $\Omega = 4.93 \times 10^{-9}$ rad^2).

Upon substituting the given values into Eqs. (9.3) and (9.4), we obtain directly that $B_\nu = 2.29 \times 10^{-16}$ W m^{-1} rad^{-2} Hz^{-1} and $S_\nu = 1.13 \times 10^{-24}$ W m^{-2} Hz$^{-1} = 113$ Jy.

As expected from Eq. (9.4), the thermal radiation of most Solar System bodies is proportional to the frequency squared; see Fig. 9.4.

Equation (9.4) between radiation temperature and flux is, of course, strictly valid only for blackbody radiation, but it has been customarily used to define an *equivalent*

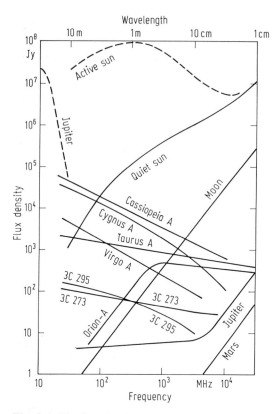

Fig. 9.4. The flux density of some strong radio sources as a function of frequency in MHz. *Dashed lines* show typical intensities in radiation bursts. For solar system bodies, the mean distance is assumed and phase effects are neglected.

radiation temperature. Often the flux contributions of several sources superpose without the individual fluxes and the solid angles of every source being separated. The equivalent temperature $\overline{T}_b$ averaged over the main beam of the antenna is then used; its value formally represents a flux density per main beam.

How is this scale of surface brightness or flux density established? In 1928, H. Nyquist found from thermodynamical considerations that an Ohmic resistance of temperature T_a emits the same noise power as would be received by an antenna placed in a black cavity of the same temperature. The relation between T_a and the resistance noise of the receiver can be extrapolated to lower values and into the range of radio astronomy observations. For calibration, the receiver is circuited with a resistor heated to various temperatures T_a. Recently, the thermal calibration has often been obtained by surrounding the feed horn of the receiver with a metal container, where absorbing material is cooled to the temperature of liquid gas (e.g., nitrogen (N), argon (Ar), helium (He)). The cooled absorber makes the container a black cavity with the radiation temperature of the boiling gas. As the evaporation temperatures of these gases are all very low, only a small extrapolation of the calibration will be needed. In the black cavity, $T_a = T_b$.

Often the source is not much larger in angular diameter than the antenna beam. In such cases,

$$T_a = \eta_b T_b, \tag{9.5}$$

where η_b is the beam efficiency; a typical value is about 75%.

Hydrogen clouds near early-type stars are almost completely ionized, as the high-energy, UV-photons emitted by the hot stars dissociate the electrons from the atoms. Free electrons passing close to protons are decelerated in their electric fields and emit *thermal bremsstrahlung*. As the electrons are bounded neither before nor after the collision with the protons, and are not tied to certain energy levels, the continuous radiation formed in this manner is also called *free–free radiation*.

The absorption of an ionized gas can be calculated. For the optical depth τ, the integrated absorption along the line of sight, Mezger and Henderson [9.5] give the following approximation:

$$\tau = 0.08235 \, T_e^{-1.35} \nu^{-2.1} \int_0^l N_e^2 \, ds, \tag{9.6}$$

where T_e (K) is the electron temperature, ν (GHz) the frequency, N_e (cm^{-3}) the electron density, and l (pc) the length of the path along the line of sight. The integral of Eq. (9.6) is called the *emission measure*:

$$E = \int_0^l N_e^2 \, ds [\text{pc cm}^{-6}]. \tag{9.7}$$

The brightness temperature T_b of the ionized hydrogen is:

$$T_b = T_e \left(1 - e^{-\tau}\right). \tag{9.8}$$

For large values of τ (i.e., at long wavelengths), $T_b = T_e$, while for small values of τ, $T_b = \tau T_e$. Converted to flux densities with Eq. (9.4), we obtain a radio spectrum similar to that of Orion A in Fig. 9.4: at low frequencies, with $\tau \gg 1$, $S_\nu \propto \nu^2$, while at high frequencies, with $\tau \ll 1$, $S_\nu \propto \nu^{-0.1} \approx$ constant.

Example: Estimate l (pc) and N_e (cm^{-3}) of Orion A. The measured (deconvolved) values at 5 GHz are: $T_b = 273$ K, $l = 4'$. Assuming $T_e = 7000$ K, the optical depth is $\tau = 0.039$. Using Eqs. (9.6) and (9.7), $E = 2.2 \times 10^6$ pc cm^{-6}. At 0.5 kpc, the size is $l = 0.5 \tan(4') = 0.6$ pc; we assume the same size in the line of sight. If N_e is constant, Eq. (9.7) becomes $E = N_e^2 \, l$ and thus $N_e = 1900$. The estimate of the volume and the mass is straightforward.

An electron losing energy by multiple collisions with protons and emitting it as free–free radiation may recombine with a proton to form a neutral atom. The electron drops from high energy levels, cascading to lower ones while emitting *recombination lines*. The prediction of intensities of recombination and many other spectral lines in the radio range is difficult, as the gas clouds are often not in thermal equilibrium, and stimulated or *maser emission* may occur.

9.2.2 Nonthermal Radiation

Radiation of an inexplicably high equivalent temperature ($\gg 10^6$ K) or with a wavelength dependence not represented by a thermal spectrum, is appropriately called *nonthermal radiation*. A useful criterion is the *spectral index* α, defined by the relation $S \propto \nu^\alpha$. A detailed study of thermal spectra shows that they must comply with the condition $\alpha \geq -0.1$, and hence $\alpha < -0.1$ necessarily implies radiation of nonthermal origin.

The commonly assumed mechanism for nonthermal radiation is *magnetobremsstrahlung*, also called *synchrotron radiation*, generated when a relativistic electron moving in a magnetic field is forced into a spiral path perpendicular to the field lines. It emits polarized radiation, whose intensity and frequency depends on the speed and the magnetic field, into a narrow cone tangential to the spiral. The combined effect of many relativistic electrons of various energies generates the observed continuous spectrum; for most nonthermal sources, the flux increases with wavelength. A typical spectrum of this kind is given in Fig. 9.4, that of the supernova remnant Cassiopeia A.

Even at low electron speeds due to the thermal motion of the gas, a low-frequency gyromagnetic radiation at the *Larmor frequency* occurs; this is observed in the kHz-range in the ionosphere.

Other nonthermal mechanisms of radiation are *plasma oscillations*, caused by a departure from the equilibrium state in an ionized medium, and *Cerenkov radiation*, through the deceleration of high-energy particles in a plasma. Both kinds of radiation are narrow band, and are seen only occasionally in the solar spectrum.

9.3 Atmospheric Influences

Depending on the wavelength, radio waves usually propagate predominantly either by surface waves or space waves. In the long- and medium-wavelength radio ranges (LF, MF), the surface wave, which follows the Earth's surface over hundreds of kilometers, is utilized. In the shortwave (HF) realm, however, the surface wave damps out after only a few kilometers, and so the space wave, which permits worldwide communication via reflection in the high atmosphere (specifically the ionosphere) and at the ground, is used instead. For even shorter wavelengths (VHF, UHF, EHF), the ionosphere is virtually transparent; since the propagation of these waves is by space wave, transmitter and receiver must be practically within sight of each other.

9.3.1 The Ionosphere

At altitudes above 50 km, the powerful UV-radiation from the Sun ionizes some of the oxygen and nitrogen molecules. The *ionosphere* thus generated may reach heights of well over 500 km. It is composed of various layers which represent ionization maxima, and are designated by letters: D at 50 km, E_1 at 100 km, E_2 at 150 km, F_1 at 190 km, and F_2 over 250 km. The intensity of ionization depends directly on the strength and duration of UV-radiation at that level in the ionosphere, or, more specifically, on the altitude of the Sun, the time of day, the season, and the solar activity, the latter often being expressed in terms of the relative sunspot number. There is also a dependence on the geographic position and on the Earth's magnetic field, which renders the ionosphere bi-refracting. All of these factors conspire to make a quantitative prediction of the ionospheric structure at a given moment, and thence of the propagation of short waves, exceedingly difficult. The structure of the ionosphere is constantly monitored by nearly 60 stations scattered over the Earth by measuring the altitude of reflection as a function of frequency: the square root of the limiting frequency at the point of onset of reflection is proportional to the electron density. These measurements show that the F_1- and F_2-layers are separated only in the daytime but merge at night, that the E-layer in twilight is located 50–80 km higher than at noon, and so on. The limiting frequency also gives the limit above which cosmic radio radiation is received on Earth. Extreme values for this limit occur in winter: the limiting frequency is over 12 MHz at noon, and around 2 MHz at night (at respective electron densities 2×10^6 and 5×10^4 cm^{-3}). Quantitative radio astronomical measurements near the limiting frequency are not reliable since the refraction may amount to several degrees, the inhomogeneities of the ionosphere make the sources scintillate, and the extinction is difficult to assess. Earth-based observations are possible above about 20 MHz. Refraction and extinction by the electron gas are proportional to the square of the wavelength, and thus their influence on cm-wavelengths is negligible.

The ionosphere is also subject to occasional disturbances. These disturbances are the subject of the following several subsections.

9.3.1.1 The Mögel–Dellinger Effect. This effect is almost always associated with the occurrence of chromospheric eruptions (i.e., flares), and causes a strong increase in the atmospheric noise in the kHz range and the cessation of ionospheric echoes of short waves. The increased UV-radiation of the Sun enhances the ionization of the lower atmosphere, leading to a strong shortwave absorption. The durations range from a few minutes to several hours.

9.3.1.2 Ionospheric Storms. These are caused by solar particle radiation triggering violent motions in the ionosphere. The height of the F_2-layer increases greatly, the limiting frequency diminishes, and the reflections weaken. A strong storm can raise the height of the F_2-layer to 1000 km and suppress the reflection at this layer via strong damping. Such storms are ordinarily accompanied by polar aurorae and geomagnetic storms, and they occur about 20 to 40 hours after chromospheric eruptions. Their time scale is on the order of several hours. Ionization occurs in the vicinity of the aurorae and is so strong that it extends the range of radio communications in the VHF-band.

9.3.1.3 Ionization by Meteors. Meteors vaporizing in the upper atmosphere can leave ionization trails strong enough to reflect VHF waves. A single meteor may cause a radio reflection lasting 0.1 to a few seconds; during meteor showers, up to 100 such reflections per minute may occur.

9.3.2 The Troposphere

While the "radio window" of the atmosphere is limited at the long-wavelength end of the electromagnetic spectrum by the ionosphere, the other boundary in the millimeter-range is caused by, numerous spectral absorption lines of water vapor and oxygen. At wavelengths of 2 cm and above, only a weak, continuous absorption by these gases occurs, typically less than 1% at the zenith. Since the atmosphere is in thermal equilibrium, this damping results in the production of weak atmospheric radiation with τ_0 as the absorption at the zenith, and T_{At} as the mean temperature of the atmosphere. The atmospheric radiation can be estimated in analogy to Eq. (9.8):

$$T_b = T_{At}\left(1 - e^{-\tau_0 \sec z}\right), \tag{9.9}$$

where $\sec z$ is an approximation to the air mass traversed at zenith distance z. The absorption increases strongly at shorter wavelengths. The absorption constant τ_0 can be computed from the observed intensity ratios of a source at various zenith distances, and, since the temperature T_{At} of the atmosphere is readily found, its emission can therefore be calibrated. This reasoning can of course be reversed: with the calibration known, the atmospheric radiation permits the determination of the absorption constant τ_0, which is comprised of contributions of oxygen, which is essentially constant, and of water vapor along the line of sight through the atmosphere, which is strongly variable. Thus, the measure of the atmospheric radiation allows the determination of the total amount of water vapor, a quantity whose determination would otherwise have to be left to the more direct—and expensive—technique of balloon ascents.

It is normally assumed that optical and radio refractions are equal. More precisely, the optical refractions by dry air and by water vapor are almost identical; the radio refraction is somewhat less for dry air and much greater for water vapor than it is for light waves, and it does not depend on wavelength. The high refractive index of water vapor favors the possibility of total reflection of radio waves by inversion layers in the lower atmosphere. Multiple reflections on the ground together with those from the inversion layer conduct the radio wave as in a waveguide. This causes propagation beyond the horizon, which can lead, for instance, to interference in television reception, when a signal from a distant station is superimposed onto that of a local station. The same effect can often be observed with the radio sun at sunrise and sunset as a "Fata Morgana," named for the legendary sister of King Arthur who could create towering castles in the air; it is in reality a type of mirage which occurs most frequently over extended water surfaces.

9.3.3 Artificial Interference and Protected Frequencies

In addition to local and distant thunderheads, electrical and electronic equipment of all kinds generate radio interference. Even when measures are taken to screen out such unwanted signals, the remaining interference level is often too high to permit radio astronomical observations at longer wavelengths; in such cases, a suitably remote observing site needs to be found. In order to maintain an organized worldwide broadcast communications network, the International Telecommunications Union (ITU) has distributed the frequency bands to the various services; radio astronomy has been recognized as one of the services since 1959. In early 1992 the World Administrative Radio Conference (WARC) of the ITU reviewed the allocations; all frequencies reserved for radio astronomy remain protected. The frequency bands presently allocated to radio astronomy are listed in Table 9.1.

But radio astronomy cannot limit itself to the assigned bands because it could not then search for new molecules and redshifted lines of known atoms and molecules.

9.4 Instrumentation

9.4.1 Antennas

At a given wavelength of observation, the sensitivity of a radio telescope is determined by the area of its collecting surface, while the angular resolution depends upon its effective diameter. For a given geometric area A_g, the area elements may be arranged in various ways, for instance, in a full circle, as in the 100-m Effelsberg telescope (Fig. 9.3), or in a rectangle, as in the 40 × 200-m telescope in Nancay, or in a narrow ring, such as in the 600-m radio telescope in the Special Astrophysical Observatory in the Caucasus mountains (Fig. 9.5), or in many individual telescopes working together as a coordinated unit, as is the case for the fourteen 25-m telescopes of the Westerbork interferometer, which has a base length of 2.7 km.

The maximum angular resolution of a telescope is estimated from its diameter using Eq. (9.1), but the value thus obtained is not necessarily a yardstick of the worth of the instrument. For instance, the radio telescope at Effelsberg is a multipurpose instrument particularly suited for spectroscopic studies, while the Westerbork Synthesis Telescope is specialized toward securing the highest possible resolution. The superb Very Large Array (VLA), a y-shaped interferometer of 27 km diameter near Socorro, New Mexico, can be used like a single dish (phased array) for VLBI measurements or as a synthesis instrument for continuum or spectroscopic mapping. The wide variety of telescope systems is typical of radio astronomy, and reflects how strongly the design is adapted to the intended use.

Consider first an antenna with a parabolic reflector whose surface consists of a conducting and connected metal layer. A plane wavefront incident along the parabola axis on the reflector is converged after reflection in phase in the focus. A receiving element (dipole, helix, or horn) located at the focus forwards the coherent radiation to the receiver proper. When there are many small, statistically distributed deviations from the parabolic shape, the reflecting dish conveys the wave to the focus with some components of the wave out of phase with the rest of the wave; the antenna beam is thereby broadened and the received intensity reduced. Investigating the accuracy requirements of antennas, J. Ruze derived an efficiency factor η_s which expresses the diminished power compared with the ideal telescope. For a root-mean-square (rms) deviation σ from the paraboloid and a wavelength λ, the efficiency factor is given by

$$\eta_s = e^{-(4\pi\sigma/\lambda)^2}. \tag{9.10}$$

Table 9.1. Radio bands allocated to radio astronomy.

Band	Protection	Use
13.36– 13.41 MHz	(2)	Continuum
25.55– 25.67 MHz	(1)	Continuum
37.50– 38.25 MHz	(2)	Continuum
73.00– 74.60 MHz	(1)	Continuum
150.05– 153.00 MHz	(1)	Continuum
322.00– 328.60 MHz	(2)	D and Continuum
406.10– 410.00 MHz	(2)	Continuum
608.00– 614.00 MHz	(1)	Continuum
1330.00–1400.00 MHz	(2)	Redsh. H and Continuum
1400.00–1427.00 MHz	(1)	H
1610.60–1613.80 MHz	(2)	OH
1660.00–1670.00 MHz	(2)	3 lines OH
1718.80–1722.20 MHz	(2)	OH
2655.00–2690.00 MHz	(2)	Continuum
2690.00–2700.00 MHz	(1)	Continuum
3260.00–3267.00 MHz	(2)	CH
3332.00–3339.00 MHz	(2)	CH
3345.80–3352.50 MHz	(2)	CH
4800.00–4990.00 MHz	(2)	H_2CO + Continuum
4990.00–5000.00 MHz	(2)	Continuum
10.60– 10.68 GHz	(2)	Continuum
10.68– 10.70 GHz	(1)	Continuum
14.47– 14.50 GHz	(2)	H_2CO
15.35– 15.40 GHz	(1)	Continuum
22.01– 22.50 GHz	(2)	H_2O
22.81– 22.86 GHz	(2)	NH_3
23.07– 23.12 GHz	(2)	NH_3
23.60– 24.00 GHz	(1)	NH_3 + Continuum
31.20– 31.30 GHz	(2)	Continuum
31.30– 31.80 GHz	(1)	Continuum
36.43– 36.50 GHz	(2)	H^+
42.50– 43.50 GHz	(2)	SiO + Continuum
48.94– 49.04 GHz	(2)	CS
51.40– 54.25 GHz	(1)	Continuum
58.20– 59.00 GHz	(1)	Continuum
64.00– 65.00 GHz	(1)	Continuum
72.77– 72.91 GHz	(2)	H_2CO
86.00– 92.00 GHz	(1)	CO and other lines
93.07– 93.27 GHz	(2)	HN_2^+
97.88– 98.08 GHz	(2)	CS
105.00– 116.00 GHz	(1)	CO

Protection: (1) no broadcast permitted, (2) limited broadcast that does not interfere with radio astronomy. See refs. [9.32], [9.33] for details.

Fig. 9.5. The ring structure of the 600-m radio telescope of the Special Astrophysical Observatory located in the Caucasus mountains of Ukraine. The telescope can be used in several different modes: as an annular telescope with a maximum resolution of $3''$ at 1 cm, looking toward the zenith; or with each of the four sectors working as independent telescopes; or using the two southern sectors with the flat mirror near the center, working as a Kraus or Nancay type telescope.

It is expected that, in the construction of new telescopes, the lower wavelength limit λ_0 to be observationally used will still have an efficiency $\eta_s = 67\%$ corresponding to an rms deviation $\sigma = \lambda_0/20$. Even deformations as small as $\sigma = \lambda_0/10$ reduce η_s to 20%, and the antenna beam is then so broadened (or substantial side lobes appear) that useful observations will be hampered. Differential heating by sunshine can degrade the surface accuracy. Therefore, major telescopes are protected by a special paint, which reduces temperature differences to below $3°C$. This is not sufficiently small, however, for millimeter telescopes; consequently, the James Clerk Maxwell Telescope in Hawaii is built into a dome, and the IRAM 30-m radiotelescope on Pico Veleta is itself temperature controlled to temperature differences below $1°C$.

Some additional concepts will help to characterize the telescopic properties. The *gain G* is the ratio of the solid angle of the entire sphere (symbolizing an isotropic radiator) to the solid angle of the antenna Ω_A at the same received energy:

$$G = \frac{4\pi}{\Omega_A}. \tag{9.11}$$

The effective collecting area A_{eff} of the telescope is less than the geometric area A_g, owing to the imperfect surface and the blockage by the feed support structure; the ratio of these areas is called the *aperture efficiency* η_a. Effective area and the resulting solid angle are related through the simple formula

$$A_{\text{eff}} = \frac{\lambda^2}{\Omega_A}, \tag{9.12}$$

which can be verified by consideration of proportionality. Combining Eqs. (9.11) and (9.12) gives

$$G = \frac{4\pi A_{\text{eff}}}{\lambda^2}. \tag{9.13}$$

For example, at $\lambda = 3$ cm the 100-m Effelsberg telescope has a gain of $G \approx 5 \times 10^7$. The gain is customarily expressed in logarithmic form in decibels (db):

$$G \text{ [db]} = 10 \log G \approx 10 \log(5 \times 10^7) \approx 77 \text{ db.}$$

The effective area is an important characteristic for rating a telescope; point sources of known flux are used to measure it. Combining Eqs. (9.4) and (9.12), the radiation flux for the system calibrated in T_a is

$$S_\nu = \frac{2kT_a}{A_{\text{eff}}} = \frac{2kT_a}{\eta_a A g}. \tag{9.14}$$

For observations above the limiting wavelength, the antenna efficiency is around 50%. With the effective area known, the antenna temperature can be estimated from the flux.

Example: Calculate the ratio of antenna temperature (K) and flux density (Jy) for the 100-m telescope and compare it with that for an amateur telescope of diameter 1 m for a wavelength of 3 cm, where $\eta_s \approx 1.0$ and $\eta_a = 0.47$.

The large telescope gives

$$\frac{T_a}{S_\nu} = \frac{\eta_a A}{2k} = \frac{0.47 \times 7854 \text{ (K m}^2\text{)}}{2 \times 1.38 \times 10^{-23} \text{ (Joule)}} = 1.34 \ (10^{26} \text{ K/Jy)}.$$

For the small telescope, this ratio is 0.00013!

A radio source of 1 Jy then gives an antenna temperature of 1.3° in the 100-m telescope. The same source would, in a 1-m reflector, result in a signal of 0.0001°, which illustrates that the sensitivity limit is quickly reached when using smaller antennas.

The two most preferred receiving antennas are the *Yagi antenna* in the VHF and UHF bands and the shortwave range, and the *paraboloid antenna* for the decimeter and centimeter wavelength ranges; both kinds are commercially available. Other inexpensive antenna types such as corner reflectors and dipole arrays can be built according to instructions given in, for example, the *Antenna Book* of the American Radio Relay League [9.6].

9.4.1.1 Paraboloid Reflectors. These antennas are now mass produced with diameters from 0.4 to about 2 meters for television reception, via satellite, of signals near 11.5 GHz. Antenna drive systems with a limited positioning range (azimuth 180°, elevation 40°) are also commercially available. Off-axis parabolic satellite antennas have recently become popular; they have a higher efficiency (no blockage by feed support legs, etc.) and allow lower system temperatures (less stray radiation and diffraction); hence, they are more sensitive. It is because of these attributes that the decision was made to construct the new 100-m telescope of the National Radio Astronomy Observatory (NRAO) in Green Bank as an off-axis telescope.

9.4.1.2 Yagi Antennas. These are the typical TV antennas. They have the advantage of being easily mounted and adjusted, and the drawbacks of a rather small antenna surface and a limited wavelength range. Manufacturers of these antennas often state the gain G in db relative to a $\lambda/2$-dipole. These G-values are to be increased by 2.1 db in order to be comparable with the isotropic radiator assumed in Eq. (9.11).

9.4.1.3 Composite Antennas and Interferometers. One obvious way to increase the antenna area is through the rigid combination of several Yagi or paraboloid antennas, combining the signals of the elements correctly in phase; thus, an effectively large antenna is synthesized.

In an interferometer, each antenna is individually mounted. For every moment and each antenna, the computer calculates and inserts a time delay in order to make the signals arrive simultaneously. The signals from all pairs of antennas are then correlated with an alternating delay of one-half wavelength (phase switching) to suppress disturbances. Every pair of antennas measures a correlated signal only when the source is smaller than the resolution calculated by Eq. (9.1). The antenna pairs thus measure the angular components of the brightness distribution along the baseline. As the baseline rotates with the Earth relative to the source, the latter can be completely measured and thus the brightness distribution deduced or an image of the source generated. This technique is called *aperture synthesis*. Its technical and computer demands are evidently high. Simpler interferometers used by amateurs are transit instruments with or without phase switching. The monograph by R.A. Thompson et al. [9.7] is the textbook for radio interferometry.

9.4.2 Radio Receivers

The features of receivers are better understood when the properties of cosmic radio radiation are visualized; unlike broadcast radio, cosmic radiation is neither modulated by nor associated with a fixed, coherent carrier frequency. It corresponds to a noise which is randomly and statistically distributed in time and frequency. Recall the definition of antenna temperature: the thermal motion of the electrons in the resistance generates the noise power. On the way from the collecting part (e.g., a dipole) over a conductor to the receiver, the radiation suffers resistance losses at the various parts; each part then adds to the signal a noise temperature corresponding to the losses, the largest contribution coming from the internal noise of the receiver. The equivalent of the total of noise contributions is called the system temperature T_s, and it substantially determines the sensitivity of the receiver.

A modern radio telescope usually employs a superheterodyne receiver, the principle of which is also used in broadcast receivers: the signal from the antenna with frequency ν_s is amplified in a low-noise preamplifier; the still very weak signal is blended in a mixer with a signal of frequency ν_0 from a *local oscillator*, resulting in a signal at intermediate frequency $\nu_s - \nu_0$ proportional to the signal from the antenna. The amplified intermediate frequency then enters a quadratic rectifier (a broadcast receiver would have here a demodulator). For direct recording, the rectifier is followed by a low-frequency amplifier and an integration unit (to set integration time or time constant; see also Sect. 9.6.4) followed by a printer or, alternatively, an analog digital converter which renders the current from the rectifier computer-readable.

The superheterodyne principle has two distinct advantages: 1. the major signal amplification occurs at the intermediate frequency which is technically easier to handle, and 2. for a series of receivers, the intermediate frequency can be chosen such as to be the same for all. This greatly simplifies the operation, as all processing instruments such as spectrometers, polarimeters, digital back-end, etc., are connected with the one intermediate frequency. To change the wavelength of observation, only the preamplifier with mixer and local oscillator is replaced. Preamplifiers used in the centimeter-range are *cooled parametric amplifiers, masers, FETs,* or *HEMTs,* which are technically very demanding and expensive; the latter will not be discussed here. The significance of preamplifiers for a receiving system may be illustrated with an example of an amplifying train composed of two preamplifiers, and a mixer with an intermediate frequency amplifier, each with a noise temperature T_i and an amplifying factor V_i of the ith component. The system temperature is then estimated as follows:

$$T_s = T_1 + \frac{T_2}{V_1} + \frac{T_3}{V_1 V_2}. \tag{9.15}$$

For a sufficiently large V_1 of the first preamplifier, the system temperature is in practice determined by its noise temperature.

The demands made upon a receiver in commercial and amateur broadcasting are quite different from those in radio astronomy. The numbers stated by manufacturers to express quality and sensitivity differ also. One often used quantity is the *noise factor F*. It is related to the system temperature via the relation

$$T_s = (F - 1)T_0, \tag{9.16}$$

with T_0 as the air temperature (290 K). An ideal receiver without a noise contribution has the noise factor 1. If F is expressed in db, the noise number is

$$F\,[\text{db}] = 10 \log F.$$

The sensitivity of the receiver is defined as the root-mean-square variation of the noise:

$$\text{rms}\,[°T_a] = T_s(t\Delta\nu)^{-0.5}, \tag{9.17}$$

where t stands for the integration time. Table 9.2 compares the sensitivities per second of receivers at the 100-m Effelsberg telescope, calculated for radio astronomy purposes, with those of amateur and commercial broadcast receivers. The sensitivity of a radio telescope may be improved by increasing the bandwidth, or, in the case of a television receiver, by adding (if need be) a preamplifier.

Assessing the sensitivity of radio and TV receivers may be difficult due to the lack of product information. Also, the commercially available antenna amplifiers sometimes have a high noise temperature, since their purpose is usually not to increase the sensitivity but rather to compensate for the cable losses. Equation (9.15) can be used to estimate the effect of a preamplifier on the sensitivity: a broadband antenna amplifier of noise number $F = 2.5$ and a power amplification $V = 16$ would double the sensitivity of the 470 MHz receiver, while at a noise number $F = 5$ the sensitivity would have already started to deteriorate.

The sensitivities calculated from Eq. (9.17) are rarely reached owing to instabilities and fluctuations of amplification, which are caused, for instance, by changes in the ambient temperature or in the net voltage. Stabilizing the temperature and voltage does not suffice to reach the sensitivity limit, but *Dicke switching* has proven successful. An

Table 9.2. Sensitivities per second of receivers at the 100-m Effelsberg radio telescope and of various amateur and commercial broadcast receivers.

λ (MHz)	T_S (K)	$\Delta\nu$ (MHz)	rms (K)	Remarks
30	5500	0.01	55.0	amateur station
470	1100	5.00	0.5	TV receiver, Ch. 21
790	2000	5.00	0.9	TV receiver, Ch. 60
2 321	100	2.00	0.07	amateur sat., GaAs FET
11 325	220	100.0	0.02	satellite receiver
1 400	40	30.0	0.015	cooled FET, 2 channel
2 700	50	80.0	0.006	cooled FET, 3 channel
4 600	40	500.0	0.002	cooled param. amplifier
23 000	35	100.0	0.004	maser
31 000	210	1000.0	0.007	cooled mixer

automatic switch precedes the receiver and periodically alternates between the antenna and a comparison signal (load switching). Since the comparison signal is affected the same way as the antenna signal by amplifying fluctuations, these changes are eliminated in forming the difference. The condition for this to work is of course to have a Dicke switching frequency higher than that of the amplifying fluctuations. Experience has shown that switch frequencies between 1 and 50 Hz are most acceptable. The price for this advantage is the diminution of the theoretical sensitivity by a factor 2 through halving the integration time and through forming the differences of signals.

A variant of the Dicke circuitry is "beam switching," that is, alternating between two horns in the focal plane separated by about 3 half-power beam widths. In the near field of the antenna, both lobes pass through practically identical paths through the atmosphere. In the far field, the lobes are separated by the 3 half-power beam widths. Forming the difference then cancels the amplification variation as well as atmospheric fluctuations. This and other observing methods have been quantitatively discussed by G. Evans and C.W. Leish in the *RF Radiometer Handbook* [9.8].

9.4.3 The Computer

The presentation so far may leave the reader with the mistaken impression that telescope size and receiver quality alone determine angular resolution and sensitivity, and that the computer is merely an ornate embellishment. In fact, a modern telescope cannot be operated without it. The computer reads the atomic clock, calculates sidereal time, calculates instantly the needed telescope position, controls the tracking, regulates the receiver functions, and transfers the data obtained to an analog printer or screen, or to storage. The advantages of computer control are exemplified by a "beam-switching" measurement of 3C120, graphed in Fig. 9.6.

The Dicke radiometer circuit operates at a frequency of 31.25 Hz, alternating at 16 ms intervals between signals from the main horn and the reference horn, which are stored separately. The telescope was moved so that one horn after the other

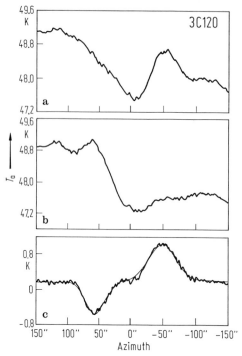

Fig. 9.6a–c. A "beam-switching" observation at $\lambda = 1.3$ cm with the 100-m Effelsberg telescope in May 1987. See explanation in the text.

passed across the calibrating source 3C120. Figure 9.6(a) shows the signal recorded by the main horn, Fig. 9.6(b) that from the reference horn. Both horns synchronously "see" a variable signal from the scatter by clouds, as has been described previously. Superposed on it is the signal from the source, displaced in the two channels by an amount corresponding to the separation of the horns. The difference of the two channels, graphed in Fig. 9.6(c), shows that atmospheric fluctuations and amplification variations are well suppressed. Main and reference lobes are well approximated by Gaussian curves.

The computer can also be used to improve the sensitivity. From Eq. (9.17), one expects the rms to be proportional to $t^{-0.5}$. In practice, this relation is fulfilled only for short integration times of up to a few seconds. Above this time, the rms does not decrease further. The computer makes it possible to use the good short-time stability: the radiation is recorded along a line traversing the faint source with a moderately low speed. With a computer, this measurement can be repeated accurately and as often as desired. It is possible to average the sections point-by-point and thereby to arrive at a large integration time. Figure 9.7 demonstrates this in a measurement of α Orionis (Betelgeuse) obtained in the same fashion as the calibrating measure from Fig. 9.6. α Orionis is a very weak object in the radio range. Ten scans do not yet show a signal (a). Averaging 40 observations (b) yields a trace of a signal. After 160 averaged scans (c), corresponding to 2 hours of monitoring time, the radio radiation

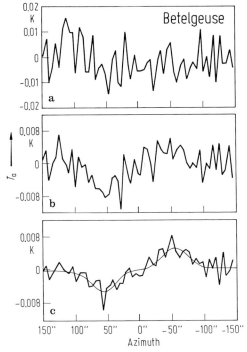

Fig. 9.7a-c. Observation of a radio star (Betelgeuse) using the "beam-switching" method at $\lambda = 1.3$ cm following the calibration in Fig. 9.6. See text for an explanation.

is distinctly recognized. Also shown in Fig. 9.6(c) are the position and amplitude of the radio star as found by the computer.

As with measurements of continuous radiation, digital techniques make it possible to improve the sensitivity for line observations by increasing the integration time. The average price of a middle-of-the-line personal computer (and the plethora of accessories which can be attached to it) has today fallen to such an affordable level that its use for data processing and, if need be, for the guiding of the observation, is a matter well worth considering.

9.5 The Objects of Observation

The following survey is oriented toward the instrumental facilities of amateur astronomers. Most modern trends of research, such as *very-long-baseline interferometry* (VLBI), aperture synthesis, millimeter observations, molecular spectroscopy, and others will be excluded.

9.5.1 Continuous Radiation

9.5.1.1 The Milky Way. Jansky's discovery led him to the noteworthy conclusions that the radio radiation which reaches maximum intensity in the direction of the galactic center is so strong that the building of more sensitive receivers is not worthwhile. The radiation is illustrated in the radio map of the entire sky at 150 MHz by T.L. Landecker and R. Wielebinski [9.9], as reproduced here in Fig. 9.8.

The observed radiation is indeed distinctly concentrated toward the galactic equator with a maximum at the galactic center. The hypothesis that this radiation originates in the spiral arms is supported by its narrow distribution in galactic latitude, and by the intensity gradients in longitude, which partly mark the boundaries of the inner spiral arms. High-latitude radiation is contributed to by the local spiral arm, by numerous extragalactic sources which have been smeared by the antenna lobe, and possibly by the galactic halo, a hypothetical spheroidal region of radiation surrounding the Milky Way galaxy.

Superposed on this large-scale distribution are a few loops, whose ring shapes suggest their identification as remnants of ancient supernova explosions. Most noticeable is Loop I centered at $l = 329°$, $b = 17°.5$, and with a diameter of $116°$; it contains the north polar spur, which points toward $l = 30°$ from the galactic equator to the north galactic pole.

The spectral index of the extended continuum is $\alpha \approx -0.8$. The radio map permits an estimation as to which radiation temperatures may be expected at other frequencies:

$$\overline{T}_b = \overline{T}_{b(150 \text{ MHz})} \left(\frac{\nu}{150 \text{ MHz}}\right)^{\alpha - 2.0}, \tag{9.18}$$

where the value -2.0 in the exponent arises from the conversion from the flux to

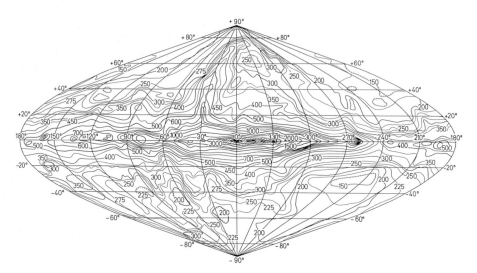

Fig. 9.8. Radio map of the entire sky at 150 MHz in galactic coordinates. The isophotes are labeled by radiation temperatures in K (after T.L. Landecker and R. Wielebinski).

radiation temperature (Eq. (9.4)). Extrapolating the radiation temperatures with this relation from $\lambda = 2.0$ m down to centimeter waves, the radiation should be unobservable except for the inner portions of the Milky Way. Extrapolated the other way to 20 MHz ($\lambda = 15$ m), the mean temperature of the sky becomes higher than 30 000 K. Every shortwave antenna thus "sees" at least this temperature, and the system temperature cannot be lower. This signifies, first, the confirmation of Jansky's find that the sensitivity for terrestrial shortwave radio is limited by cosmic radio radiation, and, second, that this radiation, because of its intensity, is easily observable, if not masked by interference. The much-cited radio map of the sky at 200 MHz by F. Dröge and W. Priester [9.10] was obtained with a 5×5-m dipole array.

9.5.1.2 Discrete Galactic Sources. The radiation field at the galactic plane, when observed in more detail at shorter wavelengths and with better angular resolution, is resolved into numerous discrete sources which superpose on an extended background of radiation, which in turn consists of a thermal and a nonthermal component. (It is the nonthermal part which determines the concentrated, long-wavelength radiation at the galactic equator, as shown in Fig. 9.8.)

Most of the discrete sources are HII regions, which, as they are excited only by early-type stars, indicate places of star formation. Nearly all HII regions are located within the spiral arms, and many of them are invisible from Earth owing to strong absorption by intervening dust. They are readily identified from radio spectra and their physical parameters derived. At the time of writing, about 1200 HII regions have been catalogued.

Table 9.3(a) lists 10 radio-bright objects which can be viewed from the northern hemisphere of the Earth. The radio name is the number in the Westerhout Catalogue (W) or in the Third Cambridge Catalogue (3C), or it is the constellation code with a following letter that indicates the relative brightness. Ori A, the strongest radio source in Orion, is often considered a "model" HII region. A small number of emission nebulae are found upon study to be in reality supernova remnants (SNRs), which emit nonthermal radiation. They often have, both in the optical and radio images, a ring structure like that of the the Cygnus Veil Nebula or Cassiopeia A, but this is not an unambiguous characteristic. The Rosette Nebula, for instance, is a ring-shaped HII region.

Other characteristic features of SNRs are the nonthermal spectral index and the absence of recombination lines. Table 9.3(b) lists 10 especially strong SNRs out of about 170 currently catalogued.

About three supernova events per century are expected in our galaxy; this frequency could suffice to explain the observed cosmic rays and the nonthermal radio background. Such statistical expectations are derived from the frequency of events in other galaxies and are, of course, uncertain, but this value is probably not too far off the mark. Although no galactic supernova has been observed now for 280 years[2], it is quite possible that one or more have occurred during that period but at such vast distances as to be obscured from view by interstellar dust.

---

2 The recent supernova SN 1987A occurred in the Large Magellanic Cloud, a nearby satellite galaxy, and not in the Milky Way proper.

Table 9.3. Fluxes and redshifts of typical strong radio sources.

Name	R.A. [2000.0]	Decl. [2000.0]	Flux[a] [Jy]	z[c]	Calibr.[d]	Identification/Remarks
(a) HII Regions/Emission Nebulae						
W 3	$02^h25^m42^s$	$+62°05'30''$	61			IC 1795
Ori A	$05^h35^m18^s$	$-05°23'46''$	382			NGC 1976, 3C145
Ori B	$05^h41^m44^s$	$-01°54'19''$	54			NGC 2024, 3C147.1
W 28	$18^h00^m38^s$	$-24°04'46''$	23			M 20
W 29	$18^h03^m45^s$	$-24°23'10''$	85			M 8, NGC 6523
W 33	$18^h14^m10^s$	$-17°56'06''$	50			IC 4701
W 37	$18^h18^m44^s$	$-13°46'16''$	108			M 16, NGC 6611
W 38	$18^h20^m29^s$	$-16°10'54''$	535			M 17, NGC 6618
W 49A	$19^h10^m17^s$	$+09°05'59''$	50			NRAO 598
W 51	$19^h23^m41^s$	$+14°30'22''$	117			3C400
(b) Supernova Remnants (SNR)						
3C10	$00^h25^m36^s$	$+64°07'28''$	21			SNR of Tycho's Star (1572)
3C58	$02^h05^m27^s$	$+64°52'20''$	27			
Tau A	$05^h34^m32^s$	$+22°01'00''$	687		A	M 1, Crab Nebula SNR of 1054
3C157	$06^h17^m10^s$	$+22°42'29''$	75			IC 443
3C358	$17^h30^m43^s$	$-21°30'17''$	7			SNR of Kepler's Star (1604)
W 41	$18^h34^m20^s$	$-08°54'48''$	24			
3C391	$18^h49^m18^s$	$-00°52'56''$	10			NRAO 583
W 44	$18^h56^m08^s$	$+01°18'58''$	149			3C392
W 49B	$19^h11^m07^s$	$+09°05'13''$	20			3C398
Cas A	$23^h23^m27^s$	$+58°49'16''$	1084		A	3C461, SNR of ca. 1700.
(c) Radio Galaxies[b]						
3C84	$03^h19^m48\!.\!^s2$	$+41°30'42''$	46.3	0.018	V	NGC 1275, Seyfert Galaxy
3C123	$04^h37^m04\!.\!^s4$	$+29°40'15''$	16.5	—	R	
3C218	$09^h18^m06\!.\!^s0$	$-12°05'45''$	13.5	0.053	R	D Galaxy
3C270	$12^h19^m23\!.\!^s9$	$+05°49'21''$	8.8	0.007		E Galaxy
Vir A	$12^h30^m49\!.\!^s6$	$+12°23'21''$	71.9	0.004	R	M 87, E Galaxy
3C295	$14^h11^m20\!.\!^s7$	$+52°12'09''$	6.4	0.460	R	D Galaxy
Her A	$16^h51^m08\!.\!^s3$	$+04°59'26''$	11.8	0.157	R	3C348, D Galaxy
3C353	$17^h20^m29\!.\!^s5$	$-00°58'52''$	21.2	0.031	R	D Galaxy
3C390.3	$18^h42^m17\!.\!^s1$	$+79°45'56''$	4.4	0.057		N Galaxy
Cyg A	$19^h59^m28\!.\!^s2$	$+40°45'42''$	459	0.057	A	3C405, D Galaxy
(d) Quasars						
3C48	$01^h37^m41\!.\!^s3$	$+33°09'35''$	5.2	0.367	R	
3C138	$05^h21^m09\!.\!^s8$	$+16°38'22''$	4.1	0.759		
3C147	$05^h42^m36\!.\!^s1$	$+49°51'07''$	8.0	0.545	R	
3C196	$08^h13^m36\!.\!^s0$	$+48°13'02''$	4.4	0.871		
3C273	$12^h29^m06\!.\!^s7$	$+02°03'09''$	5.8	0.158	V	
3C279	$12^h56^m11\!.\!^s2$	$-05°47'22''$	14.9	0.536		
3C286	$13^h31^m08\!.\!^s3$	$+30°30'33''$	7.3	0.849	R	
3C309.1	$14^h59^m07\!.\!^s5$	$+71°40'20''$	3.4	0.904		
3C380	$18^h09^m31\!.\!^s7$	$+48°44'46''$	6.2	0.691		
3C454.3	$22^h53^m57\!.\!^s7$	$+16°08'54''$	17.4	0.859	V	

[a] For wavelengths of 6 cm.
[b] Galaxy classification in the Yerkes system: E are elliptical galaxies, D are dumbbell galaxies, and N galaxies have small, unusually bright nuclei.
[c] The redshift is defined as $z = \Delta\lambda/\lambda_0$.
[d] The standard sources for intensity and constancy are given in absolute (A) and relative (R) calibration. The strongly variable sources (V) are unsuitable as standards. For details and other frequencies, see Baars et al. [9.11].

The Crab Nebula may well be the most exotic SNR. Its shell has an unusually high expansion speed of 30 000 km s^{-1}, its light is strongly polarized—apparently as optical synchrotron radiation; in the center of the explosion is a radio pulsar with the remarkably short period of 0.033 s, which is also observable optically. (It is in fact the only pulsar for which the pulses can be detected at visible wavelengths.) The radio radiation of this pulsar is exceedingly variable: at 430 MHz, the intensity level rises from practically zero to over 20 000 Jy in just one pulse. This is sufficient to be detected as interference on an ordinary television receiver! Unfortunately, such strong pulses are rare. Even at the largest radio telescopes, most of the 500 known pulsars were recognized only after averaging many pulse periods. These neutron stars, as well as the nearly 350 known radio stars, are within reach of the sensitivity of only the largest instruments.

9.5.1.3 Extragalactic Sources. Reber succeeded in confirming the radio emission from the Milky Way, but failed to detect the Andromeda Nebula as a radio source. The attempts of other observers to detect optically known objects in the radio range were similarly unfruitful. The inverse method, to first find radio sources and then identify them optically, met with early success when in 1954 the source Cyg A was identified with a faint type-D galaxy. To distinguish them from the "normal" galaxies, such as M31, M33, etc. (at that time not detectable at radio wavelengths), the newly identified objects were called *radio galaxies*. A large percentage of sources remained unidentified.

After 1960, positions with accuracies of a few arcseconds were obtained and the coincidence of some sources with star-like objects was revealed. These so-called *quasars*, optically distinguished by blue colors and UV-excesses, show redshifts which are indicative of high recessional velocities, some approaching the speed of light. According to Hubble's law, they are the most distant objects known. Table 9.3(c,d) lists ten each of the strongest radio galaxies and quasars, respectively. A comparison of the redshifts of radio galaxies and quasars suggests a gradual transition between these two classes. The largest redshift currently known is $z = 4.90$, which was measured in the quasar PC 1247+3406.

The current flurry of interest in extragalactic sources stems from the expectation that they may aid in solving cosmological problems, especially since radio radiation is not impeded by cosmic dust. The foremost question concerns the structure of the universe, whether static or expanding, infinite or finite. These inquiries may in principle be answered by source counts dependent on intensity: If sources are distributed homogeneously in space, their number N is proportional to the cube of the distance R (i.e., to the volume of the sphere), whereas the measured flux is proportional to the inverse-square of the distance. These facts combine to give an expression for the number of sources per steradian,

$$N = C S_\nu^{-1.5}, \tag{9.19}$$

where the constant C, which depends on frequency, is to be found by observation. A static universe would comply with this formula for all fluxes; in other world models, systematic deviations for low fluxes (i.e., at large distances) should occur. The underlying cause is that the very distant objects were all relatively young when emitting the radiation which is just now being measured. If the sources evolve with time—say, from high-energy quasars to lower-energy radio galaxies—the premises inherent in the source statistics break down.

Though the correct model of our universe cannot be determined in this way, such measurements are suited to estimate the number of observable sources. The constant C equals 60 for $\lambda = 6$ cm, 150 for $\lambda = 20$ cm, and 2000 for $\lambda = 170$ cm. For example, at 20 cm there are in total 1885 sources in the sky with fluxes ≥ 1 Jy; at $\lambda = 6$ cm, Gregory and Condon [9.12] recently measured 54 579 radio sources stronger than 25 mJy between declinations $0°$ and $75°$; this is the most comprehensive catalogue thus far for the northern sky.

9.5.1.4 The Sun. The Sun is the strongest discrete source over the major part of the radio spectrum. Four basic components are distinguished in its radiation:

1. radiation of the quiet Sun;
2. a slowly variable sunspot component;
3. noise storms in the meter-range;
4. outbursts.

The quiet Sun component is the constant part which comes from the corona and photosphere. At frequencies under 200 MHz, the corona is optically thick, and its radiation temperature is found to be 10^6 K. At higher frequencies, the corona becomes optically thin; the photosphere radiates through and is the only effective emitter above 10 GHz under a very low coronal component. The quiet Sun flux graphed in Fig. 9.4 shows a blackbody spectrum above 10 GHz and below 150 MHz.

The slowly varying component originating in sunspot regions is superposed on the quiet Sun, and is observed only in the transition range between 150 MHz and 10 GHz, with a maximum around 2 GHz. The spot component may here exceed the total quiet Sun radiation by a factor of about 5. It is strongly correlated with the sunspot number; the ionospheric limiting frequency and variations of the geomagetic field also influence its recorded intensity. Noise storms in the meter-range consist of many individual radiation impulses, each pulse with a duration only on the order of seconds and with a bandwidth of some MHz. The overall activity of a storm is limited to a frequency range less than 100 MHz, with durations ranging from a few hours to several days. Storm centers are observed only near the central meridian of the Sun; the radiation appears to be beamed perpendicular to the surface. The measured radiation temperatures may exceed 10^7 K and the radiation is strongly polarized, both facts suggesting a nonthermal mechanism. The correlation between noise storms and sunspots is weak; every storm relates to an active region, but not every region necessarily relates to a storm. During sunspot maxima, the occurence of noise storms can be expected more than 10% of the time.

Radiation bursts from the Sun cover the entire radio range and are usually related to visible eruptions (flares or subflares). The radiation flux in cm-waves may rise by a factor 10 to 20, in the m-range by as much as 4 orders of magnitude above the thermal radiation, including the spot component. The dashed line labeled "Active Sun" in Fig. 9.3 indicates the possible increase of radiation. The burst may last from

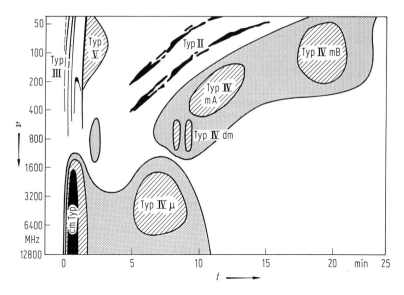

Fig. 9.9. Schematic time development of radiation bursts with respect to frequency. After O. Hachenberg [9.13].

a few seconds to several hours. The dynamical radio spectrum of a strong burst is schematically graphed in Fig. 9.9 after O. Hachenberg [9.13].

The division into six types serves to distinguish individual features of radiation bursts:

- cm-bursts (without number designation);
- the pulses or noise-storm bursts as mentioned above (Type I);
- slowly drifting bursts (Type II);
- fast-drifting bursts (Type III);
- those with continuous emission over the entire radio range (Type IV);
- those with continuous emission in the meter range (Type V).

The first phase of the burst follows the optical flare: Around 500 MHz, several Type III bursts with bandwidths of around 5 MHz appear, their frequency drifting by 20 to 50 MHz to smaller values, and a cm-burst occurs simultaneously. They are quickly superceded by Type V bursts with long waves and Type IV in short waves. For smaller events, this may terminate the observed effect.

In larger outbursts, another phase ensues: Type II bursts appear around 300 MHz with bandwidths around 50 MHz, drifting slowly at rates of 0.3 MHz per second toward lower frequencies. At this time, the Type IV burst gains strength, reaching its maximum first in microwaves (Type IVμ), then in decimeter and finally in meter waves (cf. Fig. 9.9). The radiation temperature may reach 10^{11} K in individual bursts, a fact which argues for an unambiguously nonthermal origin.

The 11-year sunspot cycle, which greatly influences the radio intensity, reached its most recent minimum in 1986.7. The most recent maximum occurred in 1991.

Detailed information on solar radio radiation appears in the monographs by M. Kundu [9.14] and A. Krüger [9.15]. See also Chap. 13 on the Sun.

9.5.1.5 Moon and Planets. The radiation of the Moon and the planets at centimeter wavelengths is explained as their thermal radiation; the effective radiation temperature results from the equilibrium of solar radiation and re-radiation. The Moon, Mercury, and Mars (the latter with and the other two without a transparent atmosphere) radiate from a depth of the order of the wavelength. At $\lambda = 1$ mm, the actual surface temperature is measured; it nearly coincides in phase with the optical measurement. Owing to the low heat conductivity of the surface crust of these bodies, the temperature propagates inward only slowly. Toward longer wavelengths, the radio phase is delayed compared with the optical one and diminishes in amplitude. Above about $\lambda = 20$ cm, no radio phase is found.

In Fig 9.10 the variation of the average brightness temperature $\overline{T}_b$ and the flux density of the Moon are given as functions of the optical phase of the Moon and of the observing wavelengths. The values are calculated after Kuzmin and Solomonovich [9.16]. The mean brightness temperature and the mean flux density as functions of wavelength are tabulated in Table 9.4.

Venus has an equilibrium temperature much higher than would be expected from its distance from the Sun. This can be explained by the so-called "greenhouse effect" of the atmosphere which reduces the amount of heat re-radiated into space. Venus shows no phases in the radio range.

A nonthermal radiation component from Jupiter appears in the decimeter range of wavelengths, and at high angular resolution is found to emanate from a double source outside the planet and located in its equatorial plane. Evidently, relativistic electrons trapped in Jupiter's magnetic field (similar to Earth's van Allen belts) emit synchrotron radiation. Figure 9.4 shows this component to be relatively weak. The longwave radiation from Jupiter is very complex and is limited primarily to the range

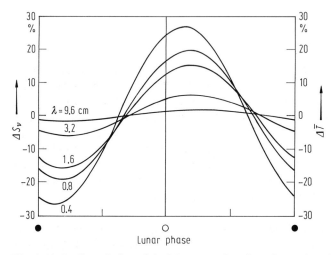

Fig. 9.10. Radio radiation of the Moon as a function of wavelength and optical phase.

Table 9.4. Lunar brightness temperatures and flux densities at various wavelengths.

λ (cm)	$\overline{T}_b$ (K)	S_ν (Jy)
0.4	203	2.25×10^6
0.8	205	5.68×10^5
1.6	207	1.43×10^5
3.2	210	3.64×10^4
9.6	218	4.19×10^3

below 30 MHz. The intensity of the measured radiation depends on the frequency, on the central meridian of Jupiter, on the Earth's latitude as seen from Jupiter, and on the position of the satellite Io. The optical rotation period of Jupiter depends on latitude; for radio radiation, the rotation system III is defined with a period $09^h55^m29\overset{s}{.}7$. Some sources occur periodically at the following longitudes in this system: A at 240°, B at 145°, C at 325°, and D at 58°. The radiation arrives in noise storms containing two kinds of bursts:

1. L-bursts lasting about 1 second.
2. S-bursts lasting about 10 ms.

The frequent and almost periodic occurrence of bursts in a noise storm is one of the radio characteristics of Jupiter. Source B shows the strongest bursts, but Source A is the most frequently active, while Source D is only very rarely so; the bursts from B and C are usually predictable. A detailed survey of the exploration of Jupiter has been edited by T. Gehrels [9.17]; a more recent compilation has been written by J.D. Kraus [9.1].

9.5.2 Spectral Lines

As has been mentioned before, atomic and molecular radio astronomy is the subject of current research. So far about 60 molecules (not counting isotopes) have been radio-identified; these include organic molecules such as formaldehyde, formic acid, and methyl alcohol. The observed, different transitions or lines total over 2300, not counting the numerous recombination lines. The fact that most lines are located in the millimeter range and are also very weak restricts their observation to elaborate spectrometers. Radiation from strong lines (e.g., from neutral hydrogen (HI; see below) and water vapor (H_2O)) and its distribution can be observed with a narrow-band receiver tuned to the line-frequency-like continuum emission.

9.5.2.1 The Neutral Hydrogen Line. Atomic hydrogen is the most abundant and widespread component of the interstellar medium. As early as 1951, H.I. Ewen and E.M. Purcell found the "forbidden line" predicted by Van de Hulst, using a horn pointed out of their laboratory window. This showed the intensity and wide distribution of the HI radiation. The central frequency of the line is $\nu_0 = 1420.406$ MHz; typical radiation temperatures are between 50 and 100 K, with a line width of 8 kHz, corresponding to the temperature.

Neutral hydrogen is concentrated in the spiral arms, and by observing the shift of frequency caused by the Doppler effect, the velocity of spiral arms relative to the Sun and thus the dynamics of the Milky Way can be inferred. Charting the Milky Way from such data demonstrates its spiral structure quite convincingly. Such measurements are already successfully being performed by amateurs with homemade spectrometers.

9.5.2.2 The Line of Water Vapor. The longest-wavelength line of H_2O at $\nu_0 = 22.235$ GHz had been known for decades through laboratory and atmospheric measurements when it was found in the Milky Way in 1969 by A.C. Cheung, D.H. Rank, C.H. Townes, and W.J. Welch. Characteristics of this radiation are high observed fluxes, extremely small source diameters, and strong variability. The source W 49A, for instance, has a flux of about 10^5 Jy, an apparent extended diameter of about $0.''001$, and thus an equivalent radiation temperature of around 10^{15} K. Of course, this suggests a nonthermal mechanism of radiation, and the *maser effect* is suspected. The radiation originates in dark clouds near compact HII regions or infrared sources. Other strong maser emissions originate in the molecules SiO and OH. Maser radiation is suspected to delineate the stellar nurseries where O- and B-type stars are born; hence the current interest in observing dark and molecular clouds.

9.6 Tested Observing Systems

Radios and television receivers are such everyday commodities that it is easily forgotten that they are also highly sensitive measuring instruments. If they were tuned regularly to some distant transmitter, and the received signal intensity recorded as a function of meteorological data, time, and date (season), some very interesting records on wave propagation in the atmosphere would be obtained. Amateurs with shortwave receivers practice such systematic observations.

Publications on the subject of amateur radio astronomy have multiplied in number during the past decade. A comprehensive survey cannot be attempted here. Books have been published by F.W. Hyde [9.18], J. Heywood [9.19], D. Heiserman [9.20], J.P. Shields [9.21], and R.M. Sickels [9.22], the last being preferred by this author. G.W. Swenson published an interesting series of articles [9.23], which were later combined into a booklet [9.24]. There are two relevant monthly journals: *Radio Astronomy*, the journal of the Society of Amateur Radio Astronomy, Inc. (SARA), a worldwide nonprofit society, and *The Radio Observer*, a magazine devoted to radio astronomy amateurs. The former is edited by J.M. Lichtman [9.25], the latter by R.M. Sickels [9.26].

From many published suggestions, a few tested and used designs are presented in the following subsections.

9.6.1 Solar Flare Monitoring

Solar activity and the state of the ionosphere are directly related; the appearance of a flare increases the noise in the lower atmosphere at the longest wavelengths (VLF) by a factor of about 2. This noise enhancement has been used for decades to indirectly identify flare activity. As early as 1959, P.J. del Vecchio [9.27] described a tested receiver system at 27 kHz with exemplary measuring results. Building instructions for such simple receivers have been published, for instance, by C.H. Hossfield [9.28], R.C. Maag [9.29], and R.M. Sickels [9.22]. The author's experience from his own observations is less satisfying. Local disturbances were too strong to permit direct identification of flares, and only by comparison with simultaneous measurements at a distant location, or with optical observations, is it possible to distinguish flares from normal interference. A network of observers in the U.S. has joined in the "Very Low Frequency Experimenter's Group." An alternative way to record flares has been described by J. Hudak [9.30] and R.M. Sickels [9.22], using the fact that a flare ionizes the ionosphere more and causes shortwave "fadeout" (SWF).

9.6.2 Jovian Bursts

As mentioned, the radiation from noise storms and bursts on Jupiter is limited to frequencies under 30 MHz, with a maximum intensity between 10 and 20 MHz, which is the range in which shortwave radio sets operate. In AM-mode and with amplification regulation switched off (AGC), such devices operate like radio astronomy receivers, and can be used directly to observe Jovian bursts, recording the radiation on tape or printer. The sensitivity of any broadcast receiver (i.e., the system temperature) should certainly suffice. It has already been noted that the strength of cosmic radiation at these wavelengths is strong enough to render worthless an improvement in the sensitivity. In order to be certain to have observed a Jovian burst, it is again advisable to compare with measurements made at a different station.

9.6.3 A Model Interferometer

G.W. Swenson [9.23,24] has designed a phase-switch interferometer for a frequency of 73.8 MHz and published the schematics for the circuitry and the construction; it is an operational model of a professional interferometer. Since many essential components are to be made up from parts which must be procured first, this project demands much time and some knowledge of electronics. Yet the end product promises to be a useful and sensitive instrument. The low effective antenna area permits observations of only a few bright sources.

9.6.4 FM and Television Receivers

A quarter of a century ago, D. Downes [9.31] described the modifications of an FM-receiver needed to make simple measurements of the Sun. These changes involve switching to AM operation and the switch-out of the automatic amplification. Modern miniature circuitry may make such modifications difficult.

For many years R.M. Sickels [9.22] has supplied a receiving system composed of a 2.1-m antenna with dipole and preamplifier (or with several paraboloids for an interferometer), a mixer and tuner for the range 500 to 750 MHz with intermediate-frequency amplifier and detector, a DC-amplifier with several different time constants, and an analog recorder. To the author's knowledge, this is the only commercially available receiver in amateur radio astronomy. The system can be homebuilt at low cost by following the instructions in the book by Sickels [9.22]. In 1982, the author built the Sickels system as a single telescope as well as an interferometer, and with it has successfully observed the Sun. Lacking space for a permanent mount, the instrument had to be set in place anew for each observation. The parabolic reflectors were unwieldy and difficult to adjust. They were replaced by a Yagi-antenna which has a low-noise preamplifier for television channel 38 (USV-G from SSB Electronics, Iserlohn) with restriction to one fixed frequency and with some sacrifice of antenna gain. The improved noise number from 2 db to 0.9 db barely compensated the lesser gain and brought about a satisfactory result. With this system, the quiet Sun and, of course, the active Sun with noise storms and various flares can be observed, as well as 5 to 10 other sources. In contrast to the model interferometer, this instrument is sensitivity limited.

9.6.5 Satellite Receivers

Low-noise receivers and mass-produced parabolic reflectors exist for the amateur bands around 1.3 GHz, 2.3 GHz, and 10 GHz, and for the television satellites at 11 GHz. According to specifications, the available receivers should be suited to amateur radio astronomy. Preliminary test measurements of a 2.3 GHz and an 11 GHz system showed the sensitivity to be sufficient to observe the Sun and the thermal radiation of the Moon; presumably the strongest point sources can also be "discovered." In Fig. 9.11, some (redrawn) analog recordings of the author's test measurements are shown. This range of wavelengths is interesting for solar observations because the radio emission is here correlated with sunspots and with the sunspot numbers.

Receiver technology will in the near future make the range under 2 cm accessible to radio amateurs. It will then be possible to perform the meteorological measures suggested above, or to record planetary radiation.

9.6.6 Evaluation

The foregoing showed some observational possibilities which are accessible to amateur astronomers or to professional astronomers working in a small college environment. The author feels that such observing systems are particularly useful for didactic purposes, such as an introduction to radio astronomy. It was noted above that monitoring programs of the Sun and Jupiter, for example, will be of scientific value if the measurements are executed regularly and carefully calibrated and evaluated.

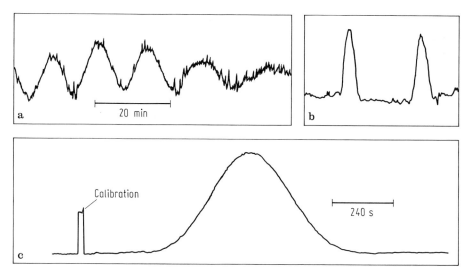

Fig. 9.11 a–c. Measurements of the Sun. **a** Driftscan made on 1982 May 15 with the interferometer at a wavelength $\lambda = 50$ cm. The baseline was 14 m. **b** Manual scan in the azimuth direction of a single antenna (Yagi) at $\lambda = 50$ cm on 1982 May 25. **c** Driftscan, observed with a parabolic reflector of diameter 1.2 m and with a satellite receiver at $\lambda = 3$ cm on 1987 August 21.

Considering the short time span of organized amateur radio astronomy, it is too early to review its contribution to science. But there are possibly some startling results such as the detection of high-energy pulses (HEPs) and some radiation anomalies observed simultaneously by several amateurs. These findings still await confirmation.

The author wishes to thank his colleagues K.H. Hoesgen and Dr. D. Graham for their help and advice, and R.M. Sickels of Bob's Electronic Service for valuable information.

9.7 Amateur Radio Astronomy Groups

Persons interested in becoming a member of the SARA group should write to them at the following address: The Society of Amateur Radio Astronomers, Inc., SARA Membership Services, 247 N. Linden St., Massapequa, NY 11758, USA.

References

9.1 Kraus, J.D.: *Radio Astronomy*, 2nd edn., Cygnus Quasar Books, Powell 1986.
9.2 Rohlfs, K.: *Tools of Radio Astronomy*. Springer, Berlin 1986.
9.3 Verschuur, G.L.: *The Invisible Universe Revealed*, Springer, New York 1987.
9.4 Verschuur, G.L., Kellerman, K.I. (eds.): *Galactic and Extragalactic Radio Astronomy*, Springer, Berlin 1988.
9.5 Mezger, P.G., Henderson, A.P.: *Astrophys. J.* **147**, 471 (1967).
9.6 American Radio Relay League: *The A.R.R.L. Antenna Book*, Newington, 1988.

9.7 Thompson, R.A., et al.: *Interferometry and Synthesis in Radio Astronomy*, Wiley, New York 1986.
9.8 Evans, G., Leish, C.W.: *RF Radiometer Handbook*, Artech House, Ottawa 1977.
9.9 Landecker, T.L., Wielebinski, R.: *Austr. J. Phys. Suppl.*, No. 16 (1970).
9.10 Dröge, F., Priester, W.: *Zeitschrift für Astroph.* **37**, 125 (1955).
9.11 Baars, J.W.M. et al.: *Astron. Astrophys.* **61**, 99 (1977).
9.12 Gregory, P.C., Condon, J.J.: N.R.A.O. preprint 90/91 (1991).
9.13 Hachenberg, O.: in *Landolt-Börnstein, NS*, VI/1, Springer, Berlin 1965.
9.14 Kundu, M.R.: *Solar Radio Astronomy*, Wiley, New York 1965.
9.15 Krüger, A.: *Introduction to Solar Radio Astronomy and Radio Physics*, Reidel, Dordrecht 1979.
9.16 Kuzmin, A.D., Salomonovich, A.E.: *Radioastronomical Methods of Antenna Measurements*, Academic Press, New York 1966.
9.17 Gehrels, T. (ed.): *Jupiter*, University of Arizona Press, Tucson 1976.
9.18 Hyde, F.W.: *Radio Astronomy for Amateurs*, Lutterworth Press, London 1962.
9.19 Heywood, J.: *Radio Astronomy: and How to Build Your Own Telescope*, ARC Books, New York 1964.
9.20 Heisermann, D.: *Radio Astronomy for the Amateur*, TAB Books, Blue Ridge Summit 1975.
9.21 Shields, J.P.: *The Amateur Astronomer's Handbook*, Crown Publl, New York 1986.
9.22 Sickels, R.M.: *Radio Astronomy Handbook*, Bob's Electronic Service, Ft. Pierce 1987.
9.23 Swenson, G.W.: *Sky and Telescope* **55**, 385, 475 (1978) and **56**, 28, 114, 201, 290 (1978).
9.24 Swenson, G.W.: *An Amateur Radio Telescope*, Pachart Publ. House, Tucson 1980.
9.25 Lichtman, J. (ed.): *Radio Astronomy*. 1425 Parkmont Dr., Roswell, GA 30076, USA.
9.26 Sickels, R.M. (ed.): *The Radio Observer* (serial), 7605 Deland Ave., Ft. Pierce, FL 34951, USA.
9.27 Del Veccio, P.J.: *Sky and Telescope* **45**, 392 (1973).
9.28 Hossfield, C.H.: *Sky and Telescope* **37**, 254 (1969).
9.29 Maag, R.C.: *Sky and Telescope* **45**, 392 (1973).
9.30 Hudak, J.: *Sky and Telescope* **68**, 452 (1984).
9.31 Downes, D.: *Sky and Telescope* **24**, 75 (1962).
9.32 Pankonin, V.: *Sky and Telescope* **61**, 308 (1981).
9.33 Pankonin, V.: *IEEE Transactions on Electromagnetic Compatibility*, Vol. EM-23, 308 (1981).

10 Modern Sundials

F. Schmeidler

10.1 Introduction

Modern sundials have become a favorite ornament in houses and gardens. Their design and construction are not difficult with some knowledge of spherical astronomy, and suggestions for improved constructions appear occasionally in the literature [10.1].

The basic parts of a sundial are the shadow-casting *style* or *gnomon*, and the dial plate. In principle, the dial can be designed for any surface. Some important and simple cases of plane dials are considered in the following. The gnomon always lies in the plane of the meridian, and is tilted so as to point toward the celestial pole. It is inclined to the horizontal plane by the same angle as the geographic latitude φ. Some special sundial forms with the gnomon not parallel to the Earth's axis are detailed by H. Lipold [10.7].

Any sundial shows directly the *true solar time*, which is obtained simply from the shadow cast by the actual Sun. True solar time is a *local time* and differs from the internationally used system of *standard time* (ST). The latter is the mean local time of the standard meridian, for instance, the 75th parallel west of Greenwich for Eastern Standard Time (EST) in the U.S. In principle, two different corrections must be applied to observed sundial readings:

1. The equation of time which reduces the true solar time to a mean, uniform time (mean solar time, or MST).
2. The correction for longitude to the standard meridian, which converts the thus obtained mean solar time to standard time. In contrast to the equation of time, the longitude correction at a given location of the dial is a constant to be applied to every reading. This suggests that the longitude correction should already be included in the marking of the dial; see Sect. 10.6 below.

It is not practical, as is sometimes suggested, to find the accurate orientation from an architect's blueprints, since such directions are often not precise enough. Nor should it be obtained from compass readings, which are influenced not only by the magnetic declination but also by unknown local magnetic disturbances which may affect the reading beyond the required precision. An astronomer can and should find the directions only by astronomical observations.

A warmly recommended reference for the actual construction of a sundial is a book by H. Schumacher [10.2]; it details problems occurring in the construction and contains numerous design graphs and other figures.

10.2 The Equinoxial Dial

The simplest type of dial is the *equinoxial* or *polar dial*, designed as a kind of armillary sphere. The shadow of the style pointing to the celestial pole is read on a circle in the equatorial plane, which is divided uniformly from 0^h to 24^h, with the 12^h mark at the shadowed position at true noon (i.e., when the true Sun is at the meridian). For the geographic longitude correction of the armillary sphere, the scale is correspondingly shifted. Such equinoxial dials are best placed atop a pillar in the garden.

10.3 Horizontal Dials and Vertical East–West Dials

Also rather simple are the sundials with faces lying in a horizontal plane or on an east–west oriented wall.

10.3.1 Computations

Figure 10.1 graphs three planes: the horizontal plane, the east–west plane oriented at right angles to it, and the plane parallel to the celestial equator and passing through the intersection $\overline{AB}$ of the other two. The formulae for dividing the faces in the horizontal plane and in the vertical east–west plane are then readily derived. The style $\overline{NS}$ is perpendicular to the equatorial plane, intersecting it in the point O. At a given hour angle h the shadow on an equinoxial dial meets the east–west line $\overline{AB}$ or its extension at the point P, then $\overline{SP}$ is the shadow on the horizontal plane referring to this hour angle, and $\overline{NP}$ that on the vertical east–west plane. The style is inclined by the geographic latitude, and for an hour angle $h = 0^h$, the shadow makes the lines $\overline{SC}$ and $\overline{NC}$ on the two primary planes. The three triangles $\triangle OCP$, $\triangle SCP$, and $\triangle NCP$ have the line $\overline{CP}$ and the right angle at C in common.

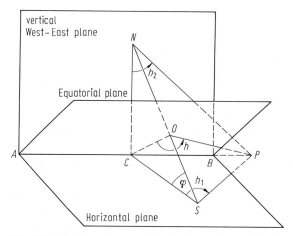

Fig. 10.1. Relations between the shadows of an equinoxial, a horizontal, and an east–west dial.

Therefore, the important time-dependent angles are

$\angle COP = h =$ hour angle on the equinoxial dial,
$\angle CSP = h_1 =$ corresponding angle in the horizontal plane,
$\angle CNP = h_2 =$ corresponding angle in the vertical east–west plane.

We then obtain

$$\begin{aligned}
\text{in } \triangle OCP && \overline{CP} &= \overline{OC} \cdot \tan h, \\
\text{in } \triangle SCP && \overline{CP} &= \overline{SC} \cdot \tan h_1, \\
\text{in } \triangle NCP && \overline{CP} &= \overline{NC} \cdot \tan h_2.
\end{aligned} \qquad (10.1)$$

And, by division of the equations,

$$\tan h_1 = \frac{\overline{OC}}{\overline{SC}} \cdot \tan h,$$

$$\tan h_2 = \frac{\overline{OC}}{\overline{NC}} \cdot \tan h. \qquad (10.2)$$

Also, in the right triangles $\triangle COS$ and $\triangle CON$,

$$\frac{\overline{OC}}{\overline{SC}} = \sin \varphi,$$

$$\frac{\overline{OC}}{\overline{NC}} = \cos \varphi, \qquad (10.3)$$

and hence

$$\begin{aligned}
\tan h_1 &= \sin \varphi \tan h, \\
\tan h_2 &= \cos \varphi \tan h,
\end{aligned} \qquad (10.4)$$

These are the two basic formulae which give for any hour angle h the corresponding angles h_1 and h_2, respectively, on horizontal and east–west dials.

10.3.2 Graphical Construction

The graphical solution, if preferred, is also readily obtained (Figs. 10.2 and 10.3).

Draw a circle with radius $\overline{MC} = 1$ and extend the radius to a point O so that $\overline{OC} = \sin \varphi$. Graph a semicircle around O and the joint tangent through C. Mark the angles $15°$, $30°$, $45°$, etc. on the smaller circle, beginning at $\overline{OC}$, corresponding to whole hours. The desired subdivisions may be added at this point or later. The radii from O are extended to intersect the tangent; the points of intersection connected with M show the dial markings of the horizontal on the larger circle. Label C with 12^h and the other points with increasing hour numbers in the clockwise direction. The style is anchored in M and forms the angle φ with the direction 12^h.

The dial plate is adjusted horizontally by a level and with 12^h pointing exactly toward north. The easiest way to do this is to calculate the transit of the Sun for any day and rotate the dial accordingly. The longitude correction is not yet allowed for. Figure 10.2 shows the construction of such a dial for the latitude $\varphi = 50°$.

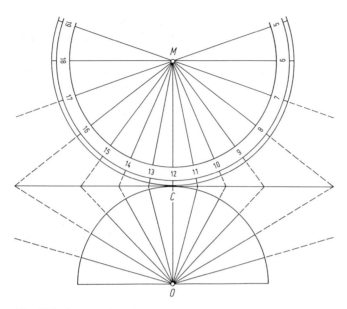

Fig. 10.2. Construction of a *horizontal* sundial for the geographic latitude $\varphi = 50°$.

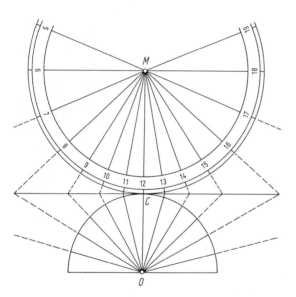

Fig. 10.3. Construction of a *vertical* sundial in the east–west direction for the geographic latitude $\varphi = 50°$.

The vertical east–west dial is constructed in the same way, except that $\overline{OC} = \cos\varphi$ (Fig. 10.3) and the marking is counterclockwise. The 6^h–18^h line is parallel to the tangent.

The adjustment of the vertical dial in the east–west plane requires an observation. The true solar time corresponding to the position of the Sun at any instant, for example, can be calculated. Adjusting the dial for grazing incidence (at the moment when the Sun crosses the prime vertical) gives the best accuracy, but this can be done only in spring and summer.

10.4 The Vertical Deviating Dial

Usually, the vertical sundial is to be installed at a given wall which in general does not lie exactly in the east–west plane. A vertical dial that is not in this plane is called a *vertical deviating dial*. The first task is to then determine the azimuth of the wall.

10.4.1 Determination of the Azimuth of the Wall

It is probably safe to assume in the construction that the wall is perfectly vertical, although, as was said previously, the azimuthal direction cannot be relied upon. This direction is best found by observing the Sun when its light is just grazing (i.e. parallel with) the wall surface. More specifically, gaze along the plane of the wall toward the east (morning) or west (afternoon). Using a suitable glass filter, observe with one eye the instant when half of the solar disk is covered by the wall. Record this time precisely with a dependable watch which has been synchronized with a time signal. With some experience, the instant mentioned can be found without instruments to about ± 10 s; the observation becomes more precise with repeated trials on several days. The recorded times will differ, because the instant of passage through the plane of the wall is affected by the continuous changes of the solar declination and of the equation of time. Each observation gives the azimuth of the wall from the well-known relation,

$$\tan A_\odot = \frac{\sin h}{\sin\varphi \cos h - \cos\varphi \tan\delta}, \qquad (10.5)$$

where

φ = geographical latitude,

h = hour angle of the Sun at the instant of observation,

δ = declination of the Sun at the instant of observation,

$A_\odot$ = azimuth of the Sun and of the wall to be found.

Weather permitting, the readings can be made in the morning or in the afternoon, but, for obvious reasons, only during the spring and summer seasons. With a little practice, the resulting azimuths may differ by about $20'$, so that just a few observations will give a sufficiently precise average.

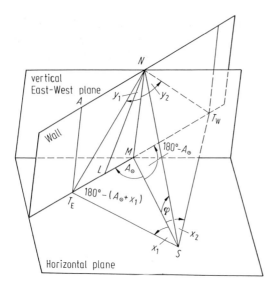

Fig. 10.4. Calculation of the face of an vertically deviating dial

10.4.2 Calculation of the Dial at the Wall

The formulae required for this analysis are more complicated than in the previous cases, but can be deduced from Fig. 10.4.

Let $\overline{NS}$ again be the style parallel to the Earth's polar axis, meeting the horizontal plane at the point S and the vertical east–west plane, as well as the plane of the wall, at the point N. Besides horizontal and wall planes, the east–west plane, which intersects the wall in a vertical through N, is graphed. The north–south plane through S meets the two other planes at the point M. The direction of the wall is given by the intersection with the horizontal. The angle $< SMT_E = A_\odot$ also gives the azimuth of the wall as defined and determined above.

The points T_E on the morning side (hour angle east) and T_W on the afternoon side (hour angle west) are, for the moment, arbitrary on the line of intersection between wall and horizontal. Then,

$$< SMT_E = A_\odot \quad \text{and} \quad < SMT_W = 180° - A_\odot.$$

In the horizontal plane, let

$$< MST_E = x_1 \quad \text{and} \quad < MST_W = x_2.$$

The angle $< NSM$ again equals φ, and x_1 and x_2 are the shadow directions of the style in the horizontal plane at hour angles h_1 and h_2, respectively. The angles at N in the plane of the wall,

$$< MNT_E = y_1 \quad \text{and} \quad < MNT_W = y_2,$$

correspond to x_1 and x_2 at the rotation by the angle $A_\odot$. $\overline{NT_E}$ and $\overline{NT_W}$ are the corresponding shadow directions at the wall for the hour angles h_1 and h_2. To find them, it is necessary only to calculate the segments $\overline{NT_W}$ and $\overline{NT_E}$ or the angles y_1, y_2.

The triangle $\triangle MST_E$ gives

$$\overline{MT_E} = \frac{\overline{MS} \sin x_1}{\sin(A_\odot + x_1)} = \frac{\overline{NS} \cos \varphi}{\cos A_\odot + \sin A_\odot \cot x_1}. \tag{10.6}$$

Because of the relation

$$\cot x_1 = \frac{\cot h_1}{\sin \varphi}, \tag{10.7}$$

as in the horizontal dial, Eq. (10.6) becomes

$$\cot y_1 = \frac{\overline{MN}}{\overline{MT_E}} = \frac{\overline{NS} \sin \varphi}{\overline{NS} \cos \varphi} \left(\cos A_\odot + \frac{\sin A_\odot}{\sin \varphi} \cot h_1 \right). \tag{10.8}$$

The length $\overline{NS}$ of the style cancels, with the result

$$\cot y_1 = \cos A_\odot \tan \varphi + \frac{\sin A_\odot}{\cos \varphi} \cot h_1, \tag{10.9}$$

or, in a more convenient form,

$$\tan y_1 = \frac{\cos \varphi}{\cos A_\odot \sin \varphi + \sin A_\odot \cot h_1}. \tag{10.10}$$

This equation holds for any true solar time in the morning hours.

For western hour angles, which occur in the afternoon, $A_\odot$ is replaced by $180° - A_\odot$ and x_1 by x_2, which gives

$$\tan y_2 = \frac{\cos \varphi}{-\cos A_\odot \sin \varphi + \sin A_\odot \cot h_2}. \tag{10.11}$$

Since the position of the points T_E and T_W on the intersection was not specified (except that one is to the left of M and the other to the right), the point at any instant of time can be determined on the line $\overline{T_E T_W}$.

Since the horizontal line $\overline{T_E T_W}$ cannot be arbitrarily long, one turns to the perpendicular to it at a suitable point, for instance, in T_E. Let this be $\overline{AT_E} = \overline{NM}$. The angle $< ANT_E$ equals $90° - y_1$. To find the corresponding points of the time scale on $\overline{AT_E}$, simply change to the co-functions.

10.4.3 Construction of the Calculated Dial Face at the Wall

The next step, which is to transfer the calculated dial markings to the wall, may be executed in the following way.

First mark the foot of the center of the style at point N. Define the vertical with a plumb bob through N, and then graph it. It will generally suffice to make the vertical exactly 1.000 m long to point M (see Fig. 10.4). At M, the horizontal is determined precisely by a level and graphed toward both sides, also precisely 1.000 m long. At the end points, the vertical upward direction is determined with the plumb bob. Thus, on both sides of the central line $\overline{NM}$, two squares each of 1.0 m side are bisected whose side pairs are oriented exactly horizontal and vertical, respectively.

Graph the points characterized as $\tan y_1$ and $\tan y_2$ from the equations above, horizontally from M to the left and to the right, and do the same with the values of $\cot y_1$ and $\cot y_2$ on the respective verticals ($\tan y_1 = \tan y_2 = 0$ lies at the point M, while $\cot y_1 = 0$ and $\cot y_2 = 0$ are diagonally opposite to M).

In this way, all computed points for the dial are at the wall, and subdivisions, perhaps half-hours or even 10 to 10 minutes, can be made according to preference and the size of the dial.

The labeled band can have any arbitrary shape. To obtain the time points on it, merely connect the corresponding points on the rectangular frame of Fig. 10.4 with the foot of M, where the style will be inserted. Then record the numbers where these lines pass through the labeled band.

Later on, after construction is completed and the wall is painted, the rectangular frame with the original time scale can be removed.

10.4.4 Inserting the Style

The insertion of the style must be performed as precisely as the construction of the dial scale. The accurate fit of the style determines how precise the time readings at the dial are. Two matters in particular should be attended to:

1. When cementing the style, its center should meet the wall as accurately as possible at the point N, since that was the point to which the dial scale referred (cf. again Fig. 10.4.).
2. The style should lie exactly in the plane of the meridian and should form the angle φ with the horizontal.

The second requirement above can best be met as follows: Imagine a plane which passes through the style and can rotate around it as an axis, and consider the angle which the intersection of this plane with the wall forms with the style. This angle will vary with the rotating plane, but reaches a minimum when the rotating plane is vertical to the wall. The intersection of the rotating plane is called a *substyle*, and is the segment $\overline{NL}$ in Fig. 10.4.

The produced directions of $\overline{NM}$, style $\overline{NS}$, and substyle $\overline{NL}$ point to three different points on the sphere, and form a spherical triangle $\triangle NSL$ with a right angle at L (Fig. 10.5). The arc $MS = 90° - \varphi$, the angle $< LMS = A_\odot$, and the arc SL marked with the symbol γ is the angle between style and substyle. Formulae for the right spherical triangle MSL give

$$\sin \gamma = \cos \varphi \sin A_\odot \tag{10.12}$$

and

$$\tan \psi = \cot \varphi \cos A_\odot, \tag{10.13}$$

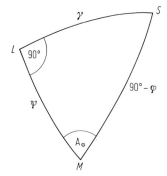

Fig. 10.5. The determination of the substylar line.

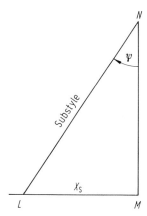

Fig. 10.6. Determination of the position of substyles.

where ψ is the angle formed by the substyles with the perpendicular $\overline{MN}$ in the plane of the wall (see Fig. 10.6).

Call L the point where the substyle meets the horizontal through M, and X_S the distance $\overline{ML}$, and make $\overline{MN}$ again equal to 1.000 m. Then

$$X_S = \tan \psi = \cot \varphi \cos A_\odot. \qquad (10.14)$$

When $A_\odot < 90°$, X_S is to the left (i.e., west) of M. Otherwise, it is to the right (or east) of M. The positions of the substyles are thus known through X_S.

To insert the style correctly, have a moderate-sized right triangle made of plywood and with one of the acute angles equal to γ. When the style of the desired length is to be inserted, this plywood triangle is placed with one of its sides perpendicular to the substylar line—to support the style—whereby the angle γ comes to lie between the style and substylar line.

If the plywood style has a hypotenuse about as long as the style, the latter can then be placed on the triangle and, if need be, lightly fastened to it. Often, the style is slightly conical, a feature which can be accommodated by a slight correction of the angle γ in the wood triangle, or by a wedged support between triangle and style. The triangle is held in place by the weight of the style or else by a small nail, but the perpendicularity of the wooden triangle with the wall should be exactly maintained. The style can now be cemented, and will be in the proper position when the wood triangle is removed two days later.

10.5 Designs for Higher Accuracy

The methods mentioned so far permit the construction of sundials whose readings may depart from standard time by up to a quarter or half hour. Higher accuracy may be reached by making additional corrections as described in the subsections below.

10.5.1 Correction for Geographic Longitude

This correction, if desired, can be incorporated into the labeling of the scale. For instance, if the sundial is located at a longitude 5° or 20 minutes west of the meridian to which the standard time refers, then the addition of 20 minutes on all marks of the dial will correct for longitude, because standard time is by this average amount ahead of local solar time. The label 12^h, located on the vertical dial directly below the style foot, is now displaced to the left, and below the foot will now be the mark $12^h 20^m$. A sundial corrected for longitude can be immediately recognized by its having 12^h reading sidewise displaced from directly below the style.

10.5.2 Correction for the Equation of Time

The changes of the equation of time from day to day, and from one year to the next, are below the reading precision of a sundial, which is at best 1 minute. To neglect the leap days creates a maximum error of one day, and the fastest change of the equation of time occurring in late December is only 1/2 minute per day. It is thus adequate to neglect February 29 and to use one permanent graph which reduces apparent to mean solar time for every day (Fig. 10.7). The date is read on the figure-of-eight, and the corresponding correction of the sundial reading on the top or bottom scale to obtain standard time.

Attempts have recently been made to incorporate this correction in sundials by suitable construction measures. Some successful attempts in this direction have been reported in *Sky and Telescope* [10.3]. In each case, the principle is to curve the gnomon or the dial or both in such a fashion that the readings include the correction for the equation of time.

Also worth reading is an article in the *Journal of the British Astronomical Association (J.B.A.A)* [10.4], which discusses under what conditions the dial is equiangular in the sense that equal time intervals are represented by equal labeling intervals. Evidently, a sundial complying with this condition is more precise and convenient to read than would otherwise be the case. On the other hand, the construction of such a dial is necessarily more complex than that of an ordinary sundial. For a simplified theory of such dials, consult H. Lippold [10.7].

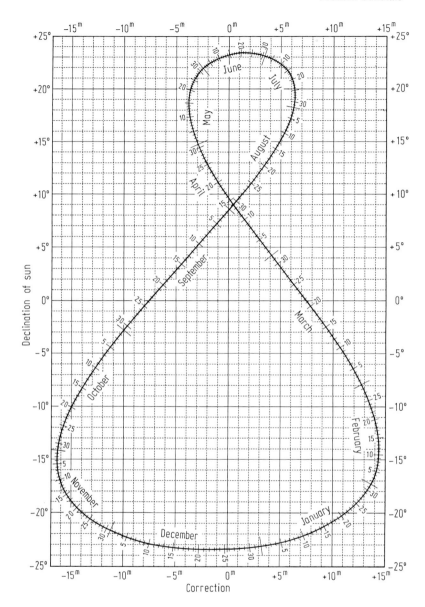

Fig. 10.7. Correction of sundial readings for the equation of time.

References

10.1 Brunner, W.: Neuartige Sonnenuhr-Konstruktionen. *Orion* **33**, 44 (1975).
10.2 Schumacher, H.: *Sonnenuhren*, Callwey, München 1973.
10.3 *Sky and Telescope* **32**, 256 (1966).
10.4 *Journal of the British Astronomical Association* **86**, 7 (1976).
10.5 Peitz, A.: *Sonnenuhren, Tabellen und Diagramme zur Berechnung*, Callwey, München 1978.
10.6 Hanke, W.: Ermittlung der Wandrichtung für eine deklinierended Vertikal- (Süd-) Sonnenuhr durch Sonnenzeitazimutbeobachtung. *Astronomie und Raumfahrt* **3** (1975).
10.7 Lippold, H.: Zur Theorie der homogenen Sonnenuhr. *Die Sterne* **61**, 228 (1985).

11 An Historical Exploration of Modern Astronomy

G.D. Roth

11.1 Introduction

There were two principal factors which motivated the people of ancient times to concern themselves with the stars: the link of celestial objects with religious concepts, and the everyday need to determine the time.

The instruments and observing methods in astronomy remained basically the same throughout and past the Middle Ages. The study of the motions of celestial objects was paramount to ancient and medieval astronomers. With the invention of the telescope early in the 17th century, progress in astronomy underwent a spectacular advance: a detailed physical exploration of celestial bodies could now be pursued.

The evolution of astronomy into a highly specialized science began in the 19th century, as the introduction of spectral analysis, photometry, and photography made astrophysics a major part of astronomy. The limitations of optical astronomy were overcome in the 20th century, when radio telescopes and space-borne instruments made the entire spectrum of electromagnetic waves accessible.

11.2 The Heliocentric System

The era of "modern" astronomy can be said to have come into existence in 1543 with the appearance of *De Revolutionibus* by Nicolaus Copernicus (1473–1543). In this work, the Polish canon introduced a model of the heavens in which the center of the planetary motions is not the Earth but the Sun. This model is known as the Sun-centered, or *heliocentric*, system. In a nutshell, his thesis was that the observed annual motion of the Sun about the Earth is actually due to the motion of the Earth; the rotation of the Earth about its axis is reflected in the diurnal rotation of the stellar sky, and properties of planetary motion can be understood only by the taking into consideration that observations are made from the moving Earth.

The German astronomer Johannes Kepler (1571–1630) studied the planetary motions in more detail. In particular, he examined carefully the observations of Mars which had been made by the Danish astronomer Tycho Brahe (1546–1601) and, as a result, discovered the motion of planets in elliptical orbits about the Sun. (Copernicus had adhered to the venerable tradition of assuming that circular motions are appropriate for celestial bodies.) Kepler's three laws of planetary motion about the Sun are seminal documents of the laws of nature which now pervade all of astronomy:

1. Each planet moves in an ellipse about the Sun, which lies at a focus of this ellipse.
2. The speed of orbital motion is higher in the orbital sections near the Sun than far away from it. The speed changes such that the radius vector from Sun to planet sweeps equal areas in equal times.
3. The ratio between the squares of the periods and the cubes of the mean distances from the Sun is the same for all planets.

Whereas Kepler arrived at his laws without telescopic observations, his contemporary, the Italian scientist Galileo Galilei (1561–1642), began using the telescope in 1610 for astronomical observations and for providing arguments in favor of the heliocentric system. Galileo discovered the four bright satellites of Jupiter and observed the phases of Venus. His advocation of the Copernican system, however, led to serious conflicts with the Church. The Roman Inquisition initiated the trial of Galileo in which he was labeled a representative of what seemed to Church authorities to be a preposterous and heretical doctrine.

One objection, in particular, against the heliocentric system remained for nearly three centuries: If the Earth did move about the Sun, then stars should exhibit annual displacements or *parallaxes* in response to this orbital motion. Critics of the system pointed to the absence of the requisite stellar displacements, while supporters argued that such displacements did exist but were immeasurably minute owing to the presumably enormous distances to the stars. The apparent absence of observable stellar parallaxes became an incentive for observers and telescope builders to design and use better measuring instruments in order to obtain a positive result. The goal was not reached, however, until 1838, when Friedrich Wilhelm Bessel in Königsberg determined the parallax of 61 Cygni, and, almost simultaneously, Wilhelm Struve in Pulkova found the parallax of α Lyrae and Thomas Henderson at the Cape Observatory that of α Centauri.

By that time, the heliocentric system had already gained wide acceptance. But the first stellar parallax measures were not merely the final word in an old dispute; they also marked the beginning of a new advance in the exploration of space with more modern measuring techniques.

11.3 Evolution of the Theory of Motions

The notion that some kind of force originates in the Sun and influences the motions of the planets was already implicit in Kepler's writings. The English mathematician and physicist Isaac Newton (1643–1727) formulated a mechanical principle which is not restricted to planetary motions only; it is universally valid:

> Two bodies attract each other with a force proportional to the product of their masses and inversely proportional to the square of the distance between them.

This *law of gravitation* permitted an understanding of motions of bodies in the solar system. Newton could write down the solution of the *two-body problem* (the motion of one planet about the Sun), but the mathematical treatment of the *many-body problem* was considerably more difficult. The latter problem occupied many famous mathematicians of the 18th century: Leonard Euler (1707–1783) in Switzerland, and Joseph Lagrange (1736–1813) and Pierre Simon Laplace (1749–1827) in France. With the law of gravitation as its base, celestial mechanics developed in the 18th and 19th centuries into a highly sophisticated science.

The discovery of the planet Uranus by Sir William Herschel (1738–1822) in 1781 and of the first minor planet by the Italian G. Piazzi (1746–1826) in 1801 also stimulated the study of motions of solar system bodies. Carl Friedrich Gauss (1777–1855) published his "Theory of Motions of Celestial Objects which Revolve about the Sun in Conic Sections" in 1809, thereby furnishing a convenient method for calculating the ephemerides of, for instance, a minor planet from only a few observations. The method developed by the German physician Wilhelm Olbers (1758–1840), "To Compute the Orbit of a Comet from Some Observations," as a parabolic approximation also found wide application.

The disoveries of asteroids and planetary bodies during the late 18th and early 19th centuries spawned further searches for planets. The discovery of Neptune in 1846 by Johann Gottfried Galle (1812–1910) is considered the crowning achievement of the celestial mechanics experts Urbain Leverrier (1811–1877) in France and John Couch Adams (1819–1892) in England, who had independently employed their mathematical skills to successfully predict the position of the unknown planet. In 1930, the planet Pluto was discovered by Clyde Tombaugh.

11.4 Cataloguing the Stellar Sky

Some much-used star catalogues date back to the 17th and 18th centuries:

1603 *Uranometria*, by Bayer;
1661 *Sternverzeichnis*, by Hevelius;
1679 First southern star catalogue, by Halley;
1725 List of all stars down to 7th magnitude, visible in northern Europe, by Flamsteed;
1762 Star catalogue, by Bradley.

The list of 103 nebulous "M" objects compiled by Charles Messier (1730–1817) in 1784 and two catalogues by Sir William Herschel in 1786 and 1789, each containing 1000 newly found nebulae and clusters, also date to before 1800. The latter had by 1782 already compiled a catalogue of 269 double stars.

Astronomical research relies heavily on the accumulation of a sufficiently large base of observed data, so that the conclusions will be statistically significant. For this reason, sky surveys, data collections, and catalogue publications occupied the substantial part of the astronomical working capacity in the 19th and 20th centuries. The prototype of large surveys was the *Bonner Durchmusterung*, prepared in 1852–59 and directed by Friedrich Wilhelm Argelander (1799–1875) with co-workers Adalbert Krüger (1832–1896) and Eduard Schönfeld (1828–1891).

Improved instruments and observing techniques triggered advances in positional astronomy during the 19th century, notably through the efforts of the German astronomer Friedrich Wilhelm Bessel (1784–1846). The positions of about 32 000 stars which he determined in the years 1821–35 laid the foundation of many subsequent

Table 11.1. Number of known variable stars as a function of year.

Year	1786	1844	1890	1896	1912	1970	1983
Number of Variables	12	18	175	393	4000	22 650	28 450

studies. Better quality and larger telescope apertures also promoted the development of special areas of observation such as double stars. In this regard, an especially assiduous observer was Friedrich Wilhelm Struve (1793–1864) of Dorpat (Russia), who discovered double stars and also edited catalogues containing their data.

Collection and compilation of data were quickly undertaken in the new fields of photometry and spectroscopy, which were to form the basis for astrophysics. Visual work characterized only the beginning of this "new astronomy," which was to be taken over wholly by photographic methods in the latter half of the 19th century. Visual photometric methods, however, were still used to generate some celebrated catalogues:

- *Cordoba Durchmusterung of the Southern Sky*, begun in 1885, published in 1892;
- *Harvard Revised Photometry*, begun in 1879, published in 1907;
- *Potsdammer-Durchmusterung*, begun in 1886, published in 1907.

The *Göttinger Aktinometrie* in the years 1904–1908 by Karl Schwarzschild (1873–1916) formed the basis for photographic stellar photometry.

There is another special subject which has gained much importance in the wake of these compilations, namely the discovery and description of variable stars. The number of known variables has rapidly increased in the past 200 years, as can be inferred from Table 11.1 [11.1].

The advent of spectroscopy led to spectral classifications and catalogues of stars. The first list by the Italian Father Angelo Secchi (1818–1878) appeared in 1867 and contained 316 stars. The monumental spectral survey at the beginning of the 20th century with 225 300 stars is contained in the *Henry Draper Catalogue*, which was published under the auspices of Harvard College Observatory in the United States during the period 1918–1924. For more information on charts and maps, the reader is directed to Sect. 1.3 and also to the Supplemental Reading List for this chapter.

11.5 Astrophysics

11.5.1 Stellar Photometry

In the latter half of the 19th century, the apparent magnitude system of stars evolved from a foundation of somewhat crude, purely estimated values into a sophisticated, calibrated basis. The transition began in 1861, when Karl Friedrich Zöllner (1834–1882) built the first visual photometer in Berlin [11.2]. Photographic plates and photoelectric cells dramatically increased the measuring precision, as is revealed in Table 11.2 [11.3]:

Table 11.2. Precision of stellar magnitudes by various methods.

Year	Method	Precision
ca. 1850	Visual estimates	$0.^{\mathrm{m}}2$
ca. 1870	Visual measures	$0.^{\mathrm{m}}1$
ca. 1912	Photoelectric measures	$0.^{\mathrm{m}}01$
ca. 1970	Photodetector measures	$0.^{\mathrm{m}}001$

Stellar magnitudes thus came to be very accurately measured in certain standardized wavelength ranges from the ultraviolet into the infrared and have been employed to determine various color indices.

11.5.2 Spectroscopy of the Sun and Stars

The groundwork of modern spectral analysis was laid with the discovery by Gustav Robert Kirchhoff (1824–1887) and Robert Bunsen (1811–1899) of the characteristic spectrum of each chemical element. The amassing of spectral data on stars and other celestial objects characterized the first phase of astronomical spectroscopy. With the invention of the photographic process, permanent records of spectra of the Sun and stars were soon made [11.4]. The famous principle discovered by Doppler [11.5] led to another application in measuring radial velocities of stars. The first photographic measures of this kind were obtained in 1888 by Hermann Carl Vogel (1841–1907).

Among several spectral classification schemes of stars, the Harvard system by Edward Charles Pickering (1846–1919) and Annie Cannon (1863–1941) was the most successful, and was officially adopted by the International Astronomical Union in 1922.

Detailed studies of the solar spectrum continue with unabated intensity to this day. The invention, in 1892, of the *spectroheliograph*, a device which is used to photograph the solar surface in the light of individual spectral lines, was a major achievement in solar physics. Another milestone was reached when the hybrid technique of spectrophotometry was created by combining the methods of photometry and spectroscopy to measure the intensity distribution in spectra.

11.5.3 Astronomical Photography

The first photograph of the Moon by J.W. Draper on daguerre plates dates from 1841. The oldest scientifically useful photograph is that of the total solar eclipse on 1851 July 18 [11.6], while the first photographs of star images were obtained by W.C. Bond in 1857. But it was the invention of the dry plate by R.L. Maddox in 1871 that signaled the real breakthrough for astrophotography.

The enormous power of the photographic technique was quickly documented by scores of impressive photographs taken of the Milky Way, star clusters, and nebulae. Much of this pioneering work was performed by the American astronomer Edward Emerson Barnard (1857–1923) and the German Max Wolf (1863–1932). Photography

Table 11.3. Size and year of construction of some famous observatories.

Observatory	Aperture	Year of Construction
Mt. Wilson Observatory, USA	152 cm	1908
Mt. Wilson Observatory, USA	254	1917
Mt. Palomar Observatory, USA	508	1948
Selenchukskaya, CIS	610	1976

rapidly established itself in all areas of astronomical research, often in connection with photometric and spectroscopic methods.

11.5.4 Large Telescopes

Concurrent with the development of advanced observing techniques was the construction of larger and more powerful telescopes. Progress in astronomical research over the last century is well correlated with the practicalities and limitations of instrument design. The development of the achromatic objective by Fraunhofer in the 19th century initiated the production of refractors with increasingly larger apertures [11.7]. These range from the moderate-sized Dorpat refractor (objective diameter 24.3 cm, focal length 4.11 m) built by Joseph von Fraunhofer (1787–1826) in 1824, to the immense Yerkes Observatory refractor (objective diameter 102 cm, focal length 19.4 m) constructed by Alvin Clark and Warner and Swasey in 1897. The latter remains the largest refractor in the world.

After 1900, although special photographic refractors were still being constructed [11.8], the demand for larger, more powerful telescopes favored the building of reflectors with considerable dimensions. The aperture diameter and year of construction for a few of the most famous instruments are presented in Table 11.3.

A large field and large aperture ratio are combined in the optical system invented by Bernhard Schmidt (1879–1935) in the years 1930 to 1931 (the *Schmidt camera*). The largest Schmidt instrument with an aperture of 134 cm went into operation in 1960 in the Karl Schwarzschild Observatory in Tautenburg, Germany.

From the photographic plate to the most recent charge-coupled devices (CCDs), the accessories used to measure and record the radiation emitted by celestial bodies are a vital part of large-telescope research. Detectors of higher sensitivity have improved the information output over shorter recording times. Photoelectric methods in stellar photometry were introduced by Joel Stebbins (1878–1966) and Paul Guthnick (1879–1947). Large telescopes have been designed for specialized purposes (e.g., the 3.8-m Infrared Telescope on Mauna Kea in Hawaii, completed in 1980).

The majority of large telescopes are located on climatically favorable mountaintop sites. Some observatories have been built for the express purpose of observing the southern skies. One of the first to be established was the Cape Observatory in South Africa (1820). Of more recent construction are the European Southern Observatory (ESO), completed in 1969, and the Cerro Tololo Inter-American Observatory (CTIO), built in 1965, both in Chile.

11.6 Stellar Evolution and Stellar Systems

11.6.1 Stellar Evolution

The matchless discoveries in the areas of stellar structure and evolution are products exclusively of the 20th century [11.9]. The recognition of nuclear processes as the source of stellar energy and the identification of the specific nuclear reactions in question paved the way for the present evolutionary models of stars. The major stepping stones along this path were:

1906 Theory of equilibrium applied to solar atmosphere by K. Scharzschild; publication of book *Gaskugeln* by R. Emden; 1906–1912 Discovery of the relation between surface temperatures and luminosities of stars (the Hertzsprung–Russell diagram);
1923 Discovery of mass–luminosity relation by E. Hertzsprung;
1925 Publication of *Stellar Atmospheres* by C.H. Payne;
1926 Publication of *The Internal Constitution of the Stars* by A.S. Eddington;
1930 Theory of convective currents in stellar atmospheres first proposed by A. Unsöld;
1934 Hypothesis of neutron stars first proposed by W. Baade and F. Zwicky;
1938 Nuclear reactions as stellar energy sources proposed by H. Bethe and C.F. von Weizsäcker;
1951 Formation of carbon in stellar interiors investigated by E.J. Öpik, E.E. Salpeter;
1955 Beginning of computer modeling of stellar evolution, allowing for changing interior composition due to nuclear processes, by M. Schwarzschild, F. Hoyle.

11.6.2 Stellar Systems

With the opening of observatories with large reflectors during the first half of the 20th century, the exploration of the Milky Way and other galaxies progressed swiftly. A rich and detailed picture of the Milky Way, further refined with radio astronomical observations since about 1951, began to emerge. The resolution of extragalactic systems (e.g., the Andromeda Nebula) into stars confirmed the fact that most objects commonly found in the Milky Way (e.g., Cepheids, RR Lyrae stars, gaseous nebulae) exist in these distant galaxies as well. Results of investigations into the nature of spiral galaxies and of the Milky Way were merged into a single picture [11.10].

Some important stepping stones of exploration of stellar systems are:

1895 Initial studies of the structure of the Milky Way system using the methods of stellar statistics (J. Kapteyn, H. von Seeliger);
1904 Interstellar reddening and absorption found (J. Kapteyn, H. Shapley);
1918 Determination of photometric distances to systems within the Milky Way using Cepheids (H. Shapley);
1924 Resolution of the outer parts of the Andromeda and other spiral galaxies into stars (E. Hubble);

1926 Theory of galactic kinematics and dynamics developed (B. Lindblad, J. Oort);
1929 Redshift of spectra of extragalactic systems found to be in proportion to their distances (E. Hubble);
1937 Interstellar CO_2 absorption bands found in spectra;
1943 Unusual spectra of certain extragalactic systems found not to be explained solely by thermal emission (C.K. Seyfert);
1951 21-cm radio radiation of neutral hydrogen first detected (Ewen and Purcell);
1952 New distance scale of extragalactic systems (W. Baade) removes size discrepancy between Milky Way and other galaxies;
1958 Theory of the explosive origin of the Milky Way from a compact protogalaxy proposed (V.A. Ambartsumian);
1963 First quasar discovered and redshift measured (M. Schmidt);
1964 Density-wave theory proposed to explain spiral structure in the Milky Way and other galaxies (Lin and Shu).

11.7 Observations at Invisible Wavelengths and Space Exploration

Until about 1950, only the visible radiation from celestial bodies reaching the Earth's surface determined the state of astronomical knowledge. This is the so-called "optical window" in the electromagnetic spectrum, which encompasses ultraviolet, X-rays, and γ-rays in one direction, and infrared, microwave, and radio waves in the other. Most of these ranges cannot be observed from the Earth's surface because of atmospheric absorption. The observations made with the aid of radio telescopes, high-altitude flights, and later space-based instruments have broadened optical astronomy in the past 40 years into a science of *all* wavelengths [11.11]. The exploration of the solar system via space missions then led to numerous discoveries regarding the planets and their satellites. The major milestones along this road are:

1931 Radio radiation from the Milky Way discovered (K.G. Jansky);
1939 Observation of concentrated radio radiation in the galactic plane and toward the galactic center (G. Reber);
1942 Discovery of extragalactic components in radio radiation (J.S. Hey, J. Southworth);
1954 Discovery of radio galaxies;
1957 First artificial satellite (Sputnik I, USSR) launched;
1960 X-ray radiation from the solar corona found by an aerobee rocket;
1961 First unmanned mission to Venus (Venera 1, USSR);
1965 Discovery of cosmic X-rays (E.T. Byram, H. Friedman, T.A. Chubb);
1967 Discovery of pulsars (J. Bell and A. Hewish);
1969 First men land on the Moon (Apollo 11, USA);
1970 160 cosmic X-ray sources discovered (UHURU X-ray satellite, USA);
1976 Unmanned landings on Mars (Viking I and II, USA);
1979 Flyby of Jupiter and satellites (Voyager 1 and 2, USA);
1981 Voyager 2 flyby of Saturn;

1986 Voyager 2 flyby of Uranus;
1989 Voyager 2 flyby of Neptune.

The extensive space programs of the USA and Russia (former USSR) since 1957 as described in special publications [11.12] are beyond the scope of this chapter.

11.8 Research in Historical Astronomy

11.8.1 Research Problems

The fascination with the history of astronomy can sometimes lead from the mere perusal of historical presentations to the direct pursuit of historical records. The study of the general history of astronomical research [11.13,14] and the description of special sections requires some background in both history and science. Biographical presentations often reach beyond purely scientific aspects, and so the historically inclined researcher is encouraged to approach the life work of an astronomer and cast it into a literary form [11.15,16,17].

There are many books devoted to extensive historical subjects, but magazine articles on historical events and persons often provide the best source of information for a report on the past of astronomy (including amateur astronomy) in both research and education. The report may include material on observational work performed or public activities provided by particular societies, groups, or individuals [11.18,19,20].

The subject can be approached from various angles. There is, on the one hand, the technical, scientific investigation, and, on the other, one which produces a publication to inform a larger audience on various subjects from astronomy history. The latter may also deal with such matters as community or local school histories.

11.8.2 Sources

Any historical study depends on the amount and quality of the information available in the accessible sources.

11.8.2.1 Primary Sources. The focal point of work for the historian is the study of primary sources, which refer to the immediate documentation of the events, and generally date from the time of the events. These include personal documents, foundation documents, diaries, and contemporary reports of all sorts (descriptions of institutes, company pamphlets, newspapers, etc.).

Primary sources are often accessible in public and private archives and museums, and also in community public and governmental libraries.

11.8.2.2 Secondary Sources. Secondary sources include mainly published studies on a particular subject. These include dissertations and theses (on history), book publications and papers, and various manuscripts. These publications generally refer to the primary sources used and include citations from them. In addition to public libraries, university and college libraries may hold important collections of secondary sources.

Topics on astronomy history after 1800 are, apart from books, also documented in serials of a technical or general nature (e.g., *Popular Astronomy* since 1893, *Sirius/Die Sterne* since 1868, *Monthly Notices of the Royal Astronomical Society* since 1831, and *Astronomische Nachrichten* since 1821).

11.8.2.3 General Historical Sources. When searching libraries for material, it may be helpful to use the following groups as a guide:

Bibliographic works (e.g., *Astronomischer Jahresbericht*); Introductions (e.g., compendia on history and specifically astronomy history); Source studies (e.g., literature on historical sources); Document and source compilations (e.g., printed letters, documents, papers, and also collected works); Books (e.g., handbooks, textbooks, comprehensive presentations of a particular period in time or of a technical subject, such as spectroscopy); Genealogies (e.g., of a family); Serials (e.g., *Journal of the History of Astronomy*); Dictionaries; Biographies and Reference Books (e.g., *Allgemeine (ADB)* and *Neue (NDB) Deutsche Biographie*).

11.8.3 Processing

The purpose of a historical investigation, whether for scientific study or for more journalistic general interest information, largely determines the kind of research and source studies to be undertaken. All of the general methods of investigation such as interviews, excursions, literature studies, and surveys of local papers may have to be tapped to gain the desired information. Particularly in amateur astronomy, the trails, if any, are difficult to find. Biographical descriptions may be quite general, but cannot be more detailed for lack of primary sources [11.21].

For the processing, it is advisable to proceed from the accessible and well-known to the obscure and less-known:

Preparatory work (Inspection of local chronicals, documents, family diaries, etc.); Library work (Study of technical and documentary literature); Archive work (in local or governmental archives); Processing (Compilation of the previously obtained results in card or computer files).

Any presentation gains by offering the reader well-selected contemporary statements and documents cited, quoted, or reproduced [11.22]. Footnotes and lists of references are found primarily in scientific studies. A general readership will undoubtedly appreciate some outline of the contemporary historical situation.

Regardless of the particular area in astronomy history, the interrelations of scientific, technological, and societal evolution will ultimately emerge as the work progresses. This interaction of all societal forces has always influenced astronomers, observatories, and research programs, a fact which must be kept in mind in any evaluation or narrative [11.23].

References

11.1 Roth, G.D.: Kosmos Astronomie-Geschichte. S.72, Franckh'sche Verlagshandlung, Stuttgart 1987.
11.2 Herrmann, D.B.: Karl Friedrich Zöllner und die "Potsdamer Durchmusterung." Versuch einer Rekonstruktion. *Die Sterne* **50**, 170 (1974).
11.3 Op. cit. [11.1], p. 82.
11.4 Seitter, W.: Aus der Geschichte der astronomischen Spektroskopie (2. Teil). *Sterne und Weltraum* **8**, 231 (1969).
11.5 Herrmann, D.B.: *Geschichte der modernen Astronomie*, p. 102 f. VEB Deutscher Verlag der Wissenschaften, Berlin 1984.
11.6 Op. cit. [11.1], p. 35.
11.7 Van Helden, A.: Telescope Building, 1850–1900. p. 40 f., in: *The General History of Astronomy Vol. 4, Part A: Astrophysics and Twentieth-Century Astronomy to 1950*. Cambridge University Press, Cambridge 1984.
11.8 King, H.C.: *The History of the Telescope*, Dover, New York 1979.
11.9 Lang, K.R., Gingerich, O.: *A Source Book in Astronomy and Astrophysics, 1900–1975*, Harvard University Press, Cambridge MA 1979.
11.10 Unsöld, A.: *The New Cosmos*, p. 202. Springer, Berlin Heidelberg New York 1967.
11.11 Op. cit. [11.9], p. 110.
11.12 Engelhardt, W.: *Planeten, Monde, Ringsysteme. Kamerasonden erforschen unser Sonnensystem*. Birkhäuser, Basel 1984.
11.13 Zinner, E.: *Astronomie, Geschichte ihrer Probleme*. Verlag Karl Alber, München 1951.
11.14 Hearnshaw, J.B.: *The Analysis of Starlight. One Hundred and Fifty Years of Astronomical Spectroscopy*. Cambridge University Press, Cambridge 1986.
11.15 Brück, H.A., Brück, N.T.: *The Peripatetic Astronomer. The Life of Charles Piazzi Smyth*, Adam Hilger, Bristol 1988.
11.16 Evans, D.S.: *Under Capricorn. A History of Southern Hemisphere Astronomy*, Adam Hilger, Bristol 1988.
11.17 Wright, H.: *James Lick's Monument. The Saga of Captain Richard Floyd and the Building of the Lick Observatory*, Cambridge University Press, Cambridge 1987.
11.18 Barthalot, R.: Astronomy in the French Revolution. *Sky & Telescope* **78**, 21 (1989).
11.19 Lankford, J.: Astronomy's Enduring Resource. *Sky & Telescope* **76**, 482 (1988).
11.20 Williams, T.R.: A Galaxy of Amateur Astronomers. *Sky & Telescope* **76**, 484 (1988).
11.21 Ashbrook, J.: J.N. Krieger: The Moon Half-Won. (see [11.22].) *Note:* Further research attempts into the life of Krieger (by the author) have been unsuccessful.
11.22 Ashbrook, J.: Astronomical Scrapbook. Sky Publishing Corporation, Cambridge MA 1984.
11.23 McCutcheon, R.A.: Stalin's Purge of Soviet Astronomers. *Sky & Telescope* **78**, 352 (1989).

12 Astronomy Education and Instructional Aids

H. J. Augensen

12.1 Introduction

The field of astronomy serves not only to perform research on matters of astronomical and astrophysical relevance, but also to administer an invaluable didactic function: that of guiding and contributing to the scientific education of the general public as well as of future astronomers, physicists, and other scientists. In this regard, instructors affiliated with public and private educational institutions and also amateur astronomers can participate in educating interested persons in this most fascinating of subjects. Astronomy is—particularly at the introductory level—arguably the most approachable of all the sciences.

This chapter[1] will also survey the broad spectrum of astronomy education in the United Kingdom, Canada, and the United States, with the intent of providing the reader with information regarding:

1. how students who will become future astronomers are trained within the formal education systems in the U.K., Canada, and the U.S. Several possible career paths will also be mentioned;
2. facilities, such as planetariums and observatories, and services, such as lectures, which are available to school groups and the general public throughout most of the English-speaking countries in the northern hemisphere;
3. some currently available (1992) sources of instructional materials such as slides, videodiscs, films/videocassette tapes, and computer software, as well as books and other printed materials for all age groups. Also listed are some of the distributers of mechanical instruments and exhibit items such as globes, planetariums, telescopes, and clocks.

---

1 This chapter has been extensively rewritten from the original German version, authored by A. Kunert, in order that it be applicable to the principal English-speaking countries in Europe and North America. Those readers who live in the southern hemisphere, particularly in Australia and New Zealand, will undoubtedly extract some useful information too.

12.2 Formal Astronomy Education[2]

The subject of education in astronomy is a complex one, especially when one considers the fact that most professional astronomers do not begin formal training in the subject until they enter graduate school. Indeed, the majority of persons who call themselves astronomers received their first university degrees (e.g., B.A., B.Sc.) in physics, mathematics, engineering, computer science, chemistry, or some other technical field. On the other hand, very few students enrolled in an undergraduate course in introductory astronomy are destined to pursue the subject beyond that point. The explanation of this paradox lies in the nature of astronomical research: it is chiefly *quantitative*, and moreover many of the principles and techniques used to obtain planetary and stellar distances, sizes, temperatures, masses, luminosities, and other relevant quantities are founded in physics or mathematics.

The popular notion that an astronomer is a man who spends every night peering through a telescope at the Moon or looking for "new" stars is far more fantasy than fact. In reality, the percentage of time actually spent at the telescope is relatively small. Moreover, very few modern research astronomers perform purely visual observations; instead they attach a plate camera, spectrograph, photometer, micrometer, or other device at or near the eyepiece position of the telescope, and record their results photographically or electronically.[3] Some astronomers observe with radio telescopes, which can be operated during the daytime, and still others observe from a ground-based control room with satellite telescopes such as the *Hubble Space Telescope* and the *International Ultraviolet Explorer (IUE)*. Furthermore, most of their time is spent in *reducing* and *analyzing* the data that they accumulate during a few nights of observing at the telescope. Thus, on a typical day an astronomer is likely to be found sitting at a computer terminal, or browsing through journals in the library, or writing up the results of research into a publishable paper, or—for astronomers who are affiliated with a college or university—teaching and advising students. Finally, astronomy (as well as physics, mathematics, and the other sciences) is for women as well as for men, and today the opportunities for women to enter the field are greater than ever before.

It is only at the introductory level that astronomy can be successfully presented in a descriptive fashion (as would be a course in art, for example) and thus is suitable for students who wish to know something of the stars and planets but without getting into an extensive mathematical treatment. A science-bound student taking such a course will have received only the "frosting" on a large, multilayered cake. If, on the other hand, the student has formed a solid undergraduate background in physics or mathematics, then he/she has also in effect been prepared for advanced work in astronomy or astrophysics.

---

2 Much of the information in this section is extracted from the career pamphlets/brochures *Becoming a Professional Astronomer* [12.1], which is available from the Association for Astronomy Education or from the London Planetarium, *Astronomer* [12.2], which is available from the Guidance Centre of the University of Toronto, and *Understanding the Universe: A Career in Astronomy* [12.3], put out by the American Astronomical Society.

3 In contrast to the other devices mentioned, micrometer data are usually recorded with pencil and paper.

The preparation for and demands of a career in astronomy vary somewhat, especially at the secondary school (pre-college) and college levels, depending upon the particular country or province where one receives his/her schooling. Nevertheless, there are so many similarities that a general presentation of the requirements at the pre-college, undergraduate, and graduate (sometimes called postgraduate) levels, as well as possible avenues in postdoctoral research and teaching can be attempted here. In fact, the further along the ladder one rises en route to becoming a professional astronomer, the less evident are these seeming dissimilarities. Fortunately for science, there is no fundamental difference in the manner in which research is performed and published by professional astronomers working on opposite sides of the Atlantic Ocean.

Not all courses in astronomy are designed for those who intend to pursue astronomy as a career; some (primarily at North American institutions) are intended for persons whose future occupations will most likely not be in science, and yet whose curiosity for the subject has been whetted by newspaper or television reports of exciting astronomical events and discoveries (e.g., the passage of Halley's comet in 1986, the appearance of Supernova 1987A in the Large Magellanic Cloud, the recent *Voyager* spacecraft encounter with Neptune). The needs of this latter group must be met just as successfully as those of future astronomers, as much of the funding to support astronomy programs, including the recent space missions, comes from government agencies (such as the Science and Engineering Research Council (SERC) in the U.K. and the National Science Foundation (NSF) in the U.S.); hence, the fate of any program is ultimately dependent upon taxpayers from all walks of life. Unless society has been adequately informed of developments in the field and convinced of the need for funding for research, support for that research will most likely not be continued.

12.2.1 The Education of Professional Astronomers: An Overview

The subsections which follow are designed to provide the reader with a very general idea as to what goes into the education of a professional astronomer, with the emphasis on the education systems in the United Kingdom, Canada, and the United States. As the terminology used to describe pre-university and university school systems in the U.K. differs somewhat from that used in Canada and the U.S., the former will be treated separately from the latter two in this subsection. As the level of education progresses to the advanced undergraduate and graduate levels, the apparent dissimilarities in education systems become insignificant. It is not unheard of for an American student, for instance, to obtain his or her undergraduate degree in the United States and the doctorate in the United Kingdom, and vice versa.

Irrespective of which country the student is educated in or what the structural details of the particular education system are, the underlying theme during the years of preparation from pre-college level to the Ph.D. (Doctor of Philosophy) degree in astronomy is a firm background in physics and mathematics. The theoretical and experimental tools acquired in the process of learning these two subjects are essential for advanced work in astronomy. Needless to say, persons with aspirations to become astronomers but without the necessary strengths in these two areas will find university studies very rough going, if not impossible.

The student who wishes to become a professional astronomer is strongly encouraged to pursue advanced studies leading to the Ph.D. While it it is not absolutely necessary to "go all the way" in order to become a professional astronomer, it must be noted that the vast majority of the available jobs in astronomy in North America and Europe do require the Ph.D. just to be considered for the post (see Sect. 12.2.6). Nevertheless, those with an undergraduate major in astronomy or physics are needed in support positions at national observatories, national laboratories, federal agencies, and sometimes in large astronomy departments at universities. An undergraduate degree in astronomy is considered excellent preparation for secondary school science teachers, laboratory technicians, and computer programmers, and it can be the basis for higher degrees in fields other than astronomy (e.g., medicine or law). In addition, there are a small number of professional positions in planetariums and science museums, and a few in science journalism and science writing, many of which do not require an advanced degree.[4]

Students who intend to pursue graduate studies should be prepared to pass one and often two written and oral examinations during their graduate years before obtaining the Ph.D. At some point the student will be expected to prepare and deliver a paper at an astronomical society meeting, or publish the results of his/her thesis work in an astronomy journal. In either case, the work will be subject to critical scrutiny by seasoned professionals from other institutions, and the student will be required to answer any queries about one or more aspects of his/her research.

Once the Ph.D. is obtained, the road to learning has not ended. After securing a position as a teacher, researcher, government employee, or some other post, the education of an astronomer continues at an advanced level. Many astronomers keep a hand in research, continuing to publish papers on ongoing research, giving invited colloquia at colleges and universities, and attending meetings of various local, national, or international astronomical societies. Some take time to write textbooks, monographs, and popular books for the edification of future generations of beginning and advanced students. Those individuals who successfully reach this stage have gone beyond merely learning about the body of knowledge which encompasses modern astronomy; they are actively contributing to it.

12.2.2 Pre-College/University Preparation

12.2.2.1 Pre-University Preparation in the United Kingdom.[5] Any young person who sincerely desires to become a professional astronomer should realize that usually six or seven (or more) years at the university level are mandatory. To ensure a strong

---

[4] It is worth noting that the recent supernova SN 1987A was discovered by a technician with a B.Sc. degree only.

[5] The source for much of this information is the pamphlet *Becoming a Professional Astronomer* by Couper [12.1], which is available from the Association for Astronomy Education or from the London Planetarium.

preparation for this formidable undertaking, careful preparation is virtually a must, beginning with judicious selections of courses at the GCSE- and A-levels.

General Certificate of Secondary Education (GCSE). The decision as to which GCSE-level courses to take depends upon one's own interests—art, music, history, and even needlework are all quite acceptable. Of course, one should not hesitate to enroll in a GCSE-level astronomy course if it is offered, but even if it is not, physics, mathematics, and chemistry are strongly recommended at this early stage. Mastering one or more languages is also a good investment of time, as professional astronomers are as a rule well-traveled. Finally, a mastery of the *English* language is indispensible, as professional astronomers regularly write and publish papers in order to communicate the results of their work to colleagues and the public. A grade of C or better should be achieved in all GCSE subjects, as these will ultimately form the basis of further study at A-level and/or be credited toward entrance to a university.

A-levels. To ensure the best possible chance of getting into a university, the student will need to take three A-level subjects: 1. mathematics (essential), 2. physics (also essential), and 3. a choice of another branch of mathematics (applied or statistics), or chemistry, biology, geography, or other science course. Note that the N.U.J.M.B.'s Physics A-level includes an astrophysics option. Two appropriate AS-level subjects may be offered in place of the third A-level.

12.2.2.2 Pre-University Preparation in Canada and the U.S. The student of secondary school age can begin preparing for a career in astronomy by taking as many courses in science, mathematics, and English as possible, the last being necessary to ensure the development of good writing skills, which are indispensible in any field of science. The science courses usually include one year each of biology, chemistry, and physics. Mathematics courses should be taken during each of the four years, and should include analytical geometry, trigonometry, and, if offered, calculus. It should be noted that the recommendations at this level are basically sound ones for a student who intends to pursue advanced work in *any* of the sciences or in engineering.

It is also desirable to attain a mastery of at least one foreign language, such as French, German, Russian, or Spanish, but even more important is familiarity with computer programming using a language such as BASIC, PASCAL, C, or FORTRAN. Computer skills are now virtually mandatory at all levels, but this is especially true if one intends to pursue advanced work in science and engineering.

During the junior year, the student should begin to consider various colleges and universities for undergraduate training. Early in the senior year, he/she should arrange to take the Scholastic Aptitude Test (SAT) and select and apply to colleges.

12.2.3 Astronomy at the College/University Level

12.2.3.1 University-Level Training in the United Kingdom: First Degree.[6] University entrance requirements vary widely, and applicants are usually provisionally offered

---

6 *Note added in proof:* See also the article by Dworetsky [12.33].

places subject to their "A-level" results being satisfactory. As a first step in deciding which course and which university to apply for, the *U.C.C.A. Handbook* (P.O. Box 28, Cheltenham, Gloucester) should be consulted. A degree in astronomy, physics, mathematics, or a combination of these will enable the student to choose from among a wide variety of research topics ranging from radio astronomy to planetary geology. A degree in one of the other sciences may place a constraint on the future choice of graduate studies; a degree in electrical engineering, for example, would normally restrict the graduate to research in astronomical instrumentation.

A great deal of care and planning should precede one's choosing a university. Some of the following questions should be considered when making the final decision: 1. Is the college or university located in the city or in the country? 2. Is the school close to home or quite distant? 3. What are the course offerings in astronomy at the school? 4. What are the research interests of the astronomy or physics department in the school? 5. Do they have an observatory on or near campus?

During the years spent at college or university, the student will probably find numerous astronomical activities to become involved in "after hours." The long vacation between academic years provides other opportunities, and some students spend the vacation between their second and third years working at an observatory in order to become familiar with the manner in which astronomers conduct research. In fact, the Royal Greenwich Observatory, now at Cambridge, and the Royal Observatory at Edinburgh run special eight week long "vacation courses" to cater to this need, but competition for these places is keen.

As the student nears the end of his/her studies toward the first degree, he/she should keep in mind that the prerequisite for acceptance into a doctoral degree program is a first degree of the highest possible grade—first or upper second class honors. If the first degree grade is not up to standard, on the other hand, it is still possible to go on. Students with upper and lower second-class honors may usually proceed straight to a one-year M.Sc. course after their first degree. After the M.Sc. has been obtained through examination or thesis, he/she may then apply for a Ph.D. place, though at the price of losing one year of SERC studentship support.

12.2.3.2 University-Level Training in Canada and the United States. In Canada and the U.S., "introductory" astronomy generally manifests itself at the university level in two tiers:

1. A one-semester survey course in astronomy, with or without an associated laboratory and/or telescope viewing sessions, designed for students who are not science-oriented; breadth rather than depth is the goal here.
2. A one- or two-semester course in general astronomy (usually with laboratory) aimed at those students with good mathematical skills and/or who are majoring in one of the sciences or engineering. This is a more rigorous introduction to astronomy than 1., and frequently numerical problems will be assigned as homework. Much of the same material is covered as in 1., but with a more quantitative, in-depth approach.

Few four-year colleges offer advanced astronomy courses beyond this, although some do have a course on advanced topics which undergraduates can enroll in, and occasionally a course on the history of astronomy is available. There do exist a small number of colleges and universities in North America which offer very fine undergraduate programs in astronomy, but the paucity of such institutions will mean that many astronomy-bound students must obtain their initial training in physics or some other area (see below).

Perhaps the best general advice that can be given to the college student who aspires to be an astronomer is to focus on the curriculum in physics and/or mathematics. The careers of physicists and astronomers are much more alike than they are different, and usually a solid undergraduate preparation in physics, even without the inclusion of a course in general astronomy (although admittedly taking such a course at this stage would certainly do no harm and might save some time later on) lays a reasonably solid foundation for first graduate work and ultimately a career in astronomy or astrophysics. According to the *1986–87 Graduate Student Survey* [12.4] which was conducted by the American Institute of Physics (AIP), approximately one-half of the students enrolled in graduate school in astronomy had been physics majors as undergraduates and one-fifth had majored in a related field such as mathematics or engineering. Only 28% of the sample had majored in astronomy.

As noted in *Enrollments and Degrees* [12.5], a survey taken by the AIP, "students interested in astronomy benefit from a greater number of options than their physics counterparts, because they can begin to concentrate on their major either at the undergraduate or graduate level and they have a choice of two types of departments, the separate astronomy departments or the combined physics and astronomy departments."

Though requirements will naturally vary from school to school, the following generic curriculum for a student majoring in physics will give the reader an idea as to what is usually required at this level. During the first year (and often into the second year) of undergraduate studies, the student will be taking introductory calculus along with the following sequence of general physics courses:[7]

- General Physics I. Introductory Mechanics, Structures, Fluid Motions, and Wave Motion (with laboratory).
- General Physics II. Heat and Thermodynamics, Electricity and Magnetism, Circuits (with laboratory).
- General Physics III. Radiation, Optics (with laboratory).

These are generally followed by more intensive work in many of the areas covered in these courses, but also with new, advanced material not previously dealt with. These include:

- Modern Physics (Atomic and Nuclear Physics, Relativity)
- Solid State Physics
- Classical Mechanics
- Thermodynamics and Statistical Mechanics
- Electricity and Magnetism
- Optics and Wave Motion

---

[7] The term "general" here connotes an introductory course for students who are oriented towards science or engineering.

– Quantum Mechanics

It is assumed that the student will be taking concurrent advanced coursework in mathematics, the most essential being differential and integral calculus, differential equations, complex variables, numerical methods, statistical analysis, and linear algebra, although others may also be included. In addition, at least one year of chemistry is advisable.

The student will also need to enroll in a minimum of one semester of a computer science class in which at least one specific computer language, such as FORTRAN, is taught and writing programs to solve problems is stressed. It is not necessary that all of the inner workings of the computer be comprehended, but rather that the essential programming skills needed to solve problems in physics and mathematics be mastered.

The student would also be well advised to get some career-related job experience during the summer months, especially if research is not an implicit part of the undergraduate program. Government institutions (such as NASA), many universities (e.g., Harvard), and some observatories (e.g., Kitt Peak National Observatory, the Maria Mitchell Observatory) offer summer jobs in astronomy for students.

By the end of the first three of the (usually) four years of mandatory undergraduate training, the student will likely have a clearer idea of whether he or she wants to pursue astronomy in earnest as a career. If the answer is negative, then the student should not feel that the education he/she has received during their college days has been for naught. On the contrary, the many skills in mathematics, computer science, physics, and astronomy that he/she has learned will be invaluable in any one of a number of nonastronomy professions. These would include in particular computer programming and teaching science at the secondary school level (much in demand in the U.S. at this time). There may also exist openings in astronomy for such posts as observing assistant at an observatory or lecturer at a planetarium.

If, on the other hand, the answer to the astronomy career question is affirmative, then preparation should now begin for the next step in the education chain: graduate school. The school chosen for graduate studies may be the same one where the undergraduate degree was attained, but usually it is not. Most graduate schools in Canada and the U.S. require at least a B or better grade point average in college and an undergraduate major in physics or astronomy. Most schools also require, or at least strongly recommend, that students take the Graduate Record Examination (GRE) administered at a local school at regular intervals during the fall, winter, and spring. The GRE tests, which are created under the aegis of the Educational Testing System in Princeton, NJ, consist of a General Test measuring verbal, quantitative, and analytical abilities, and an Advanced Test in one of Physics, Mathematics, Engineering, Computer Science, and other areas. The student should check carefully whether any or all of these are required by the particular university. The results are usually sent direct to the graduate school of the student's choice prior to the deadline for receipt of applications. The student should be aware of the fact that many schools base financial aid on the results of the GRE Advanced Test. In addition, students from non-English-speaking countries are required to take the Test of English as a Foreign Language (TOEFL).

For more information on GRE or TOEFL exams, write: ETS, Box 899, Princeton, NJ 08541, USA.

12.2.4 Astronomy at the Graduate Level

The in-depth training of an astronomer for his or her profession is often said to begin in graduate school. There he or she will, over a time interval of approximately two years, be required to take a sequence of core courses, each containing one or more elements of the foundation for advanced observational and/or theoretical research. In addition, there are more specialized courses and seminars aimed at a particular aspect of the subject.

Outstanding students who wish to achieve a doctorate should not be discouraged by the perceived financial burden to be incurred in this endeavor. In the U.K., SERC and other foundations will normally fund much or all of graduate education of those who have achieved first or upper second class honors for their first degree. In Canada and the U.S., individual universities provide financial aid in the form of teaching assistantships, research assistantships, and full fellowships in order to attract qualified students into their doctoral programs. These awards often provide for all or part of the tuition, plus a stipend for living expenses.

Course offerings at the graduate level will, of course, vary somewhat from institution to institution. The following might constitute a typical sequence of courses, a substantial subset of which would be taken over a roughly two-year period.

- Astronomical Observing Methods I. Photometry
- Astronomical Observing Methods II. Spectroscopy
- Astrophysics I. Stellar Atmospheres
- Astrophysics II. Stellar Interiors and Evolution
- Interstellar Medium and Galactic Structure
- Galaxies and Cosmology
- Seminar: Special Topics in Astronomy/Astrophysics
- Independent Study/Research

It should be noted that some universities offer only the terminal master's degree in astronomy or physics, while others offer both master's and Ph.D. degrees. This can be checked by referring to the *1990–91 Graduate Programs in Physics, Astronomy, and Related Fields* [12.6] (or the most recent edition), which lists in some detail the graduate programs of universities on the North American continent, to the *Directory of Graduate Programs Vol. C* [12.7], or to *Understanding the Universe: A Career in Astronomy* [12.3], a brochure available from the American Astronomical Society. Students in the U.K. should consult *Postgraduate Opportunities in Astronomy and Geophysics* [12.8], published by the Royal Astronomical Society (Burlington House, Piccadilly, London W1V 0NL), or the aforementioned booklet *Becoming A Professional Astronomer* [12.1].

Following the completion of graduate coursework, the student usually selects a research project which has promise of developing into a Ph.D. dissertation. The dissertation is carried out under the guidance of one or more faculty members in his/her department, the thesis *adviser(s)*. The project selected will depend on the student's propensity toward a specific aspect of observational and/or theoretical astronomy, and also upon the diversity of astronomical research done at that institution. Therefore, if a student applying to graduate schools has a strong preference for a particular research area, then he/she should be certain that the graduate program has at least one faculty member involved in that research.

Most of the major categories of astronomy research are listed below. These groupings are not necessarily mutually exclusive, and in some cases the overlap is considerable.

Solar System
- Planetary Atmospheres
- Planetary Interiors, Magnetic Fields
- Solar–Terrestrial Interactions, Aurorae
- Meteorites and Tektites
- Comets and Asteroids
- Solar Atmosphere and Composition
- Solar Flares, Prominences, Corona, and Solar Wind
- Solar Interior, Neutrinos

Stars and Interstellar Matter
- Positional Astronomy, Stellar Parallaxes and Motions
- Stellar Spectra/Atmospheres
- Stellar Interiors and Evolution
- Low Luminosity Stars, Degenerate Stars
- Binary Stars
- Star Clusters and Associations
- Interstellar Matter, Dust
- Molecular Clouds and Star Formation
- Young Stars, T Tauri Stars
- HII Regions
- Intrinsic Variable Stars
- Novae and Supernovae
- Planetary Nebulae
- Supernova Remants, Pulsars, Black Holes
- Cosmic Rays, X-ray Sources

The Milky Way Galaxy
- Kinematics and Dynamics of Stars in the Galaxy
- Spiral Arm Structure and Motions
- Galactic Halos
- Stellar Populations
- The Galactic Center

Galaxies
- Normal Galaxies
- Active Galaxies, Quasars
- Clusters and Superclusters of Galaxies

Cosmology
- Cosmology, Early Universe
- Missing Mass

– Relativistic Astrophysics

Astronomical Instruments and Techniques
– Telescope Design and Tracking
– New Technology Telescopes
– Optical Interferometry
– Spectroscopic, Photometric, and Photographic Techniques
– Charge-Coupled Devices (CCDs)

History of Astronomy
– Archaeoastronomy
– Old Astronomical Records and Techniques
– Noted Astronomers of the Past

The distribution of astronomers engaged in research in these various subfields is by no means uniform. According to the *1986–87 Graduate Student Survey* of the AIP [12.4], cosmology/extragalactic objects is currently the biggest attractor of graduate students—one-quarter of the graduate student population. The second and third largest enrollments are in stellar atmospheres and interstellar matter, respectively.

Once the dissertation research has been embarked upon, the adviser will carefully monitor the student's progress, having frequent meetings to discuss the latest results. The student should be prepared to spend three or more years on the dissertation, especially if the project is one which involves numerous observations. According to [12.4], the median number of total years spent in graduate studies (this includes both coursework and dissertation) in the U.S. was 5.4 years for 1986–87 graduates. One-third of those students took six years to complete their studies and one-quarter took seven. While these figures may seem daunting, it must be noted that students in the humanities and social sciences take even longer to complete graduate studies.

The numbers quoted above can in large part be attributed to the originality and thoroughness demanded of the dissertation research, and also to the amount of time needed to accumulate and reduce the requisite data. Not everyone is suited for the rigors of graduate school, and the conferring of a Ph.D. degree upon someone means that that individual has demonstrated fitness to be a doctoral astronomer.

It should be borne in mind, however, that although the road to the Ph.D. can be a long one, the research in which the student is involved is often quite exciting, and will sometimes involve visiting major observatories, such as Kitt Peak National Observatory in Arizona or the Canada–France–Hawaii Observatory in Hawaii, for the purpose of gathering observational data or using computer reduction facilities. It means travel to conferences and meetings in different parts of the country or of the world to present the results, whether preliminary or final, of the dissertation and to meet and engage in stimulating discussions with professional astronomers from other institutions. It can mean the excitement of being the first to discover some bit of astronomical knowledge which no one had known before. Thus, the progression toward the doctoral degree should be regarded as a genuine adventure, not a prison term!

12.2.5 Postdoctoral Positions

Although theoretically it is possible for an astronomer with a freshly minted Ph.D. to obtain a permanent position, this option is not usually available. Instead, many astronomers accept temporary (1–3 year) "postdoc" positions involving research (usually in the same area of astronomy in which the Ph.D. work was done) and occasionally some teaching as well. In fact, according to the *AIP Employment Survey 1987* [12.9], some three-quarters of new astronomers in the U.S. traditionally accept postdoctoral positions. Many employers, especially the larger research universities and laboratories, stipulate a prerequisite of one or more years of "postdoc" experience before consideration for a permanent teaching or research post.

12.2.6 Vocational Opportunities

After the receipt of the Ph.D., there are several possible routes to follow, such as college/university teaching, civil service (government), business/industry, and public service. Some of these possibilities are discussed below.

12.2.6.1 Colleges and Universities. Many astronomers will acquire at least some experience in teaching during their years of graduate training, and most will ultimately secure teaching posts at the college/university level. That is not to say that the astronomer hired to teach is absolved of research duties. On the contrary, an astronomer who teaches at a university, particularly one with a Ph.D. program, will often be engaged in a vigorous research program. In this case, his or her students can benefit from the direct experience of a teacher who is also a research astronomer. Job advertisements from 4-year colleges sometimes state the importance of the ability of the candidate to involve undergraduates in his/her research.

Once a "permanent" college/university post has been obtained, usually at the assistant professor level, the candidate will most likely be required to attain *tenure* within three to eight years after having signed a contract. During that period of time, the astronomy faculty member must demonstrate excellence usually in one or more of the areas of teaching, research, and service to the college or university. The years during which tenure is strived for can be challenging, as the astronomer must juggle the duties in these three areas, not to mention family responsibilities. Once achieved, however, the astronomer's position is considered "secure" essentially for life.

12.2.6.2 National Observatories and Government. A number of professional astronomers (about one-third in the U.S., for example) are employed by the federal government directly or by federally supported national observatories and laboratories. A Ph.D. is nearly always a prerequisite for such posts, and a form of tenure is granted after a certain period. A few of the more prominent institutions are listed below:

National Observatories
– La Palma Observatory (U.K.)
– Royal Greenwich Observatory (U.K.)
– Royal Observatory Edinburgh (U.K.)

- Dominion Astrophysical Observatory and Dominion Astrophysical Radio Observatory (Canada)
- Canada–France–Hawaii Observatory (Canada)
- James Clerk Maxwell (mm-wave) Telescope (Canada/U.K./Netherlands)
- Cerro Tololo Inter-American Observatory (U.S./Chile)
- Kitt Peak National Observatory (U.S.)
- National Radio Astronomy Observatory (U.S.)
- Space Telescope Science Institute (U.S.)

Government Agencies
- Science and Engineering Research Council (U.K.)
- Herzberg Institute of Astrophysics of the National Research Council (Canada)
- National Aeronautics and Space Administration (U.S.)
- Naval Research Laboratory (U.S.)
- U.S. Naval Observatory (U.S.)

12.2.6.3 Business and Private Industry. A small number of astronomers (fewer than ten percent in the U.S., for example) are employed in business or private industry, but there are indications that this percentage has increased. Some industries, especially in the aerospace field, hire astronomers to do astronomical research which can increase the competitiveness of the the company, and a number of consulting firms supply astronomical talent to the government for specific tasks. Still other companies, while not engaged in astronomical research, hire astronomers in order to benefit from their acquired talents in instrumentation, remote sensing, spectral observations, and computer applications to unusual problems. Industrial employment also offers a wide variety of nontechnical career paths. A Ph.D. is normally not required for such positions.

12.2.6.4 Public Service. A very small number of astronomers choose to work in planetariums, science museums, or other public service positions where they provide an invaluable interface between the technical world of professional astronomy and society in general. Such jobs require not only a knowledge of astronomy, but also the ability to interact well with the general public.

There are also some jobs available teaching physics, earth sciences, or mathematics in secondary schools and community colleges. While a doctorate is usually not a prerequisite, such positions generally require teaching certificates or practical experience, and the salary levels and job security in these positions are generally lower than those in government and university employment. In the U.S. especially there is an urgent need for qualified persons to fill these posts.

A few jobs in science writing and science journalism are also available, and these almost never require an advanced degree (although it certainly would do no harm). Job security is not as a rule very high, but the satisfaction gained by translating research discoveries into exciting and cogent material for the public can be an important asset in this position.

12.2.7 Interdisciplinary Approaches to Astronomy

While astronomy may be considered most akin to physics and mathematics, ties with other, often seemingly unrelated, disciplines have also been demonstrated. In a two-part article appearing in *Mercury* magazine and reprinted in the information packet *Interdisciplinary Approaches to Astronomy* (see Sect. 12.4.5.3), Fraknoi [12.10] has compiled a bibliography to various crossovers between astronomy and the following other subjects: literature, music, art, anthropology, archeology, energy (solar power), history (supernovae), law, meteorology, molecular biology, philately, philosophy, psychology (including UFOs), and society (light pollution). Those teaching introductory astronomy will find this reference list especially valuable as students are often fascinated to learn that astronomy has real-world ties.

12.2.8 Comments on Astrology

No survey of astronomy education would be complete without at least a mention of the "science" of astrology. The term astrology means literally "science of the stars" and was in ages past used interchangeably with astronomy. Today, however, there is a major distinction. According to *Webster's Dictionary*, astronomy is that science which treats the nature, distribution, magnitudes, motions, distances, periods of revolution, eclipses, etc. of the heavenly bodies. Astrology, on the other hand, is defined as a pseudo-science which claims to foretell the future by studying the supposed influence of the relative positions of the Moon, Sun, and stars on human affairs. While most modern scientists are well aware of this distinction, the terms astronomy and astrology are frequently still equated in the eyes of the general populace. This is perhaps not surprising given the fact that many astrologers use astronomical terminology and charts of planetary positions and motions, thus leading the uninformed public to believe that astrology is merely applied astronomy (Kruglak and O'Bryan [12.11]). On more than one occasion has a professional astronomer been misintroduced to an audience as a noted "astrologer" from the — Observatory, or quoted in a newspaper or radio interview as an "astrology" professor at — University.

Rigorous statistical tests have been applied time and again to astrological forecasts, but the conclusion has always been the same: there is simply no empirical support for them (Kelly [12.12] and Pasachoff [12.13]). Yet horoscope columns are fixtures in many newspapers, and some astrologers even appear on local television to provide the daily horoscopes for astrology devotees. According to a 1975 Gallup poll of a sample of the U.S. population[8], 22% of those surveyed of ages 18 and older said they believed in astrology, and 23% of those polled said they read astrology columns regularly. The percentages were even higher for a subsample in the age range 18–24: 38% and 26%. These and other daily experiences point to a distinct fact: that the predictions of astrologers are still taken seriously by a non-negligible fraction of the populace. Consequently, many professional astronomers, particularly those who teach

---

[8] Quoted from Table 2 in the article by Kruglak and O'Bryan [12.11], who compared these percentages with ones taken from a sample of students attending Western Michigan University.

introductory level courses or have contact with the public, are continually faced with the task of refuting the tenets of astrology.

There are numerous papers and books which deal critically with the subject of astrology and the scientific arguments against it; the reader is directed to the listing of books and articles assembled by Robbins and Fraknoi on p. 29 in [12.14] or to the information packets *Astronomy Versus Astrology* and *Debunking Pseudoscience* listed in Sect. 12.4.5.3. To these can be added two recent articles by Fraknoi: "Your Astrology Defense Kit," in *Sky & Telescope*, August 1989, p. 146, and "Scientific Responses to Pseudoscience Related to Astronomy," in *Mercury*, September/October 1990, p. 145.

12.3 Facilities and Services Available to Schools and the General Public

The best general references for this section are the *International Directory of Professional Astronomical Institutions* (A. Heck, 1989) and the *International Directory of Astronomical Associations and Societies* (A. Heck and J. Manfroid, 1988). These are published by Centre de Donnees de Strasbourg, Observatoire Astronomique, II, Rue de l'Universite, F-67000 Strasbourg, France.

For North American astronomers, the American Association for the Advancement of Science has recently published the *AAAS Science Education Directory 1989*, which is designed to be a resource for those involved in science, mathematics, or technology education. The *Directory* contains addresses and telephone numbers of principal executives, directors, administrators, and policy makers who are leaders in associations, scientific academies, museums, educational resource centers, educational laboratories, and state and federal agencies. A copy of the *Directory* may be obtained free of charge by writing to Barbara Walthall, AAAS, Office of Science and Technology Education, 1333 H Street, NW, Room 1139, Washington DC 20005, USA.

Educational Affairs Division: Programs and Services, PED-102, is a special NASA publication which gives overviews of all the educational programs and services provided by the agency's Educational Affairs Division. Included are listings of resources showing how to receive additional information. It can be obtained from Educational Affairs Division, Code XEP, NASA, Washington, DC 20546.

Other references will be found in the individual subsections.

12.3.1 Planetariums, Museums, and Exhibits

A visit to a public planetarium can be a marvelous experience for children and adults alike, especially if it is a large one with extensive facilities and adjoined by a museum and/or observatory. A sophisticated planetarium projector can simulate all the wonders of the night sky, including sunrise, sunset, twilight, the nightly motion of the stars, the seasonal motions of the Sun and planets, and the presence of the Milky Way (see also Sect. 12.4.1.2). Receptive visitors will often find that their awareness of the real night sky has thereafter been heightened. It should be noted that many of the larger planetariums are connected with observatories which provide public viewing sessions.

A listing of the major planetariums and museums in Great Britain and North America is provided in Appendix Sect. A.1. Most of the larger planetariums offer a full range of services and facilities, including exhibits, tours, lectures, classes, public telescope observing sessions, planetarium shows, special children's programs, and bookstore/giftshop. Detailed information regarding hours, fees, and services can be obtained by contacting the planetarium or museum direct. Readers can also obtain more information on individual planetariums from the *Directory of the International Planetarium Society* [12.15], from which some of the items in the aforementioned list in Appendix Sect. A.1 have been extracted, or by writing the International Planetarium Society, c/o Dr. Terrence P. Murtagh, The Planetarium, College Hill, Armagh, Northern Ireland BT61 9DB.

Those readers living in the U.K. will find a great deal of information on planetariums, observatories, and other interesting places to visit in the most recent edition of *Handbook for Astronomical Societies* [12.16], which is available from Ken Marcus, 5 Cedar Gardens, Brighton, East Sussex BN1 6YD, England. This is a 104-page guide to amateur resources in the U.K., and also lists clubs, societies, periodicals, equipment, suppliers, speakers, and sources of astronomical software. The cost is currently £3.50.

Readers in North America should consult *The Observer's Handbook 1991* [12.17], published by the Royal Astronomical Society of Canada and which contains essentially all of the major Canadian planetariums, *America's Planetariums & Observatories (A Sampling)* by R.L. Beck and D. Schrader [12.18], which is a collection of short histories and descriptions of major astronomical sites in the United States, or Kirby-Smith's *U.S. Observatories: A Directory and Travel Guide* [12.19], which lists major centers as well as many small college museums and planetariums in the United States. General listings are also provided by the *1989 Directory of Observatories, Planetariums, and Museums* [12.20] in the May 1989 issue of *Astronomy* magazine, and by the *Astronomy Resource Guide* [12.21] in the September 1991 issue of *Sky & Telescope* magazine.

12.3.2 Observatories and Research Laboratories

While a visit to a planetarium can be rewarding, a trip to a major observatory where astronomical research is carried out can be even more so, particularly if viewing sessions are permitted. The sight of the Moon, a planet, or a star cluster through even a modest instrument can be quite inspirational, especially for a young person.

There are multitudes of observatories in North America and several in Great Britain; many offer evening viewing hours for the interested public and school groups. The names and addresses of the more prominent ones can be found in Appendix Sect. A.2. These institutions should be contacted direct in writing or by telephone for information and current hours. As with planetariums and museums, readers can also obtain some basic information (e.g., visiting hours) from the *Handbook of Astronomical Societies* [12.16] (U.K.), from the *The Observer's Handbook 1991* [12.17] (Canada), from *America's Planetariums & Observatories (A Sampling)* [12.18] (U.S.), and from *U.S.*

Observatories: A Directory and Travel Guide [12.19] (U.S.), as well as [12.20] and [12.21] (Canada and U.S.).

12.3.3 Lectures

The following lectures are available in the U.S. and Canada:

- *Harlow Shapley Visiting Lectureships in Astronomy* (Canada and the U.S.). The American Astronomical Society (AAS) sponsors the Visiting Professors Program, in which professional astronomers make two-day visits to colleges and universities which do not offer a degree in astronomy. During the visit, the astronomer will give usually two (or more) lectures, from public talks to graduate-level seminars. The lectures are designed to strengthen and stimulate interest in astronomy and related sciences, and to enhance the understanding of this discipline by the general public. All costs incurred by the visiting astronomer (food, transportation, lodging) are borne by the AAS, and, in return, the participating institution is asked to make a contribution of $250 in support of the program. For a brochure with application and a list of participating astronomers, contact: Dr. Mary Kay Hemenway, AAS Education Officer, University of Texas, Department of Astronomy, Austin, TX 78712-1083. Tel: 512-471-1309.
- *Morrison Public Lectures*. The Astronomical Society of the Pacific sponsors public lectures by noted astronomers, explaining new discoveries and ideas in astronomy. Some of the lectures are offered in conjunction with amateur astronomy groups and colleges; others are presented at larger astronomical conferences. The host club or institution pays for the speaker's travel expenses, the ASP pays the honorarium. Also, the ASP each year presents the Bart Bok Memorial Lecture, which is non-technical, free, and open to the public. For more information, contact: Astronomical Society of the Pacific, 390 Ashton Avenue, San Francisco, CA 94112, USA; Tel: 415-337-1100.
- *R.A.S.C. Lectures* (Canada). The Royal Astronomical Society of Canada sponsors the vast majority of astronomy lectures in Canada. Queries should be addressed to: R.A.S.C., 136 DuPont St., Toronto, Ontario M5R 1V2.
- *Sigma Xi Lectures* (Canada and the U.S.). Sigma Xi, The Scientific Research Society, has chapters at most major universities and colleges in North America. Each year it designates an official group of scientists (often including an astronomer or astrophysicist) who are available for speaking engagements. For more information, contact: Sigma Xi, The Scientific Research Society, 345 Whitney Ave., New Haven, CT 06511-2316, USA; Tel: 203-624-9883 or 1-800-243-6534.
- *Society of Physics Students (SPS) Speakers* (Canada and the U.S.). The SPS is an organization designed explicitly for students. There are local chapters of SPS at most major universities and colleges in North America. Each academic year they publish *Speakers, Tours, and Films: A Program Resource for SPS Chapters* [12.22]. Copies may be obtained from: National Office, Society of Physics Students, American Institute of Physics, Second Floor, 2000 Florida Ave., N.W., Washington, DC 20009.

The following institutions in the U.K. have identified themselves to the Association for Astronomy Education (AAE) as being willing to provide speakers for schools. Usually the school is expected to reimburse the the visiting speaker for travel and other expenses.

- Hatfield Polytechnic Observatory
- University of London Observatory
- Lancashire Polytechnic, Preston
- Mills Observatory, Dundee
- Glasgow University Observatory

More information can be obtained from the *Handbook of Astronomical Societies* [12.16], or by contacting the AAE Secretary, 34. Aeland Crescent, Denmark Hill, London SE5 8EQ.

12.3.4 Workshops and Other Programs

The following workshops and activities are a few of the more notable ones in astronomy:

- *Astronomer for a Day Program* (Canada and the U.S.). This program is sponsored by the AAS and is designed to bring high school science teachers into contact with professional astronomers. For more information, contact Dr. Mary Kay Hemenway, Education Officer of the AAS, at the address given in the previous section.
- *ASP-Sponsored Workshops on Teaching Astronomy* (Canada and the U.S.). The ASP organizes workshops on teaching astronomy throughout North America. Approximately 100–200 teachers attend each workshop and take back information, resources, and teaching activities to their home districts. For more information, contact the ASP at the address given previously.
- *Astronomy Day*. Astronomy Day has been celebrated in North America in many ways, including display techniques, activities, and recruiting by various astronomical societies. These are described in *The Astronomy Day Coordinator's Handbook*, published by the Astronomical League. To obtain a copy, contact: Gary Tomlinson, Astronomy Day Handbook, Chaffee Planetarium, 54 Jefferson SE, Grand Rapids, MI 49503.
- *GEMS*. GEMS (Great Explorations in Math and Science) provides activities from preschool to high school, not only in astronomy but also in Earth science and many other fields. Contact: GEMS, Lawrence Hall of Science, University of California, Berkeley, CA 94720.
- *STAR*. Project STAR (Science Teaching through its Astronomical Roots) is oriented toward grades 10 through 12. It will produce the first high school astronomy textbook. Both STAR and GEMS use simple, inexpensive materials and focus on fundamental concepts. Contact: Philip Sadler, Center for Astrophysics, 60 Garden St., Cambridge, MA 02138.

12.3.5 Where to Find Information on Astronomy as a Career

Young people who wish to find out more about pursuing astronomy as a career should write for one of the available pamphlets or brochures (all of which have been referred to earlier in this chapter) which pertains to his or her locale.

- *In the United Kingdom.* The Old Greenwich Observatory produces a leaflet called *Becoming A Professional Astronomer* which may be obtained by writing to: Dr. R. Lopes, Old Greenwich Observatory, National Maritime Museum, London SE22. In addition, a little booklet called *Astronomy* and booklets on opportunities for astronomical research in the U.K. are published by: Royal Astronomical Society, Burlington House, London W1V 0NL. Finally, the British Astronomical Association (same address as the RAS) publishes various leaflets which can be obtained by reference to the BAA.
- *In Canada.* Write for the pamphlet *Astronomer* at the following address: Guidance Centre, Faculty of Education, 252 Bloor St. W., 2nd Floor, University of Toronto, Toronto, Ontario M5S 1V5.
- *In the United States.* Write for the brochure *Understanding the Universe: A Career in Astronomy* from: AAS Education Office, University of Texas, Department of Astronomy, Austin, TX 78712-1083.

For young women in particular, there is the booklet *Space for Women*, which is published by the Harvard-Smithsonian Center for Astrophysics, 60 Garden St., Cambridge, MA 02138. Also, the Astronomical Society of the Pacific publishes an information packet entitled *Women in Astronomy*, which is reprinted from the Jan/Feb 1992 issue of *Mercury* magazine; see Sect. 12.4.5.3.

For general reading, *The Astronomers* by Donald Goldsmith (St. Martin's Press, 1991), *Visit to a Small Universe* by Virginia Trimble (American Institute of Physics, 1992), *How to be an Astronomer* by Robin Scagell (Macdonald, 1980), and the career-oriented article "How We Became Astronomers" by Graham-Smith and Lovell in their book *Pathways to the Universe* (Cambridge University Press, 1988) should be especially inspirational to young persons. Also highly recommendable are the autobiographies of successful modern astronomers: *In Quest of Telescopes* by Martin Cohen (Cambridge University Press, 1980), *On the Glassy Sea, An Astronomer's Journey* by Tom Gehrels (American Institute of Physics, 1988), and *Astronomer by Chance* by Bernard Lovell (Basic Books, 1990). A recently published biography, *Clyde Tombaugh: Discoverer of Planet Pluto* by David H. Levy (University of Arizona Press, 1991), and an older but still quite serviceable book, *How to Become an Astronomer* by Freeman Miller (Sky Publishing Corp., 1969), can also be recommended.

12.3.6 Films on Astronomy as a Career[9]

The following are just a few of the several available films and/or videotapes dealing with astronomy (as well as science in general) as a career:

---

[9] A more complete discussion of the available books, films, and videotapes on astronomy will be dealt with in Sect. 12.4.

- *The Astronomers* (1991, Six 60 min tapes) — A major PBS production funded by the W.M. Keck Foundation, and broadcast as a six-part PBS series. The companion book of the same name (by Donald Goldsmith) is mentioned in Sect. 12.3.5. Each episode deals with a current area of astronomical research, and highlights the lives of some of the astronomers involved in that research. Available from KCET Public Broadcasting Service, PBS Home Video, Beverly Hills, CA.
- *The Sun and Beyond* — Highlights the work of husband-and-wife team Aden and Marjorie Meinel. It is available in both videotape and 16-mm formats from Educational Materials and Equipment Co., Old Mill Plain Road, Danbury, CT 06811, USA; Tel: 800-848-2050, in Connecticut 203-798-2050.
- *The Observatories* (27 min) — Astronomers from some of the major observatories in North and South America describe the functions of optical and radio telescopes, the discoveries being made, and the prospects for newly developed computerized equipment. Produced by the National Science Foundation. Available from Media Design Associates, Inc., Dept. P, P.O. Box 3189, Boulder, CO 80307-3189, USA; Tel: 800-228-8854 or 303-443-2800.
- *Science—Woman's Work* (27 min) — Discusses why young women are turned off by mathematics and science, and explores the lives of women who are already established in their scientific careers. Produced by the National Science Foundation. Also available from Media Design Associates, Inc.
- *A Private Universe: Misconceptions That Block Learning* (1989, 18 min) — Deals with the subject of scientific illiteracy among American children. The film was produced by Matthew H. Schneps of Project STAR (Science Teaching through its Astronomical Roots), a curriculum-development program at the Harvard-Smithsonian Center for Astrophysics. The tape may be purchased or rented from Pyramid Film & Video, Box 1048, Santa Monica, CA 90406. See p. 586 of the June 1989 issue of *Sky & Telescope* for a somewhat more detailed description of this film.
- *Space Women* (1984, 57 min) — Astronaut Sally Ride discusses the physical and mental rigors of being an astronaut, and the responsibility of role models. Available from Ambrose Video Publishing, Inc., 381 Park Avenue South, Suite 1601, New York, NY 10016; Tel: 800-526-4663. In New York: 212-696-4545.
- *Where the Galaxies Are* (1991, 8 min) — Astronomer Margaret Geller of the Harvard-Smithsonian Center for Astrophysics briefly describes the research she is performing with colleague John Huchra to map the three-dimensional structure of galaxy clusters. Available from Astronomical Society of the Pacific, 390 Ashton Ave., San Francisco, CA 94112. See p. 139 of the August 1991 issue of *Sky & Telescope* for more details.

12.3.7 Astronomical Societies and Clubs

Those who want to benefit from contact with other observers and share their mutual enthusiasm for astronomy should consider becoming a member of a local astronomy club. Residents of Canada and the U.S. should consult the *1989 Directory of Astronomy Clubs* [12.23] published in the May 1989 issue of *Astronomy* magazine or the *Astronomy Resource Guide* published in the September 1991 issue of *Sky & Telescope*. In the

U.K., there is a complete listing of the major astronomical societies in the *Handbook for Astronomical Societies* [12.16]. Information on societies can also be obtained by writing to: Astronomy Section, Old Royal Observatory, National Maritime Museum, London SE10 9NF, England.

In addition to local clubs, there are also national and international organizations, some of which have local chapters, to which an amateur or professional astronomer can belong. While some of these have been mentioned in the various chapters of this book, a more complete listing with the addresses is provided in Appendix Sect. A.3. A somewhat more extensive listing is provided by Robbins and Fraknoi in *The Universe at Your Fingertips* [12.14]. It should be noted that nearly all of the "national" societies welcome members from other countries.

12.3.8 Inns and Travel Tours

The following companies arrange travel tours to view eclipses and other astronomical events:

- *East Africa Travel Consultants, Inc.*, 574 Parliament St., Toronto, Ontario M4X IP8, Canada; Tel: 416-967-0067. Solar eclipse tours.
- *Explorers Travel Club Ltd.*, 5 Queen Anne's Court, Windsor SL4 1DG, England; Tel: 0753 842184.
- *Hanssen Tours*, 3705 NASA Rd. 1, Seabrook, TX 77586, USA; Tel: 713-326-7643.
- *Scientific Expeditions*, 1832 Quail Lake Dr., Venice, FL 34293; Tel: 800-344-6867. Organizer of specialized astronomical tours in association with *Sky & Telescope* magazine.
- *Virginia Roth's Scientific Expeditions, Inc.*, 227 W. Miami Ave., Suite #3, Venice, FL 34285; Tel: 813-484-3884 or 800-344-6867.
- *World of Oz Ltd.*, 20 E. 49th St., Suite 500, New York, NY 10017; Tel: 800-248-0234. Arranges tours to astronomical events, especially solar eclipses, in association with *Sky & Telescope*.

The following inn is quite possibly unique: it caters to astronomers:

- *Star Hill Inn*, located near Santa Fe, New Mexico at an altitude of 7200 feet. Astronomers can bring their own telescope or rent the Inn's 24-inch Cassegrain. (A wide-angle photograph of the summer Milky Way taken from this site appears on page 328 in the September 1989 issue of *Sky & Telescope*.) For free information, contact: Star Hill Inn, Sapello, NM 87745, USA; Tel: 505-425-5605.

12.3.9 Astronomy Books for the Visually Handicapped

Those who are visually impaired can learn about the wonders of the sky from Braille-printed astronomy books. For example:

- *Touch the Stars*, by Noreen Grice, a 44-page astronomy textbook which includes 11 tactile illustrations. Text pages are printed in Braille and large print.

- *Tactile Illustrations*, a set of 20 astronomy pictures with Braille captions.

Both of these books can be ordered from Museum of Science, Public Outreach Department, Science Park, Boston, MA 02114-1099; Tel: 800-729-3300.

Numerous works on astronomy are now available on audiocassette tapes; see also Sect. 12.4.2.1. The best sources are the following:

- *Recording for the Blind* — A nonprofit organization with studios in 16 states in the U.S. It maintains a central library of audiocassette tapes of approximately 80 000 books, with 4000 new books being recorded each year for the use of 20 000 borrowers. The total collection on astronomy numbers around 150 entries. The cassettes are loaned free of charge to students and include in addition to the written text complete descriptions of all illustrations and associated tables and appendices. For more information contact: Recording for the Blind, Inc., 20 Roszel Road, Princeton, NJ 08540, USA.
- *ASP* — offerings discussed in Sect. 12.4.2.1.

12.4 Educational Resources in Astronomy

The 1980s were extremely fruitful years for the development of resources in astronomy, especially in the realm of videocassette tapes and computer software. The balance of this section is devoted to the plethora of educational materials which are available at the time of writing (1992). The lists of distributors of these various products are meant to represent a generous, but not exhaustive, cross section. Completeness is not the objective and indeed is an impossibility within the scope of this chapter, as the number of distributors for slides, for instance, must total several hundred in North America alone.

Except where noted, the author does not endorse any particular product or company, but merely gives a liberal sampling of the ones that have appeared in advertisements during recent months. Note that there is a great deal of overlap in the sense that a particular product is often carried by several different companies. Prices for various items are not as a rule given here, as these will doubtlessly change during the lifetime of this *Compendium*. Any purchases of products listed here or in the distributors' catalogues should be made without haste and using common sense. The reader should telephone or write the particular company for a catalogue and/or current price information.

Readers in the U.K. can obtain more information on resources from the previously mentioned *Handbook for Astronomical Societies* [12.16], or by contacting the Association for Astronomy Education, the London Planetarium, or the London Schools Advisory Centre (addresses given earlier) for resource lists. Those in North America should consult the following:

- *The Air and Space Catalogue* by J. Makower (ed.), Vintage Books, 1989 — Extensive listings cover resources, products, planetariums, clubs, places to visit, and other useful information.
- *The Astronomer's Sourcebook* by B. Gibson, Woodbine House, 1992 — A guide to astronomical equipment, publications, planetariums, organizations, events, and history.

- *Directory of Science Education Suppliers* – Lists equipment, other firms, educational services, media producers, and publishers. It is available from the National Science Teachers Association (NSTA), 1742 Connecticut Ave. NW, Washington, DC 20009.
- *Equipment Directory 1989* [12.24] – An advertising supplement to the October 1989 issue of *Astronomy* magazine. It contains a sizable number of sources for a wide variety of observational and educational items.
- *Astronomy Resource Guide* [12.21] – A supplement to the September 1991 issue of *Sky & Telescope* magazine, It lists planetariums, observatories, museums, clubs, computer bulletin boards, and equipment manufacturers.
- *The Universe at Your Fingertips: An Educator's Desktop Reference of Astronomy Education Materials in English* [12.14] — A very comprehensive document (up to 1985) assembled under the auspices of the International Astronomical Union's Commission 46 (Education) by R. Robert Robbins and Andrew Fraknoi. It is available from the ASP. A 1988 update is available from the IAU and is published in the Teachers' Guide to the introductory astronomy textbooks by J. Pasachoff.
- *Universe in the Classroom* by A. Fraknoi (W.H. Freeman & Sons) — A highly recommended teachers' guide.
- *The Universe at Your Desk* — A resource guide for the beginning teacher of high school astronomy, and developed by Project STAR, Center for Astrophysics, 60 Garden St., Cambridge, MA 02138.

A wide variety of educational materials can also be obtained from CORE, the Central Operation of Resources for Educators. CORE was established for the national and international distribution of NASA produced educational materials, including videocassette programs, slides, computer software, and filmstrip programs that chronicle NASA's state-of-the-art research and technology. For a catalogue, write to NASA CORE, Lorain County JVS, 15181 Route 58 South, Oberlin, OH 44074; Tel: 216-774-1051 x293 or 294.

12.4.1 Mechanical Models and Exhibit Items

There are seemingly innumerable kinds of astronomy items that can be used productively in the classroom or simply as display pieces. The most common ones will be discussed in the following subsections; a general list of companies which distribute these materials can be found in Appendix Sect. A.4.

12.4.1.1 Globes. The globe or sphere is generally used to represent the terrestial and celestial spheres with true-to-scale angles and areas. There are Earth globes, lunar globes, Mars globes, celestial globes, and so on. The classic Earth globe can be modified by additional equipment and spheres so as to explain the origin of the seasons and demonstrate the subsolar point on the Earth's surface, zones of twilight, and other phenomena.

The celestial globes are designed so that the constellations can be shown in one of two ways: 1. as they are seen in the sky, so that the globe has the function of a star map, or 2. in the reverse sense, where the viewer must imagine he is observing the sky from the center of the sphere in order to "see" the patterns of constellations correctly. Some globes are transparent (or at least translucent) and can be illuminated from the inside. There are also "blackboard" globes with or without gradation on which chalk can be used.

Some of the best sources of globes are Farquhar Globes, Carolina Biological Supply Co., Central Scientific (CENCO), Edmund Scientific, Fischer Scientific, Hubbard Scientific, MMI Corporation, Schoolmasters Science, and Wards Natural Science Establishment, Inc. (addresses given in Appendix Sect. A.4).

12.4.1.2 Telluria and Planetariums. A *tellurian* is an apparatus designed to demonstrate how the Earth's position and movement (diurnal rotation, annual revolution, etc.) causes day and night, the phenomenon of the seasons, lunar phases, and eclipses. A *planetarium* is a device designed to show the motions of a variety of celestial objects, but especially planets, the Sun and the Moon. Typical museum planetariums are miniature models of our solar system which display the motions of celestial bodies by accurately calculated machinery; these devices are often referred to in the English language as *orreries*. In Germany, for instance, a planetarium is generally understood to be a *projection* planetarium, where the celestial globe is in fact projected (via multitudes of tiny, individual light beams) onto the dome of the presentation room so that the "stars" appear in a more or less natural-looking fashion.

Projection planetariums are offered nowadays in a great variety of sizes and also with different facilities; for example the largest instruments are models V and VI of the Carl Zeiss Corporation in Oberkochen, Germany. Small- and medium-sized planetariums are also available with many of the facilities of the larger ones. Some suppliers are listed later in this subsection.

A totally new development is the Digistar Planetarium from the Evans & Sutherland Computer Corporation in Salt Lake City, Utah. It employs a powerful computer (VAX 11/730) which steers via a graphic processor an ultrabright cathode-ray tube with an area 18 cm × 18 cm. The images from an 8000 × 8000 pixel surface are projected onto a dome with a superwide angle objective at videospeed (25 images per second). The images of stars, particularly those at the horizon, appear fainter and less sharp than in the traditional planetarium projectors. Although Digistar's representation of the celestial view is less than perfect, it offers many new possibilities. For instance, the proper motions of stars in the past and future can be shown with different time lapses, and even travel by spacecraft through certain constellations, where the distances of the stars have been stored in the computer, can be simulated. Currently the only European facility of this type is in The Hague in the Netherlands, but in North America there are several.

For use in schools, colleges, and public observatories, there is available a type intermediate between projection planetarium and orrery: a transparent star globe containing a tellurian. With this device, the following can be shown: day and night, the seasons, representations of lunar revolution, phases, and eclipses, the tides, universal time, sidereal time, solar time, and the motion of the lunar nodes. The relation between equator and ecliptic, seasonal nighttime views, the dependence of circumpolar stars on geographic latitude, the position of satellite orbits, the phases of Venus, the platonic year, and the precession of the Vernal Equinox can also be shown.

The Baader Planetarium in Germany, for instance, contains such a device. This instrument has an easily exchangable projection screen or dome in two different sizes, and can also be used for smaller projection planetariums. One instrument in particular corresponding to the old orreries is the *Helios Planetarium*, which contains a plexiglass dome with a northern and southern hemisphere, and inside a representation of the planetary system (but not to scale). This instrument can be used as a tellurian.

The following is a list of the names and addresses of companies which distribute orreries and planetariums.

Projection Planetariums
- Evans & Sutherland Computer Corporation, Salt Lake City, UT, USA.
- Goto Optical Mfg. Co., 4-16 Yazakicho, Fuchu-Shi, Tokyo 183, Japan; Tel: 0423 (62) 5311.
- Minolta Camera Co., Osaka Kokusai Bldg. 30-2, Chome, Azuchi, Osaka 541, Japan.
- MMI Corporation, 2950 Wyman Parkway, P.O. Box 19907, Baltimore, MD 21211, USA; Tel: 301-366-1222.
- Schoolmasters Science, 745 State Circle Box 1941, Ann Arbor, MI 48106, USA; Tel: 800-521-2832.
- Seiler Instruments & Manufacturing Co., INC., 170 East Kirkham Ave., St. Louis, MO 63119-1791, USA; Tel: 314-968-2282 or 1-800-444-7952.
- Spitz Space Systems, Chadds Ford, PA, USA.
- Starlab Planetarium Systems, 59 Waldenstreet, Cambridge, MA 02140; Tel: 800-537-8703.
- Viewlex Audio-Visual, Broadway Ave., Holbrook, NY, USA.

Other Planetariums
Most of the companies which supply globes and telluria also supply model planetariums. See the list in Sect. 12.4.1.1 and also the following possible source:
- Cochranes of Oxford Ltd., Fairspear House, Leafield, Oxford OX8 5NY, England: Tel: 099-387641. Helios planetariums.
- Learning Technologies Inc., Starlab Planetarium Systems, 59 Walden Street, Cambridge, MA 02140, USA; Tel: 800-537-8703, in Massachusetts 617-547-7724. Telex: 5106008300 LRNG TEC.
- Orbic Systems, Inc., 109 Eltingville Blvd., Staten Island, NY 10312, USA; Tel: 718-356-2328.

Accessories
- Alan Castelman, c/o Carl Zeiss Inc., 1 Zeiss Dr., Thornwood, NY 10594, USA; Tel: 914-747-1800.
- Sky-Scan, P.O. Box 3832, Rochester, NY 14610, USA.
- Talent, Inc., 1010 Marietta Way, Sparks, Nevada 89431, USA.

Planetarium Shows
- Maryland Science Center, Allan C. Davis Planetarium, 601 Light St., Baltimore, MD 21230; Tel: 301-685-2370 x440.

Literature on Planetariums
- Fernbank Science Center: Physical Science Teacher's Guide, *The Planetarium*.
- Various authors—NASA: *The Planetarium* (An Elementary School Teaching Resource, 1966).
- Berendzen, Richard (editor and conference chairman): International Conference on Education in and History of Modern Astronomy Part II. *Annals of the New York Academy of Sciences*, Vol. **198** (August 1974).
- Jettner, Frank C., and Soroka, John J.: The Planetarium in Modern Science Education. p. 178 in Berendzen ref.
- Branley, F.M.: Education in Major Planetariums. p. 192 in Berendzen ref.

Journals and Magazines
- *The Planetarian*, put out by the International Planetarium Society. Contact: Griffith Observatory, 2800 E. Observatory Rd., Los Angeles, CA 90027, USA, Telephone 213-664-1181.
- *Planetarium Sourcebook*, a new resource guide published by the Great Lakes Planetarium Association. It lists companies and organizations offering products and services of interest to the planetarium community. Cost is $5 ($7 outside the U.S.). Write: Gary E. Sampson, Wauwatosa West High School Planetarium, 11400 W. Center St., Wauwatosa, WI 53222, USA.
- *Publication of the Planetarium Association of Canada*, (Manitoba Museum of Man and Nature, 147 James Ave., Winnipeg 2, Manitoba), in Vol. **4**, 1972.
- *Planetarium News*, Carl Zeiss, D-7082 Oberkochen, Germany.

12.4.1.3 Telescopes and Accessories. The consumer is well-served for telescopes and observing equipment these days. During the latter half of the 20th century, the number of companies which market telescopes and accessories has increased dramatically (as have the prices for these items), and there is currently fierce competition to satisfy the demands of amateur and professional astronomers. Unfortunately, the cost of even a good 4-inch reflecting telescope[10] with equatorial mount and clock drive, for example, may dissuade all but the most hard-core astronomy buffs from making the purchase.

As with audio-visual equipment, the number of companies which sell telescopes, binoculars, and all the standard accessories such as eyepieces, filters, finderscopes, etc. is overwhelming. Some are well-known national and international suppliers which advertise in the popular astronomy magazines, but anyone seeking to purchase a telescope would be well advised to check local suppliers also. Two very useful guides are the *Astronomy Resource Guide* [12.21], a special insert within the September 1991 issue of *Sky & Telescope*, and the 1989 Guide to Telescopes [12.25], a special section in the October 1989 issue of *Astronomy* magazine. In the latter can also be found the *Equipment Directory 1989* [12.24], which lists major North American distributers of telescopes and accessories as well as the requisite tools and supplies for telescope making. Readers in the U.K. can find some distributers in the indespensible *Handbook for Astronomical Societies* [12.16]. These lists will not be reproduced here, but the

---

[10] 4 inches is usually considered the mininum aperture with which useful observations can be made of planets and stars with this type of scope.

names of companies which market certain specialty items or perform repairs etc. are given in Appendix Sect. A.6.

12.4.1.4 Astronomical Clocks. Astronomy is, of course, *the* science of reference for timekeeping. Below are given a few of the more distinctive astronomical clocks which are currently on the market.

- *Geochron clock.* This device is an electronic map which shows the time of day and amount of sunlight anywhere in the world. The geochron reveals how the distribution of sunlight over the Earth's surface is affected by the changing seasons. Source: American Weather Enterprises, P.O. Box 1383, Media, PA 19063, USA; Tel: 215-565-1232.
- *Lunatime clock.* A lunar clock which shows the position of the moon at all times, while following the circadian day of $24^h 50^m$.
- *Solar/Lunar clock.* Mechanism shows exactly the positions of both Sun and Moon in relation to the Earth. Gives phases of the Moon, lunar time, conventional time, and times of sunrise and sunset for each day.
- *Startimer.* Reads celestial time and illustrates the rotation of the Earth in relation to the stars.

The latter three items are available from most of the sources in the general list in Appendix Sect. A.4.

Astronomical timepieces are also manufactured by:

- *Astrotech*, 39 Periwinkle Lane, Dunstable, Beds. LU6 3NP, England; Tel: 0582 605464.
- *Gemini*, 8930 Blue Smoke Dr., Gaithersburg, MD 20879, USA; Tel: 301-921-0157.
- *Willman-Bell Inc.*, P.O. Box 35025, Richmond, VA 23235, USA; Tel: 804320-7016.

12.4.1.5 Sundials. Sundials provide a traditional and ornate way of telling the time. They come in many sizes, and are usually permanently mounted, although this need not be the case. For a full discussion of sundials and instructions on how to build one, see Chap. 10.

A few sources of sundials include:

- *American Weather Enterprises*, P.O. Box 1383, Media, PA 19063, USA; Tel: 215-565-1232.
- *Carolina Biological Supply Company*, 2700 York Road, Burlington, NC 27215, USA; Tel: 919-584-0381.
- *Wind and Weather*, The Albion Street Water Tower, P.O. Box 2320, Mendocino, CA 95460.

12.4.1.6 Early Scientific Instruments. The following offer antique scientific instruments (and/or reproductions of such) and early associated books. They should be contacted for information and a catalogue:

- *Greybird Publishers*, 11824 Taneytown Pike, Taneytown, MD 21787. Publisher of bimonthly advertising journal of antique scientific instruments.
- *Historical Technology, Inc.*, 6B Mugford Street, Marblehead, MA 01945.

- *Paul MacAlister & Assoc.*, 280 Arden Shore Rd., Lake Bluff, IL 60040, USA.
- *G.B. Manasek Inc.*, P.O. Box 961, Hanover, NH 03755, USA; Tel: 603-643-2227.
- *Renaissance Instrument Co.*, 635 6th St., Myrtle Point, OR 97458, USA; Tel: 503-572-2595.
- *St. James House Co.*, 3010 W. Montrose Ave., Chicago, IL 60618; Tel: 312-267-0400.
- *Tesseract*, Box 151, Hastings-on-Hudson, New York, NY 10706, USA; Tel: 914-478-2594.

12.4.1.7 Meteorites and Tektites. Meteorites are available in limited quantities to astronomers, schools, and collectors. The following are possible sources:

- *Astrosystems*, 1536 Meeker Dr., P.O. Box 1183, Longmont, CO 80501; Tel: 303-678-5339. œ
- *Bethany Trading Company*, P.O. Box 3726-S, New Haven, CT 06525, USA; Tel: 203-393-3395. Catalogue $2 (U.S.).
- *Fireball Electronics*, 246 E. 52nd St., Odessa, TX 79762, USA; Tel: 915-366-4802.
- *Robert A. Haag — Meteorites*, 2990 E. Michigan St., Box 27527, Tucson, AZ 85726, USA. Catalogue available.
- *R.A. Langheinrich*, 326 Manor Ave., Cranford, NJ 07016, USA.
- *Minerological Research*, 15840 E. Alta Vista Way, San Jose, CA 95127, USA; Tel: 408-923-6800.
- *James M. Williams*, 4017 W. 10 St., Odessa, TX 79763, USA; Tel: 915-381-0249.

12.4.1.8 Optics/Spectroscopy Demonstrations. The many properties of light can often be demonstrated in the classroom using a wide variety of "blackboard optics" equipment which has been available over the past several years. These devices make it especially easy to demonstrate the principles of refraction and reflection. Most of the companies in the general list of sources in Appendix Sect. A.4 supply them.

A spectroscope can be a very simple device for dispersing the radiation from a light source into its constituent colors. Several instructive experiments can be done with a spectroscope, including making observations of various gas discharge tubes, incandescent lamps, and the Sun. It is usually helpful also to have on hand a large spectrum analysis chart for identifying the various elements in the spectrum. Again, suppliers can be found from amongst the companies in the general list provided in Appendix Sect. A.4.

12.4.1.9 Maps, Posters, Starcharts, and Postcards. Various wall posters, charts, and maps are available in abundance, and the following list gives just a few of the numerous suppliers. Most of the companies in the general list in Appendix Sect. A.4 also supply miscellaneous charts, maps, and posters.

United Kingdom
- *Armagh Planetarium.* Offers a wide selection of all of these materials.
- *Broadhurst Clarkson & Fuller Ltd.*, Telescope House, 63 Farrington Road, London EC1M 3JB. Tel: 01-4052156.
- *Daily Telegraph*, 135 Fleet Street, London EC4 or bookshops. Offers *Sky at Night* map.

- *George Philip*, 59 Grosvenor Street, London W1X 9DA; Tel: 01-493-5841. Publishes atlases.
- *London Planetarium*, Marylebone Road, London NW1 5LR; Tel: 01-486-1121.
- *MacMillan Educational*, Houndmills, Baskingstoke, Hants.
- *Pictoral Charts Educational Trust*, 27 Kirchen Road, London W.13.
- *The Royal Astronomical Society*, Burlington House, London, W1V 0NL. Limited quantities of posters.
- *Spacecharts*, Newton Tony, Salisbury, Wilts.

Canada, United States
- *ASP*, 390 Ashton Avenue, San Francisco, CA 94112, Tel: (415) 337-1100. Posters of the Moon, Space Shuttle, Earth at Night, Solar System, Galaxies, An Explorer's Guide to Mars, and Astro-Posters set.
- *Astronomy Magazine*, Order Dept., 21027 Crossroads Circle, P.O. Box 1612, Waukesha, WI 53187-1612, USA; Tel: 414-796-8776. Posters of Space Shuttle by NASA, Uranus from Voyager, Map of the Universe, Solar System Exploration.
- *Black Forest Observatory*, 12815 Porcupine Lane, Colorado Springs, CO 80908; Tel: 719-495-3828. Has star charts and many unique products in astronomy.
- *Dover Publications, Inc.*, 31 East 2nd Street, Mineola, NY 11501. Puts out a series of postcards on space exploration from the archives of NASA.
- *Haddon's Posters*, Box 9514, Station A, Halifax NS, B3K 5S3, Canada. Solar System poster showing all nine planets.
- *Hansen Planetarium*, 15 South State Street, Salt Lake City, UT 84111, USA; Tel: 800-321-2369. Has wide variety of outstanding charts, posters, and postcards; see their catalogue.
- *Magrath Photography*, 2202 Willett, #45, Laramie, WY 82070. Sky photography and calendars.
- *Nightsky Enterprises*, Box 278, Pomona, NJ 08240, USA; Tel: 800-NITESKY. Offers giant $7' \times 7'$ mural of the night sky.
- *Schoolmasters Science*, 745 State Circle Box 1941, Ann Arbor, MI 48106, USA; Tel: 800-521-2832.

12.4.1.10 Calendars. Astronomical calendars abound these days. Below are a few of them which were available for 1992 (and presumably will also be on the market in succeeding years), along with the companies which distribute them.

- *The Astronomical Calendar*, available from Astronomical Calendar, Department of Physics, Furman University, Greenville, S.C. 29613, USA; Tel: 803-294-2208. In Europe and the U.K., write to: Geoffrey Falworth, 12 Barn Croft, Penwortham, Preston, PR1 0SX, England.
- *The Astronomical Companion*, same sources as for the *Astronomical Calendar*.
- *The British Astronomical Calendar*, from RAS Calendar Offer, Calendar Division, Thomas Forman Ltd., Hucknall Road, Nottingham NG5 1FE; Tel: 0602-608151 x262 or x272.
- *Moon Phase Calendar*, from ASP.
- *Wonders of the Universe Calendar*, from ASP.

12.4.1.11 Stamps. Stamps with celestial and space themes have been issued over the years in many different countries. Collectors interested in such stamps should contact: The Astronomy Study Unit of the American Topical Society, P.O. Box 630, Johnstown, PA 15907. They publish *Astronomy & Philately*, a handbook listing astronomical stamps, stationary, and cancellations (order number HB90), and *Astrofax*, a quarterly bulletin.

12.4.2 Audio-visual Media

In recent years there has been a veritable explosion in both the quantity and quality of resources available in the audio-visual realm. Nowadays there are multitudes of science companies promoting astronomy by selling slides, books, videocassettes, and other items. This is due in part to the broadening of astronomy to include observations made in the gamma-ray, X-ray, ultraviolet, and infrared regions of the electromagnetic spectrum, as well as to improved imaging (often computer-enhanced) of visible sources. It has also certainly been spawned by the spectacular images obtained from recent space missions to the planets, by technological advances in audio and video products, and by the general public's unceasing appetite for astronomical knowledge.

The most common audio-visual products include phonograph records, audiocassette tapes, still photographs, slides, transparencies, filmstrips, filmloops, films (both silent and sound), videocassette tapes, and videodiscs. These will be discussed in the following subsections, along with a few of the companies which sell or rent them; a general source list of audio-visual products is provided in Appendix Sect. A.5.

12.4.2.1 Audiocassette Tapes. While not the most popular audio-visual resource in astronomy, audiocassette tapes can supply a great deal of verbal information which is especially useful for beginners. Some cassettes provide descriptive information on a number of popular celestial objects in the night sky, while others contain recorded lectures on various subjects. Still others are condensed readings from published books on astronomy. Some of the available tapes include:

– *First Moon Landing—Flight of the Eagle*—Includes audio segments from the Apollo 11 flight to and from the Moon in July 1969. Also includes the famous speech by President Kennedy stating the goal of putting a man on the Moon during the 1960s. Available from Spacetapes (see address below).
– *The Hubble Space Telescope*—Features a 60-minute talk about the HST by Dr. Stephen Maran of the NASA/Goddard Spaceflight Center. Describes HST's history, instruments, capabilities, scientific projects, and safety features. Available from the ASP.
– *The Sky at Night—A Guided Tour of the Constellations*—by Patrick Moore. The 60-minute tape has an introduction followed by four sections for each of the seasons. Available from George Philip, London.
– *Tapes of the Night Sky*—A series of four half-hour guided "tours" (on two cassettes) of the night sky, one tour for each season. The *Tapes* can be purchased either from the ASP or from Sky Publishing Corporation.

– *Telescopes of the World*—A self-contained introduction to modern telescopes and astronomical observing. Can be used as a synchronized narration to the ASP slide set of the same name.

The following books and excerpts from books are available on audiocassette from the ASP:

– *A Brief History of Time*—Complete reading of the best-selling popular book by Stephen W. Hawking. (4 cassettes, approximately 6 hours.)
– *Coming of Age in the Milky Way*—Selections from the book of that name by Timothy Ferris, read by the author. Tells the story of how the western world arrived at its conception of the universe. (2 cassettes, 168 minutes.)
– *The Discovery of the Pulsar*—Selections from the book by Antony Hewish and Jocelyn Bell-Burnell. Recorded by the BBC, this tape includes interviews with the authors, who give a first-hand account of their discovery. (1 cassette, about 56 minutes.)
– *To Space and Back*—From the book by Sally Ride with Susan Okie. Written especially for young audiences, it tells the story of America's first woman in space, and answers questions frequently asked by space enthusiasts of all ages. (1 cassette, 54 minutes.)

Selected astronomy textbooks and other printed materials have been transcribed in their entirety to cassette tapes, and are available for loan to the visually handicapped; see Sect. 12.3.9.

The following companies also sell various cassette tapes:

– *Federation of Astronomical Societies*, 1 Tal-y-bont Road, Ely, Cardiff CF5 5EU, Wales. See Slide/Audiocassette Sets.
– *Informastron Company*, P.O. Box 262, Belle Vernon, PA 15012, USA. Cassette tapes providing information on the various objects being viewed (e.g., Pleiades, Full Moon, Sun). Free brochure available.
– *Spacetapes*, 539 Telegraph Canyon Rd., #777 Chula Vista, CA 92010; Tel: 800-777-6382.
– *Terra Firma Cassettes*, 55 Bolingbroke Road, London W14, England.

12.4.2.2 Phonograph Records. Owing to the recent popularity of audiocassette tapes, very little astronomy-related material is available in this medium with the exception of slide or filmstrips with audio narration on record. However, most of these sets now come with cassettes instead of records.

12.4.2.3 Still Photographs. A simple photograph may in some instances be preferable to motion pictures since in the former the viewer has time to mull over its meaning. There are standard prints and now high quality laser prints. Sources are listed below, including libraries which loan prints.

United Kingdom
– *Armagh Planetarium*, College Hill, Armagh, Northern Ireland BT61 9DB; Tel: 0861-524725. Sells a wide variety of laser scan prints.
– *Earth and Sky*, 21A West End, Hebden Bridge, West Yorkshire HX7 8UQ.

- *Edinburgh Astronomy Teaching Packages*, Royal Observatory, Blackford Hill, Edinburgh, EH9 3HJ, Scotland, UK; Tel: 031-668-8325/8100; Fax: 031-668-8264. Several packages to choose from, some containing suggested exercises for undergraduate students.
- *Mary Evans Picture Library*, 1 Tranquil Vale, Blackheath, London SE3 0BU; Tel: 01-318-0034.
- *Science Museum Library*, Photo Orders Service, South Kensington, London SW7 5NH; Tel: 01-938-8220.
- *Science Photo Library*, 112 Westbourne Grove, London W2 5RU; Tel: 01-727-4712 or 01-229-9847; Fax: 01-727-6041.
- *Space Frontiers Ltd.*, 30 Fifth Avenue, Havant, Hampshire PO9 2PL; Tel: 0705-475313.
- *Spaceprints*, 17A High Street, Norton, Stockton on Tees, Cleveland; Tel: 0642-584440.

Canada, United States
- *Carolina Biological Supply Company*, 2700 York Road, Burlington, NC 27215, USA; Tel: 919-584-0381. Also Box 187, Gladstone, OR 97027, USA. Tel: 503-656-1641.
- *Central Scientific Company* (CENCO), 11222 Melrose Ave., Franklin Park, IL 60131-1364, USA; Tel: 708-451-0150 or 800-262-3626.
- *Fisher Scientific*, Educational Materials Division, 4901 W. LeMoyne Street, Chicago, IL 60651, USA; Tel: 800-621-4769.
- *Orbital Productions*, P.O. Box 22202, San Diego, CA 92122; Tel: 619-457-0953.

12.4.2.4 Slides. Slides are by far the most used format in the audio-visual realm. The vast volume of material in this medium makes even a partial listing of offerings prohibitive. Instead, the reader is referred to the selection of companies given in the general list in Appendix Sect. A.5, as virtually all of them distribute slides. Also, some of the available slide sets are listed in [12.14].

12.4.2.5 Videodiscs. Videodiscs or *laserdiscs* will quite possibly supplant the standard 35-mm slides and 16-mm films (and even videocassettes) in the not-too-distant future. This new technology, which is available for a wide variety of subject areas, provides at one's fingertips a variety of audio-visuals, including still and motion pictures, that can be shown on a color monitor at a selectable speed, forward or backward, with high quality sound. Each disc is provided with an image directory in which every frame is described in detail. A disc can hold several hundred thousand individual images with quick random access via remote keypad.

The ASP, the MMI Corporation, and Optical Data Corporation (formerly Video Vision Associates) all distribute astronomy videodiscs, but the number of available sources should increase as this medium becomes more popular. For more information, consult the *Videodisc Compendium for Education and Training*, which is available from Emerging Technologies, Inc., P.O. Box 12444, St. Paul, MN 55112, USA; Tel: 612-639-3973.

12.4.2.6 Transparencies. Transparencies are in the form of clear plastic sheets (overlays) for use with overhead projectors. They are not as ubiquitous as slides, but they can be used to good purpose when they are available. Transparencies for astronomy are sold by a few companies, including the Science Museum Library in the U.K. and Fischer Scientific, Hubbard Scientific, Schoolmasters Scientific, and Wards Natural Science Establishment in the U.S.

12.4.2.7 Filmstrips and Slide/Audiocassette Sets. Filmstrips are essentially still pictures (slides) which have been put on a continuous roll, to be shown on a screen with or without synchronous audio narration (audiocassette tape or phonograph record). One notable astronomy set aimed at grades 7–12 is *The Universe: Frontiers of Discovery Series* (1984), which is distributed by National Geographic, Educational Services, Washington, DC 20036; Tel: 800-368-2728. In Maryland: 301-921-1330. There are also several good series of slide/audiocassettes available, including *Astronomical Observations and Computations*, which is essentially a short course in navigational astronomy available from National Audiovisual Center, and *Telescopes of the World*, distributed by the ASP.

In the U.K., the Federation of Astronomical Societies (FAS) distributes the following slide/audiocassettes sets: *Introducing Black Holes* by Iain Nicolson, *The Truth About UFOs* by Robert Scheaffer, and *Europe in Space* by Neville Kidger. The reader should check the catalogues of various suppliers for other astronomy filmstrips.

12.4.2.8 Motion Picture Films and Videocassette Tapes. Motion pictures or movies have for many years been one of most important tools for disseminating information in astronomy. They are available in three principal formats:

1. film (16-mm, 8-mm, and Super-8)
2. videocassette (VHS and Beta)
3. videodisc (discussed in Sect. 12.4.2.5 above).

Film, usually in the 16-mm format with or without soundtrack, was the *only* format in widespread use until the late 1970s. Recently, however, many films have become available in the Super-8 format. Some films are wound onto reels which are to be mounted with a blank take-up reel on a projector, while others are mounted in a self-winding plastic cassette called a film loop. There are purely silent movies and movies with sound tracks, the latter of which comprises the bulk of the resources available in this medium. 16-mm sound films in particular are rapidly being superseded by videocassette tapes.

One of the most successful industries to emerge in recent years has been the videocassette industry, now considered one of the biggest in the world. The use of videocassette recorders (VCRs) to record and play back television broadcasts increased exponentially during the 1980s, and, consequently, most sound movies are now being produced exclusively on videocassette, and some of the older conventional movie films have been transferred to this new medium. The number of available videocassette tapes on astronomy subjects is also growing rapidly.

The list of companies which distribute films is far too long to present here, and the reader should therefore check [12.14] (for materials available as of 1985) or consult one or more of the following film/video catalogues, some of which are updated annually:

- *AAAS Science Books and Films*, subscription $20 per year from AAAS Subscription Dept., 10th Floor, 1101 Vermont Ave., N.W., Washington, DC 20005.
- *A Cinescope of Physics*, published by the American Association of Physics Teachers in 1978.
- *Educational Film/Video Locator of the Consortium of University Film Centers and R.R. Bowker*, published by the R.R. Bowker Company, New York 1986.
- *Educator's Guide to Free Films*, published annually by Educators Progress Service, Inc., 214 Center St., Randolph, WI 53956, USA.
- *Films in the Sciences*, available from the AAAS.
- *Film and Video Finder*, published by the National Information Center for Educational Media, Albuquerque 1987.
- *Media Review Digest*, published by Pierian Press, Ann Arbor.
- *The Video Source Book*, published by National Video Clearinghouse, Syosset (NY).

There was also a listing (with descriptions) of astronomy films compiled in 1977 by C. Smith and published in *Effective Astronomy Teaching and Student Reasoning Ability*, but it is now out of print. A detailed subject index of several hundred astronomy films is contained in *Universe in the Classroom* by Andrew Fraknoi, published by Freeman in 1985.

12.4.3 Broadcasting and Communications

12.4.3.1 Radio. In North America, the syndicated daily program *StarDate* has achieved immense success during its decade-long run on radio. A typical segment consists of a 2-minute vignette focusing on events taking place in the night sky, as well as anecdotes from poetry or the history of astronomy, or news of the latest discoveries. More information can be found in the article by Byrd [12.26] in *Mercury* magazine.

For precision observing work, time signals from station WWV are broadcast on shortwave radio at 2.5, 5, 10, and 15 MHz. Similar signals are broadcast by Canadian station CHU at 3.330, 7.335, and 14.670 MHz. (See also Sect. 12.4.3.3.)

12.4.3.2 Television. The convenience of television viewing is undeniably a major factor in diverting much of the populace from other, more active pursuits during their leisure time. This can have serious consequences, especially for amateur astronomy, as many would-be skywatchers are tempted to tune in their favorite programs at night instead of making direct observations of the stars and planets. Nevertheless, the medium of television has, at least in principle, the potential to disseminate a broad spectrum of scientific principles and results to a significant fraction of the public.

High quality television programs, primarily from BBC and PBS stations, are currently being broadcast in much of Great Britain and North America. Probably the longest-running program on astronomy is Patrick Moore's *The Sky at Night*, which has been running monthly on BBC televison for over three decades. In addition, the PBS program *Star Hustler*, a 5-minute segment on happenings in the sky, is broadcast weekly in the U.S. and Canada, while *Earth Calling Basingstoke* is another astronomy program seen in Great Britain. Other television programs, most notably *Discover: The World of Science, Innovation, Nova,* and *Smithsonian World,* are oriented specifically toward issues and discoveries in science and technology, and individual programs will sometimes focus on astronomical subjects.

A very recent offering in educational television is *NASA Select Television,* a cable station which provides informational and educational programming on space and related topics aimed at inspiring young people to achieve, especially in mathematics and science. Broadcasts are made Monday through Friday, and include historical documentaries focusing on America's space program, scientific results from spacecraft, and programs designed specifically for classroom use, covering topics such as biology, geology, the atmospheric and Earth sciences, and math and engineering concepts. For more information, contact: NASA Select, c/o Associate Administrator for Public Affairs, NASA Headquarters/Code P, Washington, DC 20546, USA; Tel: 202-453-8425.

There are also limited series programs (with usually six or more episodes) some of which are now available on videocassette. These include the following:

- *The Astronomers*, a 6-part international PBS series describing the lives and work of professional astronomers. Available from PBS Video on six, one-hour videocassettes, accompanied by a substantive set of resource materials, including a 16-page reference and referral guide, a 32-page high-school curriculum guide, a poster, a career opportunities brochure, and a companion book.
- *Cosmos*, hosted by astronomer Carl Sagan. One-hour shows, now available on videocassette from numerous sources (e.g., the ASP).
- *The Day the Universe Changed*, hosted by James Burke.
- *Planet Earth*, an 8-part series hosted by Richard Kiley. At least two of the one-hour shows have material on astronomy. Now available on videocassette.
- *Ring of Truth*, hosted by astronomer Philip Morrison.
- *Universe in the Classroom*, a 39-part series produced by KOCE-TV in southern California which premiered on national television in the U.S. in 1979. The interested reader might want to look up the relevant article by Pierce [12.27] in a past issue of *Mercury* magazine.

12.4.3.3 Telephone News Services. Telephone "hotlines" on astronomy provide daily, weekly, and monthly reports on phenomena occurring in the sky. The ones listed below are a sampling from [12.21], which should be consulted for more detailed information and additional hotlines. Long distance rates apply.

- *Abrams Planetarium*: 517-332-7827 (East Lansing, MI).
- *Astronomical Society of the Pacific*: 415-337-1244 (San Francisco, CA).
- *Buehler Planetarium*: 305-475-6734 (Fort Lauderdale, FL).
- *Dial-A-Shuttle*: 900-909-NASA (Washington, DC). Available only during Space Shuttle Missions.
- *Griffith Observatory*: 213-663-8171 (Los Angeles, CA).
- *Hansen Planetarium*: 801-532-7827 (Salt Lake City, UT).
- *Hawaiian Skies*: 808-948-0759 (Honolulu, HI).
- *Maryland Science Center*: 301-539-7827 (Baltimore, MD).

- *NASA-Johnson Space Center*: 713-483-8600 (Houston, TX).
- *National Air and Space Museum*: 202-543-2000 (Washington, DC).
- *Pacific Science Center*: 206-443-2920 (Seattle, WA).
- *Sky & Telescope*: 617-497-4168 (Cambridge, MA).
- *The Skyline*: 602-955-7597 (Phoenix, AZ).
- *Smithsonian Astrophysical Observatory*: 617-497-1497 (Cambridge, MA) or 202-357-2000 (Washington, DC).
- *University of Illinois*: 217-333-8789 (Urbana, IL).

12.4.3.4 Electronic News Services and Computer Bulletin Board Systems (BBSs). There now exist many local and national computer networks which can be accessed with only a computer and modem. It is possible not only to read news and information, but also to communicate with other users and to run programs.

One of the best known networks is *CompuServe Information Service*, a major international network service which allows the user with a PC and modem to communicate with other astronomy enthusiasts via "Astronomy Forum." The Forums have large message boards and maintain text files and programs in extensive data libraries. To receive a free Inquiry Brochure, or to order a Subscription Kit direct, contact: CompuServe Inquiry Brochure, Dept. MA1017, P.O. Box 20212, Columbus, OH 43220, USA; Tel: 800-848-8199. In Ohio or outside the continental U.S.: 614-457-0802. CompuServe memberships are also sold by computer dealers.

A brief sampling of the available "computer bulletin board systems" (BBSs) which contain astronomy and space-related subject material is provided below. A typical BBS contains files of general interest, such as current sky information, upcoming celestial events, astronomical news, ephemerides, and astronomy computer programs. The telephone numbers given are for more information and/or to subscribe. The reader is directed to [12.21] for a more extensive list of BBSs and more information on individual BBSs.

- *Astronomical Society of the Atlantic BBS*: 404-985-0408 (Snellville, GA).
- *Canadian Space Network BBS*: 416-458-5907 (Brampton, ON).
- *The Comm-post*: 303-534-4646 (Denver, CO).
- *Enviro*: 703-524-1837 (Washington, DC).
- *GEnie*: 800-638-9636. The General Electric Network for Information Exchange.
- *NASA Spacelink*: 205-895-0028 (Huntsville, AL).
- *National Space Society BBS*: 412-366-5208 (Pittsburgh, PA).
- *Starbase One*: 44-71-733-3992 (London, England).
- *Starbase III*: 209-432-2487 (Fresno, CA).
- *Star*Net*: 612-681-9520 (Minneapolis, MN).
- *Yokohama Science Center BBS*: 81-045-832-1177 (Yokohama, Japan).

12.4.4 Computers and Software

The use of computers has pervaded virtually every segment of our society, and science in particular. The direct effect on research astronomy has been in relieving human laborers of the vast amounts of data reductions that used to be performed by hand and later with calculating machines. For didactic functions (planetarium shows, classroom exercises, etc.), the computer can be used in the following capacities:

- as a programmable computing machine
- to interpret results graphically
- to store and record measured data
- to steer telescopes and other instruments
- to simulate past, future, and hypothetical events
- to present instructional programs (tutorials)
- to present tests and exercises (often with control over the answers)
- as a database
- as a text processor.

The following are some references on the use of computers and programmable calculators:

- Boulet, D.: *Methods of Orbit Determination for the Microcomputer*, Willmann-Bell, Inc., 1991.
- Burgess, E.: *Celestial Basic: Astronomy on Your Computer*, Sybex Inc., 1982.
- Duffet-Smith, P.: *Astronomy with Your Personal Computer*, Cambridge University Press, 1985.
- Genet, R.M. (ed.): *Microcomputers in Astronomy, I and II*, Fairborn Observatory (1247 Folk Rd., Fairborn, OH 45324) 1983, 1984.
- Ghedini, S.: *Software for Photoelectric Photometry*, Willman-Bell, 1982.
- Klein, F.: *Pocket Computer Programs for Astronomers*, Klein Publications, 1983.
- Kuhn, W.: *Computer in Experiment*, Aulis Verlag Deubner & Co., 1986.
- Mackenzie, R.: *The Astronomer's Software Handbook*, Sigma Press, 1985.
- Meeus, J.: *Astronomical Algorithms* Willmann-Bell, Inc., 1991.
- Montenbruck, O, Pflager, T.: *Astronomy on the Personal Computer*, Springer-Verlag, 1991.
- Press, W.H., Flannery, B.P., Teukolsky, S.A., Vetterling, W.T.: *Numerical Recipes—The Art of Scientific Computing*, Cambridge University Press, 1986.
- Roy, A.E.: *Orbital Motion* (3rd ed.), Adam Hilger, 1988.
- Schmid, E.W.: *Theoretical Physics on the Personal Computer*, Springer-Verlag, 1988.
- Tattersfield, D.: *Orbit for Amateurs with a Microcomputer*, Stanley Thornes Ltd., 1984.
- Trueblood, M., Genet, R.: *Microcomputer Control of Telescopes*, Willmann-Bell, 1985.
- Vetterling, W.T., Teukolsky, S.A., Press, W.H., Flannery, B.P.: *Numerical Recipes Examples Book (FORTRAN)*, Cambridge University Press, 1986.

The list of software resources has grown enormously in the past decade. Some of these are given in the *Handbook for Astronomical Societies* (for readers in the U.K.) [12.16] or in the *Astronomy Resource Guide* (Canada and the U.S.) [12.21]. Mosely and Fraknoi [12.28] assembled a list of astronomy software available as of 1991, along with a list of software producers in North America. A copy can be obtained by sending a donation of $3.00 (U.S.) to ASP Software List Dept., 390 Ashton Ave, San Francisco, CA 94112.

In addition, NASA's Educational Affairs Division has produced the publication *Software for Aerospace Education*, PED-106, which identifies aerospace-related software currently available as well as addresses of software vendors and NASA Teacher Resource Centers. It can obtained by writing to the Educational Affairs Division, Code XEP, NASA, Washington, DC 20546.

12.4.5 Printed Material

12.4.5.1 Books and Magazines. The number of astronomy books (textbooks, laboratory manuals, observing guides, childrens's books, etc.) that have been published in English over the past several years is too enormous (and transient) to be treated here in any detail. A listing of some of the available magazines are provided in Appendix Sect. A.7. A comprehensive listing (up to 1985) is also given in the aforementioned *The Universe at Your Fingertips* by Robbins and Fraknoi [12.14], and thereafter in *Astronomy Educational Materials in Print 1985–1987* by Robbins [12.29]. Also, Fraknoi [12.30] provides, in the May/June 1989 issue of *Mercury* magazine, a listing of astronomy books which appeared in 1988, and highlights the dozen best nontechnical books of the year. Readers in the U.K. can obtain some information on printed materials in the *Handbook for Astronomical Societies* [12.16] and also by writing the Astronomy Section, Old Royal Observatory, National Maritime Museum, London SE10 9NF.

Perhaps the best way to keep apprised of what books have been published is to regularly check the book review sections of the various weekly, monthly, and quarterly astronomy and science magazines such as *American Scientist*, *Astronomy*, *Astronomy Now*, *Mercury*, *New Scientist*, *The Observatory*, *Physics Today*, and *Sky & Telescope*.

For those who wish to maintain a steady diet of reading in astronomy, the Astronomy Book Club provides members with substantial savings over the publisher's suggested prices for the books. There are books for both amateurs and professionals, beginners and seasoned observers. They may be contacted at: Astronomy Book Club, Dept. L-CX3/00087, 3000 Cindel Dr., Delran, NJ 08075, USA.

12.4.5.2 Teachers' Guides. For those who teach school children, especially those in grades about 4 to 12, the following teachers' guides will be found invaluable:

- DeBruin, J., and Murad, D.: *Look to the Sky: An All-Purpose Interdisciplinary Guide to Astronomy*, Good Apple Books, 1988. Provides ideas and references for teaching astronomy in grades 4–12. Includes puzzles, constellation finders, star-gazing, and time-telling activities, and observing hints for beginners.
- Druger, M. (ed.): *Science for the Fun of It: A Guide to Informal Science Education*, National Science Teachers Association, 1988. A collection of articles on science education outside the classroom includes sections on television programs, science writing, museums, zoos, science fairs, community science, science toys and gifts, and so on. The section on informal astronomy education was written by A. Fraknoi.
- Fraknoi, A.: *Universe in the Classroom: A Resource Guide for Teaching Astronomy*, W.H. Freeman, 1985.

- Gibson, B.: *The Astronomer's Sourcebook*, Woodbine House, 1992. Information on astronomical equipment, publications, planetariums, organizations, events, and history.
- Pasachoff, J.M., Percy, J.R. (eds.): *The Teaching of Astronomy*, Cambridge University Press, 1990. The report of IAU Colloquium 105, with the following sections: Curriculum, Astronomy and Culture, The Teaching Process, Student Projects, Computers, Textbooks, Teaching Aids and Resources, Conceptions/Misconceptions, High School Courses, Teacher Training, Popularization, Planetariums, and Developing Countries.
- Reynolds et al.: *Space Mathematics: A Resource for Teachers*, NASA, 1972.
- Schatz, D., Fraknoi, A., Robbins, R., and Smith, C.: *Effective Astronomy Teaching and Student Reasoning Ability*, Regents of University of California, 1978. From a workshop designed to improve the teaching of introductory astronomy at college and high school levels, especially in courses for nonscience majors. Examined are reasoning patterns of students, teaching strategies, texts and other reading materials, films, and laboratory activities.

12.4.5.3 Booklets and Information Packets on Astronomy. The following booklets are available from the Hansen Planetarium (address given in Appendix Sect. A.1) and are adapted from the texts of their star programs.

- *Footsteps: Man on the Moon* (32 pp.).
- *The People: Sky Lore of the American Indian* (24 pp.).
- *The Universe of Dr. Einstein* (32 pp.).
- *Springtime of the Universe* (24 pp.).
- *Skywatchers of Ancient Mexico* (24 pp.).

The ASP also puts out a series of information packets, some of which are reprinted from the society's *Mercury* magazine. They are designed to provide the beginner with clear, nontechnical information about certain subjects.

- *Astronomy on Computer* (8 pp.)
- *Astronomy as a Hobby* (28 pp.)
- *Astronomy Versus Astrology* (16 pp.)
- *Debunking Pseudoscience* (4 pp.)
- *Eclipses of the Sun* (8 pp.)
- *Interdisciplinary Approaches to Astronomy* (32 pp.)
- *Introduction to Black Holes* (24 pp.)
- *Learning About Quasars* (20 pp.)
- *Selecting Your First Telescope* (16 pp.)
- *Solar System Data Kit* (8 pp.)
- *Women in Astronomy* (48 pp.).

12.4.5.4 Sources for Rare and Out-of-Print Books. The best sources of old and historical books on astronomy are the following:

- Warren Blake, *Old Science Books*, 308 Hadley Dr., Trumbull, CT 06611, USA; Tel: 203-459-0820.

- *The Gemmary*, P.O. Box 816, Redondo Beach, CA 90277, USA; Tel: 213-372-5969. Mail order rare books and antique scientific instruments. Catalogue available.
- *Historical Technology, Inc.*, 6B Mugford Street, Marblehead, MA 01945, USA.
- *Ian Howard-Duff*, Highfield, Fairview Road, Headley Down, Hants GU35 8HQ, England. Dealer in secondhand and out-of-print books on astronomy, space, and related subjects.
- *Janus Publications*, P.O. Box 8705, Wichita, KS 67208, USA; Tel: 316-686-8320. Reproduction of Burritt's *Geography of the Heavens*.
- *Herbert A. Luft*, 46 Woodcrest Dr., Scotia, NY 12302, USA.
- *R.A. Marriott*, 24 Thirlestane Road, Far Cotton, Northampton NN4 9HD, England, Tel: 0604-65190. This company publishes a listing of available books, monographs, technical papers, and journals in a quarterly catalogue. Authors include such historical notables as Airy, Arago, John and William Hershel, Huygens, Wollaston, and Thomas Young.
- *G.B. Manasek Inc.*, P.O. Box 961, Hanover, NH 03755, USA; Tel: 603-643-2227.
- *Willman-Bell, Inc.*, P.O. Box 1325, Richmond, VA 23235, USA.

12.4.6 Games

There are several astronomy items available which fall under the category of "games." These include:

- *Earth from Moon Jigsaw* — Available from the London Planetarium.
- *Summer Milky Way Jigsaw Puzzle* — A 1000-piece jigsaw puzzle of the summer Milky Way in Sagittarius and Scorpius is available from either Deen Publications, P.O. Box 831991 Richardson, TX 75083, USA; Tel: (214) 231-0338, or from the ASP (Catalogue no. JP 101).
- *Vivid Rings of Saturn* — (PuzSN) A 600-piece jigsaw puzzle featuring Saturn's rings and Earth superposed to scale. Available from Science News Books, 1719 N St. NW, Washington, DC 20036, USA; Tel: 1-800-544-4565.
- *Good Heavens!* — A card game designed to teach simple facts about astronomy. Includes deck of 54 cards plus 24-page booklet. Available from Ampersand Press, 691 26th St., Oakland, CA 94612, USA.
- *Liftoff!* — A board game in which four players try to be the first to land a crew of astronauts on the Moon. Intended for ages 12 and up. Available from Task Force Games, 14922 Calvert Street, Van Nuys, CA 91411, USA.
- *Sky Challenger* — Games and activities for star gazers. Six different game wheels and a star clock. Available from Hubbard Scientific Company and also from Carolina Biological Supply Company.
- *Solarquest* — Space-age real estate game, recommended for ages 8 and up. Available from Science News Books.
- *Space Picture Playing Cards* — Available from The London Planetarium.
- *Stellar 28/Constellation Games* — Players learn constellations and stars by matching cards and game board. Includes 28 different games. May be played in group or solitaire. Available from Hubbard Scientific Company.

12.4.7 Music

Not surprisingly, the subject of astronomy has been the inspiration for some of the most beautiful and original musical compositions in both the classical and pop repertoires, and they have been frequently used as background music for films on astronomy. A few of these compositions and their composers will be listed below. For a very extensive listing, the reader should refer to an excellent series of *Mercury* articles by Fraknoi [12.31] and Ronan [12.32].

- Braheny, Kevin: *Galaxies*. This piece was composed as the soundtrack to a planetarium show that was written and directed by Timothy Ferris and based on Ferris' book of the same name. It has been reviewed by Robert Burnham in the February 1989 issue of *Astronomy* magazine. It is available on both cassette and CD.
- Holst, Gustav: *The Planets*. There must be nearly two dozen available recordings of this masterful and beguiling composition dating from around 1918. If the reader has not heard it, then he or she may want to check the record out from the nearest public or school library.
- Scriabin, Alexander: *Universe*. A less familiar composition by a Russian composer of the late 19th century. The recording is on Angel SR-40260.
- Serrie, John: *And the Stars Go with You*. Five tracks of synthesized space music, "a tribute to the triumphs and infrequent setbacks in the space program," and dedicated to Christa McAuliffe. Available from Miramar Productions, P.O. Box 15661, Seattle, WA 98115.
- Serrie, John: *Flightpath*. More synthesized space music for deep sky viewing. Available from Miramar Productions.
- Strauss, Johann: *Music of the Spheres*. The only recording of this work that the author is familiar with is the one by Karajan with the Vienna Philharmonic Orchestra, on Deutsche Grammaphon 419 616-2.
- Wright, Gordon: *Symphony in URSA MAJOR*. Composed by a University of Alaska professor. Excerpts of this work, performed by the Fairbanks Symphony Orchestra, are used to provide the sound track for the videotape *Aurora Borealis* (see Motion Picture Films and Videocassette Tapes).

There are numerous other compositions in the classical, pop, and folk realm that have astronomical undercurrents, and the interested reader is encouraged to pursue them.[11]

Acknowledgements
The author is indebted to a number of persons from whom he received copious amounts of information for this chapter, especially Miss Undine Concannon, Professor Roy Bishop, Mr. Martin Ratcliffe, Mrs. Rosaly Lopes, Mrs. Dolores Fidishun, and Dr. Alan Batten. Special thanks go to Drs. Andrew Fraknoi, Derek McNally, John Percy, Bob

---

11 An interesting sidenote is that the father of astronomer Galileo Galilei was a noted Renaissance musician, Vincenzo Galilei (ca. 1520–1591). There is on record a performance of his *Contrapuncto*, a short composition for two lutes; it is available on the Musical Heritage Society label, catalogue number 4152F (LP) or 6152H (cassette).

Robbins, Harry Shipman, and Charles Tolbert, all of whom made helpful comments and suggestions on a preliminary version of this chapter.

References

12.1 Couper, H.: *Becoming a Professional Astronomer*, brochure available from the Association for Astronomy Education, London 1987.
12.2 *Astronomer*, pamphlet published by the Guidance Centre of the University of Toronto, Toronto 1985.
12.3 *Understanding the Universe: A Career in Astronomy*, brochure published by the American Astronomical Society, Washington 1986.
12.4 Ellis, S.D.: *1986–87 AIP Graduate Student Survey (AIP Publ. No. R-207.20)*, American Institute of Physics, New York 1988.
12.5 Ellis, S.D.: *Enrollments and Degrees (AIP Publ. No. R-151.25)*, American Institute of Physics, New York 1988.
12.6 *1990–91 Graduate Programs in Physics, Astronomy, and Related Fields*, American Institute of Physics, New York 1989.
12.7 *Directory of Graduate Programs: Volume C*, published by the Educational Testing Service, Princeton, NJ.
12.8 *Postgraduate Opportunities in Astronomy and Geophysics*, published by the RAS.
12.9 Ellis, S.D., Mulvey, P.J.: *AIP Employment Survey 1987 (AIP Publ. No. R-151.25)*, American Institute of Physics, New York 1988.
12.10 Fraknoi, A.: Interdisciplinary Approaches to Astronomy, Parts I and II. *Mercury*, Sep/Oct 1977, p. 20 and Nov/Dec 1977, p. 20.
12.11 Kruglak, H., and O'Bryan, M.: Astrology in the Astronomy Classroom. *Mercury*, Nov/Dec 1977, p. 18.
12.12 Kelly, I.: The Scientific Case Against Astrology. *Mercury*, Nov/Dec 1980, p. 135; and Cosmobiology and Moon Madness. *Mercury*, Jan/Feb 1981, p. 13.
12.13 Pasachoff, J.: Some Tests of Astrology. *Mercury*, Nov/Dec 1980, p. 137.
12.14 Robbins, R.R., and Fraknoi, A.: *The Universe at Your Fingertips: An Educator's Desktop Reference of Astronomy Education Materials in English*, IAU Commission 46 Report, August 1985.
12.15 *Directory of the International Planetarium Society*, Hansen Planetarium, Salt Lake City.
12.16 Jones, B. (ed.): *Handbook for Astronomical Societies*, Federation of Astronomical Societies, 1991.
12.17 *The Observer's Handbook*, Royal Astronomical Society of Canada, published annually.
12.18 Beck, R.L., Schrader, D.: *America's Planetariums & Observatories (A Sampling)*, Sunwest Space Systems, Inc., St. Petersburg FL 1991.
12.19 Kirby-Smith, H.T.: *U.S. Observatories: A Directory and Travel Guide*, Van Nostrand Reinhold Co., New York 1976.
12.20 *1989 Directory of Observatories, Planetariums, and Museums*. In: *Astronomy*, May 1989, p. 50.
12.21 *Astronomy Resource Guide*. In: *Sky & Telescope*, September 1991.
12.22 *Speakers, Tours, and Films: A Program Resource for SPS Chapters*, American Institute of Physics, published annually.
12.23 *1989 Directory of Astronomy Clubs*. In: *Astronomy*, May 1989, p. 70.
12.24 *Equipment Directory 1989*. In: *Astronomy*, October 1989.
12.25 *Astronomy's* 1989 Guide to Telescopes. *Astronomy*, October 1989, p. 71.
12.26 Byrd, D.: Astronomy on the Air. *Mercury*, Jan/Feb 1980, p. 7.
12.27 Pierce, D.: Astronomy on Public TV. *Mercury*, Sep/Oct 1978, p. 118.
12.28 Mosley, J., and Fraknoi, A.: Computer Programs in Astronomy. *Mercury*, Jan/Feb 1985, p. 27; and Computer Software for Astronomy. *Mercury*, May/June 1991, p. 87.

12.29 Robbins, R.R.: *Astronomy Education Materials in Print 1985–1987*, IAU Commission 46 Report, August 1988.
12.30 Fraknoi, A.: Nontechnical Astronomy Books of 1988. *Mercury*, May/June 1989, p. 88.
12.31 Fraknoi, A.: The Music of the Spheres: Astronomical Sources of Musical Inspiration. In: *Interdisciplinary Approaches to Astronomy*, Astronomical Society of the Pacific, San Francisco 1987, p. 17.
12.32 Ronan, C.: Some Other Astronomical Pieces of Music. *Mercury*, Nov/Dec 1979, p. 131.
12.33 Dworetsky, M.M.: An Astronomy Degree Course in the U.K.: Syllabus and Practical Work. In: Pasachoff, J.M., and Percy, J.R. (eds.): *The Teaching of Astronomy*, IAU Colloquium 105, Cambridge University Press, Cambridge 1990.

Appendix A: Educational Resources in Astronomy

A.1 Planetariums, Museums, and Exhibits

A.1.1 Planetariums and Museums in the United Kingdom

England
- *AAC Planetarium*, Amateur Astronomy Centre, Bacup Road, Clough Bank, Todmorden, Lancs. OL14 7HW. Tel: 0706 816964.
- *British Museum*, Great Russell Street, London WC1B 3DG; Tel: 071-323 8395 ext. 395. Astronomical clocks.
- *British Museum (Natural History)*, Cromwell Road, South Kensington, London SW7 5BD; Tel: 071-938 9123. Extensive meteorite collection.
- *Caird Planetarium*, Old Royal Observatory, Greenwich, London SE 10.
- *William Day Planetarium*, Plymouth Polytechnic, School of Maritime Studies, Plymouth PL4 8AA. Tel: 0752 264666.
- *Electrosonic Ltd.*, 815 Woolwich Road, London SE7 8LT.
- *Greenwich Planetarium*, South Building, Greenwich Park, Greenwich, London SE 10. Tel: 081-858 1167.
- *William Herschel House and Museum*, 19 New King Street, Bath, BA1 2B1. Contact: Dr. A.V. Sims, 30 Meadow Park, Bathford, Bath; Tel: 0225 859529. Open Mar–Oct daily 2–5 pm, Nov–Feb Sundays only, 2–5 pm.
- *Jodrell Bank Planetarium and Visitor Center*, Lower Withington, Nr. Macclesfield, Cheshire SK11 9DL; Tel: 0477 71339.
- *Kings Observatory, Kew*, Old Deer Park, Richmond, Surrey TW9 2AZ.
- *University of Leicester, The Planetarium*, Department of Astronomy, University Road, Leicester LE1 7RH; Tel: 0533 522522.
- *Liverpool Museum Planetarium*, William Brown Street, Liverpool, Merseyside L3 8EN. Tel: 051-207 0001 ext. 225.
- *London Planetarium*, Marylebone Road, London NW1 5LR; Tel: 071-486 1121 (9:30–5:30), 071-486 1121 (recording).
- *City of London Polytechnic*, The Planetarium, 100 Minories, Tower Hill, London EC3N 1JY. 071-283 1030.
- *London Schools Planetarium*, John Archer School Building, Wandsworth Rd., Sutherland Grove, London SW18; Tel: 081-788 4253.

- *The Planetarium at Rickmansworth Masonic School*, Chorleywood Rd., Rickmansworth, Herts. WD3 4HF. Tel: 0923 773168.
- *Royal Greenwich Observatory*, University of Cambridge, Madingley Road, Cambridge CB3 0HA; Tel: 0223 337548. Formerly located at Herstmonceux Castle, E. Sussex.
- *Science Museum*, Exhibition Road, South Kensington, London SW7 2DD; Tel: 071-589 3456.
- *David W. Shepherd*, 5 Nab Wood Drive, Shipley, W. Yorkshire BD18 4HP.
- *Southend Planetarium*, Central Museum, Victoria Avenue, Southend-on-Sea; Tel: 0702 330214.
- *South Tyneside College Planetarium*, St. Georges Avenue, S. Shields, Tyne & Wear NE34 6ET; Tel: 091-456 0403 ext. 477.
- *Stonehenge*, near Amesbury, Wiltshire. Remains of an ancient temple or astronomical observatory, built in 3 stages between 2350 BC and 1350 BC. Restricted visiting. Contact: Area Custodian, Historical Buildings and Monuments Commission for England, Bridge House, Sion Place, Clifton, Bristol BX8 4XA; Tel: 0272 734472.
- *"Trillium"*, 206 White Lion Road, Little Chalfont, Bucks HP7 9NU.
- *Whipple Museum of the History of Science*, Free School Lane, Cambridge; Tel: 0223 334540. Old scientific instruments such as orreries, sundials, astrolabes, etc.
- *Yorkshire Museum*, Museum Gardens, York YO1 2DR; Tel: 0904 29745. Observatory on grounds.

Ireland
- *Birr Castle*, Birr, Co. Offaly; Tel: 353 50920056. 72-inch telescope of third Earl of Rosse and other exhibits.

Northern Ireland
- *Armagh Planetarium*, College Hill, Armagh BT61 9DB. The exhibition hall is open to the public 2 pm–4:45 pm Mon–Sat. Public star shows run on Saturday afternoons, shows for schools held during the week. Public shows also held on most school and public holidays except Good Friday, July 12, and December 25 and 26. Tel: 0861 523689 (reservations), 0861 524725 (administration); Fax: 0861 526187.

Scotland
- *Aberdeen Technical College, The Planetarium*, Gallowgate, Aberdeen AB9 1DN; Tel: 0224 640366.
- *Glasgow College of Nautical Studies, The Planetarium*, 21 Thistle Street, Glasgow; Tel: 041-429 3201.
- *Jewel & Esk Valley College, The Planetarium*, 24 Milton Road East, Edinburgh EH15 2PP; Tel: 031-669 8461.
- *Leith Hill Nautical College*, 24 Milton Road East, Edinburgh EH15 2PP; Tel: 031-669 8461.
- *Mills Observatory and Planetarium*, Balgay Park, Glamis Road, Dundee DD2 2UB; Tel: 0382 67138.

- *Royal Museum of Scotland*, Chambers Street, Edinburgh, EH1 1JF; Tel: 031-225 7534.
- *Royal Observatory, Edinburgh*, The Visitor Centre, Blackford Hill, Edinburgh EH9 3HJ; Tel: 031-668 8100.

A.1.2 Planetariums and Museums in Canada

Alberta
- *Calgary Centennial Planetarium*, Alberta Science Center, 701-11 Street S.W., P.O. Box 2100, Stn. M, Calgary, Alberta T2P 2M5; Tel: 403-264-4060 or 221-3700.
- *Edmonton Space Sciences Centre*, Coronation Park, 11211-142 Street, Edmonton, Alberta T5M 4A1; Tel: 403-451-7722 or 452-9100. Features planetarium Star Theatre, IMAX film theatre, exhibit galleries, Science Magic telescope shop and bookstore. Open daily.

British Columbia
- *H.R. MacMillan Planetarium*, 1100 Chestnut Street, Vancouver, BC V6J 3J9; Tel: 604-736-3656. Open daily.

Manitoba
- *The Lockhart Planetarium*, 394 University College, 500 Dysart Road, The University of Manitoba, Winnipeg, Manitoba R3T 2M8; Tel: 204-474-9785. By reservation only.
- *Manitoba Planetarium*, Museum of Man and Nature, 190 Rupert Avenue at Main Street, Winnipeg, Manitoba R3B 0N2. Tel: 204-943-3142 (program information recording) or 204-956-2830 (switchboard). Shows daily except some Mondays. Museum gift shop has scientific books and equipment.

Nova Scotia
- *Burke–Gaffney Planetarium*, Saint Mary's University, Department of Astronomy, Halifax, Nova Scotia B3H 3C3.
- *The Halifax Planetarium*, The Education Section of Nova Scotia Museum, Summer Street, Halifax, Nova Scotia B3H 3A6; Tel: 902-429-4610. Located in the Sir James Dunn Building, Dalhousie University. Free public shows given on some evenings at 8:00 pm and group shows can be arranged.

Ontario
- *Doran Planetarium*, Laurentian University, Ramsey Lake Road, Sudbury, Ontario P3E 2C6; Tel: 705-675-1151 ext. 2222.
- *McLaughlin Planetarium*, 100 Queen's Park, Toronto, Ontario M5S 2C6; Tel: 416-586-5736 (for show times) or 416-586-5751 (for sky information). Public shows Tues–Fri at 3:00 and 7:30. Additional shows on weekends and during summer. School shows, Astrocentre with solar telescope, and evening courses are available.
- *National Museum of Science and Technology*, 1867 St. Laurent Boulevard, Ottowa, Ontario K1G 5A3; Tel: 613-998-4566. Open daily.
- *Ontario Science Center*, 770 Don Mills Road, Don Mills, Ontario M3C 1T3; Tel: 416-429-4100. Open daily (except Christmas day) from 10:00 am to 6:00 pm.

Quebec
- *Dow Planetarium*, 1000 St. Jacques Street West, Montreal, P.Q. H3C 1G7; Tel: 514-872-4530. Live shows in French and English. Open daily.

A.1.3 Planetariums and Museums in the United States

Alabama
- *W.A. Gayle Planetarium*, 1010 Forest Ave., Montgomery, AL 36106; Tel: 205-832-2625.

Alaska
- *Marie Drake Planetarium*, 1250 Glacier Ave., Juneau, AK 99801; Tel: 907-586-3780.

Arizona
- *Barringer Meteor Crater*, Meteor Crater Enterprises, Inc., 121 East Birch, Flagstaff, AZ 86001. Tel: 602-774-8350. Site of a 24 000 year old crater resulted from a gigantic impact. With museum, gift shop, and refreshment stand. Open 8 am to sundown.
- *Flandreau Planetarium*, University of Arizona, Corner of Cherry and University Boulevard, Tucson, AZ 85721. Tel: 602-621-4515. Open Tue–Sun.

California
- *Chabot Planetarium (and Observatory)*, 4917 Mountain Blvd., Oakland, CA 94619. Very active center for resources and workshops.
- *Reuben H. Fleet Space Theater & Science Center*, P.O. Box 33303, San Diego, CA 92103; Tel: 619-238-1233.
- *Griffith Observatory Planetarium*, 2800 East Observatory Road., Los Angeles, CA 90027. Open fall–winter–spring: Tue–Fri 2–10 pm; Sat 10:30 am to 10 pm; Sun 1–10. During summer: Mon–Fri 2–10 with same Sat–Sun schedule as above. Planetarium shows daily. Considered one of the best centers for popular and amateur astronomy in the U.S.
- *Morrison Planetarium*, California Academy of Sciences, Golden Gate Park, San Francisco, CA 94118. Tel: 415-750-7141. Open daily.

Colorado
- *Denver Museum of Natural History and Charles C. Gates Planetarium*, 2001 Colorado Boulevard, Denver CO 80205. Tel: 303-370-6351. Includes many natural science exhibits, a Foucault pendulum, Space Shop, and a book shop. Open Tue–Sun.
- *Fisk Planetarium and Science Center*, University of Colorado, Boulder, CO 80309-0408; Tel: 303-492-5001. Open Mon-Fri.

District of Columbia
- *Albert Einstein Planetarium*, National Air and Space Museum, 6th St. and Independence Ave., Washington, DC 20560; Tel: 202-357-7000. Open daily.

Connecticut
- *Science Museum of Connecticut, Inc., and Gengras Planetarium*, 950 Trout Brook Dr., West Hartford, CT 06903; Tel: 203-236-2961.

Florida
- *Bishop Planetarium*, 201 10th St. W., Brandenton, FL 33505; Tel: 813-746-4132.
- *Miami Museum of Science & Space Transit Planetarium*, 3280 South Miami Avenue, Miami, FL 33129; Tel: 305-854-4242. Open daily.
- *Museum of Science and History and Alexander Brest Planetarium*, 1025 Gulf Life Dr., Jacksonville, FL 32207; Tel: 904-396-7062.
- *NASA Kennedy Space Center*, FL 32899; Tel: 305-452-2121. Open daily.
- *South Florida Science Museum and Aldrin Planetarium*, 4801 Dreher Trail N., West Palm Beach, FL 33405. Tel: 407-832-1988.
- *John Young Museum and Planetarium*, 810 East Rollins St., Orlando, FL 32803; Tel: 407-896-7151.

Georgia
- *Fernbank Science Center*, 156 Heaton Park Drive, NE, Atlanta, GA 30307; Tel: 404-378-4311. Open daily.

Hawaii
- *Bernice P. Bishop Museum*, P.O. Box 19000-A, Honolulu, HI; Tel: 808-847-3511.

Illinois
- *Adler Planetarium*, 900 E. Achsah Bond Dr., Chicago, IL 60605; Tel: 312-322-0304. Open daily.
- *Museum of Science and Industry and Crown Space Center*, 57th Street and Lake Shore Drive, Chicago, IL 60637. Tel: 312-684-1414. Open daily.

Indiana
- *SpaceQuest Planetarium*, Children's Museum, 3000 N. Meridian St., Indianapolis, IN 46208; Tel: 317-924-5431.

Iowa
- *Science Center of Iowa*, 4500 Grand Ave., Des Moines, IA 50312; Tel: 515-274-4138.

Louisiana
- *Louisiana Arts & Science Center*, P.O. Box 3373, Baton Rouge, LA 70821; Tel: 504-344-9465.

Maryland
- *Davis Planetarium*, Maryland Academy of Sciences, 601 Light St., Baltimore, MD 21230; Tel: 301-685-5225. Open daily.
- *Howard B. Owens Science Center*, 9601 Greenbelt Rd., Seabrook, MD 20706; Tel: 301-577-8718.

Massachusetts
- *Charles Hayden Planetarium*, Museum of Science, Science Park, Boston, MA 02114; Tel: 617-723-2500. Open daily.
- *Seymour Planetarium Observatory*, Springfield Science Museum, 236 State St., Springfield, MA 01103; Tel: 413-733-1194. Open Tue-Sun.

Michigan
- *Abrams Planetarium*, Michigan State University, East Lansing, MI 48824; Tel: 517-355-4676. Open daily.
- *Detroit Science Center*, 5020 John R. Street, Detroit, MI 48202; Tel: 313-577-8430. Open Tue–Sun.
- *Robert T. Longway Planetarium*, Flint Board of Education, 923 E. Kearsley St., Flint, MI 48503; Tel: 313-762-1181. Open daily.

Minnesota
- *Minneapolis Planetarium*, 300 Nicolet Mall, Minneapolis, MN 55401; Tel: 612-372-6644.
- *The Science Museum of Minnesota*, 30 E. 10th St., St. Paul, MN 55101; Tel: 612-221-9403.

Mississippi
- *Russell C. Davis Planetarium & Ronald E. McNair Space Theater*, 201 E. Pascagoula St., Jackson, MS 39201. Tel: 601-960-1550.

Missouri
- *McDonnell Planetarium*, City of St. Louis, 5100 Clayton Rd., St. Louis, MO 63110; Tel: 314-535-5810.

New Jersey
- *Newark Museum Planetarium*, 43–53 Washington St., Newark, NJ 07101; Tel: 212-596-6609. Open Tue–Sun.
- *New Jersey State Museum*, 205 W. State St., CN 530, Trenton, NJ 08625; Tel: 609-292-6333.

New Mexico
- *Space Center*, Top of N.M. Hwy. 2001, Alamogordo, NM 88311; Tel: 800-545-4021.

New York
- *Hayden Planetarium and American Museum*, 81st St. at Central Park West, New York, NY 10024; Tel: 212-769-5912. Open daily.
- *Link Planetarium-Kopernik Observatory*, 30 Front St., Binghamton, NY 13905; Tel: 607-772-0660. Open Tue-Sun.
- *Rochester Museum & Science Center*, 657 E. Ave., Rochester, NY 14603; Tel: 716-271-4320.
- *Vanderbilt Planetarium*, 180 Little Neck Rd., Centerport, Long Island, NY 11721; Tel: 516-262-7800. Open Tue-Sun.

North Carolina
- *Morehead Planetarium (and Observatory)*, University of North Carolina-Chapel Hill, Morehead Building, CB 3480, NC 27599-3480; Tel: 919-962-1236. Open daily.

Ohio
- *Cincinnati Museum of Natural History and Planetarium*, 1720 Gilbert Ave., Cincinnati, OH 45202; Tel: 513-621-3889. Open Tue-Sun.
- *Mueller Planetarium (and Observatory)*, Cleveland Museum of Natural History, Wade Oval, University Circle, Cleveland, OH 44106; Tel: 216-231-4600.
- *Ritter Planetarium (and Astrophysical Research Center)*, The University of Toledo, 2801 Bancroft, Toledo, OH 43606; Tel: 419-537-2650.

Oklahoma
- *Omniplex Science Museum and Kirkpatrick Planetarium*, 2100 NE. 52nd St., Oklahoma City, OK 73111; Tel: 405-424-5545.

Oregon
- *Oregon Museum of Science & Industry*, 4015 SW. Canyon Rd., Portland, OR 97221; Tel: 503-222-2828.

Pennsylvania
- *Buhl Planetarium and Institute of Popular Science*, Allegheny Square (north side), Pittsburgh, PA 15212; Tel: 412-321-4302. Open daily.
- *Fels Planetarium*, The Franklin Institute, 20th and the Parkway, Philadelphia, PA 19103; Tel: 215-448-1293. Open daily.

Tennessee
- *Cumberland Science Museum*, 800 Ridley Blvd, Nashville, TN 37203; Tel: 615-259-6099.
- *Memphis Pink Palace Museum*, 3050 Central Ave., Memphis, TN 38111; Tel: 901-454-5609.

Texas
- *Houston Museum of Natural Science*, #1 Hermann Circle Dr., Houston, TX 77030; Tel: 713-526-4273.
- *Lyndon B. Johnson Space Center*, Olin Teague Visitor Center, 2101 NASA Road 1, Public Services Branch, AP4, Houston, TX 77058; Tel: 713-483-4241. Open daily

Utah
- *Hansen Planetarium*, 15 South State Street, Salt Lake City, UT 84111; Tel: 801-538-2104. Open daily.

Virginia
- *Science Museum of Virginia*, 2500 W. Broad St., Richmond, VA 23220; Tel: 804-257-0211.
- *Science Museum of Western Virginia*, One Market Sq., Roanoke, VA 24011; Tel: 703-342-5710.

Washington
- *Eastern Washington Science Center*, Riverfront Park, Spokane, WA 99202; Tel: 509-359-6391. Open daily.
- *Pacific Science Center*, 200 2nd St. N., Seattle, WA 98109; Tel: 206-433-2001. Open daily.

Wisconsin
- *Milwaukee Public Museum*, 800 W. Wells St., Milwaukee, WI 53233; Tel: 414-278-2702. Open daily.

A.2 Observatories and Research Laboratories

A.2.1 Observatories in the United Kingdom

England
- *University of Cambridge*, Institute of Astronomy, The Observatories, Madingley Road, Cambridge CB3 0HA; Tel: 0223 337548.
- *Greenwich Old Royal Observatory*, Greenwich Park, London SE10 9NF; Tel: 081-858 1167. Open daily.
- *Hatfield Polytechnic Observatory*, Bayfordbury, Lower Hatfield Rd., Hertford, SG13 8LD; Tel: 0992-558451 ext. 334.
- *Jodrell Bank Radio Telescope*, Lower Withington, Nr. Macclesfield, Cheshire SK11 9DL; Tel: 0477 71339.
- *Lancashire Polytechnic Observatories*, Jeremiah Horrocks Observatory, Moor Park, Preston, Lancs. PR1 6AD.
- *Norman Lockyer Observatory*, Salcombe Hill, Sidmouth, Devon. Contact: Secretary, Sidmouth AS, Ross Meadows, 1 Green Mount, Sidmouth EX10 9DB; Tel: 03955 2928.
- *University of London Observatory*, Mill Hill Park, London NW7 2QS. Observatory Annexe, 33/35 Daws Lane, London NW7 4SD; Tel: 081-959 0421. Open Oct–Mar 1st & 3rd Fridays at 2:30 pm for schools, 6:30 pm and 7:30 pm for general public.
- *Mullard Radio Astronomy Observatory*, University of Cambridge, Physics Department. Contact: Dr. Kenderdine, Cavendish Laboratory, Madingley Road, Cambridge CB3 0HE.
- *Orwell Observatory*, Orwell Astronomical Society, 41 Melbourne Road, Ipswich IP4 5PP; Tel: 0473 271818. One of the largest amateur observatories in the country.
- *Royal Greenwich Observatory*, University of Cambridge, Madingley Road, Cambridge CB3 0HA; Tel: 0223 337548. Formerly located at Herstmonceux Castle, E. Sussex.
- *Planetary Science and Remote Sensing Group*, Dept. of Environmental Science, University of Lancaster, Lancaster LA1 4YQ; Tel: 0524 65201 ext. 4671.
- *Somerset Schools Observatory*, Education Department, County Hall, Taunton, Somerset; Tel: 0823 73451 ext. 5759.

Ireland
- *Dunsink Observatory*, Castelknock, Dublin 15: Tel: 387911/387959.

Northern Ireland
- *Armagh Observatory*, College Hill, Armagh BT61 9DB; Tel: 0861 522928.

Scotland
- *University of Glasgow Observatory, The Planetarium*, Acre Hill/Maryhill Road, Glasgow G20 0TL; Tel: 041-946 5213.
- *Mills Observatory and Planetarium*, Balgay Park, Glamis Road, Dundee DD2 2UB; Tel: 0382 67138. Open Mon–Sat.
- *Royal Observatory*, Blackford Hill, Edinburgh EH9 3HJ; Tel: 031-668 8100.
- *University of St. Andrews Observatory*, Department of Physics and Astronomy, University Observatory, Buchanan Gardens, St. Andrews, Fife; Tel: 0334 76161.

A.2.2 Observatories in Canada

Alberta
- *Devon Observatory*, Department of Physics, University of Alberta, Edmonton, Alberta T6G 2J1.
- *Rothney Astrophysical Observatory*, Physics & Astronomy Dept., University of Calgary, Calgary, Alberta T2N 1N4; Tel: 403-220-5385.

British Columbia
- *University of British Columbia Observatory*, 2219 Main Mall, Vancouver, BC V6T 1W5. Free public observing on clear Saturday evenings; Tel: 604-228-6186 (observing) or 604-228-2802 (tours).
- *Climenhaga Observatory*, Department of Physics and Astronomy, University of Victoria, Victoria, BC V8W 2Y2; Tel: 604-388-0001. Open daily.
- *Dominion Astrophysical Observatory*, 5071 West Saanich Road, Victoria, BC V8X 4M6. Open: *May–Aug*: Daily 9:15 am – 4:30 pm, *Sep–Apr*: Mon–Fri 9:15 am – 4:30 pm Public Observing: Saturday evenings, Apr-Oct; Tel: 604-388-0012.
- *Dominion Radio Astrophysical Observatory*, Penticton, BC V2A 6K3. Conducted tours: Sundays, Jul–Aug only, 2–5 pm. Visitors' Centre: Open year round during daytime; Tel: 604-497-5321.
- *Gordon MacMillan Southam Observatory*, 1100 Chestnut St., Vancouver, BC V6J 3J9. Open Fri–Sun, and statutory holidays 12 pm – 5 pm, 7 pm – 11 pm, weather and volunteer staff permitting. Extended hours during school holidays; Tel: 604-738-2855.

Nova Scotia
- *Burke–Gaffney Observatory*, Saint Mary's University, Halifax, Nova Scotia B3H 3C3. Open hours: *Oct–Mar*: Saturday evenings 7 pm, *Apr–Sep*: Saturday evenings 9 pm. Monday evening or daytime tours by arrangement. Tel: 902-420-5633.

Ontario
- *Hume Cronyn Observatory*, University of Western Ontario, London, Ontario N6A 3K7; Tel: 519-661-3183.
- *David Dunlap Observatory*, Richmond Hill, Ontario L4C 4Y6. Open Tuesday mornings 10 am throughout the year, Saturday evenings Apr–Oct by reservation; Tel: 416-884-2112.
- *National Museum of Science and Technology*, 1867 St. Laurent Blvd., Ottawa, Ontario K1A 0M8. Open: *Oct–Jun*: Group tours Mon–Thu, Public visits Fri (in French 2nd Fri); *Jul–Aug*: Public visits: Tue (French), Wed, Thu (English). Evening tours by appointment only; Tel: 613-991-3073.
- *Science North Solar Observatory*, 100 Ramsey Lake Road, Sudbury, Ontario P3A 2K3. Viewing of the solar spectrum and the Sun in hydrogen-alpha and white light in a darkened theatre. Open most days; Tel: 705-522-3701.

Quebec
- *Observatoire astronomique du mont Mégantic*, Notre-Dame-des-Bois, P.Q. J0B 2E0; Tel: 514-343-6718 (information on summer programs).

Saskatchewan
- *University of Saskatchewan Observatory*, Saskatoon, Saskatchewan S7N 0W0; Tel: 306-966-6434.

A.2.3 Observatories in the United States

Arizona
- *Kitt Peak National Observatory*, 950 North Cherry Avenue, Tucson AZ 85719; Tel: 602-620-5350. Open daily.
- *Lowell Observatory*, 1400 West Mars Hill Road, Flagstaff, AZ 86001; Tel: 602-774-2096. Open Mon–Sat.
- *Lunar and Planetary Laboratory*, University of Arizona, Tucson AZ 85721; Tel: 602-621-4861.
- *Steward Observatory*, University of Arizona, Tucson, AZ 85721; Tel: 602-621-2288.
- *U.S. Naval Observatory*, Flagstaff Station, P.O. Box 1149, Flagstaff, AZ 86002; Tel: 602-779-5132.
- *Fred Lawrence Whipple Observatory*, P.O. Box 97, Amado, AZ 85645; Tel: 602-629-6741.

California
- *Jet Propulsion Lab*, California Institute of Technology, 4800 Oak Grove, Pasadena, CA 91103. Public visits last Sun of each month, 1–4 pm.
- *Lick Observatory*, P.O. Box 85, Mount Hamilton, CA 95140; Tel: 408-274-5062. Open daily.
- *Mount Laguna Observatory*, San Diego State University, San Diego, CA 92115; Tel: 619-265-6182.
- *Palomar Observatory*, Palomar Mountain, CA 92060. Has the famous Hale 200-inch reflecting telescope.

- *Stony Ridge Observatory*, P.O. Box 874 Big Bear, CA 92314; Tel: 714-585-5486. Located on North Shore Drive, Big Bear Lake. Public visiting hours every Friday afternoon.

Connecticut
- *Van Vleck Observatory*, Wesleyan University, Middletown, CT 06457; Tel: 203-347-9411.

Colorado
- *High Altitude Observatory*, National Center for Atmospheric Research, Boulder, CO 80302.
- *Joint Institute for Laboratory Astrophysics*, University of Colorada, CB 390, Boulder, CO 80309; Tel: 303-492-6952.

Delaware
- *Bartol Research Institute and Physics Department*, University of Delaware, Newark, DE 19716; Tel: 302-451-8116.

District of Columbia
- *Derwood Observatory*, Radio Astronomy Group, Carnegie Institute of Washington, 5241 Broad Branch Rd., N.W. Washington, D.C. 20015.
- *U.S. Naval Observatory*, 34th and Massachusetts Avenue NW, Washington, DC 20392; Tel: 202-653-1541. Open Monday evenings only.

Georgia
- *Fernbank Science Center Observatory*, Dekalb County Board of Education, 156 Heaton Park Dr., Atlanta, GA 30307. Large (36-inch) research telescope plus several telescopes for visual observing; Tel: 404-378-4311.

Hawaii
- *Canada–France–Hawaii Telescope*, Mauna Kea, Hawaii 96743, USA. Arrangements to visit should be made in advance by writing to Canada–France–Hawaii Telescope Corporation, P.O. Box 1597, Kamuela, HI 96743, USA; Tel: 808-885-7944.
- *Haleakala Observatories*, University of Hawaii, P.O. Box 135, Kula, Maui, HI 96790.
- *Kauai Observatory*, operated by Radio Astronomy Branch, Goddard Space Flight Center, Greenbelt, MD 20771. Located at Kauai, Hawaii.
- *Mauna Kea Observatory*, Institute for Astronomy, 2680 Woodlawn Dr., Honolulu, HI 96822. Observatory office: 180 Kinoole St., Hilo, HI 96720; Tel: 808-935-3371.

Illinois
- *Dearborn Observatory*, Northwestern University, Evanston, IL 60201; Tel: 708-491-5633. Open Friday evenings Apr–Oct by reservation only.
- *Laboratory for Astrophysics and Space Research*, University of Chicago, Chicago, IL 60637; Tel: 312-702-7823.
- *Lindheimer Astronomical Research Center*, Northwestern University, Evanston, IL 60201; Tel: 708-491-5633. Open Sat 2–4 pm.

Indiana
- *Goethe Link Observatory*, Indiana University, Brooklyn, IN 46111; Tel: 812-335-6911.

Iowa
- *Erwin F. Fick Observatory*, Iowa State University, Ames, IA 50010; Tel: 515-294-5440.
- *Hills Observatory*, University of Iowa, Iowa City, IA 52240; Tel: 319-335-1686.
- *Space Physics Laboratory*, University of Iowa, Iowa City, IA 52242; Tel: 319-335-1918.

Kansas
- *University of Kansas Observatory*, Lawrence, Kansas 66045. Open each clear Friday night.

Louisiana
- *Louisiana State University Observatory*, Baton Rouge, LA 70803; Tel: 504-388-2261.

Maryland
- *NASA/Goddard Space Flight Center*, Visitor Center, Greenbelt, MD 20771; Tel: 301-286-8103.

Massachusetts
- *Five College Radio Astronomy Observatory*, Five College Astronomy Dept., Room 127, Hasbrouck Lab, University of Massachusetts, Amherst, MA. (Also: Amherst, Hampshire, Mt. Holyoke, and Smith Colleges.)
- *Harvard College Observatory and Smithsonian Astrophysical Observatory*, Cambridge, MA 02138. Tel: 617-495-7461. Open third Thursday evening and by reservation.
- *Hopkins Observatory*, Williams College, Williamstown, MA 01267; Tel: 413-597-2105.
- *Maria Mitchell Observatory*, Maria Mitchell Association, P.O. Box 712, Nantucket, MA 02554; Tel: 508-228-9273.

Michigan
- *Space Physics Research Laboratory*, University of Michigan, 2455 Hayward, Ann Arbor, MI 48109; Tel: 313-763-6200.

Nebraska
- *Behlen Observatory*, Dept. of Physics & Astronomy, University of Nebraska, Lincoln, NE 68588; Tel: 402-472-2770.

New Jersey
- *Princeton University Observatory*, Princeton, NJ 08540; Tel: 609-452-3801.

New Mexico
- *NASA-Langley Research Center Meteor Observatory*, Hampton, VA 23365. Located 12 miles east of Las Cruces, New Mexico.
- *Clyde W. Tombaugh Observatory*, New Mexico State University, Dept. of Astronomy, P.O. Box 4500, Las Cruces, NM 88003; Tel: 505-646-4438.

- *Tortugas Mountain Observatory*, New Mexico State University, Dept. of Astronomy, P.O. Box 4500, Las Cruces, NM 88003.
- *Very Large Array*, National Radio Astronomy Observatory, P.O. Box O, Socorro, NM 87801-0387. Open daily.

New York
- *Hartung Boothroyd Observatory*, Space Science Center, Cornell University, Ithaca, NY 14853. 607-255-4206.
- *Dudley Observatory*, 69 Union Ave., Schenectady, NY 12308; Tel: 518-382-7583.
- *C.E. Kenneth Mees Observatory*, University of Rochester, Rochester, NY 14627; Tel: 716-275-4385 for Sat appointment May through Oct only.
- *Vassar College Observatory*, Poughkeepsie, NY 12601; Tel: 914-452-7000 ext. 2060. Visits by appointment.

North Carolina
- *Morehead Observatory*, University of North Carolina, Chapel Hill, NC 27514; Tel: 919-962-2079.

Ohio
- *Perkins Observatory*, Ohio Wesleyan and Ohio State University, Deleware, OH 43015; Tel: 614-363-1257.
- *Ritter Astrophysical Research Center*, The University of Toledo, 2801 Bancroft, Toledo, OH 43606; Tel: 419-537-2650.
- *Warner and Swasey Observatory*, Case Western Reserve University, 1975 Taylor Road, East Cleveland, OH 44112. Tel: 216-451-5624.

Oregon
- *Pine Mountain Observatory*, Dept. of Physics, University of Oregon, Eugene, OR 97403; Tel: 503-382-8331. Open Fri–Sun evenings.

Pennsylvania
- *Allegheny Observatory*, 159 Riverview Avenue, Pittsburgh, PA 15214; Tel: 412-321-2400.
- *Astrophysics Computer Laboratory*, Department of Physics & Atmospheric Science, Drexel University, Philadelphia, PA 19104; Tel: 215-895-2707.
- *Sproul Observatory*, Swarthmore College, Swarthmore, PA 19081; Tel: 215-328-8272.
- *Villanova University Observatory*, Villanova, PA 19085; Tel: 215-645-4820.

Puerto Rico
- *Arecibo Observatory*, National Radio and Ionospheric Center, P.O. Box 995, Arecibo, Puerto Rico 00612. Operated by Cornell University.

Tennessee
- *Arthur J. Dyer Observatory*, Vanderbilt University, 1000 Oman Drive, Brentwood, TN 37027; Tel: 615-373-4897. Open Mon–Fri.

Texas
- *McDonald Observatory/W.L. Moody Visitor's Center*, P.O. Box 1331, Fort Davis, TX 79734; Tel: 915-426-3263. Open daily.

Virginia
- *Leander McCormick Observatory*, University of Virginia, P.O. Box 3818, Charlottesville, VA 22903; Tel: 804-924-7494. Open 1st and 3rd Fridays of each month.
- *NASA Langley Research Center*, Mail Stop 480, Hampton, VA 23365; Tel: 804-865-2855. Open daily.

Washington
- *Battelle Observatory*, Battelle-Northwest Labs, P.O. Box 999, Battelle Blvd., Richland, WA 99352. Located 20 miles west of Richland; Tel: 509-946-2383 or 942-7301 about tours.
- *Goldendale Observatory*, 1602 Observatory Dr., Goldendale, WA 98620; Tel: 509-773-3141.

West Virginia
- *National Radio Astronomy Observatory*, P.O. Box 2, Green Bank, WV 24944-0002; Tel: 304-456-2011. Open daily, mid-June to mid-August.

Wisconsin
- *Washburn Observatory*, University of Wisconsin, 1401 Observatory Drive, Madison, WI 53704; Tel: 608-262-WASH. Open Wednesday evenings.
- *Yerkes Observatory*, P.O. Box 258, Williams Bay, WI 53191-0258; Tel: 414-245-5555.

Wyoming
- *Wyoming Infrared Observatory*, Department of Physics and Astronomy, University of Wyoming, Laramie, WY 82071; Tel: 307-766-6150.

A.3 Astronomical Societies and Clubs

International
- *International Amateur–Professional Photoelectric Photometry Association* (IAPPPA) — An organization which coordinates photometric observations by serious amateurs and small colleges, with the goal of producing scientifically useful data. They publish a *Communication* which is of especial interest to smaller observatories wishing to carry out research. Contact: Dr. Douglas S. Hall, Dyer Observatory, Vanderbilt University, Nashville, TN 37235, USA; Tel: 615-373-4897.
- *International Astronomical Union* (IAU) — The primary international federation of professional astronomers. Commission 46 of the IAU deals with educational matters and supports a variety of worldwide educational activities. It also publishes a regular newsletter on international astronomy education which is sent regularly to its members. The editor of the newsletter is Professor John Percy, Dept. of Astronomy, University of Toronto, Canada.
- *International Dark-Sky Association* — Non-profit organization devoted to fighting light pollution. Contact: David Crawford, 3545 N. Stewart, Tucson, AZ 85716.
- *International Occultation Timing Association* (IOTA) — Contact: Terri and Craig McManus, 1177 Collins, Topeka, KS; Tel: 913-232-3693.

- *International Planetarium Society* (IPS) — Holds local and national conferences, publishes a comprehensive listing of planetariums, and runs a job information service. Also publishes *The Planetarian*. Contact: Hansen Planetarium, 15 S. State St., Salt Lake City, UT 84111, USA.
- *International Union of Amateur Astronomers* (IUAA) — Formed in 1969 to coordinate the activities of amateur astronomers worldwide. Membership is open to individuals and corporations. Contact: The Secretary, Mr. V. Deasy, Ard Faill, 633 Howth Road, Raheny, Dublin 5, Eire.
- *Society of Amateur Radio Astronomers* (SARA) — An international organization which seeks persons interested in radio astronomy research. Contact: SARA Membership Services, P.O. Box 6319, Long Island City, NY 11106; Tel: 718-545-3455.

Canada
- *Canadian Astronomical Society* (CAS) — A professional organization which supports many astronomy education activities. Contact: Norman Broten, CAS Secretary, 48 Pineglen Crescent, Nepean, Ontario, Canada K2E 6X9.
- *NCL CAN AM (Noctilucent Cloud Observers)* — Contact: Mark Zalcik, #2 14225 82nd St., Edmonton, Alberta T5E 2V7.
- *The Planetarium Association of Canada* (PAC) — 190 Rupert Ave., Winnepeg R3B 0N2.
- *Royal Astronomical Society of Canada* (RASC) — A Canadian society open to amateurs and professional astronomers. Sponsors a bimonthly *Journal of the RASC*, the *National Newsletter*, and the *Observer's Handbook*. Address: 136 DuPont St., Toronto, Ontario M5R 1V2.

United States
- *Amateur Satellite Observers* — Contact: Jim Hale, HCR 65, Box 261-B, Kingston, AR 72742.
- *American Association of Physics Teachers* (AAPT) — An organization of physicists who are also teachers. They have an Astronomy Committee which is very active in astronomy education. Address: Executive Office, Dept. of Physics and Astronomy, University of Maryland, College Park, MD 20742.
- *American Association of Variable Star Observers* (AAVSO) — An organization composed of both amateur and professional astronomers interested in variable stars. Contact: Janet Mattei, 25 Birch Ave., Cambridge, MA 02138; Tel: 617-354-0484.
- *American Astronomical Society* (AAS) — An organization of professional astronomers primarily in North America. Holds regular semi-annual meetings and sponsors publication of several technical journals. Main address: 1816 Jefferson Place, NW, Washington, DC 20036. Its education officer is Dr. Charles Tolbert, Leander McCormick Observatory, University of Virginia, P.O. Box 3818, University Station, Charlottesville, VA 22903-0818.
- *American Lunar Society* (ALS) — Contact: Francis Graham, P.O. Box 209, East Pittsburgh, PA 15112; Tel: 412-829-7455.
- *American Meteor Society* — Contact: David Meisel, Dept. of Physics & Astronomy, S.U.N.Y.–Geneseo, NY 14454.
- *Association of Astronomy Educators* (AAE) — Affiliated with the NSTA, it sponsors a newsletter of interest to astronomy educators, workshops and short courses

in astronomy in connection with NSTA meetings. Contact: Robert Allen, Physics Dept., Cowley Hall, University of Wisconsin, La Cross, WI 54601.
- *Association of Lunar and Planetary Observers* (ALPO) — An amateur organization specializing in solar system phenomena. Contact: Harry D. Jamieson, P.O. Box 143, Heber Springs, AR 72543; Tel: 501-362-7624.
- *Astronomical League* (AL) — A federation of local amateur astronomy clubs. Contact: Merry Edenton-Wooten, 6235 Omie Circle, Pensacola, FL 32504; Tel: 904-477-8859.
- *Astronomical Society of the Pacific* (ASP) — An association of amateur and professional astronomers committed to the goal of public education in astronomy. Publishes a technical journal *Publications of the ASP* and also a non-technical magazine *Mercury*. ASP is one of the best general sources of teaching materials (books, information packets, audio-visual aids, etc.) on astronomy education. Address: 390 Ashton Ave., San Francisco, CA 94112; Tel: 415-337-1100.
- *Aurora Alert Hotline* (AAH) — Contact: David Huestis, 57 Manley Dr., R.R. #1, Box 232A, Pascoag, RI 02859; Tel: 401-568-9370.
- *Earthwatch* — Contact: Blue Magruder, 680 Mt. Auburn St., Box 403N, Watertown, MA 02272; Tel: 617-926-8200.
- *Independent Space Research Group* (ISRG) — P.O. Box 23083, Rochester, NY 14692; Tel: 716-464-0125.
- *Meteoritical Society* (MS) — Contact: H.Y. McSween, Geological Sciences Dept., University of Tennessee, Knoxville, TN 37996.
- *National Aeronautics and Space Administration* (NASA) — A major government agency from which is available a copious amount of astronomy education material such as slides, prints, movies, and publications that have resulted from NASA activities. Publishes *Educational Briefs* and *Report to Educators* free of charge to interested educators. Write: NASA Educational Programs Division, Office of Public Affairs, Code FC-9, NASA, Washington, DC 20546.
- *National Space Society* (NSS) — Contact: Aleta Jackson, 922 Pennsylvania Ave., SE., Washington, DC 20003; Tel: 202-543-1900.
- *Niagara Frontier Council of Amateur Astronomical Associations* — Contact: Ed Lindberg, 113 Maple Dr., Bowmansville, NY 14026; Tel: 716-633-6725.
- *Piedmont Advocacy for Space* (PAS) — Contact: S.J. Redhead, P.O. Box 337, Fair Play, SC 29643; Tel: 803-972-3026.
- *Planetary Society* — Address: 65 N. Catalina Ave., Pasadena, CA 91106; Tel: 818-793-5100.
- *Problicom Sky Survey* (PSS) — Contact: Ben Mayer, 1940 Cotner Ave., Los Angeles, CA 90025; Tel: 213-478-2524.
- *Society for Scientific Exploration* (SSE) — Contact: L.W. Fredrick, P.O. Box 3818, Charlottesville, VA 22903; Tel: 804-924-4905.
- *Spaceweek National Headquarters* — Address: P.O. Box 58172, Houston, TX 77258.
- *Sunsearch–Supernova Search* — Contact: Steve Lucas, 14400 S. Kolin Ave., Midlothian, IL 60445.
- *Webb Society North America* — Contact: Ronald J. Morales, 1440 S. Marmora Ave., Tucson, AZ 85713; Tel: 602-628-1077.

- *Western Amateur Astronomers* (WAA) — Contact: Margaret Matlack, 13617 E. Baily, Whittier, CA 90601.
- *World Space Foundation* (WSS) — Contact: Jack Child, P.O. Box Y, South Pasadena, CA 91030; Tel: 818-357-2878.

United Kingdom
- *Association for Astronomy Education* (AAE) — Formed in 1981 to promote greater interest in astronomy in education. Contact: Bob Kibble, AAE Secretary, 34 Aeland Crescent, Denmark Hill, London SE5 8EQ.
- *British Association of Young Scientists* (BAYS) — Though not specifically an astronomical society, it is aimed at persons aged 11 to 18 who are interested in science and technology. Contact: BAYS Officer, British Association for the Advancement of Science, Fortress House, 23 Savile Row, London W1X 1AB.
- *British Astronomical Association* (BAA) — Founded in 1890, it is the leading national society for amateur astronomers. Contact: The Secretary, BAA, Burlington House, Picadilly, London W1V 0NL.
- *British Interplanetary Society* (BIS) — Formed in 1933, it aims to investigate the possibilities of interplanetary flight and to promote relevant engineering and scientific research. Contact: The Secretary, BIS, 27/29 South Lambeth Road, London SW8 1SZ.
- *British Meteor Society* (BMS) — Since 1969 the society has strived to further the advancement of meteor astronomy and allied sciences. Contact: The Director, BMS, 26 Adrian Street, Dover, Kent CT17 9AT.
- *British Society for the History of Science* (BSHS) — Though not specifically devoted to astronomy, this society will appeal to those with a scholarly interest in the history of astronomy. Contact: The Secretary, BSHS, 31 High Street, Stanford-in-the-Vale, Faringdon, Oxon, SN7 8LH.
- *Federation of Astronomical Societies* (FAS) — Formed in 1974 to bring together astronomical societies and groups for their mutual benefit, and to give help and advice to member societies and groups. Publishes a regular newsletter and an annual handbook which contains information on places to visit, equipment suppliers, visual aid sources, speakers, and societies, and also addresses of organizers of local societies. Contact: FAS Secretary, 1 Tal-y-bont Road, Ely, Cardiff CF5 5EU.
- *The Institute of Physics* (IOP) — Publishes *Physics Education*. Address: 47 Belgrave Square, London SW1 8QX.
- *Junior Astronomical Society* (JAS) — Intended for beginners of all ages in astronomy. The Society publishes the quarterly magazine *Popular Astronomy* and circulars with the latest news, and organizes meetings and visits. Contact: The Secretary, JAS, 10 Swanwick Walk, Tadley, Basingstoke, Hants RG26 6JZ.
- *Royal Astronomical Society* (RAS) — Founded in 1820, the major society for professional astronomers and geophysicists in the UK, but with members worldwide. Publishes the *Memoirs, Monthly Notices*, and the *Quarterly Journal*. Applications for Fellowship must be supported by three other Fellows in the Society. Application to Junior Membership is open to persons between ages 18 and 25 and needs to be supported by one Fellow of the Society. For further details, contact: Membership Secretary, RAS, Burlington House, Piccadilly, London W1V 0NL.

- *Scientific Instruments Society* (SIS) — Formed in 1983 to bring together people with a specific interest in historic scientific instruments. Contact: SIS Executive Secretary, Neville House, 42/46 Hagley Road, Birmingham B16 8PZ.
- *Webb Society* — Formed in 1967 and named after the Rev. T.W. Webb, the Society aims to unite amateur observers of double stars and deep sky objects. Contact: The Secretary, S.J. Hynes, Webb Society 8 Cormorant Close, Sydney, Crewe, Cheshire OW1 1LN.
- *William Herschel Society* (WHS) — Founded in 1978, it aims to continue maintaining Herschel's house and workshop in Bath and to further research into Herschel's life and work. Write: WHS Secretary, 19 New King Street, Bath BA1 2BL.

Australia
- *Astronomical Society of Australia* (ASA) — Contact: President A.G. Little, University of Sydney, Sydney, NSW 2006.
- *Astronomical Society of Western Australia* (ASWA)
- *Royal Society of New South Wales* (RSNSW)

India
- *Astronomical Society of India* (ASI) — Contact: Pres. J.V. Narlikar, Tata Institute, Homi Bhabha Road, Colaba, Bombay 400 005.

Germany
- *Astronomische Gesellschaft* (AG) — Contact: G. Klare, Landessternwarte, Königstuhl, D–69117 Heidelberg.
- *Vereinigung der Sternfreunde* (VdS) — Contact: Peter Stättmayer, Volkssternwarte, Anzinger Str. 1, D–81671 München.

Japan
- *Japan Astronomical Study Association* — Contact: Keiichi Saijo, National Science Museum, Ueno Park, Taito, Tokyo 110.
- *Oriental Astronomical Association* (OAA) — Contact: Ichiro Hasegawa, 2-3-11, Saidaiji-Nogamicho, Nara 631.

South Africa
- *Astronomical Society of Southern Africa* (ASSA)

New Zealand
- *Royal Astronomical Society of New Zealand* (RASNZ) — P.O. Box 3181, Wellington.

A.4 General List of Sources for Mechanical Models and Exhibit Items

United Kingdom
- *Armagh Planetarium*, College Hill, Armagh, N. Ireland.
- *Astro Instruments*, 45 Derby Road, Portsmouth, Hants.
- *Bretmain Ltd.*, 99b Hamilton Road, Felixstowe, Suffolk IP11 7BL.
- *Daily Telegraph*, 135 Fleet Street, London EC4 or bookshops.

- *George Philips*, Stanfords Ltd., Long Acre, London EC2 or bookshops.
- *London Planetarium*, Marylebone Road, London NW1 5LR; Tel: 081-486 1121.
- *London Schools Planetarium & Advisory Centre*, Wandsworth School, Sutherland Grove, London SW18 5PT. Tel: 081-788 4253.
- *MacDonald Educational*, Paulton, Bristol BS18 5BR. (Learning through Science packs).
- *MacMillan Educational*, Houndmills, Baskingstoke, Hants.
- *Mills Observatory*, Balgay Park, Glamis Road, Dundee DD2 2UB; Tel: 0382 67138.
- *Space Frontiers Ltd.*, 30 Fifth Avenue, Havant, Hants.

Canada, United States
- *Astrosystems*, 1536 Meeker Dr., P.O. Box 1183, Longmont, CO 80501; Tel: 303-678-5339.
- *Carolina Biological Supply Company*, 2700 York Road, Burlington, NC 27215, USA; Tel: 919-584-0381. Also Box 187, Gladstone, OR 97027, USA. Tel: 503-656-1641.
- *Central Scientific Company* (CENCO), 11222 Melrose Ave., Franklin Park, IL 60131-1364, USA; Tel: 708-451-0150 or 800-262-3626.
- *David Chandler Co.*, P.O. Box 309, La Verne, CA 91011, USA; Tel: 714-946-4814.
- *Edmund Scientific Co.*, 101 E. Gloucester Pike, Barrington, NJ 08007-1380; Tel: 609-573-6259.
- *Farquhar Globes*, 5007 Warrington Ave., Philadelphia, PA 19143. Celestial and other globes.
- *Fisher Scientific*, Educational Materials Division, 4901 W. LeMoyne Street, Chicago, IL 60651, USA; Tel: 800-621-4769.
- *Hubbard Scientific Company*, P.O. Box 104, Northbrook, IL 60065, USA.
- *Klinger Educational Products Corp.*, 83-45 Parsons Boulevard, Jamaica, NY 11432, USA.
- *MMI Corporation*, Dept. ST89, P.O. Box 19907, Baltimore, MD 21211; Tel: 301-366-1222. Extensive film list
- *Neptune Pacific*, 953 E. Colorado Blvd., Suite 201, Pasadena, CA 91106, USA; Tel: 818-794-4531.
- *Pasco Scientific*, 1876 Sabre Street, Hayward, CA 94545, USA; Tel: 415-786-2800. TWX: 910-383-2040.
- *Sargent-Welch Scientific Co.*, 7300 Linder Ave., Skokie, IL 60076. Educational and demonstration equipment.
- *Schoolmasters Science*, 745 State Circle, Box 1941 Ann Arbor, MI 48106; Tel: 800-521-2832; Fax: 313-761-8711.
- *Science Kit, Inc.*, Tonawanda, NY 14150. Educational and demonstration materials.
- *Sci/Space Craft International*, 953 E. Colorado Blvd., No. 201, Pasadena, CA 91106; Tel: 818-794-0177. High-resolution laser-cut educational model kits.
- *Ward's Natural Science Establishment, Inc.*, 5100 W. Henrietta Road, P.O. Box 92912, Rochester, NY 14692-9012, USA; Tel: 800-962-2660.

A.5 General List of Sources for Audio-Visual Aids

United Kingdom
England
- *Astro Art*, 99 Southam Road, Hall Green, Birmingham B28 0AB; Tel: 021-7771802. Variety of illustrative material, original paintings, prints, etc. by David A. Hardy.
- *Astro Slides*, 58 California, Nine Mile Ride, Finchampstead, Berkshire RG11 5HT. Sets of slides of planets, etc. from US originals. List available, mail order only.
- *Bretmain*, 99B Hamilton Road, Felixstowe, Suffolk IP11 7BL.
- *British Universities Film & Video Council*, 55 Greek Street, London, W.1.
- *Daily Telegraph*, 135 Fleet Street, London EC4.
- *Earth and Sky*, 21A West End, Hebden Bridge, West Yorkshire HX7 8UQ; Tel: 0422 845443. Wide range of astronomical photographs, slides, charts, books, etc., available in the shop or by mail order. Send two first-class stamps for current lists.
- *Federation of Astronomical Societies*, Mr. Ken Marcus, 5 Cedars Gardens, Brighton, East Sussex BN1 6YD. Set of 73 general slides taken by amateur observers.
- *Focal Point Audio-Visual Ltd.* 251 Copnor Rd., Portsmouth, Hants. PO3 5EE; Tel: 0705 665249. Slides and tape/slide sets.
- *The Geological Museum*, Exhibition Rd., South Kensington, London SW7 2DE; Tel: 071-938 8765. Small reference collection of NASA prints of the planets which may be available for loan. Contact Mr. Pulsford; Tel: 071-938 9035. Museum Bookshop sells 35-mm slides of space subjects; Tel: 071-938 9123.
- *Hammonds A/V & Video Services*, 60 Queens Road, Watford, Herts. WD1 2LA; Tel: 0923 39733.
- *Hatfield Polytechnic Observatory*, c/o Dr. Chris Kitchin, Bayfordbury, Nr Hertford, Herts SG13 8LD; Tel: 0992 558451 ext. 334. Slides covering all the constellations visible from the UK.
- *Jodrell Bank Visitor Centre*, Lower Withington, Macclesfield, Cheshire; Tel: 0477 71339.
- *Liverpool Museum Planetarium*, National Museums & Galleries on Merseyside, William Brown Street, Liverpool L3 BEN; Tel: 051-207 0001.
- *The London Planetarium*, Merchandise Department, London Planetarium, Marylebone Road, London NW1 5LR; Tel: 071-486 1121. Offers posters, charts, books, etc.
- *London Schools Planetarium & Advisory Centre*, Wandsworth School, Sutherland Grove, London SW18 5PT. Tel: 081-788 4253.
- *MacDonald Educational*, Paulton, Bristol BS18 5BR. Learning through Science packs.
- *MacMillan Educational*, Houndmills, Baskingstoke, Hants.
- *National Portrait Gallery*, 2 St. Martins Place, London WC2. Sells slides.
- *The Observatory Magazine*, The Editors of *The Observatory*, Rutherford Appleton Laboratory, Chilton, Didcot, Oxon, OX11 0QX; Tel: Abingdon 0235 21900.
- *George Philips*, Stanfords Ltd., Long Acre, London EC2.
- *The Royal Astronomical Society*, Burlington House, London, W1V 0NL. Limited quantities of posters.

- *Royal Greenwich Observatory*, University of Cambridge, Madingley Road, Cambridge CB3 0HA; Tel: 0223 337548. Formerly located at Herstmonceux Castle, E. Sussex. Wide range of material.
- *The Science Museum*, South Kensington, London SW7 2DD. Museum Shop sells some postcards and prints; Tel: 071-938 8186 or 071-938 8187. May also be obtained from Mail Order Department.
- *Science Museum Library*, Photo Orders Service, South Kensington, London SW7 5NH; Tel: 071-938 8220 (Photographic Order Service). Sells black-and-white photographs, black-and-white slides, and color transparencies, mainly of objects in the Museum's collections. Lists available.
- *The Slide Centre Ltd.*, Ilton 5H, Ilminster, Somerset, TA19 9BR; Tel: 0460 57151. Slide sets and some computer software for educational use.
- *Spacecharts*, Newton Tony, Salisbury, Wilts SP4 OHF. Tel: 098-064 672. Full color wall charts of planets, Halley's comet, space shuttle.
- *Space Frontiers Ltd.*, 30 Fifth Avenue, Havant, Hants.
- *Spaceprints*, 117A High Street, Norton, Stockton on Tees, Cleveland; Tel: 0642 555401. Supplier of NASA photographs, slide sets, videos, and posters. Produces Spaceprints Calendar.
- *Terra Firma Cassettes*, 55 Bolingbroke Road, London, W14.
- *Woodmansterne Ltd.*, Watford Business Park, Watford WD1 8RD. Color slides.

Northern Ireland
- *Armagh Planetarium*, College Hill, Armagh BT61 9DB; Tel: 0861 524725. Offers exceptionally wide range of materials, including slide sets on astronomy and spaceflight subjects, videotapes and laser discs, posters, and postcards.

Scotland
- *Mills Observatory*, Balgay Park, Glamis Road, Dundee DD2 2UB; Tel: 0382 67138.
- *The Royal Observatory, Edinburgh*, The Visitor Centre, Blackford Hill, Edinburgh EH9 3HJ; Tel: 031-668 8100. Books, posters slides, charts, etc. Also prints, slides and postcards from UK Schmidt Telescope Unit.

Wales
- *Drake Educational Productions*, St. Fagans Road, Fairwater, Cardiff CF5 3AE; Tel: 0222 560333. Sound filmstrips, tape/slide sets, and astronomical wallcharts.
- *DS Ltd.*, The George Building, Normal College, Bangor, Gwynedd.

Canada, United States
- *Apollo Space Systems*, 675 Station Rd., Bellport, NY 11713; Tel: 516-467-8033.
- *Astro Cards*, P.O. Box 35, Natrona Hts., PA 15065; Tel: 412-295-4126.
- *Astro-Info*, 1090 Ch. Pincourt, Mascouche, Quebec J7L 2X7.
- *Astronomical Society of the Pacific*, 390 Ashton Ave., San Francisco, CA 94112.
- *BFA Educational Media*, 468 Park Ave. S., New York, NY 10016.
- *Carolina Biological Supply Company*, 2700 York Road, Burlington, NC 27215; Tel: 919-584-0381. Also Box 187, Gladstone, OR 97027. Tel: 503-656-1641.
- *Central Scientific Company* (CENCO), 11222 Melrose Ave., Franklin Park, IL 60131-1364; Tel: 708-451-0150 or 800-262-3626.
- *David Chandler Co.*, P.O. Box 309, La Verne, CA 91750; Tel: 714-946-4814.

- *Educational Audiovisual, Inc.*, Pleasantville, NY 10570.
- *E.M.E. Science*, P.O. Box 17, Pelham, NY 1083.
- *Everything in the Universe*, 5248 Lawton Ave., Oakland, CA 94618; Tel: 415-547-6523.
- *Fisher Scientific*, Educational Materials Division, 4901 W. LeMoyne Street, Chicago, IL 60651; Tel: 800-621-4769.
- *Hale Observatories*, Caltech, Bookstore 1-51, 1201 East California, CA 91109.
- *Hansen Planetarium*, 15 South State St., Salt Lake City, UT 84111.
- *Hubbard Scientific Company*, P.O. Box 104, Northbrook, IL 60065.
- *Indiana University Audiovisual Center*, Bloomington, IN 47401.
- *Kalmia Company*, 21 West Circle, Concord, MA 01742.
- *Karol Media*, 625 From Rd., Paramus, NJ 07652.
- *Kitt Peak National Observatory*, Photo Dept., P.O. Box 4130, Tucson, AZ 85717.
- *Landmark Pictures, Inc.*, 72 Mallard Dr., Greenwich, CT 06830.
- *Learning Corp. of America*, 1350 Ave. of the Americas, New York, NY 10019.
- *Lick Observatory*, University of California, Santa Cruz, CA 95064.
- *Media Guild*, 11526 Sorrento Valley Rd., Suite J, San Diego, CA 92121.
- *MMI Corporation*, 2950 Wyman Parkway, P.O. Box 19907, Baltimore, MD 21211; Tel: 301-366-1222.
- *Modern Learning Aids*, Box 92912, Rochester, NY 14692.
- *National Audiovisual Center*, General Services Administration, Washington, DC 20409.
- *National Film Board of Canada* (Commercial Division), P.O. Box 6100 Station A, Montreal, P.Q. H3C 3H5.
- *National Optical Astronomy Observatory*, Public Inf. Office, P.O. Box 26732, Tucson, AZ 85726.
- *National Science Foundation*, Public Affairs & Public., Washington, DC 20036.
- *Optical Data Corporation* (formerly *Video Vision Associates*), 30 Technology Dr., P.O. Box 4919, Warren, NJ 07060; Tel: 800-524-2481 or 201-668-0022.
- *Optiken International*, 900 S. San Gabriel Blvd., San Gabriel, CA 91776.
- *RHR Filmedia*, 1212 6th Ave., New York, NY 10036.
- *Sargent-Welch Scientific Co.*, 7300 N. Linder Ave., Skokie, IL 60076.
- *Schoolmasters Science*, 745 State Circle Box 1941, Ann Arbor, MI 48106; Tel: 800-521-2832.
- *Science Graphics*, P.O. Box 7516, Bend, OR 97708; Tel: 503-389-5652.
- *Sky Publishing Corp.*, 49 Bay State Road, Cambridge, MA 02238. Sky maps and visual materials.
- *Space Photos*, Dept. V-3, 2608 Sunset Blvd., Houston, TX 77005.
- *Tersch Enterprises*, P.O. Box 1059, Colorado Springs, CO 80901; Tel: 719-597-3603.
- *UC Extension Media Center*, Univ. of Calif., Berkeley, CA 94720.
- *University of Illinois Visual Aids Service*, 1325 S. Oak St., Champaign, IL 61820.
- *Walt Disney Educ. Media Marketing Division*, 500 S. Buena Vista St., Santa Barbara CA 91521.
- *Ward's Natural Science Establishment, Inc.*, 5100 W. Henrietta Road, P.O. Box 92912, Rochester, NY 14692-9012. Tel: 800-962-2660.

A.6 Telescopes and Observing Equipment

A.6.1 Telescopes and Accessories

United Kingdom
- *Astro Instruments*, 45 Derby Road, Portsmouth, Hants.
- *Astro Promotions Ltd.*, 24 Old Bedford Road, Luton, Beds.
- *Beacon Hill Telescopes*, 112 Mill Road, Cleethorpes, South Humberside, DN35 8JD; Tel: 0472 692959.
- *Broadhurst, Clarkson & Fuller*, Telescope House, 63 Farringdon Road, London EC1M 3JB; Tel: 071-405 2156.
- *Norman Fischer Ltd.*, 5 Dagmar Road, London SE25 6HZ; Tel: 081-771 0477.
- *Charles Frank*, Head Office, Ronald Lane, Saxmundham, Suffolk, IP17 2NL; Tel: 0728 3506.
- *H.N. Irving & Son*, 258 Kingston Road, Teddington, Middx.
- *Kay Optical*, 89b London Road, Morden, Surrey.
- *London Planetarium*, Marylebone Road, London NW1 5LR; Tel: 071-486 1121.
- *Optical Instruments (Balham) Ltd.*, 6 Weir Road, Balham, London, SW12 0NA; Tel: 081-673 8513.
- *Opticraft Ltd.*, Unit 4, Queen Street Mill, Harlesyke, Burnley; Tel: 0282 412215.
- *Opticron*, P.O. Box 81, St. Albans, Herts. AL1 3NT; Tel: 0727 56516.
- *Orion Optics*, Unit 3M, Zan Industrial Park, Wheelock, Sandbach, Cheshire CW11 0QD; Tel: 0270 768474.
- *Osborne Optics*, 139 Dean House, Eastfield Avenue, Walker, Newcastle upon Tyne NE6 4UU; Tel: 091-263 8826.
- *Henry Wildey*, 34 Warners Avenue, Broxbourne, Herts. EN11 8LR; Tel: 0992 465886.
- *Wise Instruments Ltd.*, Unit 9, Hollins Business Centre, Marsh Street, Stafford ST16 3BG; Tel: 0785 223535.

Canada
- *Cosman & Associates*, Islington, Ontario M9B 1A8.
- *Efstonscience*, 3350 Dufferin Street, Toronto, Ontario M6A 3A4; Tel: 800-263-2935 or 416-787-4581.
- *Khan Scope Centre* (formerly *Scope City Canada*), 247 Marlee Ave., Unit 201, Toronto, Ontario M6B 4B8; Tel: 416-783-4140.
- *Northern Lites*, 801 Stanehill Place, R.R.1, Victoria, BC V8X 3W9.
- *Perceptor*, Brownsville Junction Plaza, Box 38, Suite 103, Schomberg, Ontario L0G 1T0; Tel: 416-939-2313.
- *Quasar Optics*, 7220A Fairmount Dr. S.E., Calgary, Alberta T2H 0X7; Tel: 403-255-7633.
- *Ruby Optics*, Box 2136, Kingston, Ontario K7L 5J9, Canada; Tel: 613-544-5857.
- *Science City*, 50 Bloor Street West, Toronto, Ontario.
- *Sky Instruments*, M.P.O. Box 3164, Vancouver, BC V6B 3X6; Tel: 604-270-2813.

– *Spartan Scientific Ltd.*, 116 Viceroy Rd., Unit 5, Bldg. A, Concord, Ontario L4K 2M1; Tel: 416-738-0393.
– *Star Trak Instruments*, P.O. Box 3234, Stn. C, Hamilton, Ontario L8H 7K6.
– *Telescopes TransCanada*, P.O. Box 823, Aurora, Ontario L4G 4J9; Tel: 416-939-8274.
– *Win Optic Vision Inc.*, 620 Alden Rd., Suite 102, Markham, Ontario L3R 9R7.

United States
– *Ad-Libs Astrometics*, 2401 Tee Circle, Suite 105, Norman, OK 73069; Tel: 405-364-0858 or 800-422-7876.
– *Adorama*, 42 W. 18th St. New York, NY 10011. Tel: 212-741-0052.
– *Aries Optics*, Rt. 1, Box 143G, Palouse, WA 99161. Tel: 509-878-1713.
– *Astro-Computer Control*, RD #1, Alexandria, PA 16611. Tel: 814-669-4483.
– *Astronomics*, 2401 Tee Circle, Suites 105/106, Norman, OK 73069; Tel: 405-364-0858.
– *Astro-Physics*, 7470 Forest Hills Road, Loves Park, IL 61111; Tel: 815-282-1513.
– *Astro-Systems*, P.O. Box 1183 M, Longmont, CO 80501. Tel: 303-678-5339.
– *Astro-Tech*, 101 West Main, P.O. Box 2001, Ardmore, OK 73402; Tel: 405-226-3074.
– *Astroworks*, P.O. Box 86, Cloudcroft, NM 88317. Tel: 505-682-2218.
– *Astro World*, 5126 Belair Rd., Baltimore, MD 21206. Tel: 301-483-5100.
– *Basic Concepts in Astronomy*, 11278 East Meadow Glen Way, Escondido, CA 92026.
– *Berger Bros. Camera Exchange*, 209 Broadway (Route 110), Amityville, NY; Tel: 800-262-4160 or 516-264-4160.
– *California Telescope Company*, P.O. Box 1338, Burbank, CA 91507; Tel: 800-843-4780.
– *Carolina Biological Supply Company*, 2700 York Road, Burlington, NC 27215; Tel: 919-584-0381. Also Box 187, Gladstone, OR 97027. Tel: 503-656-1641.
– *Celestron International*, 2835 Columbia Street, Torrance, CA 90503; Tel: 213-328-9560, or 800-421-1526 (Continental USA); Fax: 213-212-5835; Telex: 182471.
– *Central Scientific Company* (CENCO), 11222 Melrose Ave., Franklin Park, IL 60131-1364; Tel: 708-451-0150 or 800-262-3626.
– *Century Telescope*, 12555 Harbor Blvd., Garden Grove, CA 92640; Tel: 714-530-3861.
– *Chicago Optical*, P.O. Box 1361, Morton Grove, IL 60053. Tel: 708-827-4846.
– *City Camera*, 15336 W. Warren, Dearborn, MI 48126. Tel: 313-846-3922.
– *Cosmic Connections*, 1460 N. Farnsworth, Aurora, IL 60505; Tel: 708-851-5353, 800-634-7702.
– *Coulter Optical, Inc.*, P.O. Box K, Idyllwild, CA 92349-1107; Tel: 714-659-4621.
– *D & G Optical*, 6490 Lemon St., East Petersburg, PA 17520; Tel: 717-560-1519.
– *Denny's Astro Mart*, 832 Sydney Baker, Kerrville, TX 78028; Tel: 512-896-6377.
– *Dobbins Instrument Co.*, 5168 Lynd Ave., Lyndhurst, OH 44124; Tel: 216-449-5730 or 216-631-6611.
– *E & W Optical Inc.*, 2420 E. Hennepin Ave., Minneapolis, MN 55413; Tel: 612-331-1187.

- *Eagle Optics*, 6109 Odana Rd., Madison, WI 53719; Tel: 608-271-4751.
- *Eastern Tele-Optics*, P.O. Box 426, Springfield, PA 19064; Tel: 215-284-1725.
- *Edmund Scientific*, 101 E. Gloucester Pike, Barrington, NJ 08007-1380; Tel: 609-573-6259.
- *Fisher Scientific*, Educational Materials Division, 4901 W. LeMoyne Street, Chicago, IL 60651; Tel: 800-621-4769.
- *Focus Camera*, 4421 13th Ave., Brooklyn, NY 11219. Tel: 718-436-6262.
- *Fujinon Inc.*, 10 High Point Drive, Wayne, NJ 07470. Tel: 201-633-5600. Specializes in high quality astronomical binoculars.
- *Galaxy Optics*, P.O. Box 2045, Buena Vista, CO 81211. Tel: 719-395-8242. Supplier of mirror optics for reflecting telescopes.
- *Great Plains Instruments*, 2321 C. Court, Enid, OK 73703; Tel: 405-237-4034.
- *Handmade Telescopes*, 2906 Spooky Nook Rd., Manheim, PA 17545; Tel: 1-800-257-0702.
- *A. Jaegers*, 691S Merrick Rd., Lynbrook, NY 11563. Telescope making equipment and accessories.
- *Jena Scientific Instruments, Inc.*, 820 2nd Ave., New York, NY 10017; Tel: 212-867-3051.
- *Jim's Mobile Ind.*, 1960 County Road 23, Evergreen, CO 80439; Tel: 303-277-0304; Fax: 303-526-9140.
- *Jupiter Telescope Co.*, 818 S., U.S. Highway 1, Suite 4-237, Jupiter, FL 33477; Tel: 407-881-1365. Advanced design large telescopes.
- *Kenneth Novak & Co.*, Box 69X, Ladysmith, WI 54848. Tel: 715-552-5102.
- *Krauth Precision Instruments*, 528-30 Main St., South Weymouth, MA 02190; Tel: 617-331-3795.
- *Lewis-Michaels Engineering*, 48 Delemere Blvd., Fairport, NY 14450; Tel: 716-425-3470. Telescope-making supplies and accessories.
- *Lorraine Precision Optics*, 1319 Libby Lane, New Richmond, OH 45157.; Tel: 513-553-4999. Source of Schiefspieglers.
- *Lumicon*, 2111 Research Dr. #5S, Livermore, CA 94550. Tel: 415-447-9570. Major distributer of telescopes and supplies.
- *Meade Instruments Corporation*, 1675 Toronto Way, Costa Mesa, CA 92626; Tel: 714-556-2291.
- *F.C. Meischner Co.*, 182 Lincoln St., Boston, MA 02111; Tel: 617-426-7092, or 1-800-321-VIEW.
- *MMI Corporation*, 2950 Wyman Parkway, P.O. Box 19907, Baltimore, MD 21211. Tel: 301-366-1222.
- *National Camera Exchange*, 9300 Olson Memorial Highway, Golden Valley, MN 55427; Tel: 612-546-6831.
- *New England Astro-Optics, Inc.*, Box 834, Simsbury, CT 06070; Tel: 203-658-0701.
- *Northern Sky Telescopes*, 5667 Duluth Street, Golden Valley, MN 55422; Tel: 1-800-345-4202, or 612-545-6786, Fax 612-545-9297.
- *Ohara Corp.*, 50 Columbia Rd., Somerville, NJ 08876; Tel: 201-218-0100.
- *Optica b/c Company*, 4100 MacArthur Blvd., Oakland, CA 94619.
- *Optical Guidance Systems*, 2450 Huntingdon Pike, Huntingdon Valley, PA 19006; Tel: 215-947-5571.

- *The Optical Point*, 8016 Gonzaga Avenue, Los Angeles, CA 90045.
- *Orion Telescope Center*, 421 Soquel Ave., Dept. N, P.O. Box 1158, Santa Cruz, CA 95061; Tel: 800-447-1001, in California 800-443-1001.
- *Parks Optical*, 270 Easy St., Simi Valley, CA 93065. Tel: 805-522-6722.
- *Pauli's Wholesale Optics*, 29 Kingswood Road, Danbury CT 06811; Tel: 203-746-3579.
- *Research Instruments*, 15000 S.W. Barcelona Way, Beaverton, OR 97007; Tel: 503-641-5551.
- *Safari Telescopes*, 110 Pascack Rd., Pearl River, NY 10965; Tel: 212-621-9199. Innovative lightweight, large-aperture telescopes.
- *Scope City*, 679 Easy St., Simi Valley, CA 93065. Tel: 805-522-6646.
- *Seiler Instrument & Manufacturing Co.*, 170 E. Kirkham Ave., St. Louis, MO 63119-1791; Tel: 800-444-7952.
- *Sharpshooters Photographic*, 1034 W. Hillsborough Ave., Tampa, FL 33603, USA; Tel: 800-272-9899.
- *Sky Designs*, 4100 Felps, #C, Colleyville, TX 76034. Tel: 817-656-4326. Lightweight, rigid truss design telescopes.
- *Sky Optical*, 12428 Gladstone, Sylmar, CA 91342. Tel: 818-361-6576. Telescope making supplies.
- *Spectra Astro-Accessories*, 8726-D So. Sepulveda Blvd. Suite 441, Los Angeles, CA 90045; Tel: 818-343-1352. Free catalogue.
- *Star-Liner Co.*, 1106 S. Columbus, Tucson, AZ 85711. Tel: 602-795-3361.
- *Sunwest Space Systems*, P.O. Box 20500, St. Petersburg, FL 33742; Tel: 813-577-0629. Mail order telescopes and accessories. Publishes *Astronomy Industry Newsletter & Catalogue*.
- *Swift Instruments Inc.*, 952 Dorchester Ave., Boston, MA 02125; and P.O. Box 562, San Jose, CA 95106.
- *Tectron Telescopes*, 2111 Whitfield Park Avenue, Sarasota, FL 34243; Tel: 813-758-9890.
- *Telescopics*, P.O. Box 98, La Canada, CA 91011. Tel: 818-952-0953.
- *Texas Nautical Repair Co.*, 2129 Westheimer, Houston, TX 77098; Tel: 713-529-3551 or 529-8480.
- *T.R. Inc.*, P.O. Box 65, Mooers, NY 12958; Tel: 514-672-5697.
- *Roger W. Tuthill, Inc.*, 11 Tanglewood Lane, Box 1086, Mountainside, NJ 07092; Tel: 1-800-223-1063, in N.J. 1-201-232-1786, Fax 1-201-232-3804.
- *Unitron Inc.*, 175 Express Street, Plainview, NY 11803. Tel: 516-822-4601. Fax: 516-931-7660.
- *University Optics Inc.*, P.O. Box 1205, Ann Arbor, MI 48106; Tel: 800-521-2828.
- *USA Sport Optics*, P.O. Box 8015-304, Redondo Beach, CA 90277.
- *VernonScope & Co.*, Candor, NY 12743; Tel: 607-659-7000.
- *Ward's Natural Science Establishment, Inc.*, 5100 W. Henrietta Road, P.O. Box 92912, Rochester, NY 14692-9012. Tel: 800-962-2660.
- *Wholesale Optics of Pennsylvania*, Box 15, Sterling, PA 18463-0015; Tel: 717-842-1500 or 717-344-5217.

In addition, the following companies have clientele worldwide:

- *Kasai Trading Co.*, P.O. Box 38, Mushashino City, Tokyo 180, Japan; Tel: 81-422-55-1703.
- *Ottica e Micromeccanica SNC*, Via del Perlar, 29/B-ZA1, 37135 Verona, Italy.
- Carl Zeiss Jena GmbH, Tatzendpromenade 1a, P.O. Box 125, D–07745 Jena, Germany.

A.6.2 Telescope Mounts and Drives

Companies specializing in (but not necessarily limited to) the manufacture of computer-driven telescopes, tracking systems, or mounts are listed below:
- *A.B. Machining*, 5734 Industrial Rd., Fort Wayne, IN 46825, USA; Tel: 219-483-1418.
- *Arrick Robotics*, P.O. Box 1574, Hurst, TX 76053, USA; Tel: 817-571-4528.
- *Astro-Computer Control*, RD#1, Alexandria, PA 16611, USA; Tel: 814-669-4483.
- *Astrotech*, 39 Periwinkle Lane, Dunstable, Beds. LU6 3NP, England; Tel: 0582-605464.
- *Astro-Track Engineering*, 3900-B East Mira Loma Ave., Anaheim, CA 92807, USA; Tel: 714-630-7381.
- *Beacon Hill Telescopes*, 112 Mill Road, Cleethorpes, South Humberside, DN35 8JD, England; Tel: 0472 692959.
- *Byers Co.*, 29001 West Highway 58, Barstow, CA 92311, USA; Tel: 619-256-2377.
- *Celestial Innovations*, HCR Box 3228, Oracle, AZ 85623, USA; Tel: 602-896-9109.
- *Cheshire Instrument Co.*, P.O. Box 65, Mableton, GA 30059, USA; Tel: 404-438-9200.
- *Contraves Goerz Corp.*, 610 Epsilon Dr., Pittsburgh, PA 15238, USA; Tel: 412-967-7989.
- *Davionx*, 3535 Schafer Dr., Santa Clara, CA 95051, USA; Tel: 408-244-4660.
- *DFM Engineering Inc.*, 1035 Delaware Ave., Longmont, CO 80501, USA; Tel: 303-678-8143. Installs professional computer-controlled telescopes.
- *Dobbins Instrument Co.*, 5168 Lynd Ave., Lyndhurst, OH 44124, USA.
- *Epoch Instruments*, 2331 American Avenue, Hayward, CA 94545, USA; Tel: 414-784-0391.
- *Hollywood General Machining Inc.*, 1033 N. Sycamore Ave., Los Angeles, CA 90038, USA; Tel: 213-462-2855.
- *Jena Scientific Instruments, Inc.*, 820 Second Ave., New York, NY 10017, USA; Tel: 212-867-3051.
- *Jim's Mobile Ind.*, 1960 County Road 23, Evergreen, CO 80439, USA; Tel: 303-277-0304; Fax: 303-526-9140.
- *Optical Guidance Systems*, 2450 Huntingdon Pike, Huntingdon Valley, PA 19006, USA; Tel: 215-947-5571.
- *Thomas Mathis Co.*, 830 Williams St., San Leandro, CA 94577, USA; Tel: 415-483-3090.
- *F.C. Meischner Co.*, 182 Lincoln St., Boston, MA 02111, USA; Tel: 617-426-7092, or 1-800-321-VIEW.

- *Moore Technology*, P.O. Box 2281, Morgan Hill, CA 95038-2281, USA; Tel: 408-848-2649.
- *Optic-Craft Machining*, 33918 Macomb, Farmington, MI 48024, USA; Tel: 313-476-5893.
- *Purus Astro Mechanic*, Postfach 31, 6800 Mannheim 71, Germany.
- *Trax Instrument Corp.*, 10100 Cochiti Rd. SE, Albuquerque, NM 87123, USA; Tel: 505-292-3366.
- *Tresco Machining*, 4009 Pacific Coast Hwy., M.S. 486, Torrance, CA 90505, USA; Tel: 213-373-1427.
- *Vista Instrument Company*, 307 E. Tunnell St., Santa Maria, CA 93454, USA; Tel: 805-922-2545.
- *Vogel Enterprises, Inc.*, P.O. Box 3717, Oak Park, IL 60303, USA; Tel: 800-457-TRAK.

Astronomers living in Canada and the U.S. who are interested in purchasing or selling used equipment and supplies may want to subscribe to *The Starry Messenger* or *The Astro-Trader*. In addition, there are noncommercial advertising sections in magazines such as *Sky & Telescope* ("Sky Gazers Exchange") and *Astronomy* ("Reader Exchange") where the reader can obtain multifarious information regarding used equipment (sometimes from professional observatories) as well as job opportunities in astronomy.

A.6.3 Telescope Repairs, Cleaning, Tune-Ups

There are also firms which perform telescope repairs, restorations, cleanings, or tune-ups:

- *Cosmic Connections, Inc.*, 1460 N. Farnsworth Ave., Aurora, IL 60505, USA; Tel: 708-851-5353 or 800-634-7707.
- *Kay Optical*, 89b London Road, Morden, Surrey, England.
- *Optical Instruments (Balham) Ltd.*, 6 Weir Road, Balham, London SW12 0NA; Tel: 01-673 8513.
- *Photon Instrument Ltd./ Photon Observatory*, 30 King Street, Port Jefferson Station, L.I., NY 11776, USA; Tel: 516-331-3869.
- *Texas Nautical Repair Co.*, 2129 Westheimer, Houston, TX 77098, USA; Tel: 713-529-3551.

A.6.4 Testing of Optical Surfaces

The following firms test optical surfaces for amateur astronomers:

- *David Harbour & Assoc.*, 2321 C. Ct., Box 6081, Enid, OK 73702, USA; Tel: 405-237-4034. Testing of parabolic surfaces for amateur astronomers.
- *Orion Optics*, Unit 3M, Zan Industrial Park, Wheelock, Sandbach, Cheshire CW11 0QD, England; Tel: 0270 768474.

A.6.5 Radio Telescopes

There are at least two firms which specialize in the manufacture of custom radio receivers and observatories:

- *Astrotech*, 39 Periwinkle Lane, Dunstable, Beds. LU6 3NP, England; Tel: 0582 605464.
- *Bob's Electronic Service*, 7605 Deland Ave., Fort Pierce, FL 34951, USA; Tel: 305-464-2118. Supplies mainly schools and colleges.

A.6.6 Spectrographs, Photometers, Astrocameras, Micrometers, and Other Auxiliary Instruments

For more serious observing, starlight can be analyzed quantitatively by means of spectrographs, photometers, micrometers, and other devices which are mounted on the telescope in place of the eyepiece. Sources include:

- *Astro Link*, P.O. Box 1978, Spring Valley, CA 92077. Supplies image intensifiers and housings, digitizers for IBM AT, analog RGB monitors, CCD cameras, and Baker–Schmidt optical systems.
- *Ron Darbinian*, 1681 12th Street, Los Osos, CA 93402, USA; Tel: 805-773-0421. Sells filar micrometers.
- *Electrophysics Corp.*, 48 Spruce St., Nutley, NJ 07110, USA; Tel: 201-667-2262. Offers, among other things, an image intensifier for use with visual and photographic observations: Astrolight Viewer (Model 9100C).
- *Faraday & Wheatstone Computer Graphics*, 194 Main St., Marlborough, MA 01752, USA. Supplies videocameras for astronomy. Free catalogue.
- *KL-9 Engineering*, 935 Glenhaven, Pacific Palisades, CA 94619, USA. Mail order dealer in blink comparators.
- *Optomechanics Research, Inc.*, P.O. Box 87, Vail, AZ 85641, USA; Tel: 602-647-3332. Spectrographs.
- *Optec, Inc.*, 199 Smith, Lowell, MI 49331, USA; Tel: 616-897-9351. Stellar photometers.
- *Photometrics Ltd.*, 2010 Forbes Blvd, Tucson, AZ 85745, USA; Tel: 602-623-6992. Manufacturer of cooled CCD digital imaging systems
- *Sky Scientific*, 28578 Hwy 18, #184, Skyforest, CA 92385, USA; Tel: 714-337-3440. Astrocameras.
- *Thorn EMI Gencom Inc.*, 23 Madison Rd., Fairfield, NJ 07006, USA; Tel: 201-575-5586. Photomultiplier tubes.
- *Roger W. Tuthill Inc.*, Box 1086-A, Mountainside, NJ 07092-0086, USA. Sells StarTrap Video CCD Camera System to view stars, planets etc. on a TV set and record the images on a VCR.

A.6.7 Mirror Coatings

Companies which perform this service include:

- *Denton Vaccum Inc.*, Cherry Hill Ind. Center, Cherry Hill, NJ 08003, USA.
- *E & W Optical Inc.*, 2420 E. Hennepin Ave., Minneapolis, MN 55413, USA; Tel: 612-331-1187.
- *Evaporated Metal Films*, 701 Spencer Rd., Ithaca, NY 14850, USA; Tel: 607-272-3320.
- *David Hinds Ltd.*, Unit 34, Silk Mill, Brook Street, Tring, Herts. HP23 5EF, England.
- *Morvac Optical Coating*, 2300 Walnut, #B, Signal Hill, CA 90806, USA; Tel: 213-424-2062.
- *Orion Optics*, Unit 3M, Zan Industrial Park, Wheelock, Sandbach, Cheshire CW11 0QD, England; Tel: 0270 768474.
- *P.A. Clausing*, 8038 Monticello Ave., Skokie, IL 60076, USA; Tel: 708-267-3399. Beral coatings.
- *P.A.P. Coating Services*, 1112 Chateau Ave., Anaheim, CA 92802, USA; Tel: 714-535-4460.
- *Vacuum Coatings Ltd.*, 25 Lea Bridge Road, London E.5 9QB, England; Tel: 01-806 7335.
- *Wise Instruments Ltd.*, Unit 9, Hollins Business Centre, Marsh Street, Stafford ST16 3BG, England; Tel: 0785 223535.

A.6.8 Photographic Equipment, Film Sensitizing and Processing

Sources:

- *Advance Camera Corp.*, 15 West 46 St., New York, NY 10036, USA; Tel: 800-248-0234. Full-service photographic and optical supply house.
- *Beacon Hill Telescopes*, 112 Mill Road, Cleethorpes, South Humberside DN35 8JD, England; Tel: 0472 692959.
- *Kenmore Camera*, P.O. Box 82467, Kenmore, WA 98028, USA; Tel: 206-485-7447. Photofinishing and camera equipment sales.
- *Lakeside Photography*, P.O. Box 370027, Bearss Station #15, Tampa, FL 33697, USA; Tel: 813-968-9307. Photographic processing.
- *Lunicon*, 2111 Research Dr. #5S, Livermore, CA 94550, USA; Tel: 415-447-9570. Provides a variety of hypersensitized film as well as kits for do-it-yourself hypersensitizing.
- *Photokinesis*, 1359 Fox, Ferndale, MI 48220, USA; Tel: 313-398-1510. Custom photographic lab-film processing, prints from slides.
- *Space Technologies and Research*, 700 Seminola Blvd, Casselberry, FL 32707, USA. Dealer in hypered film and lens warmers.
- *Speedibrews*, 54 Lovelace Drive, Pryford, Woking, Surrey GU22 8QY, England; Tel: 093-23 46942. Special films and developers for astronomical use.

- *Stellar Graphics, Inc.*, 7989 Canadice Road, Springwater, NY 14560, USA; Tel: 716-426-1577. Offers full range of services from hypersensitized film to giant prints in color and B&W. Free catalogue.

A.6.9 Filters

Sources:

- *Cole Enterprises*, 40714 E. Acacia Ave., Hemet, CA 92344, USA; Tel: 714-654-8991. Chroma-Scan Pocket Filters.
- *Daystar Filter Corp.*, P.O. Box 1290, Pomona, CA 91769, USA; Tel: 714-591-4673. Sub-angstrom filters, solar and nebula filters.
- *HC Designs, Inc.*, Box 33245, Baltimore, MD 21218, USA; Tel: 301-889-0460. Dealer in shade No. 14 welder's lenses for naked-eye solar observations.
- *Edwin Hirsch*, 168 Lakeview, Dr. RR2, Tomkins Cove, NY 10986, USA; Tel: 914-786-3738.
- *Lumicon*, 2111 Research Dr. #5S, Livermore, CA 94550, USA; Tel: 415-447-9570.
- *Orion Telescope Center*, 421 Soquel Ave., P.O. Box 1158, Santa Cruz, CA 95061, USA; Tel: 800-447-1001; in California 800-443-1001.
- *Paton Hawksley Electronics Ltd.*, Rockhill Laboratories, Wellsway, Keynsham, Bristol BS18 1PG, England; Tel: 0272 862364.
- *Thousand Oaks Optical*, Box 248098, Farmington, MI 48024, USA; Tel: 313-353-6825. Full aperture glass and Mylar solar filters.

A.6.10 Observing Gear and Clothing

The winter months can make long nights of observing uncomfortable and even outright unbearable. Below are listed a few companies which sell cold-weather gear.

- *L.L. Bean, Inc.*, Freeport, ME 04033, USA. Sells a variety of outdoor gear (coats, shoes, jackets, etc.). Catalogue published quarterly; Tel: 800-221-4221 (24 hours, 7 days).
- *Damart Thermawear*, Dept. 70057, 1811 Woodbury Ave., Portsmouth, NH 03805, USA; Tel: 603-742-7420. Manufacturer of Thermolactyl thermal underwear and other cold-weather garments.
- *Spectra Astro-Accessories*, Suite 441 Los Angeles, CA 90045, USA; Tel: 818-343-1352. Sells a 3-piece, industrial-grade observing outfit for observing in sub-zero temperatures. Spectra also offers other astronomical accessories.

A.6.11 Domes

There are several companies which specialize in the manufacture and construction of observatory domes for individuals, groups, or schools. A few of these are listed below and may be contacted for a brochure or catalogue.

- *Ace Dome*, 3186 Juanita, Las Vegas, NV 89102, USA; Tel: 702-853-5790. Manufactures portable domes.
- *Ash Manufacturing Company, Inc.*, Box 312, Plainfield, IL 60544, USA; Tel: 815-436-9403. Manufactures permanently anchored domes in a variety of sizes.
- *Jena Scientific Instruments, Inc.*, 820 Second Ave., New York, NY 10017, USA; Tel: 212-867-3051.
- *Kinard Manufacturing Co.*, P.O. Box 971, Hillsboro, TX 76645, USA; Tel: 817-582-8154. Manufactures the "Hexadome" (described on p. 492 in the May 1989 issue of *Sky & Telescope*).
- *Observa-Dome Laboratories, Inc.*, 371 Commerce Park Drive, Jackson, MS 39213, USA; Tel: 601-982-3333, or 800-647-5364; Telex: 585438.
- *Stewart Research Enterprises*, 1658 Belvoir Dr., Los Altos, CA 94024, USA; Tel: 415-941-6699.

A.6.12 Miscellaneous Observing Aids

Various observing aids such as planispheres, star dials, star finders, and so on can be ordered from the following:

ASP
- SD203 The Night Sky Star Dial. For latitudes 20°–32° N.
- SD204 Same, but for latitudes 30°–40° N.
- SD205 Same, but for latitudes 38°–50° N.
- SD206 Same, but for latitudes 30°–40° S.

Astro Cards, P.O. Box 35, Natrona Heights, PA 15065.
- Index card finder charts.

Astronomy, Order Department, 21027 Crossroads Circle, P.O. Box 1612, Waukesha, WI 53187-1612, USA.
- 30097 Philips' Planisphere for 32°N.
- 30098 Philips' Planisphere for 42°N.
- 30177 Pocket Starguide Set for 32°N.
- 30178 Pocket Starguide Set for 42°N.
- 30189 Miller Planisphere for 40°N.
- 30190 Miller Planisphere for 30°N.

Basic Concepts in Astronomy, 11278 East Meadow Glen Way, Escondido, CA 92026.
- "Dewmaster" moisture controller (currently available only for Celestron 8 and Meade 8 telescopes.

Bausch & Lomb, 135 Prestige Park Circle, East Hartford, CT 06108, USA; Tel: 203-282-0768.
- Starwatcher's Decoder Set. For beginners who want to learn the constellations. Kit contains a plastic frame which is held up to the sky with transparent insert sheets that show the stars and constellations.

Carolina Biological Supply Company, 2700 York Road, Burlington, NC 27215, or Box 187, Gladstone, OR 97027, USA.

- #61-2405W Telescope–Astronomy Kit. Includes all components for a six-element lens system, 15× refractor telescope. With instruction manual.

David Chandler Co., P.O. Box 309, La Verne, CA 91750, USA.
- The Night Sky Star Dial (see ASP).

Greenwich Star Disc, P.O. Box 88, Brentford, Middlesex TW8 8PD, England.
- Greenwich Star Disc for latitudes 50°N to 58°N (British Isles and Canada).

Harris Hobbies, P.O. Box 850237, Richardson, TX 75085-0237, USA; Tel: 214-690-4943.
- "Geek-Lite" hands-free adjustable red light worn on the head.

The M31 Company, 1 Shepard Springs Ct., Durham, NC 27713, USA; Tel: 919-943-6385.
- "Light Blocker" black cape.

- *Rose Star Products*, 2913 Teague Dr., Tyler, TX 75701, USA; Tel: 214-592-5218.
- Red Lens Dark-Adaptation Astrogoggles.
- Reusable Heat Packs.

Sky Publishing Corporation, P.O. Box 9111, Belmont, MA 02178-9111, USA.
- S0008 Precision Planet and Star Locator, available for latitudes 30°, 35°, 40°, 45°, 50°, or in a variable 30°–60° design.

Smithsonian Institution, Department 0006, Washington, DC 20073-0006.
- 6080 Cosmic Constellation Viewer.
- 6032 Smithsonian Star Finder.

A.7 Printed Materials

The following is a list of selected available astronomy handbooks, textbooks, magazines, and other printed items. The list is composed mostly of recent (up to early 1992) British and American publications. Names of frequently referred-to publishers appear in abbreviated form, the coding being as follows:

Abbrev.	*Publisher*
AcdP	Academic Press, New York/London
AdHg	Adam Hilger, Bristol/Philadelphia
AdWs	Addison-Wesley, Menlo Park, CA
AltP	Athlone Press, London
ASP	Astronomical Society of the Pacific, San Francisco
BlkS	Blackwell Scientific Publishers, Oxford
CaUP	Cambridge University Press, Cambridge/New York
CoUP	Columbia University Press, New York
DovP	Dover Publications, New York
EnsP	Enslow Publ., Aldershot/Hillside
FbFb	Faber & Faber, London
FrbP	The Fairborn Press, Mesa, AZ
FrmC	W.H. Freeman and Company, New York/San Francisco/Oxford
GdBr	Gordon and Breach Scientific Publishers, New York/Paris/London
HaRo	Harper & Row, New York

HaUP	Harvard University Press, Cambridge, MA
HoMf	Houghton-Mifflin Co., Boston
JHUP	Johns Hopkins University Press, Baltimore
JoWS	John Wiley & Sons, New York/Chichester
KlwA	Kluwer Academic Publishers, Dordrecht
McMi	MacMillan Co., New York/Toronto/London
MGrH	McGraw-Hill Book Co., New York
MITP	MIT Press, Cambridge MA London
NorC	W.W. Norton and Co., New York
OxUP	Oxford University Press, Oxford
PaPH	Pachart Publishing House, Tucson
PlnP	Plenum Press, New York
PngB	Penguin Books, London/New York
PerP	Pergamon Press, Oxford
PrHl	Prentice-Hall, Englewood Cliffs
PrUP	Princeton University Press, Princeton
ReiP	Reidel Publishing (now Kluwer), Dordrecht
SaCP	Saunders College Publishing, Philadelphia
ScrS	Charles Scribner's Sons, New York
SiSh	Simon & Schuster, New York
SkyP	Sky Publishing Corp., Belmont, MA
SpVg	Springer-Verlag, Berlin/Heidelberg/New York
VNoR	Van Nostrand Reinhold Co., New York
UAzP	University of Arizona Press, Tucson
UCaP	University of California Press, Berkeley
UChP	University of Chicago Press, Chicago
UScB	University Science Books, Mill Valley, CA
UTxP	University of Texas Press, Austin
WadP	Wadsworth Publishing Co., Belmont, CA
WlmB	Willmann-Bell, Richmond, VA
YaUP	Yale University Press, New Haven/London

A.7.1 Observing Handbooks, Guides, Almanacs, Calendars, Catalogues, Atlases, Charts, and Maps

- Acker, A. (ed.): *Catalogue des étoiles les plus brillantes*, Observatoire de Strasbourg Centre de Données Stellaires, Strasbourg (France) 1984.
- Alter, D., Cleminshaw, C.H., Phillip*s, J.G.: *Pictorial Astronomy* (5th edn.), HaRo 1983.
- *Apparent Places of Fundamental Stars – 1989*, Astronomisches Rechen-Institut, 1989.
- Arp, H.C., Madore, B.F.: *A Catalogue of Southern Peculiar Galaxies*, Vols I and II, CaUP 1987.
- Asimov, I.: *Isaac Asimov's Guide to Earth and Space*, Random House, New York 1991.
- Asimov, I.: *The Universe: From Flat Earth to Quasar* (3rd edn.), Walker, New York 1980.
- *The Astronomical Almanac for the Year 19–* (annual), published by US Government Printing Office, Washington, and H.M. Stationery Office, London.
- *The Astronomical Calendar*, available from SkyP. Published yearly.

Appendix A: Educational Resources in Astronomy

- *The Astronomical Companion*, available from SkyP. Published yearly.
- Audouze, J., Israel, G. (eds.): *The Cambridge Atlas of Astronomy*, CaUP 1988.
- Bečvář, A.: *Atlas Coeli 1950.0*, Czechoslovak Academy of Sciences, Prague 1956.
- Bečvář, A.: *Atlas Eclipticalis 1950.0*, Czechoslovak Academy of Sciences, Prague 1958.
- Bishop, R.L. (ed.): *The Observer's Handbook 1991*, Royal Astronomical Society of Canada 1990. (Available from SkyP).
- *Bonner Durchmusterung (BD)* of Argelander and Schönfield, 1855.
- Brun, R., Vehrenberg, H.: *Atlas der Kapteynschen Areas* (Atlas of Selected Areas), Treugesell-Verlag, Düsseldorf 1971.
- Buil, C., Thouvenot, E.: *The Buil-Thouvenot CCD Atlas of Deep-Sky Objects*, SkyP 1991. On high-density disks.
- Chartrand, M.: *The Audobon Society Field Guide to the Night Sky*, Alfred A. Knopf, 1991.
- Clark, R.N.: *Visual Astronomy of the Deep Sky*, CaUP 1991.
- Consolmagno, G., Davis, D.M.: *Turn Left at Orion*, CaUP 1989.
- Dibon-Smith, R.: *Starlist 2000: A Quick Reference Star Catalog for Astronomers*, JoWS 1992.
- Dickinson, T.: *NightWatch: An Equinox Guide to Viewing the Universe*, Camden House/Harrowsmith, Charlotte 1989.
- Dickinson, T., Dyer, A.: *The Backyard Astronomer's Guide*, Camden House Publishing, 1991.
- Dreyer, J.L.E.: *New General Catalogue of Nebulae and Clusters* (NGC), Memoirs of the Royal Astronomical Society, London 1888.
- Duerbeck, H.W.: *A Reference Catalogue and Atlas of Galactic Novae*, ReiP 1987.
- Dunlop, S.: *Astronomy: A Step by Step Guide of the Night Sky*, Hamlyn, Feltham 1985.
- Dunlop, S. (ed.): *Atlas of the Night Sky*, Newnes, Feltham 1983.
- Eicher, D.J.: *Deep-Sky Observing with Small Telescopes*, EnsP 1989.
- Eicher, D.J.: *The Universe from Your Backyard*, CaUP and SkyP 1988.
- Enright, L.: *The Beginner's Observing Guide 1992*, The Royal Astronomical Society of Canada, Toronto 1992.
- Ekrutt, J.: *Stars and Planets*, Barron's, 1992.
- Fjermedal, G.: *New Horizons in Amateur Astronomy*, Putnam, 1989.
- Harrington, P.S.: *Touring the Universe Through Binoculars*, JoWS 1991.
- Harmann, W.K., Miller, R.: *Cycles of Fire: Stars, Galaxies, and the Wonders of Deep Space*, Workman, New York 1987.
- Hartung, E.J.: *Astronomical Objects for Southern Telescopes*, CaUP 1984.
- *Henry Draper Catalogue (HD)* (9 vols.), Harvard Annals 91–99.
- Hewison, W.: *Spaced Out—Punch Amongst the Galaxies*, Grafton, London 1987.
- Hirschfeld, A., Sinnott, R.: *Sky Catalogue 2000.0*, Vols. 1 and 2, CaUP 1982, 1985.
- Hoffleit, D., Jaschek, C.: *The Bright Star Catalogue* (4th edn.), Yale University Observatory, New Haven 1982.
- Hoffleit, D., Saladyga, M., Wlasuk, P.: *A Supplement to the Bright Star Catalogue* Yale University Observatory, New Haven 1983.
- Hunt, G., Moore, P.: *Atlas of Uranus*, CaUP 1988.
- Hynes, S.J.: *Planetary Nebulae: A Practical Guide and Handbook for Amateur Astronomers*, WlmB 1991.
- *Index Catalogue of Visual Double Stars, 1961.0*, (IDS, 2 vols.) Publ. Lick Observatory Vol. XXI, 1963.
- Jenkins, L.: *General Catalogue of Trigonometric Stellar Parallaxes*, Yale University Observatory, New Haven 1963.
- Jones, B., with Edberg, S. (ed.): *The Practical Astronomer*, SiSh 1990.
- Jones, K.G.: *Messier's Nebulae and Star Clusters*, CaUP 1991.
- Jones, K.G. (ed.): *The Webb Society Deep-Sky Handbook*, Vol. 1: *Double Stars*; Vol. 2: *Planetary and Gaseous Nebulae*; Vol. 3: *Open and Globular Clusters*; Vol. 4: *Galaxies*; Vol. 5: *Clusters of Galaxies*; Vol. 6: *Anonymous Galaxies*; Vol. 7: *The Southern Sky* EnsP 1987.

- Kals, W.S.: *Stars and Planets*, Sierra Club, 1990.
- Karkoschka, E.: *The Observer's Sky Atlas*, SpVg 1990.
- Kholopov, P.N. et al.: *General Catalogue of Variable Stars* (4th edn., 4 vols.), Nauka, Moscow 1985-90.
- Kirby-Smith, H.T.: *U.S. Observatories: A Directory and Travel Guide*, VNoR 1976.
- Kozak, J.T.: *Deep-Sky Objects for Binoculars*, SkyP 1988.
- Krisciunas, K.: *The Pictoral Atlas of the Universe*, Mallard, New York 1989.
- Laustsen, S., Maben, C., West, R.M.: *Exploring the Southern Sky*, SpVg 1987.
- Levy, D.H.: *The Sky: A User's Guide*, CaUP 1991.
- Liller, W., Mayer, B.: *The Cambridge Astronomy Guide: A Practical Introduction to Astronomy*, CaUP 1985.
- Lovi, G., Tirion, W.: *Men, Monsters, and the Modern Universe*, WlmB 1989.
- Luginbuhl, C.B., Skiff, B.A.: *Observing Handbook and Catalogue of Deep-Sky Objects*, CaUP 1989.
- Malin, D.: *A Celebration of Colour in Astronomy*, Indian Academy of Sciences, Bangalore 1991.
- Malin, S.: *The Cambridge Guide to the Planets*, CaUP 1989.
- Malin, S.: *The Cambridge Guide to Stars, Galaxies, and Nebulae*, CaUP 1989.
- Matloff, G.L.: *The Urban Astronomer: A Practical Guide for Observers in Cities and Suburbs*, JoWS 1991.
- Mayall, R.N. & M.W.: *The Sky Observer's Guide*, Western 1965. Available from SkyP.
- Mayer, B.: *Astrowatch*, Perigee, 1988.
- Menzel, D.H., Pasachoff, J.M.: *A Field Guide to the Stars and Planets*, HoMf 1983.
- Monkhouse, R.: *3-D Star Maps*, HaRo 1989.
- Moore, P.: *The Amateur Astronomer*, NorC 1990.
- Moore, P.: *Armchair Astronomy*, Patrick Stephens, Wellingborough 1984.
- Moore, P.: *Astronomers' Stars*, NorC 1989.
- Moore, P.: *Exploring the Night Sky with Binoculars*, CaUP 1989.
- Moore, P.: *The Mitchell Beazley New Concise Atlas of the Universe*, Mitchell Beazley, London 1978.
- Moore, P.: *Observers' Astronomy*, PngB 1988.
- Moore, P.: *Philip's Stargazer: The Complete Astronomy Map and Guide Pack*, George Philip, London 1991.
- Moore, P.: *The Sky at Night*, Harrap, London 1989.
- Moore, P. (ed.): *1991 Yearbook of Astronomy*, Sidgwick & Jackson, London 1990.
- Motz, L., Nathanson, C.: *The Constellations: An Enthusiast's Guide to the Night Sky*, Doubleday, 1988.
- Muirden, J.: *Astronomy with Binoculars* (2nd edn.), FaFa 1976.
- Muirden, J.: *Astronomy with a Small Telescope*, George Philip, London 1985.
- Neckel, T., Vehrenberg, H.: *Atlas of Galactic Nebulae*, Treugsell-Verlag, Düsseldorf 1990.
- Newton, J., Teece, P.: *The Guide to Amateur Astronomy*, CaUP 1988.
- Nitschelm, C.: *Catalogue of Named Stars: Names and Characteristics*, Institut d'Astronomie, Université de Lausanne, Switzerland 1989.
- Norton, A.P.: *Star Atlas and Reference Handbook*, Gall and Inglis, London 1954.
- Ottewell, G. (ed.): *Astronomical Calendar 1991*, Furman University Physics Dept., 1990. (Available from SkyP.)
- Palmer, L.: *The Trained Eye: An Introduction to Astronomical Observing*, Holt Dryden Saunders, 1990.
- Pasachoff, J.M.: *Peterson First Guides—Astronomy*, HoMf 1988.
- Peltier, L.: *Guide to the Stars—Exploring the Sky with Binoculars*, CaUP 1987.
- Porcellino, M.R.: *Through the Telescope: A Guide for the Amateur Astronomer*, Tab, 1989.
- Price, F.W.: *The Moon Observer's Handbook*, CaUP 1988.
- Raymo, C.: *365 Starry Nights*, PrHl 1986.
- Reed, G.: *Naked i Astronomy*, International Planetarium Society, Rochester 1989.
- Ridpath, I.: *American Nature Guides: Astronomy*, Gallery Books, 1990.

- Ridpath, I. (ed.): *Norton's 2000.0 Star Atlas and Reference Handbook* (18th edn.), Longman/JoWS 1989.
- Ridpath, I.: *Star Tales*, Lutterworth, Cambridge (Universe Books, New York) 1988.
- Ridpath, I., Tirion, W.: *The Monthly Sky Guide*, CaUP 1990.
- Roth, G.D. (ed.): *Astronomy: A Handbook*, SpVg 1975.
- Roth, G.D.: *The Amateur Astronomer and His Telescope* (2nd edn.), FaFa 1972.
- Rowan-Robinson, M.: *Our Universe: An Armchair Guide*, FrmC 1990.
- Sandage, A., Bedke, J.: *Atlas of Galaxies Useful for Measuring the Cosmological Distance Scale*, NASA, Washington 1988.
- Schaaf, F.: *The Starry Room*, JoWS 1988.
- Schweighauser, C.: *Astronomy from A to Z: A Dictionary of Celestial Objects and Ideas*, Illinois Issues, 1991.
- Sherrod, P.C.: *A Complete Manual of Amateur Astronomy*, PrHl 1989.
- Sidgwick, J.B.: *Amateur Astronomer's Handbook*, (4th edn.), EnsP 1980.
- Sidgwick, J.B.: *Observational Astronomy for Amateurs*, DovP 1971.
- Sinnott, R.W.: *NGC 2000.0—The Complete New General Catalogue (NGC) and Index Catalogue (IC) of Nebulae and Clusters of Stars*, CaUP/SkyP 1989. Available on diskette.
- *Sky Gazer's Almanac*, SkyP. Published yearly.
- *Smithsonian Astrophysical Star Catalog* (SAOC, 4 vols.), Washington 1966.
- Smyth, W.H.: *The Bedford Catalogue* (from a *Cycle of Celestial Objects*), WlmB 1986.
- Snyder, G.S.: *Maps of the Heavens*, Andre Deutsch, London 1984.
- Stephenson, F.R., Houlden, M.A.: *Atlas of Historical Eclipse Maps: East Asia 1500 B.C.– A.D. 1900*, CaUP 1986.
- Stott, C.: *The Cambridge Guide to Stargazing*, CaUP 1989.
- Stott, C.: *The Cambridge Guide to Astronomy in Action*, CaUP 1989.
- Sulentic, J.W., Tifft, W.G.: *The Revised New General Catalogue of Nonstellar Astronomical Objects*, UAzP 1973.
- Taylor, G. (ed.): *The Handbook of the British Astronomical Association 1991*, British Astronomical Association, London 1990.
- Tirion, W.: *Cambridge Star Atlas 2000*, CaUP 1991.
- Tirion, W., Rappaport, B., Lovi, G.: *Uranometria 2000.0—Vol. I: The Northern Hermisphere to* $-6°$*, Vol. II: The Southern Hemisphere to* $+6°$, SkyP.
- Tirion, W., Ridpath, I.: *The Night Sky*, Collins, London 1985.
- Tully, R.B.: *Nearby Galaxies Catalogue*, CaUP 1988.
- Tully, R.B., Fisher, J.R.: *Nearby Galaxies Catalog*, CaUP 1990.
- Vehrenberg, H.: *Atlas of Deep Sky Splendors* (4th edn.), CaUP 1984.
- Vehrenberg, H.: *Atlas of Selected Areas*, Treugesell-Verlag, Düsseldorf 1965.
- Vehrenberg, H.: *Atlas Stellarum 1950.0*, Treugesell-Verlag, Düsseldorf 1965.
- Vehrenberg, H.: *Photographischer Sternatlas*, Treugesell-Verlag, Düsseldorf 1977.
- Westfall, J. (ed.): *The A.L.P.O. Solar System Ephemeris: 1992*, Association of Lunar and Planetary Observers, 1991.
- Whitney, C.A.: *Whitney's Star Finder*, Alfred A. Knopf, 1989.
- *Wonders of the Universe Calendar*, SkyP. Published yearly.
- Wray, J.: *Color Atlas of Galaxies*, CaUP 1988.

A.7.2 Selected Astronomy Textbooks and Exercise Manuals

- Acker, A., Jaschek, C.: *Astronomical Methods and Calculations*, JoWS 1986.
- Birney, D.S.: *Observational Astronomy*, CaUP 1990.
- Block, D.: *Starwatching*, Lion Publishing Corp, Batavia 1988.
- Brewer, S.G.: *Do-It-Yourself Astronomy*, Edinburgh University Press (available from CoUP) 1988.

- Brück, M.T.: *Exercises in Practical Astronomy Using Photographs*, AdHg 1990.
- *Edinburgh Astronomy Teaching Package for Undergraduates* (revised and expanded version 1988), available from the Royal Observatory, Blackford Hill, Edinburgh EH9 3HJ, Scotland.
- *Edinburgh Astronomy Spectroscopic Teaching Package for Undergraduates* (1988), available from the Royal Observatory, address given above.
- Ferguson, D.C.: *Introductory Astronomy Exercises*, WadP 1990.
- Gainer, M.K.: *Astronomy Laboratory and Observation Manual*, PrHl 1989.
- Gingerich, O. (ed.): *Laboratory Exercises in Astronomy*, available from SkyP.
- Graham-Smith, F., Lovell, B.: *Pathways to the Universe*, CaUP 1988.
- Hartmann, W.K.: *Astronomy: The Cosmic Journey* (4th edn.), WadP 1989.
- Hemenway, M.K., Robbins, R.R.: *Modern Astronomy: An Activities Approach*, UTxP 1992.
- Hoff, D.B., Kelsey, L.J., Neff, J.S.: *Activities in Astronomy, 3rd edition*, Kendall/Hunt, Dubuque 1992.
- Karttunen, H., Kroger, P., Oja, H., Poutanen, M., Donner, K.J. (eds.): *Fundamental Astronomy*, SpVg 1987.
- Kaufmann, W.J.: *Discovering the Universe* (2nd edn.), FrmC 1990.
- Kitchin, C.R.: *Stars, Nebulae, and the Interstellar Medium*, AdHg 1987.
- Kleczek, J.: *Exercises in Astronomy*, KlwA 1987.
- Kutner, M.L.: *Astronomy: A Physical Perspective*, JoWS 1988.
- Moché, D.: *Astronomy — A Self-Teaching Guide*, available from SkyP.
- Moeschel, R.: *Exploring the Sky: 100 Projects for Beginning Astronomers*, Chicago Review, Chicago 1988.
- Moore, P.: *Stars and Planets*, Merehurst, London 1988.
- Pasachoff, J.M.: *Astronomy: From the Earth to the Universe* (4th edn.), SaCP 1991.
- Pasachoff, J.M.: *Contemporary Astronomy*, SaCP 1985.
- Roy, A.E., Clarke, D.: *Astronomy: Principles and Practice* (3rd edn.), AdHg 1988.
- Roy, A.E., Clarke, D.: *Astronomy: Structure of the Universe* (3rd edn.), AdHg 1989.
- Schlosser, W., Schmidt-Kaler, T., Milone, E.F.: *Challenges of Astronomy: Hands-on Experiments for the Sky and Laboratory*, SpVg 1991.
- Seeds, M.: *Foundations of Astronomy*, WadP 1990.
- Shu, F.: *The Physical Universe*, UScB 1982.
- Tattersfield, D.: *Projects and Demonstrations in Astronomy*, JoWS 1979.
- Unsöld, A., Baschek, R.B.: *The New Cosmos* (4th edn.), SpVg 1991.
- Vorontsov-Vel'yaminov, B.A.: *Essays About the Universe*, Mir, Moscow (distributed by Imported Publications, Chicago), 1985.
- Waxman, J.: *A Workbook for Astronomy*, CaUP 1984.
- Zeilik, M., Gaustad, J.: *Astronomy: The Cosmic Perspective* (2nd edn.), JoWS 1990.
- Zeilik, M., Gregory, S.A., Smith, E.: *Introductory Astronomy and Astrophysics* (3rd edn.), SaCP 1989.

A.7.3 Advanced Textbooks and Monographs

- Aller, L.H.: *Atoms, Stars, and Nebulae* (3rd edn.), CaUP 1991.
- Aller, L.H.: *Physics of Thermal Gaseous Nebulae*, ReiP 1984.
- Böhm-Vitense, E.: *Introduction to Stellar Astrophysics* Vol. 1: *Basic Stellar Observations and Data*, CaUP 1989.
- Böhm-Vitense, E.: *Introduction to Stellar Astrophysics* Vol. 2: *Stellar Atmospheres*, CaUP 1989.
- Böhm-Vitense, E.: *Introduction to Stellar Astrophysics* Vol. 3: *Stellar Structure and Evolution*, CaUP 1991.
- Bowers, R.L., Wilson, J.R.: *Numerical Modeling in Applied Physics and Astrophysics*, Jones and Bartlett, 1991.

Appendix A: Educational Resources in Astronomy

- Celnikier, L.M.: *Basics of Cosmic Structures*, Editions Frontières, France 1989.
- Chandrasekhar, S.: *An Introduction to the Study of Stellar Structure*, DovP 1939.
- Chandrasekhar, S.: *Radiative Transfer*, DovP 1950.
- Chandrasekhar, S.: *Selected Papers*, Vol. 1: *Stellar Structure and Stellar Atmospheres*, UChP 1989.
- Collins, G.W.: *The Fundamentals of Stellar Astrophysics*, FrmC 1989.
- Davies, P.C.W.: *Superforce*, Heinemann, London 1984.
- Davies, P.C.W., Brown, J. (eds.): *Super Strings—A Theory of Everything*, CaUP 1988.
- Dalgarno, A., Layzer, D.: *Astrophysical Plasmas*, CaUP 1987.
- Demianski, M.: *Relativistic Astrophysics*, PerP 1985.
- Eddington, A.S.: *The Internal Constitution of the Stars*, reissued by CaUP 1988.
- Eddington, A.S.: *Space, Time and Gravitation*, reissued by CaUP 1987.
- Encrenaz, T.: *The Solar System*, SpVg 1990.
- Gaisser, T.K.: *Cosmic Rays and Particle Physics*, CaUP 1990.
- Ginzburg, V.L.: *Physics and Astrophysics, a Selection of Key Problems*, PerP 1985.
- Goldberg, H.S., Scadron, M.D.: *Physics of Stellar Evolution and Cosmology*, GdBr 1981.
- Gray, D.F.: *The Observation and Analysis of Stellar Photospheres*, CaUP 1992.
- Gribbin, J.: *In Search of Schrödinger's Cat*, Wildwood House, 1984.
- Harwit, M.: *Astrophysical Concepts* (2nd edn.), SpVg 1988.
- Hey, T., Walters, P. (eds.): *The Quantum Universe*, CaUP 1987.
- Katz, J.I.: *High Energy Astrophysics*, AdWs 1988.
- Kippenhahn, R., Weigert, A.: *Stellar Structure and Evolution*, SpVg 1990.
- Kitchin, C.R.: *Astrophysical Techniques*, AdHg 1984.
- Kormendy, J. (ed.): *Dark Matter in the Universe* (IAU Symposium No. 117), ReiP 1987.
- Léna, P.: *Observational Astrophysics*, SpVg 1988.
- Longair, M.S.: *Theoretical Concepts in Physics*, CaUP 1984.
- Osterbrock, D.E.: *Astrophysics of Gaseous Nebulae and Active Galactic Nuclei*, UScB 1989.
- Piddington, J.H.: *Cosmic Electrodynamics*, Krieger, Florida 1981.
- Rolfs, C.E., Rodney, W.S.: *Cauldrons in the Cosmos: Nuclear Astrophysics*, UChP 1988.
- Roy, A.E.: *Orbital Motion*, AdHg 1988.
- Rybicki, G.B., Lightman, A.P.: *Radiative Processes in Astrophysics*, JoWS 1985.
- Sanchez, F., Vazquez, M. (eds.): *New Windows on the Universe* (2 vols.), CaUP 1991
- Saslaw, W.: *Gravitational Physics of Stellar and Galactic Systems*, CaUP 1985.
- Scheffler, H., Elsässer, H.: *Physics of the Galaxy and Interstellar Matter*, SpVg 1987.
- Sexl, R., Sexl, H.: *An Introduction to Relativistic Astrophysics*, AcdP 1979.
- Seymour, P.: *Cosmic Magnetism*, AdHg 1986.
- Shapiro, S.L., Teukolsky, S.A.: *Black Holes, White Dwarfs, and Neutron Stars*, JoWS 1983.
- Shapiro, S.L., Teukolsky, S.A. (eds.): *Highlights of Modern Astrophysics*, JoWS 1986.
- Shu, F.: *The Physics of Astrophysics Vol. I: Radiation*, UScB 1991.
- Shu, F.: *The Physics of Astrophysics Vol. II: Gas Dynamics*, UScB 1992.
- Stephani, H.: *General Relativity*, CaUP 1982.
- Unsöld, A.: *Physik der Sternatmosphären* (2nd edn.), SpVg 1968.
- Vangioni-Flam, E., Cassé, M., Audouze, J., Tran Thanh Van, J. (eds.): *Astrophysical Ages and Dating Methods*, Editions Frontières, France 1990.
- Wehrse, R. (ed.): *Accuracy of Element Abundances from Stellar Atmospheres*, SpVg 1990.

A.7.4 General Reference Books and Encyclopedias

- Allen, C.W.: *Astrophysical Quantities*, AltP 1973.
- *Apparent Places of Fundamental Stars – 1991*, Astronomisches Rechen-Institut, Heidelberg 1990.
- Cotardière, P. de la (ed.): *Larousse Astronomy*, Hamlyn, 1987.
- Curtis, A.R.: *Space Almanac*, Arcsoft, 1989.
- Flügge, S. (ed.): *Handbuch der Physik*, Group XI, *Astrophysik* (5 vols.), SpVg 1958-60.
- Heck, A.: *Acronyms & Abbreviations in Astronomy & Space Sciences*, Observatoire Astronomique, Strasbourg 1990.
- Illingworth, V. (ed.): *The Facts on File Dictionary of Astronomy*, McMi/SkyP 1990.
- Illingworth, V. (ed.): *The Macmillan Dictionary of Astronomy*, McMi 1985.
- *International Directory of Astronomical Associations and Societies (IDAAS)* and *International Directory of Professional Astronomical Institutions (IDPAI)*, Observatoire Astronomique, Strasbourg, France 1990.
- Landolt-Börnstein: *Numerical Data and Functional Relationships in Science and Technology*, New Series, Group VI, *Astronomy, Astrophysics, and Space Research*, SpVg 1965–1982.
- Lang, K.R.: *Astrophysical Formulae: A Compendium for the Physicist and Astrophysicist* (2nd edn.), SpVg 1980.
- Lang, K.R.: *Astrophysical Data: Planets and Stars*, SpVg 1991.
- Lang, K.R.: *Astrophysical Data: Galaxies*, SpVg 1991.
- Makower, J. (ed.): *The Air and Space Catalogue*, Vintage, 1989.
- Man, J., Wace, M. (eds.): *The Astronomer's Library* (2 vols.), W.H. Smith, New York 1989.
- Maran, S.P. (ed.): *The Astronomy and Astrophysics Encyclopedia*, VNoR 1991.
- Meyers, R.A. (ed.): *Encyclopedia of Astronomy and Astrophysics*, Academic, San Diego 1988.
- Mitton, J.: *A Concise Dictionary of Astronomy*, OxUP 1991.
- Moore, D.H.: *The HarperCollins Dictionary of Astronomy and Space Science*, Harper Collins, New York 1992.
- Moore, P.: *The Guinness Book of Astronomy*, Guinness, 1988.
- Moore, P.: *Patrick Moore's A–Z of Astronomy*, Stephens, Wellingborough 1986.
- Moore, P. (ed.): *1992 Yearbook of Astronomy*, NorC 1991.
- Qudouze, J., Israel, G., Falque, J.-C.: *The Cambridge Atlas of Astronomy*, CaUP 1988.
- Parker, S.P. (ed.): *McGraw-Hill Encyclopedia of Astronomy*, MGrH 1983.
- Plant, M.: *Dictionary of Space*, Layman, Harlow 1986.
- Rycroft, M.: *The Cambridge Encyclopedia of Space*, CaUP 1991.
- Shore, S. (ed.): *Encyclopedia of Astronomy and Astrophysics*, AcdP 1989.
- Time-Life Books, Editors of: *Voyage Through the Universe* (approx. 20 vols.), Time-Life Books, Richmond 1990.
- Tyson, N.: *Merlin's Tour of the Universe*, CoUP 1989.
- Walker, P.M.B: *Cambridge Air and Space Dictionary*, CaUP 1990.
- Williamson, M.: *Dictionary of Space Technology*, AdHg 1990.
- Zombeck, M.V.: *Handbook of Space Astronomy and Astrophysics* (2nd edn.), CaUP 1990.

A.7.5 Reviews, Overviews, and Collections

- *Annual Review of Astronomy and Astrophysics*, Annual Reviews, Palo Alto. Published annually.
- Asimov, I.: *The Secret of the Universe*, Doubleday, New York 1991.
- Beatty, J.K., Chaikin, A. (eds.): *The New Solar System* (3rd edn.), CaUP 1990.
- Beer, P. (ed.): *Vistas in Astronomy, Volume 26*, PerP 1985.

- Bertola, F., Sulentic, J.W., Madore, B.F. (eds.): *New Ideas in Astronomy*, CaUP 1988.
- Blanco, V.M., Phillips, M.M.: *Progress and Opportunities in Southern Hemisphere Optical Astronomy*, Brigham Young University Press, Provo 1988.
- Cordova, F. (ed.): *Multiwavelength Astrophysics*, CaUP 1988.
- Dunlop, S., Gerbaldi, M. (eds.): *Stargazers: The Contributions of Amateur Astronomers*, Proceedings of Colloquium 98 of the IAU, 1987 June 20–24, SpVg 1988.
- Dyson, F.: *Infinite in all Directions*, PngB 1989.
- Fabian, A.C. (ed.): *Origins: The Darwin College Lectures*, CaUP 1988.
- Gustaffsson, B., Nissen, P.E. (eds.): *Astrophysics: Recent Progress and Future Possibilities*, The Royal Danish Academy of Science and Letters, Cophenhagen 1990.
- Henbest, N. (ed.): *Observing the Universe*, BlkS and New Scientist, London 1984.
- Klare, G. (ed.): *Reviews in Modern Astronomy 1*, SpVg 1988.
- Klare, G. (ed.): *Reviews in Modern Astronomy 2*, SpVg 1989.
- Lang, K.R., Whitney, C.A.: *Exploration and Discovery in the Solar System*, CaUP 1991.
- Lightman, A.: *Time for the Stars: Astronomy in the 1990s*, PngB 1992.
- Lugger, P. (ed.): *Asteroids to Quasars: A Symposium for the 60th Birthday of William Liller*, CaUP 1991.
- McNally, D. (ed.): *Highlights of Astronomy*, Vol. 8, KlwA 1989.
- Morrison, D., Wolff, S.C.: *Frontiers of Astronomy*, SaCP 1990.
- Morrison, P.: *Philip Morrison's Long Look at the Literature*, FrmC 1990.
- National Research Council (ed.): *The Decade of Discovery in Astronomy and Astrophysics*, National Academic, 1991.
- Osterbrock, D. (ed.): *Stars and Galaxies: Citizens of the Universe*, FrmC 1990.
- Philip, A.G.D., Upgren, A.R. (eds.): *Star Catalogues: A Centennial Tribute to A.N. Vyssotsky*, L. Davis, Schenectady, New York 1989.
- Preiss, B., Fraknoi, A. (eds.): *The Universe*, Bantam, New York 1988.
- Smoluchowski, R., Bahcall, J.N., Matthews, M. (eds.): *The Galaxy and the Solar System*, UAzP 1986.
- Sutton, C. (ed.): *Building the Universe*, BlkS 1985.
- Swings, J.-P. (ed.): *Transactions of the IAU Vol. XXA (Reports 1988). Reports on Astronomy*, KlwA 1988.
- Wall, J.V., Boksenberg, A.: *Modern Technology and its Influence on Astronomy*, CaUP 1990.
- West, R.M. (ed.): *Reports on Astronomy, Transactions of the International Astronomical Union Vol. XIXA*, ReiP 1985.

A.7.6 Children's Books

There are quite a few books on the market today which are especially suitable for young children (ages 3 to 12) and teenagers. Suggested ages are given when they have been provided by the publisher. For more complete listings of books for young people, refer to R.R. Robbins and A. Fraknoi: *The Universe at Your Fingertips* (IAU Commission 46 Report, August 1985) and R.R. Robbins: *Astronomy Education Materials in Print 1985–1987* (IAU Commission 46 Report, August 1988).

- Apfel, N.: *The Moon and its Exploration*, Watts, 1982.
- Asimov, I.: *Isaac Asimov's Library of the Universe*. 4 vols.: The Earth's Moon, *The Sun, Our Solar System*, and *Our Milky Way and Other Galaxies*. For ages 8–12, available from the ASP.
- Becklake, S.: *Space—Stars, Planets & Spacecraft*, Dorling Kindersley, 1988. Ages 11–14 years.
- Berger, M.: *Bright Stars, Red Giants, and White Dwarfs*, Putnam, 1983.
- Blumberg, R.: *The First Travel Guide to the Moon*, Four Winds, 1980.

- Branley, F.: *Comets*, HaRo. Also available from ASP.
- Branley, F.: *The Sky is Full of Stars*, HaRo 1981. Ages 3–9. Available from the ASP as part of a two-book set with Branley's book on the planets (see below).
- Branley, F.: *The Planets in Our Solar System*, HaRo 1981. Ages 3–9 years.
- Branley, F.: *Sunshine Makes the Seasons*, HaRo. Ages 3–9. Available from ASP.
- Branley, F.: *Uranus: The Seventh Planet*, Comwell/HaRo 1988.
- Branley, F.: *What the Moon Is Like*, HaRo. Ages 3–9. Available from ASP.
- Burnham, R.: *The Star Book*, CaUP 1983.
- Butterfield, M.: *Satellites and Space Stations*, Usborne, 1985.
- Chandler, D.: *Exploring the Night Sky with Binoculars*, David Chandler. For older children and teens.
- Deutsch, K.: *Space Travel in Fact and Fiction*, Watts, 1980.
- Embury, B.: *The Dream is Alive: A Flight Aboard the Space Shuttle*, HaRo 1990. Ages 9 and up. Available from ASP.
- Gatland, K.: *The Young Scientist Book of Spaceflight*, Usborne, 1982.
- Goldsmith, D.: *What is a Star?*, Interstellar Media, 1979.
- Hadley, E., Hadley, T.: *Legends of the Sun and Moon*, CaUP 1983.
- Henbest, N.: *Spotter's Guide to the Night Sky*, Mayflower, 1979.
- Hirst, R., Hirst, S.: *My Place in Space*, Orchard, 1988. Available from ASP.
- Krupp, E., Krupp, R.: *The Big Dipper and You*, Morrow, 1989. Ages 5–12. Available from ASP.
- Lewellen, J.: *Moon, Sun, and Stars*, Children's, 1981.
- Maynard, C.: *The Young Scientist Book of Stars and Planets*, Usborne, 1977.
- McGowan, T.: *Album of Space Flight*, Rand McNally, 1982.
- Moché, D.: *Astronomy Today*, Kingfisher (Random House in U.S.), 1984. Ages 9–15.
- Moché, D.: *My First Book About Space*, Random House, 1982.
- Moore, P.: *Astronomy for the Under Tens*, George Philip, 1986.
- Moore, P.: *The Space Shuttle Action Book*, Random House, 1983.
- Moore, P.: *Space Travel for the Under Tens*, George Philip, 1988.
- Moore, P.: *Travellers in Time and Space*, Park Lane, 1983. Aimed at youngsters but suitable for beginners of any age.
- Myring, L., Snowden, S.: *Sun, Moon, and Planets*, Usborne, 1982.
- Neri, R.: *The Blue Planet: A Trip to Space*, Vantage, New York 1989.
- Osman, T.: *Space History*, Michael Joseph, 1983.
- Ottewell, G.: *To Know the Stars*, Furman University, 1982.
- Parker, E.: *The Universe*, CaUP/Dinosaur, 1983. For children 7 and younger.
- Pasachoff, J.: *Peterson First Guide to Astronomy*, HoMf, 1988.
- Porcellino, M.R.: *A Young Astronomer's Guide to the Night Sky*, TAB, Blue Ridge Summit, PA 1990.
- Rey, H.A.: *The Stars: A New Way to See Them*, HoMf 1975.
- Ridpath, I.: *Secrets of the Sky*, Hamlyn, 1985. For ages 11–13 years.
- Ridpath, I.: *The Young Astronomer*, Hamlyn Publ., 1981. Suitable for beginners of all ages.
- Ronan, *The Practical Astronomer*, Pan, 1981. Ages 11–14.
- Schaaf, F.: *Seeing the Deep Sky: Telescopic Astronomy Projects Beyond the Solar System*, JoWS 1992.
- Schaaf, F.: *Seeing the Sky: 100 Projects, Activities, and Explorations in Astronomy*, JoWS 1990. Ages 12 and up.
- Schaaf, F.: *Seeing the Solar System: Telescopic Astronomy Projects*, JoWS 1991.
- Seymour, P.: *Adventures with Astronomy*, Murray, 1983. A book of projects for ages 11–14 years. C.: *Let's Find Out About the Sun*.
- Simon, T.: *The Search for Planet X*, Scholastic, 1962.
- Sullivan, N.: *Pioneer Astronomers*, Scholastic, 1964.
- Taylor, G.J.: *Volcanoes in Our Solar System*, Dodd, 1983.
- Thompson, C.E.: *Glow in the Dark Constellations*, Science News, 1989. Contains glow-in-the-dark illustrations of major constellations. Ages 6 to 10.

- Usborne Explainers Series: *First Guide to the Universe*, Usborne, 1982. Actually contains three books in one volume: *Finding Out About Rockets and Spaceflight*, *Finding Out About Sun, Moon, and Planets*, and *Finding Out about Our Earth*.
- Usborne Hobby Guides: *The Young Astronomer*, Usborne.
- VanCleave, J.P.: *Astronomy for Every Kid: 101 Easy Experiments that Really Work*, JoWS 1989.
- Vautier, G.: *The Way of the Stars*, CaUP 1983.
- Whitney, C.: *Whitney's Star Finder*, Knopf, 1984.
- Zim, H., Baker, R.: *Stars*, Golden, 1975.

A.7.7 Magazines, Periodicals, and Newsletters

The following are devoted almost exclusively to astronomy (particularly the observing aspect) and astronomy education:

- *Abrams Planetarium Sky Calendar* — A one-page calendar of celestial events, backed by a star chart of the entire visible sky. Contact: Abrams Planetarium, Michigan State University, East Lansing, MI 48823, USA. Also available free with membership in the ASP (see below)
- *Astrofax* — A quarterly bulletin which details all postal items with astronomical or space-related designs. Write: George G. Young, P.O. Box 632, Tewksbury, MA 01876, USA, or Dr. Michael A. Seeds, Astronomy Program, Franklin & Marshall College, Lancaster, PA 17604-3003, USA.
- *The Astrograph* — A magazine devoted exclusively to astrophotography. Write: The Astrograph, Box 2283, Arlington, VA 22202, USA; Tel: 703-830-2229.
- *The Astronomer* — A monthly magazine devoted primarily to reporting observations and discoveries of amateurs worldwide. Write: John Colls, 177 Thunder Lane, Norwich, NR7 0JF, England; Tel: 0603 36695.
- *Astronomy* — Published monthly, provides news and information on all aspects of astronomy with emphasis on observations. Write: Astronomy, 21027 Crossroads Circle, P.O. Box 1612, Waukesha, WI 53187-1612, USA; Tel: 414-796-8776 (ask for Customer Service).
- *Astronomy Industry Newsletter & Catalogue* — Published by Sunwest Space Systems, P.O. Box 20500, St. Petersburg, FL 33742, USA; Tel: 813-577-0629.
- *The Journal of Astronomy Education* — Published by the AAE. Contact: Dr. Wayne Osborn, Physics Dept., Central Michigan University, Mt. Pleasant, MI 48858, USA.
- *Astronomy Educator* — Published 9 times a year as a supplement to *Astronomy* magazine. Provides news on astronomy and space education, a forum of ideas from astronomy educators, an activities section designed for use in the classroom, and a teachers' resource section. Write: same address and number as *Astronomy* magazine.
- *Astronomy Now* — A monthly periodical edited by astronomer Patrick Moore. Write: Astronomy Now, 193 Uxbridge Road, London W12 9RA, England.
- *The Astronomy Quarterly* — Devoted especially to astronomy education and historical astronomy. Contact: Pachart Publishing, 1130 San Lucas Circle, Tucson, AZ 85704, USA.
- *The Astro-Trader* — Publishes bi-weekly classified listings for astronomy. Write: Astro-Trader, P.O. Box 155 Casa Grande, AZ 85222, USA. Tel: 602-836-6890.
- *ATM Journal* — Published quarterly as the newsletter for the Amateur Telescope Makers Association to promote advances in telescope making and related topics. Write: Amateur Telescope Makers Association (ATMA), c/o William J. Cook, 17607 28th Ave. SE., Bothell, WA 98012.
- *Celestial Computing: A Journal for Personal Computers and Celestial Mechanics* — A new publication aimed at owners of personal computers who wish to do their own programming of orbital calculations, occultations, eclipses, etc. Published quarterly, each issue (at an additional cost) is accompanied by an IBM PC diskette containing programs discussed in the issue. Write: Science Software, 7370 S. Jay St., Littleton, CO 80123, USA.

- *Final Frontier* — Published bi-monthly. Magazine devoted to all facets of space exploration. Write: Final Frontier, P.O. Box 3803, Escondido, CA 92025-9562, USA.
- *Gnomon* — Newsletter of the AAE, sent three times yearly. Contains information on activities within the AAE and articles on pure astronomy as well as on current educational issues.
- *Griffith Observer* — Published monthly. Write: Griffith Observatory, 2800 E. Observatory Rd., Los Angeles, CA 90027, USA.
- *Guide to the Night Sky* — A free monthly guide. Write: Astronomy Department, National Museum of Science and Technology, 1867 St. Laurent Blvd, Ottawa, Ontario, Canada K1A 0M8.
- *IAU Circulars* — Postcard-sized notifications (approximately 10 per month) of information about astronomical phenomena that require prompt dissemination. Also available via "computer service" (see Sect. 28.4.3.4). Contact: Central Bureau for Astronomical Telegrams, Smithsonian Astrophysical Observatory, Cambridge, MA 02138, USA.
- *International Comet Quarterly* — Information and news on comets. Write: Mail Stop 18, Smithsonian Astrophysical Observatory, 60 Garden St., Cambridge MA 02138, USA.
- *Journal for the History of Astronomy* — Features scholarly papers, book reviews, and notes. Published three times a year by Science History Publications Ltd., Halfpenny Furze, Mill Lane, Chalfont St. Giles, Bucks HP8 4NR, England. A supplement *Archeoastronomy* is published annually.
- *Mercury* — Published bimonthly by the ASP. Write: Astronomical Society of the Pacific, 390 Ashton Ave., San Francisco, CA 94112, USA.
- *News! from the Naval Observatory* — A monthly newsletter which includes a guide to the sky. Write: U.S. Naval Observatory, 34th & Massachusetts Ave., NW, Washington, DC 20390, USA.
- *Night Skies* — An astrophotography magazine from: Hobby Publ., P.O. Box 1567, Lynwood, CA 90262, USA.
- *Observatory Techniques* — A magazine about astronomy and computer techniques. Write: Mike Otis (Dept. A1), 1710 SE 16th Ave., Aberdeen, SD 57401.
- *Odyssey* — A colorful astronomy magazine for children. Published by Kalmbach Publishing, 1027 N. 7th St., Milwaukee, WI 53233, USA.
- *The Planetarian* — Published quarterly by the IPS.
- *The Planetary Report* — Magazine available to members of the Planetary Society.
- *Radio Astronomy* — A monthly publication by SARA for radio astronomers.
- *The Reflector* — An extensive newsletter from the Astronomical League (see Sect. 28.3.8 Societies and Clubs) containing a consumer column evaluating astronomical equipment.
- *Sky & Telescope* — A monthly magazine devoted to most aspects of observational astronomy in the northern hemisphere. Write: Sky Publishing Corporation, P.O. Box 9111, Belmont, MA 02178-9111, USA.
- *Sky Views* — A monthly newsletter containing charts, calendars, letters, etc. Write: Skyviews, 130 E. Main St., Suite 168, Medford, OR 97501, USA; Tel: 503-772-8776.
- *Space Education* and *Spaceflight* — Publications of the BIS.
- *Space World Magazine* — Published in cooperation with the National Space Institute, P.O. Box 7535, Ben Franklin Station, Washington, DC 20044, USA.
- *The Starry Messenger* — A monthly publication containing advertisements for buying and selling used telescopic equipment. Write: The Starry Messenger, P.O. Box 4823-N, Ithaca, NY 14852, USA; Tel: 201-992-6865.
- *StarDate* — Published bimonthly by the McDonald Observatory of the University of Texas, and containing scientific articles as well as star charts, sky calendars, and advice on star gazing. Also explores the human aspects of astronomy. Sample copies available. Write: Stardate, The University of Texas, RLM 15.308, Austin, TX 78712, USA.
- *Sterne und Weltraum* — A monthly devoted to all aspects of astronomy (in German). Write: Verlag Sterne und Weltraum Dr. Vehrenberg GmbH, Portiastr. 10, D–81545 München, Germany

- *The Strolling Astronomer* — Published by ALPO. Contains articles, book and film reviews, etc.
- *The Universe in the Classroom* — A quarterly free newsletter on teaching astronomy in grades 3–12. Cosponsored by all three of the professional astronomical organizations in North America (the AAS, ASP, and the CAS), the newsletter is distributed to about 20,000 teachers and curriculum specialists around the world. Contact: Astronomical Society of the Pacific, Teacher's Newsletter, Dept. N, 390 Ashton Ave., San Francisco, CA 94112.

The following science periodicals frequently contain news, information, and articles on astronomy:

- *American Scientist* — Published bimonthly by Sigma Xi, The Scientific Research Society, 345 Whitney Ave., New Haven, CT 06511, USA; Tel: 203-624-9883.
- *Discover* — From Time–Life Books, Inc.
- *Mosaic* — Published by the National Science Foundation.
- *National Geographic* — Published monthly by the National Geographic Society.
- *Nature* — Published monthly.
- *New Scientist* — British weekly science magazine but containing science news from all around the world. Write: Holborne Publishing, Commonwealth House, 1–19 New Oxford St., London WC1A 1NG, England.
- *Physics Today* — Published monthly by the AIP and distributed to members of AIP societies.
- *Quantum* — A bimonthly magazine of math and science designed especially for students. Write Springer-Verlag New York, Inc., 175 Fifth Avenue, New York, NY 10010, USA.
- *Report to Educators* — A quarterly periodical of the Educational Affairs Division, Code XE, NASA, Washington, DC 20546, USA. Contact: Dr. Robert W. Brown, Director.
- *Science News* — A weekly newsmagazine for science. Almost always contains articles on astronomy as well as latest discoveries. Write: Science Service, Inc., 1719 N. St., N.W., Washington, DC 20036, USA.
- *The Sciences* — From the New York Academy of Sciences.
- *Scientific American* — Published monthly.
- *Smithsonian* — Published monthly by the Smithsonian Associates, 900 Jefferson Dr., Washington, DC 20560, USA.

Supplemental Reading List for Vol. 1

The following is a list of suggested readings to supplement the references given at the end of each chapter in Vol. 1. It is composed mostly of recent (up to early 1992) British and American bookprints, some of which may have already been referred to in the individual chapters. Older books are included as far as there is a fair chance of obtaining them by interlibrary loan. Few non-English books are included, since they are seldom obtainable; even university libraries carry few foreign books in the sciences because of the language barrier. Names of frequently referred-to publishers appear in coded form, as given in Sect. A.7.

Chapter 1

- *Astronomy and Astrophysics Abstracts* (semi-annual), SpVg.
- Jaschek, C.: *Data in Astronomy*, CaUP 1989.
- Jaschek, C., Heintz, W.D. (eds.): *Automated Data Retrieval in Astronomy*, ReiP 1982.
- Jaschek, C., Sterken, C.: *Co-ordination of Observational Projects in Astronomy*, CaUP 1988.
- Simmonds, D., Reynolds, L.: *Computer Presentation of Data in Science*, KlwA 1988.
- Werner, H., Schmeidler, F.: *Synopsis of the Nomenclature of the Fixed Stars*, Wissenschaftl Verlag 1986.

Chapter 2

- Brouwer, D., Clemence, G.M.: *Methods of Celestial Mechanics*, AcdP 1961.
- Danby, J.M.A.: *Fundamentals of Celestial Mechanics* (2nd edn.), WlmB 1988.
- Fisher, N.I., Lewis, T., Embleton, B.J.J.: *Statistical Analysis of Spherical Data*, CaUP 1987.
- Green, R.M.: *Spherical Astronomy*, CaUP 1985.
- Lieske, J.H., Abalakin, V.K. (eds.): *Inertial Coordinate System on the Sky*, ReiP 1990.
- Mihalas D., Binney, J.: *Galactic Astronomy* (2nd edn.), FrmC 1981.
- Murray, C.A.: *Vectorial Astronomy*, AdHg 1983.
- Saunders, H.N.: *All the Astrolabes*, Senecio, Oxford 1984.
- Seidelmann, P.K. (ed.): *Explanatory Supplement to the Astronomical Almanacs*, University Science Books, 1992.
- Smart, W.M.: *Textbook on Spherical Astronomy*, CaUP 1977.
- Taff, L.G.: *Celestial Mechanics*, JoWS 1985.
- Woolard, E.W., Clemence, G.M.: *Spherical Astronomy*, AcdP 1966.
- Yallop, B.D., Hohenkerk, C.Y.: *Compact Data for Navigation and Astronomy for the Years 1991–1995*, CaUP 1990.

Chapter 3

- Boulet, D.: *Methods of Orbit Determination for the Microcomputer*, WlmB 1991.
- Burgess, E.: *Celestial Basic: Astronomy on Your Computer*, Sybex, Berkeley 1982.
- Craig, I., Brown, J.: *Inverse Problems in Astronomy*, AdHg 1986.
- Duffet-Smith, P.: *Astronomy with Your Personal Computer*, CaUP 1990.
- Duffet-Smith, P.: *Practical Astronomy with Your Calculator* (3rd edn.), CaUP 1988.
- Hockney, R.W., Eastwood, J.W.: *Computer Simulation Using Particles*, AdHg 1988.

- Jaschek, C., Murtagh, F. (eds.): *Errors, Bias and Uncertainties in Astronomy* CaUP 1990.
- Klein, F.: *Pocket Computer Programs for Astronomers*, Klein, Los Altos 1983.
- Lawrence, J.L.: *Introduction to Basic Astronomy with a PC*, WlmB 1989.
- Levison, H.: *Astro-Navigation by Calculator*, David & Charles, Newton Abbot 1984.
- Lyons, L.: *A Practical Guide to Data Analysis for Physical Science Students*, CaUP 1991.
- Mackenzie, R.: *The Astronomer's Software Handbook*, Sigma, Wilmslow 1985.
- Meeus, J.: *Astronomical Formulae for Calculators*, WlmB 1982.
- Meeus, J.: *Astronomical Algorithms*, WlmB 1991.
- Menke, W.: *Geophysical Data Analysis: Discrete Inverse Theory*, AcdP 1984.
- Montenbruck, O., Pfleger, T.: *Astronomy on the Personal Computer*, SpVg 1991.
- Montenbruck, O.: *Practical Ephemeris Calculations*, SpVg 1989.
- Murtagh, F., Heck, A.: *Multivariate Data Analysis*, KlwA 1986.
- Press, W.H., Flannery, B.P., Teukolsky, S.A., Vetterling, W.T.: *Numerical Recipes—The Art of Scientific Computing*, CaUP 1986.
- Roy, A.E.: *Orbital Motion* (3rd edn.), AdHg 1988.
- Tattersfield, D.: *Orbits for Amateurs with a Microcomputer*, Stanley Thornes 1987.
- Trumpler, R.J., Weaver, H.F.: *Statistical Astronomy*, UCaP 1953.
- Tattersfield, D.: *Orbits for Amateurs with a Microcomputer*, JoWS 1984.
- Taylor, J.R.: *An Introduction to Error Analysis*, UScB 1982.
- Vetterling, S.A., Teukolsky, S.A., Press, W.H., Flannery, B.P.: *Numerical Recipes Examples Book (FORTRAN)*, CaUP 1986.
- Vetterling, S.A., Teukolsky, S.A., Press, W.H., Flannery, B.P.: *Numerical Recipes Examples Book (PASCAL)*, CaUP 1986.
- Whitney, C.A.: *Random Processes in Physical Systems: An Introduction to Probability-Based Computer Simulations*, JoWS 1990.

Chapter 4

- Bell, L.: *The Telescope*, DovP 1922.
- Berry, R.: *How to Build a Dobsonian Telescope* (2nd edn.), AstroMedia, Milwaukee 1983.
- Berry, R.: *Introduction to Astronomical Image Processing with IMAGEPRO Software*, WlmB 1991.
- Buil, C.: *CCD Astronomy: Construction and Use of an Astronomical CCD Camera*, WlmB 1991.
- Dobson, J.L.: *How and Why to Make a User-Friendly Sidewalk Telescope*, Everything in the Universe, Oakland 1991.
- Doherty, P.: *Building and Using an Astronomical Observatory*, Patrick Stephens, Wellingborough 1986.
- Hayes, D.S., Genet, D.R., Genet, R.M.: *New Generation Small Telescopes*, FrbP 1988.
- Hearnshaw, J.B., Cottrell, P.L. (eds.): *Instrumentation and Research Programmes for Small Telescopes*, ReiP 1986. available from SkyP.
- Hiltner, W. (ed.): *Astronomical Techniques* (Vol. II of *Stars and Stellar Systems*), UChP 1962.
- Horne, D.F.: *Optical Production Technology* (2nd edn.), AdHg 1983.
- Howard, E.G.: *Handbook for Telescope Making*, FaFa 1969.
- Ingalls, A.G.: *Amateur Telescope Making I, II, III*, Scientific American, New York 1953/1961.
- Jacoby, G. (ed.): *CCDs in Astronomy*, Astronomical Society of the Pacific Conference Series, San Francisco 1990.
- Jenkins, F., White, H.: *Fundamentals of Optics*, MGrH 1976.
- Kingslake, R. (ed.): *Applied Optics and Optical Engineering* (5 volumes), AcdP 1965.
- Kuiper, G.P., and Middlehurst, B.M.: *Telescopes* (Vol. I of *Stars and Stellar Systems*), UChP 1960.
- Léna, P.: *Observational Astrophysics*, SpVg 1988.
- Kitchen, C.R.: *Astrophysical Techniques* (2nd edn.), AdHg 1991.
- Manly, P.I.: *Unusual Telescopes*, CaUP 1992.

- McLean, I.S.: *Electronic and Computer-Aided Astronomy*, JoWS 1989.
- Miczaika, G.R., Sinton, W.M.: *Tools of the Astronomer*, HaUP 1961.
- Morgan, J.: *Introduction to Geometrical and Physical Optics*, MGrH 1953.
- Muirden, J.: *How To Use An Astronomical Telescope*, Linden/SiSh 1985.
- Nussbaum, A.: *Geometrical Optics: An Introduction*, AdWs 1968.
- O'Shea, D.C.: *Elements of Modern Optical Design*, JoWS 1985.
- Page, T., Page, L.W.: *Telescopes* (Vol. IV of *Sky & Telescope's Library of Astronomy*), McMi 1966.
- Parker, S.P.: *Optics Source Book*, MGrH 1989.
- Robinson, L.B. (ed.): *Instrumentation for Ground-Based Optical Astronomy*, SpVg 1988.
- Rutten, H., van Venrooij, M.: *Telescope Optics: Evaluation and Design*, WlmB 1988.
- Schroeder, D.C.: *Astronomical Optics*, AcdP 1987.
- Sobel, M.I.: *Light*, UChP 1987.
- Southall, J.P.C.: *Mirrors, Prisms, and Lenses* (3rd edn.), McMi 1933.
- Smith, G., Thomson, J.H.: *Optics* (2nd edn.), JoWS 1988.
- Taylor, H.D.: *Adjustment and Testing of Telescope Objectives* (5th edn.), AdHg 1983.
- Texereau, J.: *How to Make a Telescope* (2nd edn.), WlmB 1984.
- Twyman, F.: *Prism and Lens Making*, AdHg 1988.
- Walker, G.: *Astronomical Observations: An Optical Perspective*, CaUP 1987.

Chapter 5

- Avallone, E.A., Baumeister, T.: *Marks' Standard Handbook for Mechanical Engineers* (9th edn.), MGrH 1987.
- Ballard, J.: *Handbook for Star Trackers—Making and Using Star Tracking Camera Platforms*, SkyP 1988.
- Hayes, D.S., Genet, R.M.: *Remote-Access Automatic Telescopes*, FrbP 1989.
- Horowitz, P., Hill, W.: *The Art of Electronics* (2nd edn.), CaUP 1989.
- Timoshenko, S.: *Vibration Problems in Engineering* (3rd edn.), VNoR 1955.
- Trueblood, M., Genet, R.: *Microcomputer Control of Telescopes*, WlmB 1984.

Chapter 6

- Arnold, H.J.P.: *Night Sky Photography*, George Philip, London 1988.
- Covington, M.A.: *Astrophotography for the Amateur*, CaUP 1985.
- Gordon, B.: *Astrophotography* (2nd edn.), WlmB 1985.
- King, E.S.: *A Manual of Celestial Photography*, SkyP 1988.
- Little, R.T.: *Astrophotography*, McMi 1986.
- Marx, S., Pfau, W.: *Astrophotography with the Schmidt Telescope*, CaUP 1992.
- Wallis, B., Provin, R.: *A Manual of Advanced Celestial Photography*, CaUP 1988.

Chapter 7

- Abt, H.A., Neinel, A.B., Morgan, W.W., Tapscott, J.W.: *An Atlas of Low-Dispersion Grating Stellar Spectra*, 1968.
- Adelman, S.J., Lanz, T.: *Elemental Abundance Analyses*, Université de Lausanne 1988.
- Cannon, C.J.: *The Transfer of Spectral Line Radiation*, CaUP 1985.
- Cox, P.A.: *The Elements: Their Origin, Abundance, and Distribution*, OxUP 1989.
- Jaschek, C., Jaschek, M.: *The Classification of Stars*, CaUP 1990.
- Kaler, J.B.: *Stars and their Spectra*, CaUP 1989.
- Keenan, P.C., McNeil, R.C.: *An Atlas of Spectra of the Coolest Stars: Types G, K, M, S, and C*, Ohio State University Press, Columbus 1976.
- Parker, S.P.: *Spectroscopy Source Book*, MGrH 1989.
- Thorne, A.P.: *Spectrophysics*, Chapman and Hall, London 1974.

Chapter 8

- Carleton, N. (ed.): *Astrophysics Part A: Optical and Infrared*, Vol. 12 of *Methods of Experimental Physics*, AcdP 1974.
- Genet, D.S. (ed.): *The Photoelectric Photometry Handbook*, Vols. I and II, FrbP 1988/89.
- Genet, D.S. (ed.): *Solar System Photometry Handbook*, WlmB 1983.
- Ghedini, S.: *Software for Photoelectric Photometry*, WlmB 1982.
- Henden, A.A., Kaitchuck, R.C.: *Astronomical Photometry*, VNoR 1982.
- Hall, D.S., Genet, R.M.: *Photoelectric Photometry of Variable Stars*, International Amateur–Professional Photoelectric Photometry (IAPPP), Fairborn, Ohio 1982.
- Hall, D.S., Genet, R.M., Thurston, B.L. (eds.): *Automatic Photoelectric Telescopes*, FrbP 1986.
- Hardie, R.H.: Photoelectric Reductions. In: Hiltner, W.A.: *Stars and Stellar Systems II: Astrophysical Techniques*, UChP 1962, p. 178.
- Hopkins, J.L.: *Zen and the Art of Photoelectric Photometry*, HPO Desktop Publishing, Phoenix 1990.
- Johnson, H.L.: Photoelectric Photometers and Amplifiers. In: Hiltner, W.A. (ed.): *Stars and Stellar Systems II: Astronomical Techniques*, UChP 1962, p. 157.
- Kuiper, G.P., Middlehurst, B.M.: *Stars and Stellar Systems*, Vol. I: *Telescopes*, UChP 1960.
- Warner, B.: *High Speed Astronomical Photometry*, CaUP 1988.
- Wolpert, R.C., Genet, R.M.: *Advances in Photoelectric Photometry*, Vol. 1, Fairborn Observatory, Fairborn, Ohio 1983.

Chapter 9

- Budden, K.G.: *The Propagation of Radio Waves*, CaUP 1988.
- Christiansen, W.N., Högbom, J.A.: *Radio Telescopes*, CaUP 1985.
- Hey, J.S.: *The Radio Universe*, PerP 1983.
- Kraus, J.D.: *Radio Astronomy* (2nd edn.), Cygnus-Quasar, Powell, Ohio.
- Krüger, A.: *Introduction to Solar Radio Astronomy and Radio Physics*, ReiP 1979.
- Kuiper, G.P., Middlehurst, B.M.: *Telescopes* (Vol. I of *Stars and Stellar Systems*), UChP 1960.
- Maclean, T.S.: *Principles of Antennas: Wire and Aperture*, CaUP 1986.
- Meeks, M.L. (ed.): *Methods of Experimental Physics*, Vol. 12, Part B: *Radio Telescopes*, AcdP 1976.
- Meeks, M.L. (ed.): *Methods of Experimental Physics*, Vol. 12, Part C: *Radio Observations*, AcdP 1976.
- Orr, W.I.: *Radio Handbook* (23rd edn.), Howard W. Sams, 1986.
- Page, T., Page, L.W.: *Telescopes* (Vol. 4 of *Sky & Telescope's Library of Astronomy*), McMi 1966.
- Rohlfs, K.: *Tools of Radio Astronomy*, SpVg 1986.
- Sickels, R.M.: *Radio Astronomy Handbook*, Radio Astronomy Systems, Fort Pierce, Florida 1984.
- Swenson, G.W.: *Amateur Radio Telescopes*, PaPH 1980.
- Verschuur, G.L.: *The Invisible Universe Revealed—The Story of Radio Astronomy.*, SpVg 1987.
- Verschuur, G.L., Kellermann, K.I. (eds.): *Galactic and Extragalactic Radio Astronomy* (2nd edn.), SpVg 1988.

Chapter 10

- Jenkins, G., Bear, M.: *Sun Dials and Time Dials: A Collection of Working Models to Cut and Glue Together*, Parkwest, 1988.
- Mayall, R.N., Mayall, M.: *Sundials: How to Know, Use, and Make Them*, SkyP 1973.
- Stoneman, M.: *Easy to Make Wooden Sundials: Instructions and Plans for Five Projects*, DovP 1982.
- Waugh, A.: *Sundials: Their Theory and Construction*, DovP 1973.

Chapter 11

- Abbott, D. (ed.): *The Biographical Dictionary of Scientists: Astronomers*, Peter Bedrick, New York 1984.
- Ashbrook, J.: *The Astronomical Scrapbook*, CaUP 1985.
- Aveni, A.F. (ed.): *The Lines of Nazca*, American Philosophical Society, Philadelphia 1990.
- Aveni, A.F.: *World Archaeoastronomy*, CaUP 1989.
- Batten, A.H.: *Resolute and Undertaking Characters: The Lives of Wilhelm and Otto Struve*, ReiP 1988.
- Bauer, H.H.: *Beyond Velikovsky: The History of a Public Controversy*, University of Illinois Press, Urbana 1984.
- Beck, R.L., Schrader, D.: *America's Planetariums and Observatories (A Sampling)*, Sunwest Space Systems, St. Petersburg 1991.
- Bhathal, R., White, G.: *Under the Southern Cross: A Brief History of Astronomy in Australia*, Kangaroo Press, Kenthurst, Australia 1991.
- Blaauw, A.: *ESO's Early History*, European Southern Observatory, 1991.
- Bobrovnikoff, N.T.: *Astronomy Before the Telescope*, Vol. 2: *The Solar System*, PaPH 1990.
- Bond, P.: *Heroes in Space—from Gagarin to Challenger*, BlkS 1987.
- Bondi, H.: *Science, Churchill, and Me*, PerP 1990.
- Brashear, J.A.: *A Man Who Loved the Stars*, University of Pittsburgh Press, Pittsburgh 1988. Reprint of the 1924 autobiography.
- Brück, H., Brück, M.: *The Peripatetic Astronomer: The Life of Charles Piazzi Smyth*, AdHg 1988.
- Bühler, W.: *Gauss: Eine biographische Studie*, SpVg 1987.
- Burrows, W.E.: *Exploring Space: Voyages in the Solar System and Beyond*, Random House, 1990.
- Calvin, W.H.: *How the Shaman Stole the Moon: In Search of Ancient Prophet-Scientists from Stonehenge to the Grand Canyon*, Bantam, New York 1991.
- Chandrasekhar, S.: *Eddington: The Most Distinguished Astrophysicist of His Time*, CaUP 1983.
- Chapman, A.: *Dividing the Circle: The Development of Critical Angular Measurement in Astronomy 1500–1850*, Ellis Horwood, London 1990.
- Christianson, G.E.: *In the Presence of the Creator: Isaac Newton and His Times*, McMi 1984.
- Clark, D.H.: *The Quest for SS433*, AdHg 1986.
- Clark, D.H., Stephenson, F.R.: *The Historical Supernovae*, PngB 1977.
- Cornell, J.: *The First Stargazers: An Introduction to the Origins of Astronomy*, ScrS 1981.
- Coyne, G.V., Heller, M., Zycinski, J. (eds.): *Newton and the New Direction in Science*, Libreria Editrice Vaticana 1988.
- Crowe, M.J.: *The Exterrestrial Life Debate 1750–1900: The Idea of a Plurality of Worlds from Kant to Lowell*, CaUP 1986.
- Crowe, M.J.: *Theories of the World from Antiquity to the Copernican Revolution*, DovP 1990.
- DeVorkin, D.H.: *The History of Modern Astronomy and Astrophysics: A Selected and Annotated Bibliography*, Garland, New York 1982.
- Dickinson, T.: *Exploring the Sky by Day: The Equinox Guide to Weather and the Atmosphere*, Camden House, 1988.
- Donahue, W.H. (ed.): *Johannes Kepler's New Astronomy (1609)*, CaUP 1991.
- Drake, S.: *Galileo: Pioneer Scientist*, University of Toronto Press, Buffalo 1990.
- Dreyer, J.L.E.: *A History of Astronomy from Thales to Kepler*, DovP 1953.
- Dreyer, J.L.E., Turner, H.H. (eds.): *History of the Royal Astronomical Society*, Vol. 1: *1820–1920*, BlkS Oxford 1987.
- Duhem, P.: *Medieval Cosmology: Theories of Infinity, Place, Time, Void, and the Plurality of Worlds*, translated from the French by R. Ariew, UChP 1985.
- Durham, F., Purrington, R.: *Frame of the Universe*, CoUP 1983.
- Eelsalu, H., Hermann, D.: *Johann Heinrich Mädler*, Akademie-Verlag, Berlin 1985.
- Elliot, J., Kerr, R.: *Rings: Discoveries from Galileo to Voyager*, MITP 1984.

- Evans, D.S.: *Under Capricorn: A History of Southern Hemisphere Astronomy*, AdHg 1988.
- Evans, D.S.: *Lacaille: Astronomer, Traveler*, PaPH 1992.
- Evans, D.S., Mulholland, J.D.: *Big and Bright*, UTxP 1986.
- Fauvel, J., Flood, R., Shortland, M., Wilson, R. (eds.): *Let Newton Be!*, OxUP 1988.
- Forbes, E.G., Howse, D., Meadows, A.J.: *Greenwich Observatory 1675–1975* (3 vols.), Taylor & Francis, London 1975.
- Friedman, A.J., Donley, C.C.: *Einstein as Myth and Muse*, CaUP 1989.
- Galilei, G.: *Sidereus Nuncius or The Sidereal Messenger*, UChP 1989.
- Gamow, G.: *The Great Physicists from Galileo to Einstein*, DovP 1989.
- Gascoigne, S.C.B., Proust, K.M., Robins, M.O.: *The Creation of the Anglo-Australian Telescope*, CaUP 1990.
- Gehrels, T.: *On the Glassy Sea: An Astronomer's Journey*, American Institute of Physics, New York 1988.
- Gerlach, W.: *Johannes Kepler und die Copernicanische Wende* (2nd edn.), J.A. Barth, Leipzig 1978.
- Gingerich, O., (ed.): *Astrophysics and Twentieth-Century Astronomy to 1950*–Part A, CaUP 1984.
- Gingerich, O.: *The Great Copernicus Chase*, CaUP/SkyP 1992.
- Gingerich, O., Westman, R.S.: *The Wittich Connection: Conflict and Priority in Late Sixteenth-Century Cosmology*, American Philosophical Society, Philadelphia 1989.
- Goldsmith, D.: *The Astronomers*, St. Martin's Press, 1991.
- Hallyn, F.: *The Poetic Structure of the World: Copernicus and Kepler*, Zone (distributed by MITP) 1990.
- Hartmann, W.K., Miller, R., Lee, P.: *Out of the Cradle*, Workman, New York 1986.
- Hawking, S.W., Israel, W.: *300 Years of Gravitation*, CaUP 1987.
- Hawkins, G.S.: *Stonehenge Decoded*, Doubleday, New York 1965.
- Hearnshaw, J.B.: *The Analysis of Starlight: One Hundred and Fifty Years of Astronomical Spectroscopy*, CaUP 1986.
- Heath, T.: *Aristarchus of Samos*, DovP 1921.
- Heath, Sir T.L.: *Greek Astronomy*, DovP 1991.
- Hellemans, A., Bunch, B.: *The Timetables of Science: A Chronology of the Most Important People and Events in the History of Science*, SiSh 1991.
- Henderson, J.B.: *The Development and Decline of Chinese Cosmology*, CoUP 1989.
- Herrmann, D.B.: *The History of Astronomy from Herschel to Hertzsprung*, CaUP 1984.
- Hetherington, N.S. (ed.): *The Edwin Hubble Papers: Previously Unpublished Manuscripts on the Extragalactic Nature of Spiral Nebulae*, PaPH 1990.
- Hetherington, N.S.: *Science and Objectivity: Episodes in the History of Astronomy*, Iowa State University Press, Ames 1988.
- Hoffmann, B.: *Relativity and Its Roots*, Scientific American, FrmC 1983.
- Houk, R.: *From the Hill: The Story of Lowell Observatory*, Lowell Observatory, Flagstaff 1991.
- Howse, D.: *Nevil Maskelyne. The Seaman's Astronomer*, CaUP 1989.
- Hoyt, W.G.: *Planets X and Pluto*, UAzP 1980.
- Ilyas, M.: *Astronomy of Islamic Times for the Twenty-First Century*, Mansell, London 1989.
- Jaki, S.L.: *Olbers Studies: With Three Unpublished Manuscripts by Olbers*, PaPH 1991.
- Kelly, H.L, with McKim, R.: *The British Astronomical Association: The First Fifty Years*, British Astronomical Association, London 1989.
- Krisciunas, K.: *Astronomical Centers of the World*, CaUP 1988.
- Krupp, E.C.: *Beyond the Blue Horizon*, HaRo 1991.
- Krupp, E.C.: *Echoes of the Ancient Skies: The Astronomy of Lost Civilizations*, HaRo 1983.
- Lang, K., Gingerich, O. (eds.): *A Source Book in Astronomy and Astrophysics, 1900–1975*, HaUP 1979.
- Layzer, D.: *Constructing the Universe*, Scientific American Books, FrmC 1984.
- Levy, D.: *Clyde Tombaugh: Discoverer of Planet Pluto*, UAzP 1991.

- Lightman, A., Brawer, B.: *Origins: The Lives and Worlds of Modern Cosmologists*, CaUP 1990.
- Littmann, M.: *Planets Beyond: Discovering the Outer Solar System*, JoWS 1988.
- Lovell, B.: *Astronomer by Chance*, Basic Books, 1990.
- Lovell, B.: *The Jodrell Bank Telescopes*, OxUP 1985.
- Lovell, B.: *Voice of the Universe. Building the Jodrell Bank Telescope* (Revised edn.), Greenwood Press, London 1987.
- Luther, P.W. (compiler): *Bibliography of Astronomers — Books and Pamphlets in English by and about Astronomers*, Vol. 1: *The Spirit of the Nineteenth Century*, WlmB 1989.
- Malvill, J.M., Putnam, C.: *Prehistoric Astronomy in the Southwest*, Johnson, Boulder 1989.
- Maor, E.: *To Infinity and Beyond: A Cultural History of the Infinite*, Birkhäuser, 1987.
- Moore, P.: *Fireside Astronomy: An Anecdotal Tour Through the History and Lore of Astronomy*, JoWS 1992.
- Moore, P.: *Patrick Moore's History of Astronomy* (6th edn.), Macdonald, London 1983.
- Morton, B.: *Halley's Comet, 1755–1984: a Bibliography*, Greenwood, Westport/London 1985.
- Murdin, L.: *Under Newton's Shadow: Astronomical Practices in the Seventeenth Century*, AdHg 1983.
- Murray, C., Cox, C.B.: *Apollo: The Race to the Moon*, SiSh 1989.
- Neugebauer, O.: *Astronomy and History: Selected Essays*, SpVg 1983.
- Neugebauer, O.: *A History of Ancient Mathematical Astronomy* (3 vols.), SpVg 1975.
- Osterbrock, D.E.: *James E. Keeler: Pioneer American Astrophysicist*, CaUP 1984.
- Osterbrock, D.E., Gustafson, J.R., Unruh, W.J.S.: *Eye on the Sky: Lick Observatory's First Century*, UCaP 1988.
- Overbye, D.: *Lonely Hearts of the Cosmos*, Harper Collins, New York 1991.
- Pannekoek, A.: *A History of Astronomy*, George Allen and Unwin, London 1961 (available from DovP and SkyP).
- Porter, R. (ed.): *Man Masters Nature. 25 Centuries of Science*, BBC, London 1987.
- Roth, G.D.: *Joseph von Fraunhofer: Handwerker – Forscher – Akademiemitglied 1787–1826*, Wissenschaftliche Verlagsgesellschaft, Stuttgart 1976.
- Rossi, B.: *Moments in the Life of a Scientist*, CaUP 1990.
- Ruggles, C.L.N.: *Megalith Astronomy (A New Archaeological and Statistical Study of 300 Western Scottish Sites)*, B.A.R. British Series, Oxford 1984.
- Ruggles, C.L.N. (ed.): *Records in Stone: Papers in Memory of Alexander Thom*, CaUP 1988.
- Sabino Maffeo, S.J.: *In the Service of Nine Popes: 100 Years of the Vatican Observatory*, Vatican Observatory, Rome 1991.
- Schmeidler, F.: *Nikolaus Kopernikus*, Wissenschaftliche Verlagsgesellschaft, Stuttgart 1970.
- Schove, D.J., Fletchet, A.: *Chronology of Eclipses and Comets*, Boydell & Brewer, Woodbridge, Suffolk 1984.
- Schröder, W. (ed.): *Historical Events and People in Geosciences*, Peter Lang, Frankfurt 1985.
- Sekido, Y., Elliot, H.: *The Early History of Cosmic Ray Studies*, ReiP 1985.
- Sesti, G.M.: *The Glorious Constellations: History and Mythology*, Harry N. Abrams, New York 1991.
- Sheehan, W.: *Planets and Perception: Telescopic Views and Interpretations, 1609–1909*, UAzP 1988.
- Shipman, H.L.: *Humans in Space: 21st Century Frontiers*, PlnP 1989.
- Shklovsky, I.: *Five Billion Vodka Bottles to the Moon: Tales of a Soviet Scientist*, NorC 1991.
- Smith, C., Wise, M.N.: *Energy and Empire: A Biographical Study of Lord Kelvin*, CaUP 1989.
- Smith, R.W.: *The Expanding Universe: Astronomy's Great Debate*, CaUP 1982.
- Stephenson, F.R., Houlden, M.A.: *Atlas of Historical Eclipse Maps*, CaUP 1986.
- Sugimoto, K.: *Albert Einstein: A Photographic Biography*, Schocken, 1989.
- Sullivan, W.T. III (ed.): *The Early Years of Radio Astronomy: Reflections Fifty Years after Jansky's Discovery*, CaUP 1984.

- Swarup, G., Bag, A.K., Shukla, K.S.: *History of Oriental Astronomy, The Proceedings of IAU Colloquium 91*, CaUP 1987.
- Taton, R., Wilson, C. (eds.): *Planetary Astronomy from the Renaissance to the Rise of Astrophysics, Part A: Tycho Brahe to Newton*, CaUP 1989.
- Tayler, R.J. (ed.): *History of the Royal Astronomical Society, Vol. 2: 1820–1920*, BlkS 1987.
- Thom, A.: *Megalithic Sites in Britain*, Clarendon, Oxford 1967.
- Thoren, V.E.: *The Lord of Uraniborg: A Biography of Tycho Brahe*, CaUP 1991.
- Thrower, N.J.W. (ed.): *Standing on the Shoulders of Giants*, UCaP 1990.
- Tucker, W., Giacconi, R.: *The X-ray Universe*, HaUP 1985.
- Tucker, W., Tucker, K.: *The Cosmic Inquirers: Modern Telescopes and Their Makers*, HaUP 1986.
- Ulansey, D.: *The Origins of the Mithraic Mysteries*, OxUP 1989.
- Véron, Ph., Ribers, J.C.: *Les comètes d'antiquité à l'ève spatiale*, Hachette, Paris 1979.
- Wali, K.C.: *Chandra: A Biography of S. Chandrasekhar*, UChP 1991.
- Wattenberg, D.: *Johan Gottfried Galle. Leben und Wirken eines deutschen Astronomen*, J.A. Barth, Leipzig 1963
- Wayman, P.A.: *Dunsink Observatory 1785–1985*, Royal Dublin Society, 1987.
- Webb, G.E.: *Tree Rings and Telescopes: The Scientific Career of A.E. Douglass*, UAzP 1983.
- Westfall, R.S.: *Essays on the Trial of Galileo*, Vatican Observatory, distributed by University of Notre Dame Press, Notre Dame 1989.
- Whitrow, G.J.: *Time in History*, OxUP 1988.
- Wright, H.: *James Lick's Monument: The Saga of Captain Richard Floyd and the Building of the Lick Observatory*, CaUP 1987.
- Yataro, S., Elliot, H. (eds.): *Early History of Cosmic Ray Studies* (Astrophysics and Space Science Library Vol. 118), ReiP 1985.
- Zinner, E.: *Regiomontanus: His Life and Work*, Elsevier Science Publishers, New York 1990.

Chapter 12

- Abrams, B., Moore, P.: *Extending Science 17: Astronomy, Selected Topics*, Stanley Thornes, Cheltenham 1989.
- Brown, R.A. (ed.): *An Education Initiative in Astronomy*, Space Telescope Science Institute, Baltimore 1990.
- DeBruin, J., Murad, D.: *Look to the Sky: An All-Purpose Interdisciplinary Guide to Astronomy*, Good Apple, 1988.
- Druger, M. (ed.): *Science for the Fun of It: A Guide to Informal Science Education*, National Science Teachers Association 1988.
- Fraknoi, A.: *Universe in the Classroom: A Resource Guide for Teaching Astronomy*, FrmC 1985.
- Gibson, B.: *The Astronomer's Sourcebook*, Woodbine House, 1992.
- Pasachoff, J.M., Percy, J.R. (eds.): *The Teaching of Astronomy*, Proceedings of the 105th Colloquium of the International Astronomical Union, CaUP 1990.
- Reynolds et al.: *Space Mathematics: A Resource for Teachers*, NASA, Washington 1972.
- Schatz, D., Fraknoi, A., Robbins, R., Smith, C.: *Effective Astronomy Teaching and Student Reasoning Ability*, Regents of University of California 1978.

Index to Volume 1

aberration
    astigmatic 70, 98
    chromatic 75f, 87, 114
    comatic 70
    curves of 76
    of light 23
    spherical 70, 89, 96
    wave 84
absorption, interstellar 327
achromat 77, 86, 430
accumulator 210f
admittance, electric 170
air mass, photometric 342
Airy disk 104
almanac, observing 514
alt-azimuthal mounting 138
anastigmat 73, 99
angle encoder 184
annus fictus 28
anomaly
    eccentric 50
    mean 50
    true 51
antenna (see radio telescope)
antifriction bearing 162
antihalation backing (photography) 247
apochromat 86
astigmatism 69f
astrocamera, suppliers 509
astrology 450
astrometry, photographic 49
Astronomical Ephemeris 6
astronomy, history 425
Astronomy and Astrophysics Abstracts 3
atlas, sky 514
atmosphere
    Mie scattering 385
    optical window 432
    radio absorption and emission 390
    Rayleigh scattering 385
    refraction 61, 390
    scattering by clouds 385, 398
audio-visual aids 466, 500

axial force 160
axis adjustment 202

ball bearing 158
Balmer series 290
band drive 178
Barlow lens 115, 224
base fog (photography) 249
battery 207f
bearing 153f
bearing clearance 163
Besselian day numbers 25
blackbody radiation 288, 385
blaze grating 303, 309
Boltzmann equation 292
burning (photography) 271

calendar, observing 514
cantilever loading 139
Cassegrain telescope 91
center loading 139
characteristic curve (photography) 248
charge-coupled device (CCD) 132, 220, 306, 326
chart, star 514
chopper (current) 190
chromatic aberration 75f
clutch 177
cold camera 261f
color index 329
color temperature 316
coma 71, 90
comets
    photometry of 342, 363
    photography of 236, 245
    spectroscopy of 299, 308
composite printing 272
computer 38f, 472
    at radio telescope 383, 397
    bulletin board 472
    software 473
constellation 5
construction material 154f
continuous spectrum 288

contrast enhancement 272
control circuit (of drive) 184f
coordinates
    ecliptic 15
    galactic 16
    geographic 9
    spherical 10
    transformation of 14
coronograph 101
counterweight 139
criteria, mechanical 147f

data center 4
data line 29
DC motor 177f
dead time, photometer 373
declination 13
declination axis 138, 159
deformation constant 67
density curve, photographic 311, 323
detector
    photoelectric 132
    photographic 129
    semiconductor 132
dewcap 214
    heating 215, 231
Dewey Decimal System 3
dial (see sundial)
dies reductus 28
differential gear 180
diffraction disk 71, 104
diffraction grating 302
digression, largest 20
dispersion
    angular 301f
    chromatic 76f
displacement (mounting) 147
distortion, optical 73
Dobsonian mounting 138
dodging (photography) 271
dome, construction of 511
Doppler effect 314
drive motor 176f
Durchmusterung 5, 428

Ebert mounting 305
echelle grating 303
eclipse, photography 233, 237
eclipsing binary
    light curve analysis 367
    minima prediction 366
    photometry 365
ecliptic 15
education, astronomy 437
    college level 441
    games 476

    graduate level 445
    music 477
    postdoctoral 448
    software 472
equipment, suppliers of 503
elasticity modulus 147, 150
electron density 387
electron temperature 387
elements, orbital 50
emission measure 387
emulsion, photographic 246f, 323
encoder 184
English mounting 139
entrance pupil 64, 108, 372
equatorial mounting 138
equinox, vernal 13
equivalent width 316
errors, propagation of 40, 172
    random 39, 172
    systematic 39, 172
excitation of atoms 292
exit pupil 64, 108, 112, 126f
extinction, atmospheric 327, 343
    ionospheric 389
eye
    contrast step 322
    quantum efficiency 322
eyepiece 113f

Fabry lens 371, 373
field curvature 70f, 93, 96, 114
field lens 66
field of view 108
field rotation 228
film speed 250, 308
filter
    photographic 250
    transmission curves 252
finding chart 336
flash spectrum 298, 309
flexure 151
flux density 384, 402
focal length 63, 75, 79, 107
focal ratio 68f
focusing screen 221
Folly mounting 140
fork mounting 139
foundation, telescope 164f
frequency allocation, radio 391
friction bearing 162
friction clutch 178
friction wheel drive 178

gamma value, emulsion 249
gear ratio 174, 178
gegenschein, photography of 242

German mounting 138
grating constant 302, 312
Gregorian calendar 26
ground vibration 170
guide telescope 120, 227
guiding head, sensor 196
guiding, photoelectric 195

half-step motor 181
halo, atmospheric 235
handbook, observing 514
Hartmann test 81
heliocentric system 425
Hertzsprung-Russell diagram 295
high-energy pulse 411
history
    library and archive 434
    research 433
    sources 433
    topics 434
horizon 11
horseshoe mounting 140
hour angle 14,138
hour circle 138
hydrogen atom 290
hypersensitization 254f

IC module 185f
illuminance 321
image processing, digital 346
imaging scale 63
impedance, mechanical 169f
increment encoder 205
inhibit circuit 191
integration, numerical 48
interference, radio
    broadcast 385
    radar 385
    thunderstorm 381
interferometer 85, 93, 106
interpolation 45
intrinsic color 332
ionization 292
    ionospheric 389
ionosphere, layers of 388
iris photometer 345

jansky (Jy) 384
Julian calendar 26
Julian date 28, 349
Jupiter, radio radiation from 386

Kepler's equation 50f
Kepler's laws 425
knife-edge test 83, 167

latent image 247
latitude

ecliptic 15
galactic 16
geographic 10f
law of gravitation 426
laws of cosine and sine 32
least squares, method of 9, 39, 41
light pollution 238
line spectrum 289
Littrow mounting 305
longitude
    ecliptic 15
    galactic 16
    geographic 10f
low-pass filter 169
luminance 320
luminous flux 320
luminosity 320
Lyman series 290

magnification 63, 107f
magnitude
    absolute 321
    apparent 319, 329
    at mean heliocentric distance 356
    limiting 112, 131, 275
    photographic 329
Maksutov camera 99
meteor
    photometry 363
    photography 232
    spectroscopy 299, 308
microdensitometer 345
micrometer, filar 121
microprocessor 185, 190
microstep motor, circuit 181, 189
minimum light (variable star)
    predicting 366
    timing 351
minor planet
    guiding 236
    photography 236
    spectroscopy 299, 308
Mögel-Dellinger effect 389
Moon
    eclipse photography of 233, 237
    photometry of 361
    photography of 243
nebular filter 239

Newton's equation (optics) 63
nova, spectra of 297
nutation 21

objective grating 312
objective prism 303, 307
obliquity of ecliptic 15

observatories 430, 452
observatory 488
occultation, lunar 56
off-axis guiding 195, 227
operational amplifier 194, 209
operation torque 177
optics
    aplanatic 72, 99
    geometrical 60
orbit, elements of 50
overmagnification 108

parallax
    diurnal 24
    stellar (annual) 25
paraxial field 62
periodical 2
phase angle 53, 360
photography 217f
    equipment 510
    grades and codes (of photo paper) 269, 270
    history of 429
photodiode 199, 325, 369
photometer 509
    commercial 369
    construction 370
    photoelectric 337, 369
    photographic 313
    two-beam 374
photometry
    absolute 338
    accuracy of 328
    differential 337
    of Moon 361
    of occultations 339
    photographic 345
    of planets 357, 361
    point 334
    of solar system 356
    of Sun 362
    surface 334, 341, 368
    visual 334, 428
photomultiplier tube 324f, 369
pixel 346
planetarium 437, 451, 481
planetary moons
    photography of 235, 245
    spectra of 298
planets
    photometry of 361
    photography of 243
    radio radiation from 383f, 407
    spectra of 298
plate constants 49

Platonic year 21
plywood material 153
Pogson's method 351
polar alignment 228
pole, galactic 17
polynomial fit 352
positioning, digital 204
positive projection 242
precession 21f
prime vertical 20
principal ray 65
projection screen 124
proper motion 21
Purkinje phenomenon 322

quiet sun 297

radiation bursts, radio 386
radiation flux 321, 384
radiation mechanisms, radio
    blackbody radiation 385
    Cerenkov radiation 388
    free-free radiation 387
    Larmor frequency 388
    magneto-bremsstrahlung 388
    plasma oscillations 388
    synchrotron radiation 382
    thermal bremsstrahlung 387
radio sources 381f
    catalogs 401
    Milky Way 381, 400
    planets 383, 407
    pulsars 383
    quasars 383, 402
    Sun 383, 404
    variable 383, 402
radio telescope 381f
    antenna 391
    aperture efficiency 393
    aperture synthesis 395
    beam efficiency 387
    calibration 386
    gain 393
    home-made 409
    interferometer 395, 409
    resolution 384
    Ruze formula 391
    suppliers of 509
    VLBI 391
Rayleigh criterion 104
Rayleigh-Jeans's law 289
receiver, radio 395, 409
    antenna temperature 395
    beam switch 398
    Dicke switch 396
    load switch 397

maser 396
 noise factor 396
 parametric amplifier 396
 sensitivity 396
 system temperature 396
reciprocity failure, low-intensity 254
redshift (radio) 402
reference ellipsoid 10
reflectivity (coatings) 111
reflector
 Cassegrain 91
 Newtonian 88
 tube 118
refraction
 ionospheric 389
 law of 300
 optical 12, 61
 radio 390
refractive index 300
refractive power 63
refractor 65, 86, 118, 430
resolving power 104
 of emulsion 131, 256
 of eye 125, 323
 of grating 303
 of prism 301, 307
 radio 384
right ascension 13
Ritchey-Chrétien system 93f
roller bearing 162
Ronchi test 86
Rowland circle 306

Saha equation 292
sandwich method (photography) 272
satellite
 geosynchronous 234
 trail 234
scale, imaging 63
Scheiner method (adjustment) 202, 228
schiefspiegler 94
scintillation 105f
Schmidt camera 95, 220, 430
Schmidt-Cassegrain camera 98, 220
Schwarzschild exponent 255f
seeing 7, 105, 224
Seidel theory 66
selected area 345
semidiurnal arc 20
sensitivity
 of the eye 126
 of a radio receiver 396
sensitometer 312
setting circle 203
Shapley lens 224

sidereal time 13f
societies, astronomical 494
solar telescope 100
space exploration, history of 432
spectra
 analysis of 287, 429
 classes of 294f, 314
 comparison lines 311
 molecular 291
 secondary 219
 theory of 287
 width 307
spectral instrument
 detector 306
 slit 304, 309
 slitless 239, 303
spectral line
 forbidden 291
 hydrogen 21 cm 382, 407
 maser effect 408
 profile 317
 radio 382
 recombination 387
 telluric 287
 water vapor 407
spectroscope 509
standard, photometric 332
standard deviation 7, 39
stars
 catalogs of 5, 427
 designation of 5
 flare 365
 luminosity classes of 296
 photometry of 342
 spectra of 293
star trail 230
Stefan–Boltzmann law 289
stellar evolution (history) 431
step-estimate method 334
stepping motor 177f, 188
stiffness 147f
stop
 aperture 64
 field 64f
summing gear 180
Sun
 chromosphere 298
 corona 298
 eclipse photography of 233, 237
 photography of 243
 photometry of 361
 radiation bursts from 386, 404
 radio 383
 spectrum of 297
sundial

    equinoxial 414
    horizontal 414
    style 413, 420
    vertical 417
surface photometry 368
synchronous motor 177f
systems, stellar, history 413

teleconverter lens 224
telescope (see under types) 86f
    adjustment 201f
    drive 176, 224
    illumination 212f
    suppliers 462, 503
tellurian 460
three-color composite 275
time
    ephemeris 30f
    equation of 18, 413, 422
    heliocentric 349
    international atomic (TAI) 30
    sidereal 13f
    solar 17, 413
    terrestrial dynamical (TDT) 30
    universal (UT) 31, 349
time signals 125
torque, on motor 177, 181
tracking accuracy 178, 226
tracking head 195
transit, meridian 20
transmission
    atmospheric 327, 390
    optical 109

tumbling error 172f

UBV system 330
Universal Decimal Classification (UDC) 3
unsharp-masking method (photography) 273

variable stars
    minimum light, time of 351
    period 353
    photometry 365f
velocity, radial 314
vibration, telescope 168f
videocassette 468
vignetting 65, 218
    correction in photos 272

Weber-Fechner law 322
wide-angle adaptor 220
Wien's law 289
wood, construction by 146, 154
worm drive 178
Wright-Schmidt camera 220
Wright-Väisälä camera 220

year
    anomalistic 27
    Gregorian (civil) 27
    Julian 27
    sidereal 27
    tropical 19, 27

zenith distance 11
zodiacal light, photography of 242